Mathematical Methods in Economics

Using Mathematica

Mohammad R. Safarzadeh
California State Polytechnic University-Pomona

Bent Tree Press

Printed in the United States of America.

ISBN 978-1-4652-8790-8

Bent Tree Press

59 Damonte Ranch Parkway, Suite B284 • Reno, NV 89521 • (800) 970-1883

www.benttreepress.com

Address all correspondence and order information to the above address.

to Simin, Pooya, Pasha and Sasha

CONTENTS

PREFACE

Economics is a science with many diverse fields and topics of interest such as the theory of firm, consumer theory, welfare economics, international trade, etc. It is also a science with topics of interest that change from time to time. Economic issues of interest today are quite different from those in the 1970s, 1980s, and 1990s. The diversity of the fields and the changing nature of the economic issues make the study of economics a challenging, and at times, a laborious task. However, most of these seemingly diverse fields and subjects have strikingly similar and common elements, especially if they are cast into simple mathematical models.

As a tool, mathematics can be used to facilitates a clear and organized presentation and test of economic theories and problems. It provides a basis for setting up a model, making explicit assumptions about the model, finding optimal solutions, and extending the results by varying the parameters and assumptions of the model. A large gain in clarity and economy of effort can be achieved by incorporating mathematics into economic analysis. Today, it is critical for students of business and economics to be fully equipped with major mathematical tools. Once a student has learned the mathematical tools, s/he can apply them in analyzing many diverse problems in the subject.

Unfortunately, many students entering colleges and universities do not have enough preparation in mathematics to allow them to manipulate mathematical symbols, to do numerical calculations and to graph the mathematical relationships. These techniques are basic tools of economic analysis and students lacking them, often, do poorly in economic studies. In many schools, the lack of preparation by students has resulted in cancellation of quantitative parts of the economic curriculum, lowering the course standards, and omitting some of the advanced topics. Even if students had enough preparation in mathematical methods, presenting certain topics in the class room using the traditional methods is often very time consuming and sometimes tedious. To deal with these problems, many colleges and universities have incorporated use of computers and software into their curriculum to enhance the efficiency of teaching and learning quantitative economics.

I had written the manuscript, Mathematical Methods in Economics, for the mathematical economics course that I teach at Cal Poly-Pomona. During the fall quarter, 1997, I experimented with the use of the *Mathematica* in teaching this course. The results were beyond my expectations. The use of *Mathematica* resulted in covering more chapters of the book than I was ever able to cover in the past. Students enjoyed the course more than ever and their progress and positive response were outstanding. In 1998, in consultation with the Wolfran Research, the publishers of the *Mathematica* software, I incorporated *Mathematica* into the manuscript, revised several chapters, added new material and changed the name of the manuscript to Mathematical Methods in Economics Using *Mathematica.*

I find *Mathematica* to be a comprehensive software package that helps to remove students handicap in mathematical analysis and allows them to concentrate more on the economic concepts, economic theories, and the structure of economic optimization rather than worrying about details of numerical calculations and symbolic manipulations. *Mathematica* allows students to experiment with economic theories and concepts using a very powerful and comprehensive software without being tied up by time consuming mathematical manipulations and calculations. Students can experiment dynamics of economic theories using this powerful, and user friendly software which can handle dynamic and stochastic economic models. The two-dimensional and three-dimensional graphic features of the program brings economic theories of dynamic market analysis to life. With *Mathematica*, I believe that for the first time economic students have access to a wealth of laboratory to experiment economic theories and models.

The unifying and simplifying property of the mathematical methods together with the power and facilitating features of the *Mathematica* are the basic reason for writing this book. I have attempted to present a number of mathematical methods which are useful in analyzing various business and economic problems and use *Mathematica* to show how easily they can be handled and solved using a current computer software.

Each chapter begins with a tool box presenting all the Mathematica commands used in that chapter. To practice the commands, click on the hyperlink key of the software attached to the book. This will provide you with a practice notebook that contains several examples. As well, the Magenta-colored hyperlinks link you directly to the description of the command in the Help menu of the *Mathematica*. It is highly advisable to practice the commands by changing the parameters and the assumptions of the models and by rerunning the programs.

Each chapter of the book starts with a clear statement of pertinent definitions, principles and theorems. Then a number of examples in applications of the concepts to the most common business and economic problems in the theory of price, markets, and the allocation of resources are provided. These examples help to illustrate and amplify the theory. The manual solution to the examples are followed by the *Mathematica* commands and solutions which can be run as you read the material from the book or CD. You can always do more practice by changing the parameters in the example and test for the consistency of the results or experiment with the theory. At the end of each chapter, there are a number of numerical exercises for further practice by the students.

This book is written for intermediate undergraduate and beginning graduate students in business and economics. The requirements include college algebra and principles of economics. An attempt is made to keep the exposition of mathematical theorems brief and as simple as possible. Proofs of theorems are given whenever they are instructive and are not too tedious to follow. The emphasis is on the application of the mathematical theorems and techniques to problem solving in business and economics.

The book is essentially divided into four parts. Chapters 1 through 5 are the beginning chapters that review the fundamental concepts necessary for developing the materials in other chapters. I intentionally made the introduction to *Mathematica* and the *Mathematica* commands in Chapter 1 so brief. I believe that the *Mathematica* commands can be learned gradually as students study chapters of the book and solve the examples. For students who would like to learn *Mathematica* commands at a faster pace than the chapters, I refer them to the *Mathematica*, The Student Book, by Stephen Wolfran, Addison Wesley Publishing Company. Other concepts in chapter 1 are the number system, sets, and mathematical relations. Chapter 2 introduces the concept of function and different types of functions and their graphs. Chapter 3 deals with fitting equations to sets of data. Chapter 4 concentrates on the elements of linear algebra. Chapter 5 discusses some of the advanced topics in linear algebra. Teaching this chapter or parts of it may be optional in the undergraduate courses. The concepts in the beginning chapters 1-5 are needed as working knowledge throughout the book. Chapters 6 through 8 are concerned with optimizing techniques, unconstrained and constrained. These techniques are the fundamental tools of many economic analysis and problem solving. Chapters 9 and 10 are devoted to linear programming and its applications to input-output systems and the theory of games. Chapters 11 through 14 concentrate on dynamic analysis including integral calculus, differential equations, and difference equations.

I would like to thank the many colleagues and students who have given me the benefit of their comments and suggestions. My special thanks to Dr. David Park (California State Polytechnic University-Pomona) who set the cornerstone of this book and encouraged me to create it. Dr. Maureen Burton (California State Polytechnic University-Pomona), and Osman Yurekli (University of North Carolina, Wilmington) read the first draft and made many useful comments and suggestions. Professor James E. Sutton (California State Polytechnic University-Pomona) generously devoted his time to reading chapters and making many valuable contributions to the development of this book Sousan Badiei and Ehab Ghaly, graduate students at the California State Polytechnic University-Pomona, helped me with homework problems and were the source of many suggestions. I especially acknowledge my debt to Dr. Daniel C. Vandermuelen's (Claremont Graduate School) lectures in mathematical economics and Dr. Thomas D. Willett's (Claremont McKenna College) lectures in economics.

How To Use The CD

Mathematical Methods in Economics Using *Mathematica* is designed to help students to learn economics by hands-on practice of economic theories and economic models. Each chapter presents a number of mathematical concepts as tools for economic analysis. Using the tools, each chapter represents economic theories of the consumer behavior, theory of firms, markets, and others through simple mathematical models for further practice. The models are then programmed into Mathematica commands. Each program provides students with a laboratory of economic analysis and practice. Run a program to observe the results. Change the parameters and the assumptions of the model and re-run the program. Compare the results under different assumptions. Test for the extreme cases using the extreme values and compare your results with what the economic theory would predict under different circumstances.

No pre-knowledge of Mathematica is required for using this book or CD. There are many ways to write a program in Mathematica. I have attempted to keep the Mathematica programs and the Mathematica commands as simple as possible. In every program the main body of the commands is printed in the normal-size characters. The parts of the program that deal with "fancy" presentation of outputs such as the thickness of the graph, labeling of the graphs or axes, colors, and others are printed in small size characters. Some programs may seem long and may contain too many commands simply because they may contain too many of those fancy features. If you are not familiar with Mathematica, you may skip reading or learning those fancy parts. For example, to graph y = x^2, the program may be written as:

(Note: To run the program, place the cursor in front of the first line of the program and press Shift Enter keys. To open or close the file, double-click on the cell bracket to the right.)

```
Clear[x, y]
y = x^2;
Plot[y, {x, -10, 10}]
```

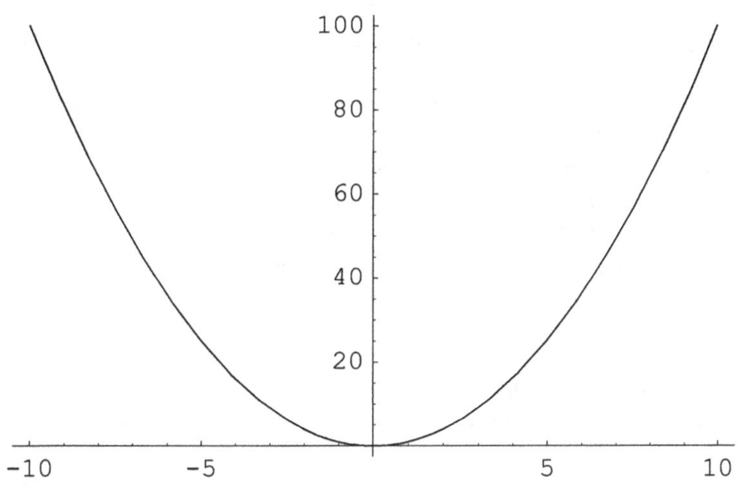

This simple program will do the job and will graph y = x^2 in the range x = -10 to 10. However, the same graph may be done by adding several fancy features such as labels and colors as presented below,

```
Clear[x, y]
y = x^2;
Plot[y, {x, -10, 10}, AxesLabel -> {"X", "Y"},
  PlotStyle -> {AbsoluteThickness[3], RGBColor[1, 0, 1]}, PlotLabel -> "Plot of a Parabola"]
```

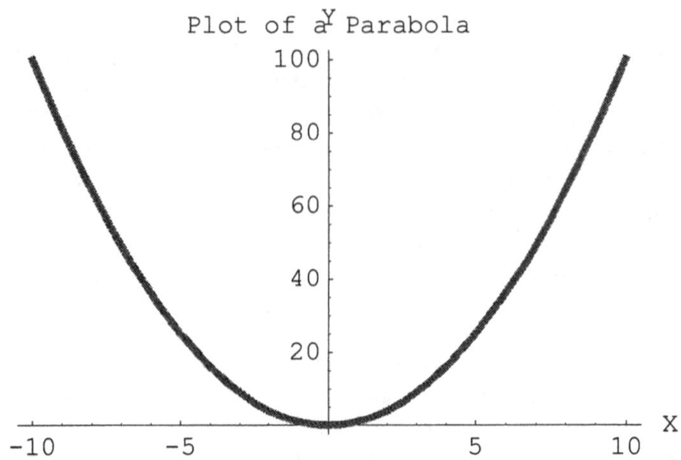

Run every program from the first line of the program which starts with "Clear[ ]" command. This command removes any values assigned to variables in the earlier programs and clears all definitions for functions. If a program doesn't run smoothly and gives error message, go back to "Clear" command and re-run the program. If the problem persists, highlight the second line of the program, go to "Cell" in the menu bar of Mathematica and use "Divide Cell" to divide the cell and re-run the program. If the problem still continues, do the dividing of the cells for all the program lines and run the program line by line.

Although you do not need to know the Mathematica commands or programming to be able to use this book or CD, it is highly recommended to learn them as you run the models and practice with the system Mathematica is a very powerful program that is used in every science that deals with numbers and computations. Its commands are very simple and very naturally correspond to the common mathematical terminology and logic. For example, to solve an equation, say $x^2 + 3x - 4 = 0$, the Mathematica command is Solve[x^2 + 3x - 4==0, x]. This command naturally corresponds to the mathematical terminology and the mathematical logic of solving an equation for the unknown variable, here x. The command reads exactly as what it does; solve $x^2 + 3x - 4 = 0$ for x. The only special programming feature that you will need to learn is that the command "Solve" should be capitalized, followed by the square brackets, and that the equality sign in the Solve command is a double equal sign "= =". Don't try to memorize the Mathematica commands. Only a handful of Mathematica commands are used in each chapter, with each command being repeated in several places in other chapters. Therefore, if you pay attention to the logic and format of the programs, you will learn the commands as you proceed practicing with them. At some point, you will be able to write programs of your own in Mathematica.

Hints On Running The Programs: Run every program from the first line of the program command which starts with "Clear[ ]". This command removes any values assigned to variables in the earlier programs or clears all definitions for functions. If a program doesn't run smoothly and gives error message, go back to the Clear command and re-run the program. In some cases, especially is some programs of difference and differential equations, the Kernel may print a list of error messages. These messages may be too long or they may run through loops. In such cases, click on the "Kernel" on the Mathematica menu bar, click on the "Quit Kernel", and click on the "Local". This will stop the Kernel from running. Go back to the Clear[ ] command and run the program again. You may have to change the Clear[ ] command to ClearAll[ ] command or vise versa to run the program.

CHAPTER 1

PRELIMINARY CONCEPTS

The three main topics studied in this book are the optimizing techniques, linear economic theory, and dynamic economic analysis. However, it is difficult to discuss these subjects before making certain concepts clear. The definitions and discussions in this chapter are a review of elementary concepts in college algebra or introductory calculus. The chapter begins with an introduction to *Mathematica* and continues with the number system, sets, mathematical relations, logarithms, and progressions. The economic applications require a knowledge of microeconomics and macroeconomics at the principles of economics level.

TOOL BOX

Mathematica Commands	Function
Abs[n]	Absolute value of n
N[n]	Numerical value of n
Sqrt[n]	Square root of n
Solve[f(x) == 0, x]	Solves equation f(x) = 0 for x
t = {5, 3, 6, 8}	t is a list of data
t = {{5, 2}, {1, 0}, {6, 8}, {7, 9}}	t is a list of paired observations on x & y
u = {a, b, c}	u is a set with elements a, b, c
Random[ ]	gives a uniformly distributed pseudorandom in the range 0-1.
Random[Integer,{a, b}]	generates a random integer in the range from a to b.
Union[u, v]	Finds union of two sets u and v
Intersection[u, v]	Finds intersection of two sets u and v
Complement[u, v]	Finds complement of two sets u and v
Log[b, n]	Finds log base b of n
Table[i^2, {i, 0,5}]	Makes a table of values for i = 0 to 5
Sum[i^2, {i, 0, 5}]	Sums a table of values of i^2 for i =0 to 5
Out[n]	Value of the n^{th}output
%	Value of the preceding output
/.	Replace or Evaluate
%/. p->2	Evaluate the preceding output as p ->2

; at the end of command statements suppresses the output

I. Command Structure of the *Mathematica*

Mathematica is a general computer software system and language for mathematical and other applications. *Mathematica* has a wide range of applications in quantitative and logical analysis. Its use in every area of science is growing fast. In this book, we will use the numerical, symbolic, programming and graphical capabilities of *Mathematica* in solving some common quantitative business and economic problems. The simplest way to use *Mathematica* is to use it like a calculator. You type in inputs, *Mathematica* prints output. However, the computation capabilities of *Mathematica* is enormous. In fact, it is the sophistication of the *Mathematica* that makes its use so simple. *Mathematica* has a large number of built-in functions that can be combined with programming language and other features of *Mathematica* to perform some of the most advanced mathematical calculations with a few simple commands. Following a few rules can save you a lot of time and frustrations.

Commands: All the commands in *Mathematica* start with a capital letter. For example, "Solve", "Fit", and "Plot" are commands for solving equations, fitting equations to sets of data, and plotting functions, respectively. For the convenience of the users, *Mathematica* commands in the CD attached to this book are hyperlinked to the *Mathematica*'s description of the commands in the help section of the program. For more information on the structure, formulation, and the arguments of the commands double click on the command names, wherever they appear in **Magenta** color. As well, many calculus commands and operators can be copied from the "palettes" in the pull-down window of the "BasicInputs" or other pull down menus in palettes. You can locate the palettes by clicking on the "File" in the toolbar. For example, to find square root of 25 you may write the command as

```
In[1]:=  Sqrt[25]

Out[1]=  5
```

or you may use the "BasicInputs" menu of the palettes to copy the operator $\sqrt{\square}$ and fill in for the number "25" in the space provided.

In[2]:= $\sqrt{25}$

Out[2]= 5

In this book, I have used the two types of commands and operators alternately to expose the students to both methods. Once you have established your own style and feel comfortable with, you may use the one that you prefer.

Variables: Except for few characters which are reserved for certain functions in *Mathematica*, any other character or combination of characters can be used as a variable name. The reserved characters are used for commands or for some other purposes in *Mathematica*. For example, "D" is a command used for finding derivative of functions, "N" is used for finding numerical values, and "E" is reserved for natural number. These characters can not be used as variable names. Also, x and X are considered as two different variables in *Mathematica*. That is, the lower case of a letter and its capital count as two different variables. Considering that the reserved characters in *Mathematica* are all capital letters and commands start with capital letters, it is highly advisable not to use capital letters for variable name. In combining letters, be careful that xy counts as one variable "xy". However, x y, that is, x space y, means multiplication of two variables, x*y.

Brackets: Brackets play an important role in command structure of *Mathematica*. Four types of bracketing are used in *Mathematica:*

1- Square brackets "[ ]": All the *Mathematica* commands are followed by a square bracket. The argument of a command should be enveloped in square brackets. As well, the arguments of functions should be enveloped in square brackets. For example, to plot y = Sin x, the *Mathematica* command is

Plot[Sin[x]].

2- Curly brackets "{ }" are used for enveloping lists. For example, if the function y = Sin x is plotted for the range -10 to 10, the plot command is written as Plot[Sin[x], {x, -10, 10}].

3- Parentheses are used for grouping of arithmetic and algebraic terms.

4- Double square brackets are used for indexing. For example, v[[i]] returns the value of the ith element of a list.

Running The Programs: Run every program from the first line of the program command which starts with "Clear[ ]". This command removes any values assigned to variables in the earlier programs or clears all definitions for functions. To run the program clik on the first line of the program to set the cursor and push the **Shift** and **Enter** keys of the keyboard together. If a program doesn't run and returns an error messages, go back to the Clear command and re-run the program. In some cases, especially in some programs of difference and differential equations, the Kernel may print a list of error messages. These messages may be too long or they may run through loops. In such cases, click on the "Kernel" on the toolbar, click on the "Quit Kernel", and click on the "Local". This will stop the Kernel from running. Go back to the Clear[ ] command and run the program again. You may have to change the Clear[ ] command to ClearAll[ ] to run the program.

Five Tips To Avoid Frustration When Working With *Mathematica:*

1- Always start the programs with the "Clear" command to clear the existing numerical values of the variables off the memory.

2- Never capitalize the variable names. *Mathematica* reads x and X as two different variables.

3- Not every character is accepted as variable name. Letters such as D, I, and C are reserved and can't be used as variable names.

4- Note that *Mathematica* reads xy as one variable. If your intention is to multiply x by y, you should write it as x y (x space y) or x*y.

5- Learn the commands as your needs arise. Most of the arithmetic and algebraic calculations can be done by knowing and combining few *Mathematica* commands. Don't memorize the commands. You always have access to "Help" pull-down window. This window provides you with examples on the use of every *Mathematica* command. You can copy the commands form the example and paste it to your own notebook.

II. The Number System and Numerical Calculations: The **N** Command

The Number System: A number in general form can be written as **a + bi**, where a and b are real numbers, b≠ 0, and i is the imaginary unit having the property $i^2 = -1$. A Number in this form is called a **complex numbers.** Furthermore, a is the real part of the number and bi is the imaginary part. The following are some concepts and definitions in number systems:

Natural Numbers A natural number is a whole number used in counting objects. Numbers such as 5, 2, and 10 are examples of natural numbers. These numbers are also called positive integers.

Rational Numbers A rational number is a ratio of two integers, positive or negative. A rational number, often called a fraction, may qualify as integer, such as $\frac{9}{3} = 3$. Therefore, all integers are considered to be rational numbers, but not all rational numbers are integers. Zero is considered a rational number.

Irrational Numbers Some numbers cannot be written as the ratio of two integers. Such numbers have infinite decimal expansions which are non-repeating. For example $\sqrt{2}$ = 1.41412...., Π = 3.1415...., and e = 2.718..... Some fractions, like $\frac{1}{6}$ = .1666, also have infinite decimal expansion. However, the difference between the two types is that the decimal expansions for rational numbers are repeating, like the sequence of 6 in 1666. Whereas, decimal expansions for irrational numbers are non-repeating.

Real Numbers Natural numbers, rational numbers, and irrational numbers taken together constitute the real number system.

Purely Imaginary Numbers For the equation $x^2 + 1 = 0$, or any equation in the form $x^2 + k = 0$, where k > 0, there is no real number that will satisfy the equation. Therefore, the solution of $x^2 + 1 = 0$ is $x = \pm\sqrt{-1} = \pm i$. For $x^2 + k = 0$, the solution is $x = \pm i\sqrt{k}$. $\pm$ i and $\pm i\sqrt{k}$ are purely imaginary numbers.

Complex Numbers The real numbers and the imaginary numbers together form the complex number system.

Absolute Value of a Real Number The absolute value of a real number x, denoted by |x|, is equal to x if $x \geq 0$ and is equal to -x if x < 0. The absolute value of a number shows the magnitude, but it does not indicate whether the number is positive or negative.

Numerical Calculations: Arithmetic operations on numbers with *Mathematica* is very similar to doing arithmetic operations with a calculator. Type the input to be evaluated using the same mathematical operators as in a calculator and press **Shift** and **Enter** keys together.

Example: Do the following arithmetic operations: $10 + 3^5$, 3^4, $\frac{236}{12}$, $\frac{236}{12}$ (result in decimal).

Solution:

```
In[3]:= 10 + 3*5

Out[3] = 25

In[4]:= 3^4

Out[4]= 81

In[5]:= 236/12

Out[5]= 59/3

In[6]:= 236./12

Out[6]= 19.6667
```

Note that by putting a "." in front of the number in numerator or denominator you are requesting the results in decimal form.

The order of the arithmetic operation in *Mathematica* is similar to that of a calculator. It operates first on exponent (^), then on division (/), followed by multiplication (*), and finally addition (+) and subtraction (-). Anytime in doubt, it is highly advisable to use as many parenthesis as needed, or as many parenthesis as you like to make the operation clear. For example, a command such as 25^3/2 will raise 25 to the power of 3 and will divide the results by 2. That is, the operation will be $\frac{25^3}{2}$. However, if the intention were to have 25 raised to the power $\frac{3}{2}$, ($25^{\frac{3}{2}}$), the command should read as

25^(3/2).

Example: Find the numerical value of $\sqrt{\frac{124}{35} * \frac{13}{5} + 45^{\frac{1}{3}}}$.

Solution:

```
In[7]:=  N[Sqrt[(124/35)*(13/5) + 45^(1/3)]]
Out[7]=  3.57328
```

or you may command as:

```
In[8]:=  Sqrt[(124/35)*(13/5) + 45^(1/3)] // N
Out[8]=  3.57328
```

or you may command as:

```
In[9]:=  N[(124/35 13/5 + 45^(1/3))^.5]
Out[9]=  3.57328
```

or using palettes, you may command as:

In[10]:= $N\left[\sqrt{\frac{124}{35}\ \frac{13}{5} + 45^{\frac{1}{3}}}\right]$

Out[10]= 3.57328

Note that, to multiply two numbers, you may use "*" between the two numbers or you may leave a space between the two numbers. In the operation above a space between $\frac{124}{35}$ and $\frac{13}{5}$ is used to multiply the two numbers.

Example: For the imaginary unit i = $\sqrt{-1}$, find the values of (2 + 3i)(2 - 3i), (2 + 3i)(3 + 4i), and $(2 + 3i)^8$.

Solution:

```
In[11]:=  (2 + 3I) * (2 - 3I)
Out[11]=  13
In[12]:=  (2 + 3I) * (2 + 4I)
Out[12]=  -6 + 17 i
In[13]:=  (2 + 3I)^8
Out[13]=  -239 + 28560 i
```

Example: Find the numerical values of irrational numbers Pi, E, and $\sqrt{2}$ up to 30 digit of accuracy.

Solution:

```
In[14]:=  N[Pi, 30]
Out[14]=  3.14159265358979323846264338328
In[15]:=  N[E, 30]
Out[15]=  2.71828182845904523536028747135
In[16]:=  N[Sqrt[2], 30]
```

Out[16]= 1.41421356237309504880168872421

Example: Find the absolute values of 8, -8 and x.

Solution:

In[17]:= Abs[8]

Out[17]= 8

In[18]:= Abs[-8]

Out[18]= 8

In[19]:= Abs[x]

Out[19]= Abs[x]

Note that Abs[ ] will be left unevaluated if the expression in the square bracket is not a number.

Example: Future Value (Compounding)

Find the compounded value of $100 deposited at a rate of 6% for 5 years, if the interest is compounded annually, quarterly, monthly, and continuously.

Solution: For a discrete compounding, the formula is $S = A(1 + \frac{r}{n})^{nt}$, where S is the compounded value, A is the initial investment, r is the interest rate, n is the number of time the interest is compounded a year, and t is the number of years. If the interest is compounded continuously, the compounding formula is $S = Ae^{rt}$. Annual Compounding:

In[20]:= N[100 (1 + .06) ^ 5]

Out[20]= 133.823

Quarterly Compounding:

In[21]:= N[100 (1 + .06 / 4) ^ (4 * 5)]

Out[21]= 134.686

Monthly Compounding:

In[22]:= N[100 (1 + .06 / 12) ^ (12 * 5)]

Out[22]= 134.885

Continuous Compounding:

In[23]:= N[100 E^ (.06 * 5)]

Out[23]= 134.986

Future Value: General Formulation

You may formulate the problem in general form so that repeated calculations can be done by changing the parameters. Using the palettes, formulate $S = A(1 + \frac{r}{n})^{nt}$, with the parameters S, the compounded value, A the initial investment, r the interest rate, n the frequency with which the interest is compounded a year to be filled for given problem. For the case of preceding problem A = 100, r = .06, n = 1, and t = 5, the problem can be solved as,

In[24]:=
```
Clear[S, A, r, n, t]
A = 100 ; r = .06; n = 1; t = 5;
S = N[A (1 + r/n)^(n t)]
```

Out[24]= 133.823

Do More Practice: In the future value model above change the interest rate r to 7%, 8%, 9%, and 10%. Run the program and record the corresponding values of S. Graph the relationship between S and r on a graphing paper. What is the relationship between the two variables.

Find the compounded value of $1000 deposited at an annual rate of 6% after 10 years if:

a- the interest is compounded annually.

b- the interest is compounded quarterly.

c- the interest is compounded monthly.

d- the interest is compounded daily

e- the interest is compounded continuously (you may select n as a very large number as, n = 10000).

Example: Present Value (Discounting)

An investment is expected to have a return of $22000 at the end of the first year, $28500 at the end of the second year, $32000, at the end of the third year, $18000 at the end of the fourth year, and $12000 at the end of the fifth year. At the end of the fifth year the business is expected to have a market value of $52000. What is the present value of the future returns if the interest rate is 6%? what is the present value of the investment?

Solution: To find the present value of the expected future returns for the next n years, use the formula $PV = \frac{R1}{(1+r)^1} + \frac{R2}{(1+r)^2} + \frac{R3}{(1+r)^3} + \ldots\ldots + \frac{Rn}{(1+r)^n}$, where PV is the present value, Ri is the return in the ith year, r is the rate of return or interest rate, and n is the number of years. Therefore, $PV = \frac{22000}{(1.06)} + \frac{28500}{1.06^2} + \frac{32000}{1.06^3} + \frac{18000}{1.06^4} + \frac{12000}{1{,}06^5} = 96212.20$. The present value of the investment, PVI is calculated by adding the present value of the scrap price of the investment to PV. That is $PVI = 96212.20 + \frac{52000}{1.06^5} = 135070$.

Using *Mathematica*, the present value of the future returns is calculated as,

In[25]:=
```
PV = N[22000 / (1 + .06) + 28500 / (1 + .06) ^2 +
  32000 / (1 + .06) ^3 + 18000 / (1 + .06) ^4 + 12000 / (1 + .06) ^5]
```

Out[25]= 96212.2

The present value of the investment (PVI) is calculated by adding the present value of the scrap value or the final value of the capital to the present value of the annual returns as,

In[26]:=
```
PVI = N[22000 / (1 + .06) + 28500 / (1 + .06) ^2 + 32000 / (1 + .06) ^3 +
   18000 / (1 + .06) ^4 + 12000 / (1 + .06) ^5 + 52000 / (1 + .06) ^5]
```

Out[26]= 135070.

Present Value: General Formulation

You may formulate the problem in general form so that repeated calculations can be done by changing t

he parameters. Using the palettes, the formulation is done below, where PV is the present value, R_t is cash flow at time period t, and r is the discounting factor. For the preceding problem, the formulation is,

```
In[27]:= Clear[PV, R1, R2, R3, R4, R5, r]
         R1 = 22000;
         R2 = 28500;
         R3 = 32000;
         R4 = 18000;
         R5 = 12000;
         r = .06;
```

$$PV = \frac{R1}{(1+r)^1} + \frac{R2}{(1+r)^2} + \frac{R3}{(1+r)^3} + \frac{R4}{(1+r)^4} + \frac{R5}{(1+r)^5}$$

```
Out[27]= 96212.2
```

Do More Practice: Change the discount rate to 7%, 8%, 9%, and 10%. Find the PVs and record them. Graph the relation between discount rates and the PVs on a graphing paper. What can you say about the relationship between the two variables?

Net Present Value

Net present value, NPV is the difference between the value of the initial investment, Io and the present value of investment, PVI. If NPV > 0, the investment is profitable. If NPV < 0, the investment will have a loss. Suppose an investment is expected to have returns of \$22000, \$28500, \$32000, \$18000, and \$12000 at the end of each year for the next five years. At the end of the fifth year, the business is

expected to have a market value of MV = \$52000. Given that the initial investment is \$120000 what is the net present value of the investment, if the discount rate is 6%?

```
In[28]:= Clear[NPV, Io, MV, R1, R2, R3, R4, R5, r]
         Io = 120000;
         R1 = 22000;
         R2 = 28500;
         R3 = 32000;
         R4 = 18000;
         R5 = 12000;
         MV = 52000;
         r = .06;
```

$$PV = \frac{R1}{(1+r)^1} + \frac{R2}{(1+r)^2} + \frac{R3}{(1+r)^3} + \frac{R4}{(1+r)^4} + \frac{R5}{(1+r)^5} + \frac{MV}{(1+r)^5} - Io$$

```
Out[28]= 15069.6
```

Do More Practice: Change the discount rate to 7%, 8%, 9%, and 10%. Find the NPVs and record them. What discount rate makes the business to break even (NPV = 0)?

Keep discount rate as it is (6%). Increase Io by factor \$5000 each time. Find NPVs and record them. At what value of Io the business breaks even?

Example: Perpetuities

A Perpetuity is a security that yields a stream of annual payments at an annual rate of r forever. The value of the perpetuities are decided by the sum of the present value of the stream of payments for infinite number of years. That is, PV = $\sum_{t=1}^{\infty} \frac{R}{(1+r)^t}$ where, R is the annual payments and r is the discount factor. Suppose a perpetuity yields annual payments of \$1 forever. What is the present value of the perpetuity if discount rate is 5%?

To formulate the problem you may use the **Sum** command or you may use palettes for summation operator "$\sum_{\square=\square}^{\square}\square$" and fill in the information at the proper spaces. With the sum command the problem is formulated as,

```
In[29]:= Clear[R, t, r, PV]
         r = .05; R = 1;
         PV = Sum[1 / (1 + r) ^t, {t, 1, Infinity}]
Out[29]= 20.
```

Using palettes the problem is formulated as,,

```
In[30]:= Clear[t, i, r, pv]
         r = .05; R = 1;
```

$$PV = \sum_{t=1}^{Infinity} \frac{R}{(1+r)^t}$$

```
Out[30]= 20.
```

Do More Practice: Change the discount rate to 10%, 15%, and 20%. Find the PVs and record them. What is the relationship between discount rates and PVs?

Keep R = 1 and change the maturity date from Infinity to 1 year, 10 years, 20 years, and 50 years and find the corresponding values of PVs. What is the relationship between the PVs and the extent of the maturity?

Example: Production Function

Production function is a technical relation between inputs and output. One of the very well-established production functions in economic literature is Cobb-Douglas production function. Cobb-Douglas production function is expressed as Q = A $L^{\alpha}K^{\beta}$ where, A represents the state of the technology, L is labor input, K is capital input, Q is output, and α and β are the shares of labor and capital inputs from the output, respectively (for more information on the properties of this function see, for example, Nicholson (1998)). Suppose the relationship between labor input (L), capital input (K) and output (Q) is given as Q = $100L^{.75}K^{.25}$. Find the output level for a labor input of 50 units and capital input of 25 units.

Solution: This function which shows the physical relationship between labor input, capital input, and output is called Cobb-Douglas production function. For the given levels of inputs, the output is Q = $100(50)^{.75}(25)^{.25}$ = 4204.48.

Using Mathematica, the output can be evaluated by substituting for L and K in the function as,

```
In[31]:= Q = 100 (50) ^.75 (25) ^.25
Out[31]= 4204.48
```

Cobb-Douglas Production Function: General Formulation

You may formulate the problem in general form as,

```
In[32]:= Clear[Q, A, L, K, α, β]
         A = 100; L = 50; K = 25; α = .75; β = .25;
         Q = N[A L^α K^β]
Out[32]= 4204.48
```

Do More Practice: In the example above double, triple, and quadruple the inputs and write the corresponding values of the outputs. What is the effect of proportional change in inputs on output? What type of return to scale exists in this production function? (Note that, in this problem $\alpha + \beta = 1$).

Change the α and β coefficients to $\alpha = 1$ and $\beta = .50$, where $\alpha + \beta > 1$. Run the production function model with the initial values of labor and capital and record the output. Run the model by doubling, tripling, and quadrupling the inputs. What happens to the corresponding outputs? What is the effect of a change in scale of operation in this case? What is the economic name of the returns to scale in this case?

Problems:

1- Find the numerical value of $\sqrt[3]{\left(\frac{2}{3} + \sqrt{2}\right)^5 - \frac{4}{5}}$

2- Find the numerical value of $8\sqrt[3]{\left(\frac{29}{23}\right)^2 + 25}$

3- Find the numerical value of $\dfrac{\sqrt[3]{\frac{2}{3}}}{\sqrt[5]{5\left(\frac{1}{3}\right)^2}}$

4- What is the compounded value of \$100 invested at a rate of 8% for five years, if the interest is compounded biannually? Change the interest rate to 16% and redo the problem. What is the effect of doubling of interest rate on the compounded value?

5- In the present value problem of the text with the stream of revenues given as: R1=22000; R2=28500; R3=32000; R4=18000; R5=12000; and discount factor of r =6% find the present value. Change the interest rate to 12% and find the present value. What is the effect of doubling discount rate on the present value?

6- In the Cobb-Douglas production function $Q = 100\, L^{\alpha}K^{\beta}$ let L =50, K = 10, $\alpha = .8$ and $\beta = .2$ ($\alpha + \beta = 1$). Find the value of Q.

a- Now, double, triple, and quadruple inputs (L and K) and find Q. How does proportional changes in inputs change output?

b- Let $\alpha = .9$ and $\beta = .7$, where $\alpha + \beta > 1$. Now, double, triple, and quadruple inputs and find Q. How does proportional changes in inputs change output?

c- Let $\alpha = .5$ and $\beta = .2$, where $\alpha + \beta < 1$. Find Q. Now, double, triple, and quadruple inputs and find Q. How does proportional changes in inputs change output?

III. Sets: The Complement, Union, and Intersection Commands

A set is a collection of well-defined, distinguishable objects. A set is usually denoted by a curly bracket. It may contain a finite number of elements, such as S = {a, m, c, p} and T = {10, 2, 13, 4, 9} or it may contain an infinite number of elements, such as the set N of natural numbers T = {1, 2, 3,........}. If b is an element of set S, then we write $b \in S$. If b is not an element of S, then we write $b \notin S$.

Subset of a Set If every element of set S is also an element of set T, but $S \neq T$, we say that S is contained in T or S is a subset of T. We then write $S \subset T$. For example, if S = {b, c, d} and T = {a,b,c,d}, then $S \subset T$.

Universal Set If all sets under consideration are subsets of a set U, then U is called the universal set. For example, a deck of playing cards is a universal set. A set containing four aces or a set containing four jacks are subsets of universal set.

The Complement of a Set Let S be a subset of a universal set U. The set of elements which belongs to U but not to S is called the complement of S and is denoted by S'. For example, for a deck of playing cards which contains faces and numbers, a set containing faces is the complement of a set containing numbers.

Example: Let S = {a, b, c} and T = {a, c, d} be subsets of the universal set U = {a, b, c, d, e}. Find S' and T'.

Solution: By the definition of the complement sets above, S' = {d, e} and T' = {b, e}.

```
In[33]:= Clear[u, s, t]
         u = {a, b, c, d, e};
         s = {a, b, c};
         t = {a, c, d};
         Complement[u, s]

Out[33]= {d, e}

In[34]:= Complement[u, t]

Out[34]= {b, e}
```

Example: Let S be a set of 5 random drawings and T be a set of 10 random drawings of numbers between 0 and 9. Find S' and T'.

Solution: The universal set is U = {0, 1, 2, ..., 9}. S and T can be generated by drawing 5 and 10 random observation from a bag containing the numbers 0 through 9 or by a program that generates random numbers. Here, we use *Mathematica* to generate those numbers Using *Mathematica,* U = **Table**[x, {x, 0, 9}] generates a set of numbers from 0 to 9, S = Table[**Random**[Integer, {0, 9}], {5}] generates a set of 5 random numbers in the range 0 to 9 and T=Table[Random[Integer, {0, 9}], {10}] generates a set of 10 random numbers in the range 0 to 9.

```
In[35]:= Clear[x, U, S, T]
         U = Table[x, {x, 0, 9}]
         S = Table[Random[Integer, {0, 9}], {5}]
         T = Table[Random[Integer, {0, 9}], {10}]

Out[35]= {0, 1, 2, 3, 4, 5, 6, 7, 8, 9}

         {5, 4, 2, 3, 4}

         {8, 5, 6, 4, 1, 0, 2, 3, 9, 1}
```

The following commands find the complement of S and T.

```
In[36]:= S' = Complement[U, S]

Out[36]= {0, 1, 6, 7, 8, 9}

In[37]:= T' = Complement[U, T]

Out[37]= {7}
```

DO More Practice: Let U be the universal set of two digit numbers. Let S, T, and W be random subsets of u each containing 80, 90, and 100 random picks. Find the complements of S, T, and W. How many elements each subset contains?

Null Set If a set contains no elements it is called an empty set and is denoted by 0.

Equality of Sets Two sets are equal if and only if they contain the same elements. If set S is equal to set T, we write S = T, otherwise we write S ≠ T.

Example: Let S = {1, 2, 3}, T = {3, 1, 2} and V = {3, 3, 2}. Then, S = T, S ≠ V, and T ≠ V.

Union of Two Sets The union of two sets S and T, written S ⋃ T, is the set of all elements which are members of S or T.

Example: Let s = {1, 3, 5} and t = {7, 3, 1, 2}. Find s ⋃ t.

Solution: Union of two sets s and t is, s ⋃ t = {1, 2, 3, 5, 7}.

```
In[38]:= Clear[s, t]
         s = {1, 3, 5};
         t = {7, 3, 1, 2};
         Union[s, t]
Out[38]= {1, 2, 3, 5, 7}
```

Example: Let S be a set of 5 random drawings and T be a set of 7 random drawings of numbers between 0 and 9. Find S ⋃ T.

Solution: S and T can be generated by drawing 5 and 7 random observation from a bag containing the numbers 0 through 9 or by a program that generates random numbers. Here, we use *Mathematica* to generate those numbers Using *Mathematica,* S = Table[Random[Integer, {0, 9}], {5}] generates a set of 5 random numbers in the range 0 to 9 and T=Table[Random[Integer, {0, 9}], {7}] generates a set of 7 random numbers in the range 0 to 9. The command Union[S, T] finds the union of the two sets.

```
In[39]:= Clear[S, T]
         S = Table[Random[Integer, {0, 9}], {5}];
         T = Table[Random[Integer, {0, 9}], {7}];
         Union[S, T]
Out[39]= {0, 2, 3, 5, 6, 7, 8, 9}
```

DO More Practice: Let u be the universal set of two digit numbers. Let S, T, and W be random subsets of u each containing 80, 90, and 100 random picks. Find S ⋃ T, S ⋃ W, and T ⋃ W. Find (S ⋃ T) ⋃ W. Is this set equal to u?

Intersection of Two Sets: The intersection of two sets S and T, written S ⋂ T, is the set of all elements which are members of both S and T.

Example: Let s = {1, 3, 5} and t = {7, 3, 1, 2}. Find s ⋂ t.

Solution: Intersection of two sets s and t is, s ⋂ t = {1, 3}

```
In[40]:= Clear[s, t]
         s = {1, 3, 5};
         t = {7, 3, 1, 2};
         Intersection[s, t]
Out[40]= {1, 3}
```

Example: Let S and T be two sets of 10 random drawings of numbers between 0 and 9. Find S ∩ T.

Solution: S and T can be generated by drawing 10 random observations from each set from a bag containing the numbers 0 through 9 or by a program that generates random numbers. Here, we use *Mathematica* to generate those numbers Using *Mathematica,* S = Table[Random[Integer, {0, 9}], {10}] generates a set of 10 random numbers in the range 0 to 9 for S and T=Table[Random[Integer, {0, 9}], {10}] generates a set of 10 random numbers in the range 0 to 9 for T. The command Intersection[S, T] finds the intersection of the two sets.

```
In[41]:= Clear[S, T]
         S = Table[Random[Integer, {0, 9}], {10}];
         T = Table[Random[Integer, {0, 9}], {10}];
         Intersection[S, T]
Out[41]= {0, 5, 6}
```

DO More Practice: Let u be the universal set of two digit numbers. Let S, T, and W be random subsets of u each containing 80, 90, and 100 random picks. Find S ∩ T, S ∩ W, and T ∩ W. Find (S ∩ T) ∩ W.

Disjoint Sets: Sets S and T are called disjoint if S ∩ T = 0. That is , the sets do not have any common elements.

Example: Two sets s = {set of all odd numbers 1, 3, 5, ...} and t = {set of all even numbers 2, 4, 6, ...} are disjoint and s ∩ t = 0.

Solution: The two sets have no common elements. Therefore, s ∩ t = 0.

```
In[42]:= Clear[s, t]
         s = {1, 3, 5};
         t = {2, 4, 6};
         Intersection[s, t]
Out[42]= {}
```

DO More Practice: Let T be the universal set of one digit numbers and S be the universal set of two digit numbers. Show that the two sets are disjoint sets.

Algebra of Sets: The theory of sets, founded by the German mathematician George Cantor, is an important branch of mathematics and is a very useful tool in probability theory or in analyzing economic issues, especially in the area of consumer theory and choice. The algebra of sets is very similar to, but not identical with, the algebra of the real numbers. The following relationships compare the algebra of sets to the algebra of real numbers.

	Algebra of Sets	Algebra of Real Numbers
1-Commutative Laws:	$S \cup T = T \cup S$	$S + T = T + S$
	$S \cap T = T \cap S$	$S \times T = T \times S$
2-Associative Laws:	$S\cup(T\cup R)=(S\cup T)\cup R$	$S+(T+R)=(S+T)+R$
	$S\cap(T\cap R)=(S\cap T)\cap R$	$S\times(T\times R)=(S\times T)\times R$
3-Distributive Laws:	$S\cap(T\cup R)=(S\cap T)\cup(S\cap R)$	$S\times(T+R)=(S\times T)+(S\times R)$

$$S \cup (T \cap R) = (S \cup T) \cap (S \cup R) \qquad S + (T \times R) \neq (S+T) \times (S+R)$$

Furthermore, it should be noted that, although commutative laws hold for union and intersection, they do not hold in general for set differencing. If S and T are two sets, S - T and T - S are not in general equal. Also, set differencing is not associative; i.e., S - (T - R) is not in general equal to (S - T) + R.

Example: If S = {2, 3, 5, 7} and T = {3, 4, 5, 6, 7, 8}, S - T = {2}, whereas T - S = {4, 6, 8}. Therefore, S - T ≠ T - S.

Example: Let S = {3, 2, 5, 6}, T = {1, 2, 3, 4}, and R = {2, 5, 1, 4}. Find S - T, T - S, S - (T - R), and (S - T) + R.

Solution: S - T = {5, 6} and T - S = {1, 4}. Therefore, S - T ≠ T - S.

S - (T - R) = {2, 5, 6} and (S - T) + R = {2, 5, 6, 1, 4}, therefore, S - (T - R) ≠ (S - T) + R.

DO More Practice: Let U be the universal set of two digit numbers. Let S and T be a random subset of U each containing 10 random picks.

1- Find the complements of S and T.

2- Find intersection of S and T.

3- Find the union of S and T.

4- Run your program from the first line several times. Do you come with the same solutions each time?

Problems

1. Let S = {3, 4, 5}, T = {6, 7, 8}, V = {5, 4, 3}, and Z = {3, 2, 5, 6, 9, 1}. Which of the following statements are true?

a) $S = T$
b) $S = V$
c) $4 \in S$
d) $6 \in S$
e) $S \subset T$
f) $V \subset T$
g) $V \subset Z$
h) $S \cup T = Z$
i) $S \cup V = S$
j) $S \cup V \cup T = Z$
k) $S \cap T = 0$
l) $2 \in S \cap Z$

2. Let S = {1, 2, 3, 4, 5, 6, 7, 8, 9}. List the elements of the following subsets of S.

a) {x: x is odd}
b) {x: x is even}
c) {x: x is greater than 5}
d) {x: x is less than 10}

3. Given S = {1, 2, 3, 4, 5}, T = {2, 3, 5, 6}, and V = {1, 2, 4, 6, 7}, verify that

a) $S \cap (T \cup V) = (S \cap T) \cup (S \cap V)$

b) $S \cap (T \cap V) \neq (S \cup T) \cap V$

4. Let S be a subset of a universal set U. Show that

a) $S \cup S' = U$

b) $S \cap S' = 0$

c) $0' = U$

5. Take 0 to 9 as a universal set, U. Take your telephone number as a set, T.

   a- Find the complement of T.

   b- Find intersection of U and T.

   c- Find the union of U and T.

6. Let S = {1, 2, 3} and T = {1, 3, 6} be subsets of the universal set U = {1, 2, 3, 4, 5, 6}. Find the followings:

   a) S' b) T'

   c) S' $\cup$ T' d) S' $\cap$ T'

IV. Mathematical Relations: The **Solve** Command

Before discussing the mathematical relations, three concepts which will be used frequently in this section and later chapters will be introduced.

Ordered Pairs An ordered pair of numbers {x, y} is a set containing two elements, one of which (x) precedes the other (y). The element x is called the first coordinate, and the element y the second coordinate. The pair {x, y} is said to be "ordered" because the order in which the numbers appear is of significance. The pairs {5, 8} and {8,5} are different ordered pairs.

Variables A variable is a symbol that may represent any member of a particular set S and may take on different values.

Example: Suppose that monthly observations on income and consumption of an individual is given as,

Income: 2300, 2500, 1800, 2000, 1500, 2200

Consumption: 2040, 2350, 1740, 1850, 1450, 1960

The first row (Income) may be represented by a variable, say Y, and the second row (consumption) by a variable, say C. The variables Y and C can take any values that consumer may have in his income and consumption set.

Coordinate System Geometrically, real numbers can be represented as points on a straight line. It is customary to draw this line horizontally and associate an arbitrary point on it with the integer zero. Points to the right of the zero point will represent positive numbers and points to the left will represent negative numbers. Such a line is called the x-axis and the point corresponding to zero is called the origin. The portion of the line to the right of zero is called the positive axis and the portion to the left of origin is called the negative axis.

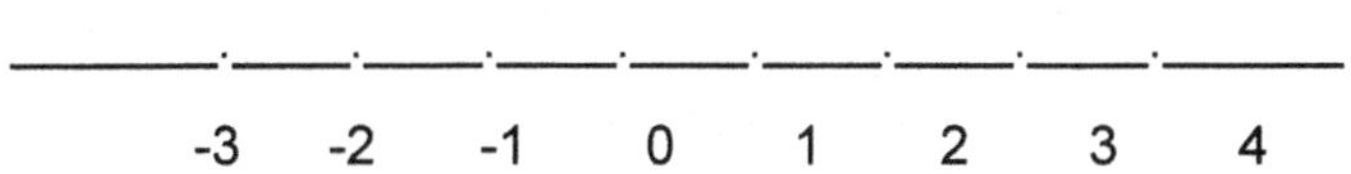

Figure 1. x-axis and origin

To present ordered pairs geometrically, draw a line perpendicular to the x-axis at the origin and call the l

ine the y-axis. It is customary to draw the positive y-axis above the origin, and to draw the negative y-axis below the origin. The plane determined by the x- and y-axes is called the xy-plane,

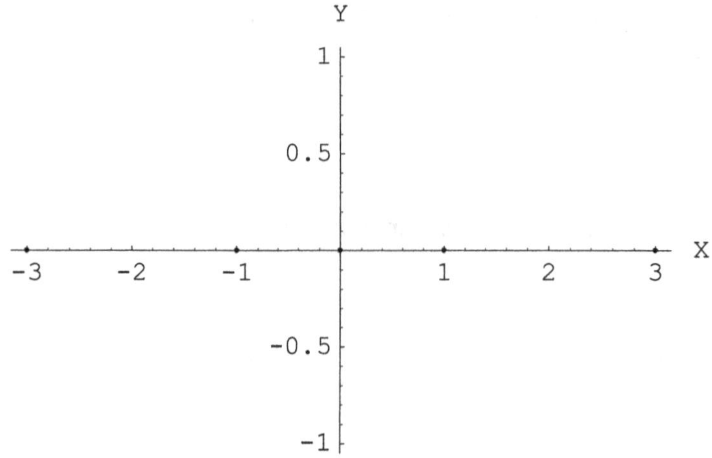

Rectangular Coordinates Given an ordered pair of real numbers (x, y), we can associate this pair with a point (x, y) on the x-y plane. Rectangular coordinates use the distance from the origin on x and y axes to locate a point on the coordinate system. For example, the ordered pairs (0, 1) and (2, -3) are shown in the figure below.

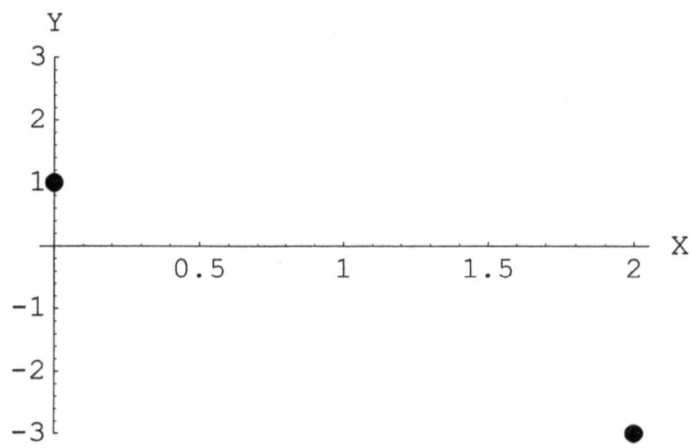

Polar Coordinates Another way to locate a point on the coordinate system is to use polar coordinates. The polar coordinates use a distance from a reference point, say origin, and an angle to describe locations. In polar coordinates, 0 is along a horizontal ray to the right of the reference point and angles are measured counterclockwise. Polar coordinates are written as ordered pairs (r, θ) where r, radius, is the polar distance and θ is the measure of the polar angle.

The relation between rectangular and polar coordinates follows the equality of the ordered pairs (x, y) = (rsinθ, rcosθ). To convert rectangular coordinates to polar coordinates take the following steps:

1- Find $r = \sqrt{x^2 + y^2}$

2- Find θ using $y = r\sin\theta$, $x = r\cos\theta$, or $\tan\theta = \frac{y}{x}$.

3- Find $\theta = \sin^{-1}(\frac{y}{r}) = \tan^{-1}(\frac{y}{x})$

Example: Find the point (3,4) on the coordinate system and convert it to polar coordinates.

Solution: $r = \sqrt{3^2 + 4^2} = 5$, $y = r\sin\theta$ or $4 = 5\sin\theta$, $\sin\theta = .8$ and $\theta = 53.13^{o}$. Therefore the polar coordinate are (5, 53.13).

Economic Relationships: A relationship between two variables can be expressed in three forms: Schedule, Graph, and Equation.

Schedule: A relation here is defined as any set of ordered pairs (x, y), where x and y are real numbers.

Example: The relationship between income (Y) and consumption (C) may be studied by listing of data and observing the relation.

(Y)	(C)	(Y, C)
2300	2040	(2300, 2040)
2500	2350	(2500, 2350)
1800	1740	(1800, 1740)
2000	1850	(2000, 1850)
1500	1450	(1500, 1450)
2200	1960	(2200, 1960)

Graph: The same information may be studied through plotting the paired observations on the coordinate system. Plotting data will be studied in Chapter 2. Here, the following graph shows the relationship between C and Y.

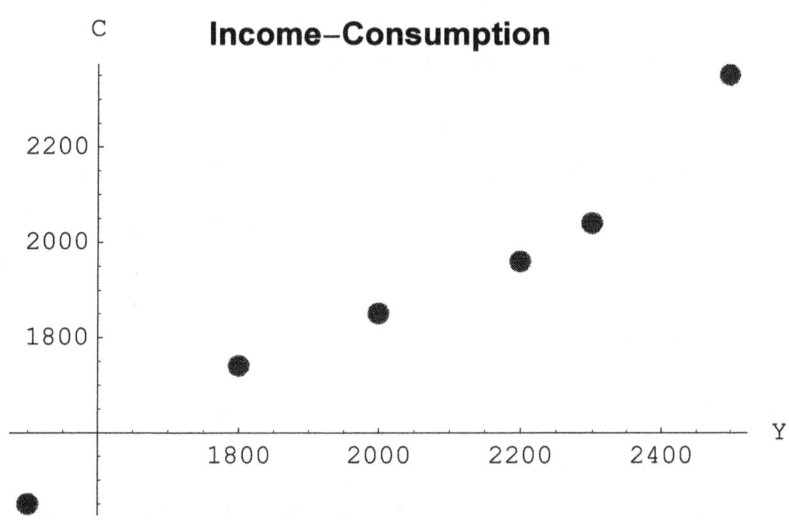

Equation: An equation is a statement of equality between two or more variables. The relationship between two or more variables my be studied by fitting an equation to the observed data and studying the mathematical relationship between the variables. For example, for the income-consumption data above, we can fit an equation of a straight line. Fitting equations to data will be studied in Chapter 3. Here, assuming that such equation is estimated as $C = 219 + 0.8Y$, the plot of the points and the equation are shown in the following graph.

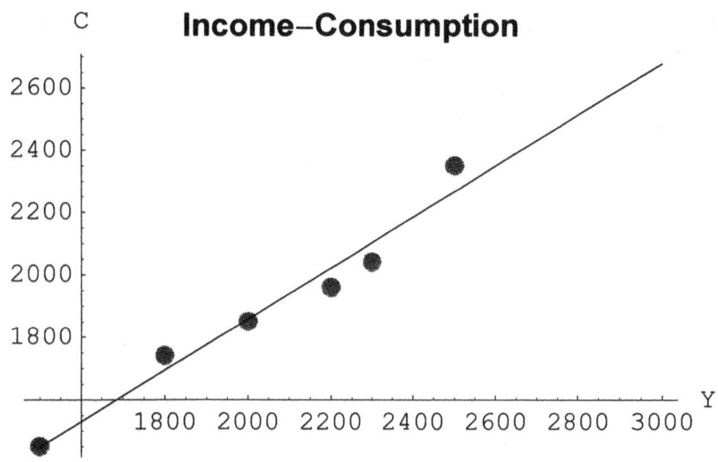

Solution of Equations: The **Solve** Command

The solution of an equation in variable x is a set of numbers or objects that satisfies the equation.

Example: Solve $x^2 - 4 = 0$

Solution: Set $x^2 = 4$ and solve for x. The solution is $x = \sqrt{4} = \mp 2$. Using *Mathematica*,

```
In[43]:= Clear[x]
         Solve[x^2 - 4 == 0, x]
Out[43]= {{x → -2}, {x → 2}}
```

Note that, equality sign "=" in Solve command of *Mathematica* is "==".

Example, Quadratic Equations: Solve $x^2 + 2x - 8 = 0$.

Solution: The general form of quadratic equation is $ax^2 + bx + c = 0$, the solution of which is $x = \frac{-b \mp \sqrt{b^2 - 4ac}}{2a}$. Therefore, by substituting the values to the formula, $x = \frac{-2 \mp \sqrt{2^2 - (4)(-8)}}{2} = -4$ & 2. Using *Mathematica*, the solution is

```
In[44]:= Clear[x]
         Solve[X^2 + 2X -8 == 0, X]
Out[44]= {{X → -4}, {X → 2}}
```

Example: Find the solution: set of $|x+3| = 4$

Solution: If $x > -3$, $|x + 3| = x + 3$. If $x < 3$, $|x + 3| = -(x + 3)$. Therefore, $|x + 3| = 4$ give rise to two equations, $x + 3 = 4$ and $-x - 3 = 4$, the solutions of which are x = 1 & -7.

Using *Mathematica*, the solution is

```
In[45]:= Clear[x]
         Solve[Abs[x+3]==4, x]

Solve::ifun :
 Inverse functions are being used by Solve, so some solutions may not be found.
```

```
Out[45]= {{x → -7}, {x → 1}}
```

Example: Find the solution set of $x^3 - x^2 + 4x - 4 = 0$.

Solution: To solve a cubic equation or equations of higher order, use a graphic equation to find the zeroes of the equation. Zeroes of an equation are the x-intercept of the equation. Alternatively, you can set values for x and through trial and error find the x-value that will make the equation equal to zero. For example, in the equation of the example, if x = 0, the value of the equation is -4. If x = 2, the value of the equation is +8. Therefore, the x-value that makes equation equal to zero lies between zero and two. By assigning x some values in this range we find that x = 1. This is only one of the solutions. Since the equation is a cubic equation, there should be two more solutions. To find the other two solutions, divide $x^3 - x^2 + 4x - 4$ by (x - 1) to get $\frac{x^3 - x^2 + 4x - 4}{(x-1)} = x^2 + 4$. The solution of $X^2 + 4 = 0$ is $x = \mp 2i$.

Using *Mathematica*, the solution is

```
In[46]:= Clear[x]
         Solve[x^3 -x^2 + 4x -4==0,x]

Out[46]= {{x → -2 i}, {x → 2 i}, {x → 1}}
```

Example: Application to Market Analysis- Demand and supply for a product are given as Qd = 20-2P and Qs = -1+ P. Find the equilibrium price and quantity.

Solution: At the equilibrium point, quantity demanded and quantity supplied are equal. Therefore, by setting Qd = Qs, solve the resulting equation for P. That is 20 - 2P = -1 + P, where P = 7 is the solution. Substitute for P in Qd or Qs to get equilibrium Q = 6.

Using *Mathematica*,

```
In[47]:= Clear[Qd, Qs, p, sol]
         Qd = 20 - 2p;  Qs=-1+p;
         sol = Solve[Qd == Qs, p]
         Qd /. sol

Out[47]= {{p → 7}}, {6}
```

Example: Application to Break-Even Analysis.

Find the break-even quantity for a business with an annual fixed cost of $24000 and average variable cost of $2 per unit of output. The product is selling at market price of $5 per unit. Find the quantity that will yield an annual profit of $18000.

Solution: Total revenue (TR) equals price times quantity. Total cost (TC) equals fixed cost plus variable cost, where variable cost equals average cost times quantity. At the break-even point TR = TC. Therefore, define TC and TR with the given information and solve the equation TR = TC for q. That is, 5q = 24000 + 2q. Solving this equation for q results in q = 8000. To find the q that results in $18000 profit, we know that Profit = TR - TC. Therefore, set TR - TC = 18000 and solve for q. This results in a quantity of q = 14000 units.

```
In[48]:= Clear[TR, TC, q, profit]
         TR = 5q;
         TC = 24000 + 2q;
```

```
Solve[TR == TC, q]
Out[48]= {{q → 8000}}
In[49]:= profit = TR-TC;
         Solve[profit==18000,q]
Out[49]= {{q → 14000}}
```

Example: Application in Finance (Internal Rate of Return).

Internal rate of return is the break-even discount rate for an investment. It is the rate at which the present value of the net future cash flow equals to the amount invested. That is, Io = $\frac{R1}{(1+r)^1} + \frac{R2}{(1+r)^2} + \frac{R3}{(1+r)^3} + ...$, where Io is the initial investment, Rt is the return at the end of the tth year, and r is the discount rate.

Suppose a \$20,000 initial investment has annual returns of \$4,500 for five years and a scrap value of \$2,500 at the end of the fifth year. The internal rate of return on the investment is calculated by setting the present value of the future returns equal to the value of investment . Therefore, formulate the problem as $20000 = \frac{4500}{1+r} + \frac{4500}{(1+r)^2} + \frac{4500}{(1+r)^3} + \frac{4500}{(1+r)^4} + \frac{4500}{(1+r)5} + \frac{2500}{(1+r)^5}$ and solve for r. Since the highest power of the equation is five, there will be five solutions to the equation. However, only one of the solutions is acceptable. To find the solution, substitute values for r and use the trial error method of the previous example to find the r-value that will make the two sides of the equation equal to each other. This value is r = 7.41%.

Using *Mathematica*, the solution is

```
In[50]:= Clear[ r, sol ]
sol = Solve[20000==4500/(1+r) + 4500/(1+r)^2 + 4500/(1+r)^3 + 4500/(1+r)^4 + 4500/(1+r)^5 +
2500/(1+r)^5, r]; N[sol]
Out[50]= {{r → 0.0740788}, {r → -1.61085 - 0.42478 i}, {r → -1.61085 + 0.42478 i},
         {r → -0.813685 - 0.744264 i}, {r → -0.813685 + 0.744264 i}}
```

Note that, since the equation is a 5th order polynomial, there will be five solutions, some with imaginary numbers. To find the real-number solution in decimals, use the N command. Here, the only real solution is r = .0740788 or a rate of return of 7.41%.

Example:

Find the internal rate of return for \$24000 investment with annual net cash flow of R1 = \$5000, R2 = \$7000, R3 = \$8500, R4 = \$6000, R5 = \$3500, and a scrap value of \$1200 at the end of the 5th year.

```
In[51] :=  Clear[R 1, R2, R3, R4, R5, Rs, Io, r, IRR]
   R1 = 5000; R2 = 7000; R3 = 8500; R4 = 6000; R5 = 3500; Rs = 1 200; Io = 24000;
IRR = N[Solve[ R1/(1+r)^1 + R2/(1+r)^2 + R3/(1+r)^3 + R4/(1+r)^4 + R5/(1+r)^5 + Rs/(1+r)^5 - Io == 0, r]]
Clear::ssym : R 1 is not a symbol or a string.
Out[51]= {{r → 0.0956087}, {r → -1.5228 - 0.369277 i}, {r → -1.5228 + 0.369277 i},
         {r → -0.920839 - 0.655767 i}, {r → -0.920839 + 0.655767 i}}
```

Do More Practice: Change the initial investment in the preceding example to \$25000, \$26000, \$27000, \$28000, \$29000, and \$30000, and find the corresponding IRRs. Graph the relationship between the initial investments and the IRRs on a graphing paper and comment on the relationship between the two variables.

Application to Finance: Investment and Growth

Suppose an initial investment of $24000 has a net return of $1000 at the end of first year and an expected continuous returns with an annual growth rate of 10% in the coming years. What is the net

present value of the firm after 5 years, 10 years, 15 years, and 20 years if the discount rate is 6%? After how many years the firm will break even?

Solution: Assuming the initial investment Io, the first year return R1, the growth rate g, and the discount rate r, the problem can be formulated as: NPV = $\sum_{t=1}^{n} \frac{R1(1+g)^t}{(1+r)^t}$ - Io. The NPV after 5, 10, 15, 20 years can be calculated by substituting for the parameters of the NPV equation. The NPV at the end of 5th, 10th, 15th, an 20th years are -18404.70, -11670.90, -3566.96, and 6185.82, respectively. The number of years needed for the firm to break-even is calculated by setting NPV = 0 and solving for n. Using *Mathematica*,

In[52]:=
```
Clear[R1, Io, g, r, NPV, n]
R1 = 1000; Io = 24000; g = .1; r = .06; n = 5;
```
$$NPV = \sum_{t=1}^{n} \frac{R1\ (1+g)^t}{(1+r)^t} - Io$$

Out[52]= `5595.34 - i o`

In[53]:=
```
Clear[R1, Io, g, r, n]
R1 = 1000; Io = 24000; g = .1; r = .06;
```
$$sol = Solve\left[\sum_{t=1}^{n} \frac{R1\ (1+g)^t}{(1+r)^t} - Io == 0,\ n\right]$$
```
N[sol]
```

```
Unique::usym : 25 is not a symbol or a valid symbol name.

Sum::write : Tag Unique in Unique[25] is Protected.

Unique::usym : 25 is not a symbol or a valid symbol name.

Solve::ifun : Inverse functions are
  being used by Solve, so some solutions may not be found.
```

Out[53]= $\{\{n \to -\frac{4\,Log[5]+Log[11]-Log[12875]}{Log[5]+Log[11]-Log[53]}\}\}$

`{{n → 16.9378}}`

Do More Practice: Change the initial investment in the preceding example to $25000, $26000, $27000, $28000, $29000, and $30000, and find the corresponding number of years that the business needs to break-even. Graph the relationship between the initial investments and the corresponding number of years on a graphing paper and comment on the relationship between the two variables.

Change the growth rate to 15%, 20%, and 25%. Keep other parameters as in the text. Find the number of years that the business needs to break-even. Graph the relationship between the growth rate and the corresponding number of years on a graphing paper and comment on the relationship between the two variables.

Change the discount rate to 8%, 10%, 12%, and 15%. Keep other parameters as in the text. Find the number of years that the business needs to break-even. Graph the relationship between the discount rate and the corresponding number of years on a graphing paper and comment on the relationship between the two variables.

4- What growth rate is needed for the firm to break-even in 10 years, if the discount rate is 6%?

Problems

1- Solve the following equations for x.

a) $x^2 + 4 = 0$

b) $x^2 - 3x - 4 = 0$

c) $x^2 - 6x + 25 = 0$

d) $x^3 + 2x^2 + 2x + 1 = 0$

e) $x^4 - 5x^2 + 4 = 0$

2- Demand and supply for a product are given as Qd = 100 - 20p and Qs = 5 + .5p, respectively. Find the equilibrium price and quantity.

3- Demand and supply for a product are given as Qd = $10 + \frac{20}{p}$ and Qs = $5p^2$, respectively. Find the equilibrium price and quantity. Find the excess supply when the market price is $2.50.

4- Find the break-even quantity for a manufacturer with a fixed annual cost of $60000 and average variable cost of $3 per unit. The product is sold at market price of $7 per unit.

5- Find the break-even price for a manufacturer that has a fixed cost of $78000 and average variable cost of $2. The manufacturer is expected to produce at full capacity of 6000 units a year.

6- A trucking company is operating with an annual fixed cost of $730,000. The company has 60 trucks. Every in-use truck has an average daily cost of $35. Every idle truck has an average daily cost of $4. Every truck is rented at an average daily price of $95.

a- Find the number of trucks that should be rented a day for the company to break even.

b- Find the number of trucks that should be rented a day for the company to have an annual profit of $350400.

7- Find the rate of return on $260,000 investment with annual returns of $44,500 at the end of the first year, $65000 at the end of the second year, $75200 at the end of the third year, $55000 at the end of the fourth year, $42000 at the end of the fifth year, and $21000 at the end of the sixth year. At the end of the sixth year, the investment has a scrap value of $25000.

V. Logarithms: The Log Command

Definition: Let y $=b^x$, b > 0. We define the exponent x as the logarithm of y to the base b and write $\log_b^y = x$. Expressions y = b^x and $\log_b^y = x$ are equivalent. The logarithms of numbers to the base 10 are called **common logarithms**, and are denoted by log. The logarithms of numbers to the base e, the natural number, is called **natural logarithm** and is usually denoted by ln. Logarithms are one of the basic tools for linearizing (log-linearizing) non-linear relationships. This property of logarithms is used frequently in economic models, data analysis, and econometrics.

Note that if the base of the logarithm is not defined *Mathematica* and most computer programs take "Log" as the natural logarithm.

Some of the properties of the logarithms are as follows,

Logarithms of products	loguv = logu + logv
Logarithms of a ratio	$\log(\frac{u}{v})$ = logu - logv
Logarithms of a power	$\log u^m$ = mlogu
Conversion of logarithm base	$\log_b^u = \log_e^u \log_b^e$
Inversion of logarithm base	$\log_b^u = \frac{1}{\log_u^b}$

Example: Find the logarithm of 356 to the base 10, 2, and e.

Solution: Use the logarithm table or a calculator to find the values. Using *Mathematica,* the values can be found as

```
In[54]:= N[Log[10, 356]]
Out[54]= 2.55145
In[55]:= N[Log[2, 356]]
Out[55]= 8.47573
In[56]:= N[Log[E, 356]]
Out[56]= 5.87493
```

Example: The relationship between the variable input (L) and the quantity of output (Q) is given as:

L: 1, 2, 3, 4, 5, 6 7 8 9 10

Q: 2, 7, 15, 29, 42, 51 57 62 64 62

a- Make a list of paired inputs and outputs and plot it.

b- Make a list of the natural logarithmic values of paired inputs and outputs and plot it.

c- Compare the two plots.

Solution: Find each pair on the x-y axes and plot them. Find the logarithmic values of each pair and plot them. The two plots will look similar. However, the scales of the two plots are different and the logarithmic plot will look more smoother than the other one.

Using *Mathematica*, the paired inputs and outputs can be listed as,

```
In[57]:= Clear[list1]
         list1 = {{1, 2}, {2, 7}, {3, 15}, {4, 29}, {5, 42}, {6, 51}, {7, 57}, {8, 62}, {9, 64}, {10, 62}}
Out[57]={{1, 2}, {2, 7}, {3, 15}, {4, 29},
         {5, 42}, {6, 51}, {7, 57}, {8, 62}, {9, 64}, {10, 62}}
```

the plot will look as follows (plotting of the functions is studies in the next chapter)

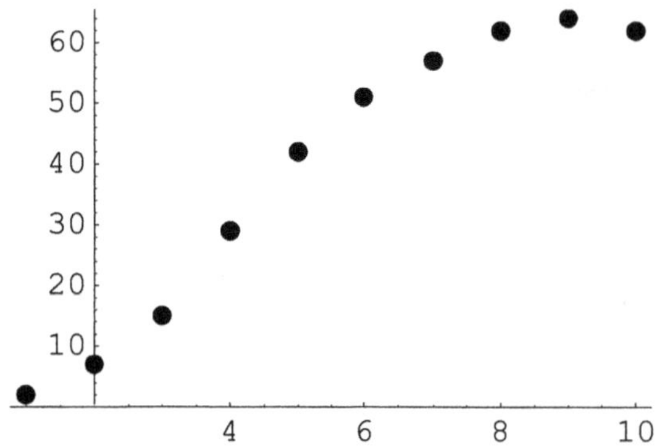

Plot of list1

to transform the numbers to logarithmic values (natural log) use the following command

```
In[58]:= list2 = N[Log[list1]]
Out[58]={{0.,0.693147},{0.693147,1.94591},{1.09861,2.70805},
{1.38629, 3.3673}, {1.60944, 3.73767}, {1.79176, 3.93183}, {1.94591, 4.04305},
 {2.07944, 4.12713}, {2.19722, 4.15888}, {2.30259, 4.12713}}
```

the plot of the logarithmic numbers is shown below

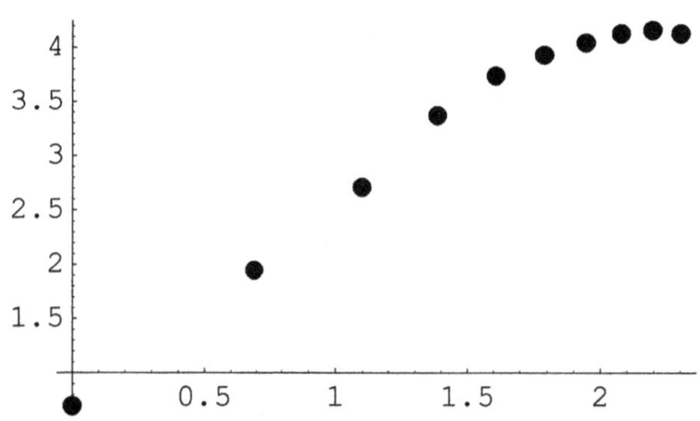

Plot of logarithmic values (list2)

Example: Log linearize the Cobb-Douglas production function $Q = AL^{\alpha}K^{\beta}$.

Solution: Using the logarithm rules, LogQ = LogA + αLogL + βLogK.

Example: Application in Business (Compounding) - How long does it take the initial value of an investment to double, if the interest rate is 8%.

Solution: Using the annual compounding formula $S = A(1 + r)^t$, the doubling of the initial investment renders $2A = A(1 + .08)^t$. Therefore, solve for t in the equation $2 = 1.08^t$ by taking the log as Log2 = tLog1.08 or $t = \frac{\text{Log2}}{\text{Log1.08}}$ = 9 years. Using *Mathematica*, the problem can be solved as

```
In[59]:= Clear[t]
         Solve[2 == 1.08^t, t]
```

```
Solve::ifun : Inverse functions are
   being used by Solve, so some solutions may not be found.
```

Out[59]= `{{t → 9.00647}}`

Do More Practice:

1- How long does it take the initial value of an investment to double, if the interest rate is 16%. (Answer: 4.67 years)

2- How long does it take the initial value of an investment to quadruple, if the interest rate is 8%. (Answer: 18 years)

Example: Growth

Show that if y grows exponentially at the rate of 20% from an initial value of 10, the natural log of y, lny, will grow at a constant rate along a line with a slope of .20. List the values of y and lny for t = 0 to 10 and plot the numbers on a graphing paper. Compare the two graphs.

Solution: $y = 10\, e^{.20t}$. The natural log of y, lny = ln10 + .20t. That is lny will change along a straight line with a slope of .20. To find the points for plotting, make a table of y and lny values for t = 0 to 10. You may use the *Mathematica* command **Table** as,

In[60]:= `Clear[y]`
`Table[10 E^(.20 t), {t, 0, 10}]`

Out[60]= `{10, 12.214, 14.9182, 18.2212, 22.2554, 27.1828, 33.2012, 40.552, 49.5303, 60.4965, 73.8906}`

The plot of the points for y is shown below

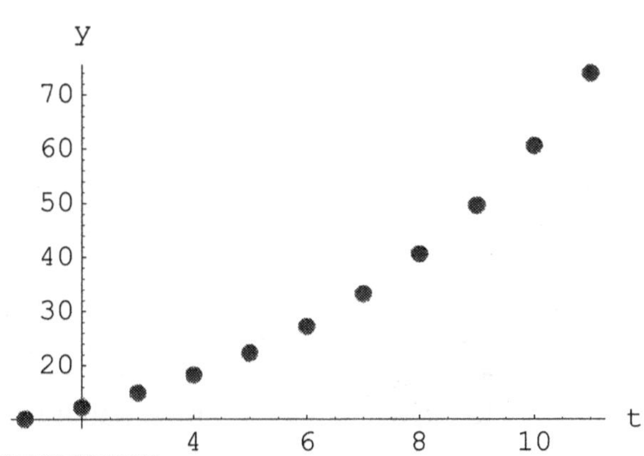

In[61]:= `Clear[y]`
`Table[Log[10] + .20t, {t, 0, 10}]`

Out[61]= `{Log[10], 2.50259, 2.70259, 2.90259, 3.10259, 3.30259, 3.50259, 3.70259, 3.90259, 4.10259, 4.30259}`

The plot of the points for lny is shown below

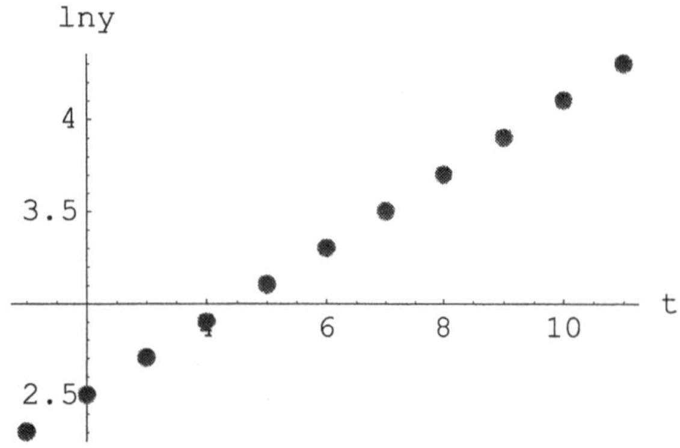

Do More Practice: Change the rate of growth of the preceding example from 20% to 5%. Find the values of y and lny for t = 0 to 10. Using graphing paper plot the values of y and lny. How do they differ from each other.

2- An investor has deposited $100 in a certificate of deposit (CD) account, earning 6% at the end of each year. Find the compounded values of the investment at the end of each year for the next 10 years and plot them on a graphing paper. Find the natural log of the compounded values of the investment at the end of each year for the next 10 years and plot them on a graphing paper. How does the shape of two graphs differ from each other?

Problems

1- Evaluate Log base 10 of 1, 10, 100, 200, and 300.

2- Evaluate log base 2 of 1, 2, 4, 8, and 16.

3- Evaluate log base 3 of 1, 3, 9, 27, and 81.

4- Write the following functions in log-linear form:

a) $Q = AL^{\alpha}K^{\beta}$

b) $S = A\ (1 + i)^{t}$

c) $Q = \frac{Ay^{\alpha}}{P^{\beta}}$

5- A constant return to scale Cobb-Douglas production function, $Q = 32L^{\alpha}K^{\beta}$, has an output of 300 units for labor input level of L = 10 and capital input level of K = 8. Find α and β. (Hint: A Cobb-Douglas production function of the form in this problem is constant return to scale if $\alpha + \beta = 1$. A constant return to scale production means that a proportional increase in inputs increases output at the same proportion).

6- The relationship between labor input (L) and output (Q) of a firm is given as.

L: 0, 1, 2, 3, 4, 5, 6, 7, 8, 9, 10, 11, 12, 13, 14, 15

Q: 0, 10, 22, 36, 54, 74, 91, 106, 117, 126, 133, 138, 142, 143, 141, 136

a) Display the relationship between labor input and output.

b) Display the relationship between the logarithm of labor input and the logarithm of output.

7- A consumer's utility (U) for a product is given as

Q: 0, 1, 2, 3, 4, 5, 6, 7, 8, 9, 10

U: 0, 5, 12, 20, 27, 33, 37, 40, 42, 43, 43

a) Display the relationship between Q and U.

b) Display the relationship between LogQ and LogU.

8- Solve the following equations:

a) $\log_{10}^{(2x+10)} = 4$

b) $\log_{x}^{(2x+8)} = 2$

c) $\log_{3}^{(x+2)} = -2$

d) $\log_{e}^{(2x+12)} = \log_{e}^{(x-3)}$

e) $\log_{x}^{(2x-x^2)} = 3$

VI. Progressions: The **Sum** Commands

Definition: A progression is a sequence of numbers any of which has a specific relation to the preceding one. The two common types of progressions are **arithmetic progressions** and **geometric progressions.**

Arithmetic Progressions is a sequence of numbers, each of which differs from its preceding number by a factor of constant. The constant number is called the **common difference**. For example, the following sequence of numbers

a, a + d, a + 2d, ..., a + (n-1)d

is a progression with an initial value of a and common difference of d. In Business and Economics we are most concerned with the sum of a progression or with the nth element of the progression. The nth element of the arithmetic progression is **a + (n - 1)d** and the sum is $\mathbf{S = n(a + \frac{d(n-1)}{2})}$.

Example: List 20 elements of a progression with an initial value of 4 and a common factor of 12. Find the 10th element and the sum of the first twenty elements.

Solution: The table follows the arithmetic progression rule which produce a table as t = {4, 16, 28, ...}. To find the 10th element and the sum, use the two formulas above. The 10th element of the progression is 4 + (10 - 1)12 = 112 and the sum is $S = 20(4 + \frac{12(20-1)}{2}) = 2360$.

In *Mathematica,* the table is generated by defining the element of the progression as 4 + 12i for i = 0, 1, ..., 19. (**Note** that 0 to 19 results in 20 elements, n = 20)

```
In[62]:= Clear[i]
         Table[4 + 12 i, {i, 0, 19}]

Out[62]= {4, 16, 28, 40, 52, 64, 76, 88, 100, 112,
          124, 136, 148, 160, 172, 184, 196, 208, 220, 232}
```

to find the 10th element of the table, use the command,

```
In[63]:= %[[10]]

Out[63]= 112
```

and to find the sum of the elements, use the command,

```
In[64]:= Sum[4 + 12 i, {i, 0, 19}]

Out[64]= 2360
```

Example: A firm producing an initial output of 100 units can increase output by 10 units a week. The management is interested in knowing how many weeks it takes to achieve an output level of 1260 units.

Solution: The increase in output follows an arithmetic progression with an initial value of 100 and a common increment factor of 10 units. That is, the ith week's production is 100 + 10i, the sum of which over an unknown period x should be 1260. To find x, set the sum of the weekly production equal to 1260 and solve for x. That is, $x(100 + \frac{10(x-1)}{2}) = 1260$. This gives x = 9 & -28, where only x = 9 is an acceptable solution. Using *Mathematica*,

```
In[65]:= Clear[i, x]
         Solve[Sum[100 + 10 i, {i, 0, x - 1}] == 1260, x]

Out[65]= {{x → -28}, {x → 9}}
```

Note that, as in the previous example (i-count 0 to 19), the i-count here is from 0 to x - 1 and not x.

Geometric Progression is a sequence of numbers, each of which is obtained by multiplying the preceding term by a constant factor. The constant factor is called the **common ratio**. For example, the following numbers

$$a,\ ar^1,\ ar^2\ ,\ ar^3,\ \ldots\ldots\ldots\ldots\ldots,\ ar^{(n-1)}$$

is an n-element geometric progression with an initial value of "a" and a common ratio of r. The nth element of the geometric progression is $\mathbf{ar^{(n-1)}}$, and the sum is $\mathbf{S = a\frac{1-r^n}{1-r}}$.

Example: List 20 elements of a geometric progression with an initial value of 5 and a common ratio of 2. Find the 10th element and the sum of the first twenty elements.

Solution: Using the geometric progression rule, the table of elements is t = { 5, 10, 20, ...}. Using the two formulas above, the 10th element is $5(2)^{10} = 2560$, and the sum is $S = 5\frac{1-2^{20}}{1-2} = 5242875$.

In *Mathematica*, define the elements of the progression as $5(2)^i$ for i = 0, 1, 2, ..., 19. The table can be generated as follows. (**Note** that, for 20 elements, i is set from 0 to 19).

```
In[66]:= Clear[i]
         Table[5*2^i, , {i, 0, 19}]

Out[66]= {5, 10, 20, 40, 80, 160, 320, 640, 1280, 2560, 5120, 10240,
          20480, 40960, 81920, 163840, 327680, 655360, 1310720, 2621440}
```

the 10th element is derived as

```
In[67]:= %[[10]]

Out[67]= 2560
```

the sum can be derived as

```
In[68] := Sum[5*2^i, {i, 0, 19}]

Out[68]= 5242875
```

Example: Application in Macroeconomics (Multiplier Effect): In a two-sector, simple Keynesian macro-model of the economy with a consumption function of C = bY, investment function of I = Io, and the income identity of Y ≡ C + I, every one dollar increase in I will increase Y by the following amounts over time periods,

Time Period: 1, 2, 3, ..., n

Increase Y: 1, b, b^2,..............., $b^{(n-1)}$

Find the total increase in output after n periods (simple multiplier.)

Solution: The progression is a geometric progression with an initial value of 1 and a common ratio of b. Using the sum formula, $S = \frac{1-b^n}{1-b}$. However, in this case b, which in macroeconomics is called marginal propensity to consume (MPC), is a number less than 1 (b < 1). Therefore, as n approaches infinity, b^n = 0 and $S = \frac{1}{1-b}$.

```
In[69]:= Clear[i]
         Sum[1*b^i, {i, 0, Infinity}]
```

Out[69]= $\frac{1}{1-b}$

Problems

1- Find the 10th and 20th terms of the following progressions

a) 4, 7, 10, ...

b) 1, $\frac{4}{3}$, $\frac{5}{3}$, ...

c) 3, $\frac{7}{2}$, 4, ...

2- Find the sum of the first thirty terms of the progressions in problem 1.

3- Find the 10th and the 20th terms of the following geometric progressions

a) 3, 9, 27, ...

b) 3, 1, $\frac{1}{3}$, ...

c) 9, 6, 4, ...

4- Find the sum of the first 3, 5, and 10 terms of the geometric progressions in problem 3.

5- What will be the total savings of a person 10 years from now if she deposits \$100 at an annually compounded interest rate of 10% a year. How long does it take to double a principal of any amount at that rate?

6- In a simple macro model of an economy C = 100 + .8Y, I = 200, and Y = C + I.

a- Assume that investment (I) increases by \$100. What is the effect on Y after 2 years, 5 years, and 10 years? What is final effect on Y?

7- In a simple two-sector macro model of an economy $C = 100 + .75Y$, $G = 500$, and $Y = C + G$.

a- Assume that government spending (G) increases by $200. What is the effect on Y after 1 year, 4 years, and 10 years? What is final effect on Y?

b- Assume that autonomous spending by consumers increases by $200. What is the effect on Y after 1 year, 4 years, and 10 years? What is final effect on Y?

c- Assume that government decreases taxes by $200. What is the effect on Y after 1 year, 4 years, and 10 years? What is final effect on Y?

Chapter 2

FUNCTIONS AND THEIR GRAPHS

The concept of functions plays a major role in mathematics. It is also an important tool in formulating and analyzing economic relationships. Economic relationships, generally, follow a certain functional form. In this chapter, after defining functions, different types of single-variable and multi-variable functions are presented, their graphs are shown, and their relevance to economic theory is discussed. The last section of this chapter presents parametric functions and their graphs.

TOOL BOX

Mathematica **Commands**	**Function**
f[x]	Defines f as an expression or a function with an argument x
f[x, y]	Defines f as an expression or as a function with arguments x & y
f[x_]: =	Defines f with any argument, named x
`Table` [x^2, {x, 0, 10}]	Makes a table of x^2 value for x-values from 0 to 10.
`ListPlot`[data, PlotJoined→True]	Plots the list of points called data and joins the points.
Plot[f(x), {x, x1, x2}]	Plots y = f(x) in the x-range, x1 to x2.
Plot3D[f(x,y), {x,x1,x2},{y, y1,y2}] 3D	Plot of a z = f(x, y) in the x-range x1- x2 and y rage y1- y2
`ContourPlot`[f(x,y),{x,x1,x2},{y,y1,y2}]	Contour plots the relation between x & y for given value of z in z = f(x, y)
ParametricPlot[{f(x), f(y)}, {t, t1, t2}	Graphs a parametric plot of f(x), y(x) for t-range t1-t2
ParametricPlot3D[{f(x),f(y),f(z), {t, ..]	Graphs a 3D Parametric Plot of f(x), f(y), f(z) for t-range t1 - t2

I. **Functions:** The **f[x]** and **f[x_]** Expressions

Let X and Y be sets of real numbers. A real-valued function *f* of a real variable x from X to Y is a correspondence that assigns to each number x in X exactly one number y in Y. The set X is called the **domain** of *f*. A subset of Y that consists of all images of numbers in X under *f* is called the **range** of the function. This relation can be denoted as **y =** ***f*****(x)**. The variable x is called **independent variable**, its image under *f*, is called **dependent variable** (y).

A function is a special kind of relation. A relation is an arbitrary set of ordered pairs. Whereas, a function is a set of ordered pairs such that no two distinct pairs have the same elements. For example, the pairs (1, 2) and (1, 3) can be pairs of a relation, but not pairs of a function.

Example: Which of the following two graphs represents a functional relationship?

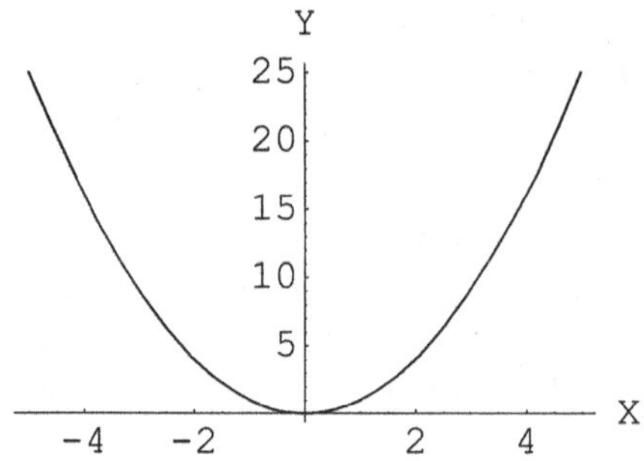

Graph 1: y = x^2

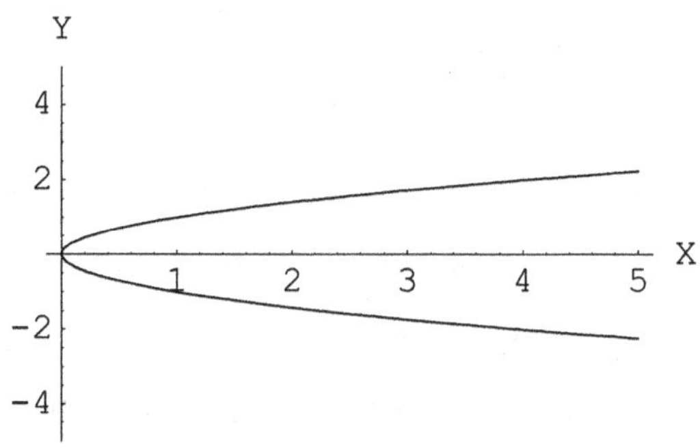

Graph 2: y^2 = x

Solution: It is apparent from graph 1 that for every value of x there is only one value of y. In other words, any vertical line will cross the graph 1 only at one point. Therefore, the first graph represents a function. Any vertical line will cross the second graph in two points. Therefore, the second graph does not represent a function.

When a variable y is a function of a variable x, the y in y = f(x) represents the value of the function. The values of a function are the numbers in its range. In *Mathematica*, a function of x is defined by "f[x_]". That is y = f(x) = x^2 + 2x is define as f[x_] := x^2 + 2x or f[x_] = x^2 + 2x. The ":" preceding equality sign suppresses the output during operation.

Example: Find the values of the function $y = f(x) = x^3 - 3x^2 + 5$ for x = 3 and x = -10.

Solution: By substituting for x in y = f(x), the corresponding values of y can be found as, $y = f(3) = 3^3 - 33^2 + 5 = 5$ and $y = f(-10) = (-10)^3 - 3(-10)^2 + 5 = -1295$.

In *Mathematica,* define the function as f[x_] = x^3 - 3x^2 + 5 and find the y values by finding f[3] and f[-10]. **Note** that, if you define the function as **f[x_] :=** the output will be suppressed during the calculation.

```
In[1]:= Clear[x]
        f[x_] = x^3 - 3 x^2 + 5
```

Output[1]= $5 - 3x^2 + x^3$

```
In[2]:= f[3]
```

Output[2]= 5

```
In[3]:= f[-10]
```

Output[3]= -1295

Implicit Functions A relation between x and y may be given in explicit form as y = f(x). A relation may be given by an equation of the form **F(x, y) = 0** where, F denotes a real function of two real variables x and y. For this equation to hold, it is clear that x and y are not independent of each other. For every value of x there is one or more value for y. The equation F(x, y) = 0 determines a relation. However, it may give rise to one or more functions. Such functions are called implicit function.

Example: The equation 2x - 3y + 6 =0 is an implicit relationship between x and y. Write the function in explicit from and find the values of y for x = 10 and x = -1.

Solution: The explicit function can be derived from this relation by solving the equation for y, $y = \frac{2}{3}x + 2$. By substituting for the given x values, $y = \frac{26}{3}$ for x = 10 and $y = \frac{4}{3}$ for x = -1.

In *Mathematica*, solve the equation for x and define a function whose argument is the solution to the equation.

```
In[4]:= Clear[x, y, sol]
        sol = Solve[2 x - 3 y + 6 == 0, y]
```

Output[4]= $\{\{y \to \frac{2(3+x)}{3}\}\}$

The preceding command solves y in terms of x and sets the results equal to sol. To express the functional from so that it can be evaluated for the given values of x , use the following command to form y = f(x) = solution of the equation (sol)

```
In[5]:= y[x_] = y /. sol[[1]]
```

Output[5]= $\frac{2(3+x)}{3}$

now that y is defined in terms of x, its values for x = 10 or x = -1 or any other number can be evaluated through y[10] and y[-1] or other values of x.

```
In[6]:= y[10]
```

Output[6]= $\frac{26}{3}$

```
In[7]:= y[-1]
```

Output[7]= $\frac{4}{3}$

Example: Using the following implicit relationship between x and y, y - xy + 2x - 10 = 0, define y as an explicit function of x . Find the values of y for x = 105 and x = 1. 5.

Solution: Factor out y as y(1-x) + 2x - 10 = 0 and solve the equation for y, $y = \frac{10-2x}{1-x}$. Substitute for the given x values to find the corresponding y values. For x = 105, $y = \frac{25}{13}$ and for x = 1.5, y = -14. In *Mathematica,* solve the equation for x and define a function whose argument is the solution to the equation. Then, find the value of the function for x = 105 and x = 1.5.

```
In[8]:= Clear[x, y, sol]
        sol = Solve[y - x y + 2 x - 10 == 0, y];
        y[x_] = y /. sol[[1]]
```

Output[8]= $\frac{2(-5+x)}{-1+x}$

```
In[9]:= y[105]
```

Output[9]= $\frac{25}{13}$

```
In[10]:= y[1.5]
```

Output[10]= -14.

Example: The equation $y^2 - xy + x^2 - 10 = 0$ gives rise to two explicit functions as shown in solution below. Write the two explicit function for y and find the two values of y for x = 3.

Solution: $y = \frac{x \mp \sqrt{x^2 - 4(x^2 - 10)}}{2} = \frac{x \mp \sqrt{-3x^2 + 40}}{2}$. By substituting for x = 3, $y_1 = \frac{3 - \sqrt{13}}{2}$ and $y_2 = \frac{3 + \sqrt{13}}{2}$.

In *Mathematica,* solve the implicit equation for x and define two functions whose arguments are the solutions to the equation. The following command solves the equation for x.

```
In[10]:= Clear[x, y, sol]
         sol = Solve[y^2 - x y + x^2 - 10 == 0, y]
```

Output[10]= $\{\{y \to \frac{1}{2}(x - \sqrt{40 - 3x^2})\}, \{y \to \frac{1}{2}(x + \sqrt{40 - 3x^2})\}\}$

The following commands sets y1[x_] and y2[x_] equals to the solution one and solution two of the equation.

```
In[11]:= y1[x_] = y /. sol[[1]]
         y2[x_] = y /. sol[[2]]
```

Output[11]= $\frac{1}{2}(x - \sqrt{40 - 3x^2})$

$\frac{1}{2}(x + \sqrt{40 - 3x^2})$

The following commands finds the values of the y1 and y2 for x = 3.

```
In[12]:= y1[3]
```

Output[12]= $\frac{1}{2}(3 - \sqrt{13})$

```
In[13]:= y2[3]
```

Output[13]= $\frac{1}{2}(3 + \sqrt{13})$

Note that, since the equation has two solutions, two different functions are produced, one for each solution. y1[x_]=y/.sol[[1]] is the function generated from the first solution and y2[x_]=/.sol[[2]] is the function produced for the second solution.

Example: Production Possibilities Frontier

Production possibilities frontier (PPF) shows the maximum amount of goods and services that a firm or a country can produce if all the available resources are used efficiently. At any point on PPF, the slope of the PPF represents the relative price or the opportunity cost of attaining an extra unit of one good at the cost of giving up some units of the other goods. Suppose that in a simple two-goods model of an economy PPF is represented by $x^2 + 9y^2 = 81$, where x and y are quantities of the two goods, food and fun, respectively. Graph the PPF. Given that the slope of the PPF is $m = \frac{-x}{9y}$, find the slope at x quantities of 1, 2, and 3. Show that when more x is produced it has higher opportunity cost in terms of the number of units of y that must be given up to produce one more unit of x

Solution: The function represents an ellipse with x intercept of 9 and y intercept of 3. Finding few other points helps to graph the function. The following graph shows the trade-off between x and y on the PPF.

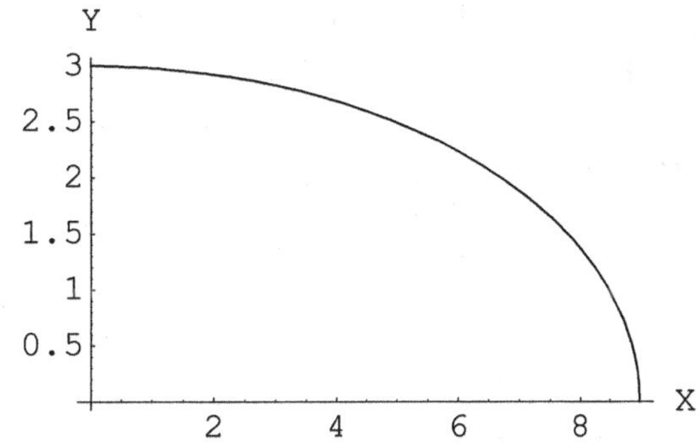

To find the slope at the given points, find the y values corresponding to the given x values and substitute for x and y in m = $\frac{-x}{9y}$. At x = 1, y = 2.98, and m = -.037. That is the opportunity cost of producing one more unit of x at this point is to give up .037 units of y. At x = 2, y = 2.92, and m = -.076. That is, at this point to have one more unit of x, .076 units of y should be sacrificed. At x = 3, y = 2.85, and m = -.118, which means at this point .116 units of y must be given up to have one extra unit of x. The changes in m show that as more and more of x is produced the opportunity cost of producing it increases. That is the PPF is an **increasing opportunity cost** PPF.

To graph the PPF using *Mathematica*, first solve y in terms of x then graph y.

```
In[14]:= Clear[x, y, sol]
         sol = Solve[x^2 + 9 y^2 == 81, y];
         y2[x_] = y /. sol[[2]];
         Plot[y2[x], {x, 0, 9}, AxesLabel -> {"X", "Y"} ]
```

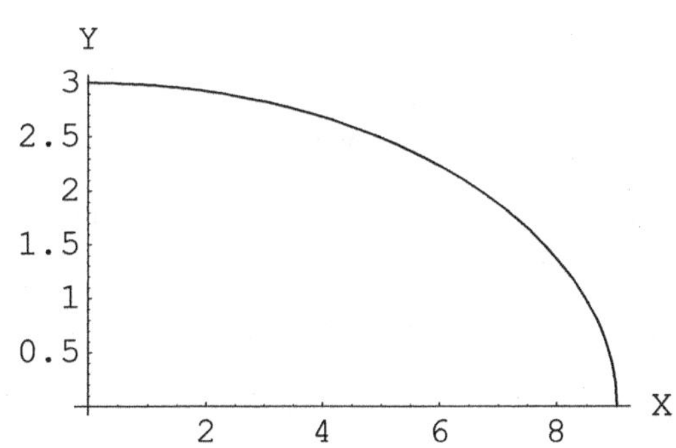

To find the y values corresponding to x = 1, 2, and 3, use the following commands

```
In[15]:= N[y2[1]]
         N[y2[2]]
         N[y2[3]]

Output[15]= 2.98142

            2.92499

            2.82843
```

To evaluate the slope m at (1, 2.98), (2, 2.92). and (3, 2.63), first define m then evaluate it at the given points.

```
In[16]:= m = -x / (9 y);
         m /. {x -> 1, y -> 2.98}
         m /. {x -> 2, y -> 2.92}
         m /. {x → 3, y → 2.83}
Output[16]= -0.0372856
            -0.0761035
            -0.117786
```

Do More Practice: Change the constant in the PPF above from 81 to 144. Find the y values corresponding to x = 1, 2, and 3. Find the slope of the PPF at x = 1, 2, 3 and their corresponding y values. Show that, although the size of the economy (represented by the PPF) is grown, the slope at the given points have not changed.

2- Change the PPF function to $9x^2 + y^2 = 81$. Graph the PPF. Find the y values corresponding to x = 1, 2, and 3. Find the slope of the PPF at x = 1, 2, 3 and their corresponding y values. Compare the results with case 1.

3- Suppose the PPF is given as x + 3y = 9.

a- Graph the PPF

b- Find the values of y for x = 1, 2, and 3.

c- How much of y is given up each time to have one more unit of x?

d- Find slope of the PPF at the given points above. Show that this PPF is a **constant opportunity cost** PPF.

Problems

1- Write consumption as a function of income. Define the domain and the range of the function.

2- Write a demand function expressing quantity demanded as a function of price. Define the domain and the range of the function.

3- Express the trade-off between inflation and unemployment in a functional form. Define the domain and the range of the function.

4. Write the following relationships between x and y in the explicit forms and evaluate y for the values of x = 1 and x = 5.

a) $3x + 2xy + 6y + 10 = 0$

b) $x^2 y + 2x - xy + 2x^2 - 4y + 10 = 0$

c) $x^2 - xy + y^2 = 10$

d) $2x + x^2 - x^2 y^2 + 3y^2 - 5 = 0$

5- Suppose that the production possibilities frontiers (PPF) of an economy is represented by $25x^2 + 9y^2 = 2025$, where x and y are quantities of the two goods, food and fun, produced given the limited resources. Graph the PPF. Given that the slope of the PPF is $m = \frac{-25x}{9y}$, find the slope at x quantities of 1, 2, and 3. Show that the PPF is an increasing opportunity cost PPF.

II. Types of Functions and Their Graphs: The **Plot** Command

Functions are mainly of two types: **Algebraic functions,** generated by algebraic operations on a variable. Examples of algebraic functions are polynomial functions, rational functions, and other explicit algebraic functions. **Transcendental functions** include logarithmic, exponential, and periodic functions.

Plotting a function without using graphic calculator or computer program requires finding the intercepts of the function (y-intercept and x-intercept), finding critical points of the function (maxima, minima, and inflection points), and finding the asymptotes of the function (convergence to a horizontal, vertical, or slant line as x or y converges to infinitely large numbers). A more systematic way of graphing functions will be discussed in later chapters. Here, we will utilize *Mathematica* to graph different types of functions to have a better understanding of the shape of the functions as their algebraic form or parameters change. To graph the relationship between x and y of y = f(x) in the x-range of x_1 to x_2, use the command Plot[y, {x1, x2}] or Plot[f(x), {x, x1, x2}]. The command "AxesLabel -> {"X", "Y"}" may be used for labeling the axes.

Polynomial Functions are the simplest of algebraic functions. Such functions have the property of being smooth and continuous. A polynomial function has the following form $y = f(x) = a_1x^n + a_2x^{n-1} + \dots\dots\dots + a_nx^{n-n}$. The highest exponent of the polynomial function defines its degree. For example, y = f(x) = 3x + 6, y = f(x) = $2x^2$ - x - 5, and y = $x^3 + 2x^2$ + 3x +10 are first degree, second degree, and third degree polynomials, respectively. In the following examples several polynomial functions are graphed and their shapes are explained.

Example: First-Degree Polynomial - Equation of a Line

Graph the relationship between x and y for y = f(x) = 3x + 6 in the x-range of -10 to 10.

Solution: A first-degree polynomial is the equation of a line. To draw the line, find two points of the line and join them. Set x = 0 to find the y-intercept (y = 6). Set y = 0 to find x-intercept (x = -2). Given the two points, draw the line.

Using *Mathematica* command "Plot" the line can be graphed as,

```
In[17]:= Clear[x, y]
         y = 3 x + 6
         Plot[y, {x, -10, 10}, AxesLabel -> {"X", "Y"}]
```

Output[17]= 6 + 3 x

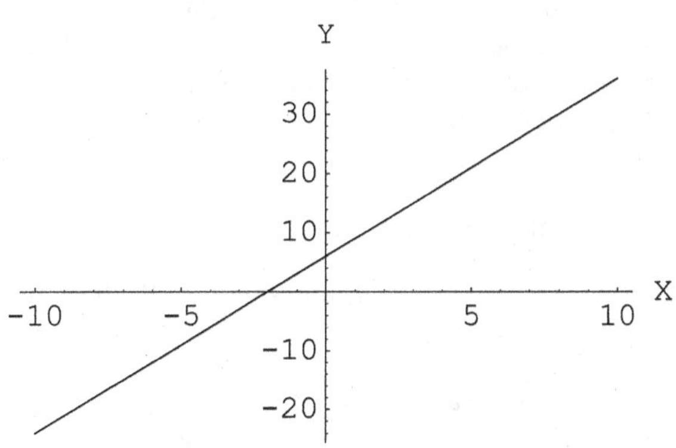

Do More Practice: Change the slope coefficient to . 3 and draw the line.

2. Change the slope coefficient to -3 and draw the line.

3. Change the intercept to zero and draw the line.

4. Change the intercept to -6 and draw the line.

Example: Linear Demand Function

An airline has been selling its round-trip tickets for flight between Los Angeles and Chicago at $450 each. At this rate the company has been selling 800 round-trip tickets a week. A survey indicates that for each $20 rebate offered, the number of round-trip travelers will increase by 50 per week. Find the demand function of the airline and graph it. Find the total revenue function for the flights between Los Angeles and Chicago.

Solution: Conventional demand function assumes price to be presented by the vertical axes and quantity by the horizontal axes. The equation of the demand can be written as P = mQ + c, where P is the price, Q is the quantity, m is the slope of the demand line and c is the P intercept. The slope of a line is the ratio of rise to run, or m = $\frac{P_2-P_1}{Q_2-Q_1} = \frac{430-450}{850-800} = \frac{-20}{50}$ = -.4. To find c, substitute in P = mQ + c for a point and m as, 450 = -.4 (800) + c or c = 770. Therefore, the demand function is P = -.4Q + 770. Total revenue (TR) is, TR = P.Q. Multiply P by Q to get the total revenue as TR = -.4Q^2 + 770Q. The demand function can be graphed by finding two points of the line and drawing the line. Below the demand function is graphed using *Mathematica.*

```
In[18]:= Clear[P, Q]
         Plot[-.4 Q + 770, {Q, 0, 1000}, AxesLabel -> {"Q", "P"}]
```

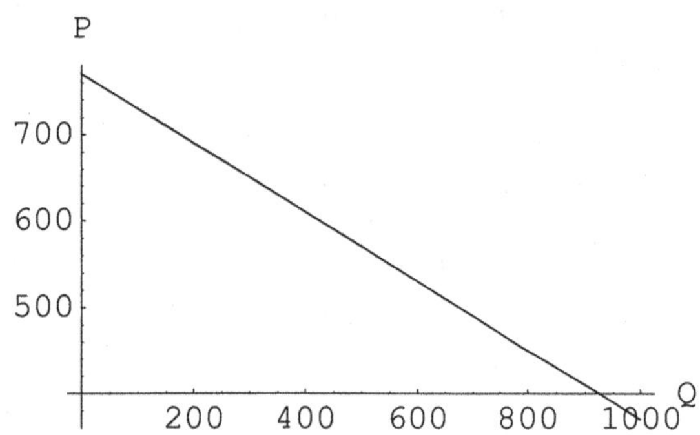

<u>Second-Degree Polynomial :</u> A second degree polynomial has a standard shape, a **parabola**. The function is either concave upward or concave downward, with a vertex and a symmetry about the y-axis or a line parallel to y- axis. The function, also called **quadratic function**, is of the general form y = f(x) = ax^2 + bx + c. This function has a y-intercept of (0, c), x-intercepts (derived by setting ax^2 + bx + c = 0 and solving for x), ($\frac{-b+\sqrt{b^2-4ac}}{2a}$, 0) and ($\frac{-b-\sqrt{b^2-4ac}}{2a}$, 0), and the vertex of ($\frac{-b}{2a}$, f($\frac{-b}{2a}$)). The vertex of the functions lies on the line of symmetry x = $\frac{-b}{2a}$. If b= 0, the y-axis is the line of symmetry. If a > 0, the parabola is concave upward. If a < 0, the parabola is concave downward. Once the intercept points and the vertex are located, the function can be graphed by joining the points and using the symmetry property of the parabola around the line parallel to y-axis and passing through the vertex x = $\frac{-b}{2a}$.

Example: Graph the relationship between x and y for y = 2x^2 - x - 5 in the x-range of -3 to 3.

Solution: For this function , the y intercept is (0, -5), the x-intercepts are ($\frac{1+\sqrt{41}}{4}$, 0) and ($\frac{1-\sqrt{41}}{4}$, 0), the line of symmetry is x = $\frac{1}{2}$, the vertex is ($\frac{1}{2}$, -5), and the parabola is concave upward.

Using *Mathematica* command "Plot", the function is graphed below for the x range -3 to +3,

```
In[19]:= Clear[x, y]
         y = 2 x^2 - x - 5
         Plot[y, {x, -3, 3}, AxesLabel -> {"X", "Y"}]
```

Output[19]= $-5 - x + 2x^2$

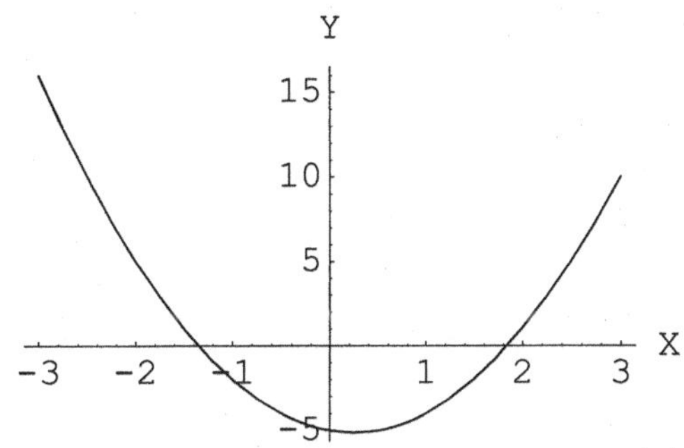

Do More Practice:

1. Change the coefficient of x^2 to -2 and graph the function. Observe the changes in concavity (the new function will be concave down).

2. Change the coefficient of x to zero and graph the function.

Example: Total Revenue Function of a Linear Demand Function

For the linear demand function of the example above (Linear Demand Function) graph the demand and total revenue functions. Graphically show that total revenue is maximum at the middle point of the linear demand line. What is the price elasticity of demand at this point? (Note: price elasticity of demand $\xi_p = \frac{\Delta Q\%}{\Delta P\%} = \frac{\Delta Q}{\Delta P} \cdot \frac{P}{Q}$)

Solution: The demand function of the example is P = -.4Q + 770 and the total revenue function is TR = $-.4Q^2$ + 770Q. The total revenue function is a second degree polynomial. To graph the total revenue, the function has a vertex and line of symmetry of $Q = \frac{-b}{2a} = \frac{770}{.8} = 962.5$. Since a < 0, the parabola is concave downward. The parabola has Q-intercepts of zero and 1925. Using these information the total revenue function can be graphed.

Below the total revenue and demand functions are graphed for the Q range 0 to 2000 using *Mathematica* command "Plot".

```
In[20]:= Clear[P, Q, TR]
         TR = - .4 Q^2 + 770 Q;
         P = - .4 Q + 770;
         Plot[TR, {Q, 0, 2000}, AxesLabel -> {"Q", "TR"}]
         Plot[P, {Q, 0, 2000}, AxesLabel -> {"Q", "P"}]
```

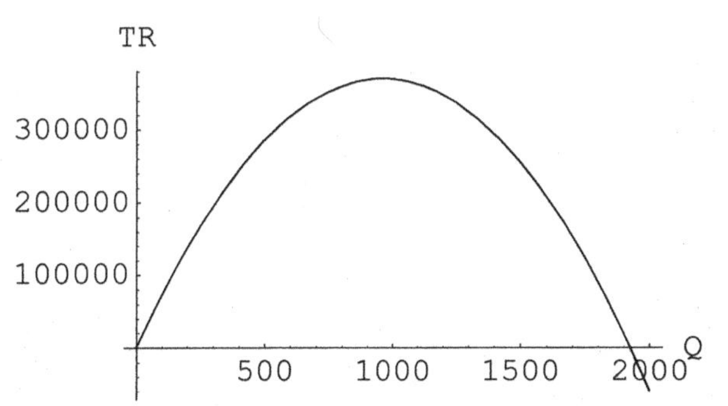

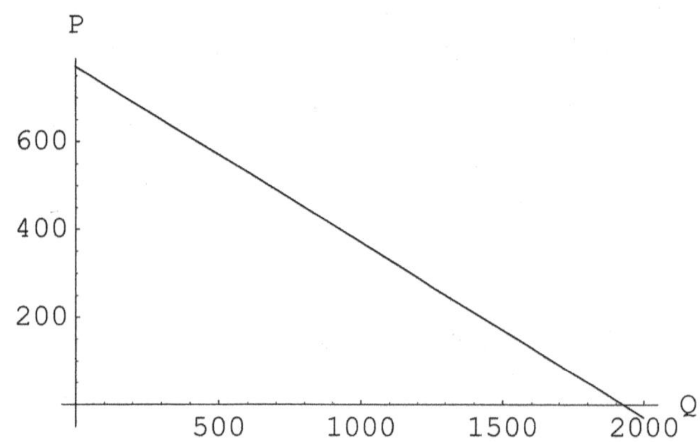

The graphs clearly show that TR is maximum at the middle of the linear demand curve. The Q-intercept of the demand line is Q = 1925 units, where $\frac{1925}{2}$ = 962.5 units. The vertex of the TR is also at Q = 962.5 units.

To find the price elasticity of demand, the revenue-maximizing quantity is Q = 962.5 and the revenue-maximizing price is P = \$385 (the mid-point between zero and P-intercept). The slope of the demand line is $\frac{\Delta P}{\Delta Q}$ = -.4 or $\frac{\Delta Q}{\Delta P}$ = -2.5. Therefore, $\zeta_P = \frac{\Delta Q}{\Delta P} \cdot \frac{P}{Q} = -2.5 \frac{385}{9625} = -1.$

Third-Degree Polynomial: Graph $y = x^3 + 2x^2 + 3x + 10$ in the x-range -5 to 5.

Solution: A third degree polynomial has a general form $y = f(x) = ax^3 + bx^2 + cx + d$, with a y-intercept of (0, d) and a rotational symmetry of 180° around $x = \frac{-b}{3a}$. The point of symmetry is $(\frac{-b}{3a}, f(\frac{-b}{3a}))$. If a > 0, the function starts from the from South-west of the third quadrant and rises up to the north-east of the first quadrant, with possible vertices in between. If a < 0, the function starts from the north-west of the second quadrant and declines to south-east of the fourth quadrant, with possible vertices in between. For the function above the y-intercept is (0, 10) and the point of symmetry is $(\frac{-2}{3}, \frac{232}{27})$. Since a > 0, the function's extreme points will be in the third and first quadrant. These information and few other points of the function are list in table below. The table helps with plotting the function.

x	$-\infty$	-1	$\frac{-2}{3}$	0	1	$+\infty$
f (x)	$-\infty$	8	8.59	10	16	$+\infty$

Using *Mathematica*, the function is graphed for the x range -5 to +5 as,

```
In[21]:= Clear[x, y]
         y = x^3 + 2 x^2 + 3 x + 10;
         Plot[y, {x, -5, 5}, AxesLabel -> {"X", "Y"}]
```

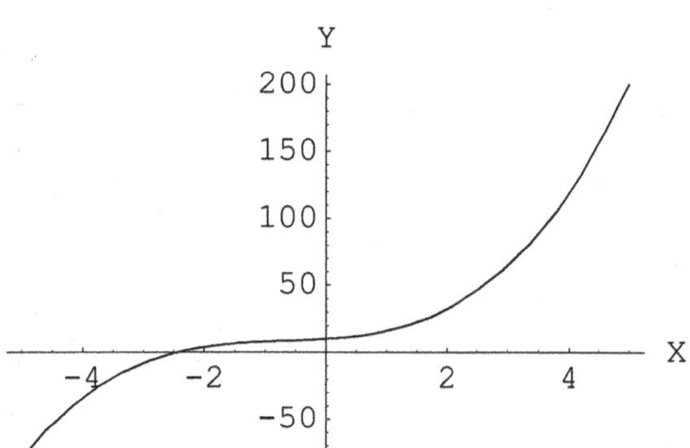

Do More Practice: Change the coefficient of x^3 to -1, keeping others coefficients the same as in the example. Graph the function and compare it to the graph in the example.

2. Change the coefficient of x to -3, keeping other coefficients the same as in the example. Graph the function and compare it to the one in the example.

3. Change the coefficient of x^2 to zero, keeping other coefficients the same as in the example. Graph the function and compare it to the one in the example.

4. Change the coefficient of x to zero, keeping other coefficients the same as in the example. Graph the function and compare it to the one in the example.

Higher-Degree Polynomials: Graph the functions $y = f(x) = 2x^4 - x^3 + 5x + 12$ and $y = f(x) = x^5 + 2x^4 - 5x^3 + 2x^2 - 10$.

Solution: Polynomials of higher degree where the highest exponent is an even number have a shape similar to second-degree polynomials. The shape of the higher degree polynomials where the highest exponent is an odd number is similar to the shape of third-degree polynomials. Business and economic problems randomly utilize polynomials of higher than third degree. The graph of the functions are as shown below using the *Mathematica* command "Plot" for the x-range -10 to +10.

```
In[22]:= Clear[x, y]
         y = 2 x^4 - x^3 + 5 x + 12;
         Plot[y, {x, -10, 10}]
```

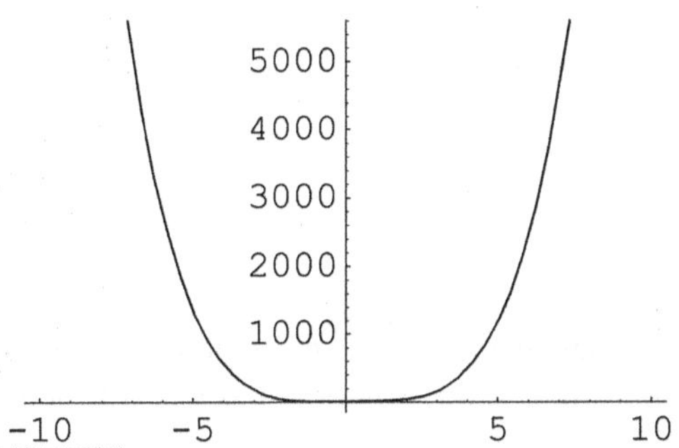


```
In[23]:= Clear[x, y]
         y = x^5 + 2 x^4 - 5 x^3 + 2 x^2 - 10;
         Plot[y, {x, -10, 10}, AxesLabel -> {"X", "Y"}]
```

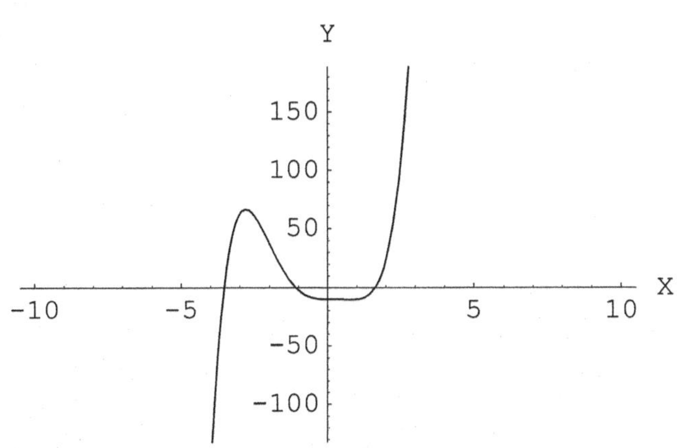

Do More Practice: Graph $y = f(x) = x^4 + 2x^3 - 3x^2 + 5x - 10$.

2. Change the coefficient of x^4 in the function of practice 1 to -1 and graph the function.

3. Graph y = f(x) = $x^5 + 3x^4 + 2x^3 - 3x^2 + 5x - 10$.

4. Change the coefficient of x^5 in the function of practice 3 to -1 and graph the function.

Polynomial functions have many applications in economic theory. For example, the relationship between variable input and output or total cost and output are, generally, best expressed by a third-degree polynomial. The relationship between total revenue and sales is usually a second-degree polynomial.

Example: Production Function

The production function, $Q = 5000L^2 - 80L^3$, expresses output (Q) in terms of labor input (L). Plot the production function and find the value of Q at L = 30.

Solution: The function is a third degree polynomial with a rotation symmetry around $\frac{-b}{3a} = \frac{5000}{2400}$ and Q-intercept of (0, 0). Finding few more points helps to have a better sketch of the graph. The general shape of the function in the L-range -100 to 100 is as follows,

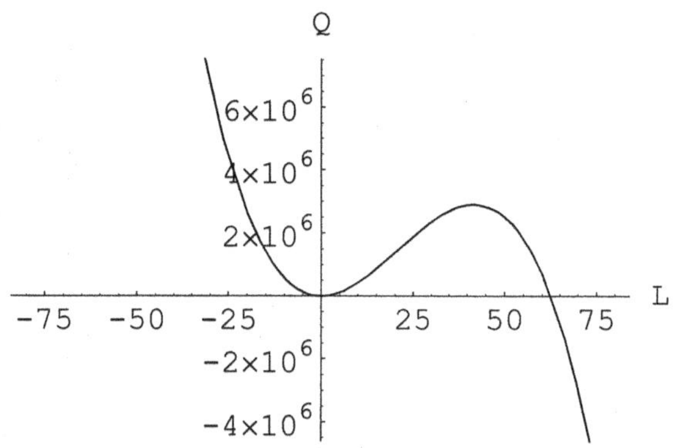

However, since Q represents a production function and and L are both positive numbers, the function should be graphed in the acceptable range of L, say the L-range of 0 to 50. The *Mathematica* solution and graph is shown below.

```
In[24]:= Clear[L, Q]
         Q = 5000 L^2 - 80 L^3;
         Plot[Q, {L, 0, 50}, AxesLabel -> {"L", "Q"}]
```

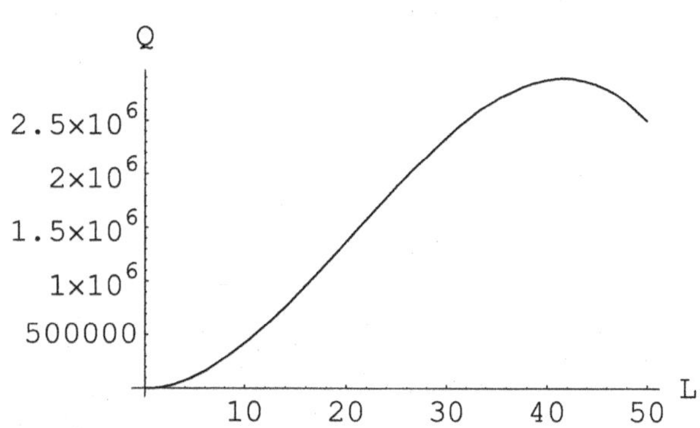

To find the value of Q at L = 3, substitute for L in the equation to find Q = 2340000. Using *Mathematica*,

```
In[25]:= Q /. L -> 30
```

Output[25]= 2340000

Example: Cost Function

Total cost (TC) of a firm as a function of output (Q) is given as TC = $.1Q^3$-3.6Q^2+ 45Q + 100. Graph the function and find the total cost of producing 30 units of output.

Solution: The function is a third degree polynomial with a TC intercept of (100, 0), rotational symmetry at $\frac{-b}{3a}$ = $\frac{3.6}{.3}$= 12, and point of symmetry (12, 294.4). Since a is positive, the curve moves up from third to first quadrant. By finding few more points, the function can be graphed.

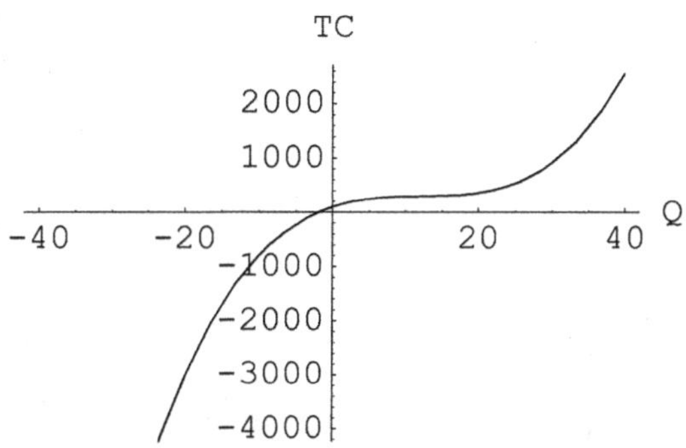

However, since Q >0, using *Mathematica* the total cost function is graphed in the acceptable range of Q as,

```
In[26]:= Clear[TC, Q]
         TC = .1 Q^3 - 3.6 Q^2 + 45 Q + 100;
         Plot[TC, {Q, 0, 40}, AxesLabe l -> {"Q", "TC"}]
```

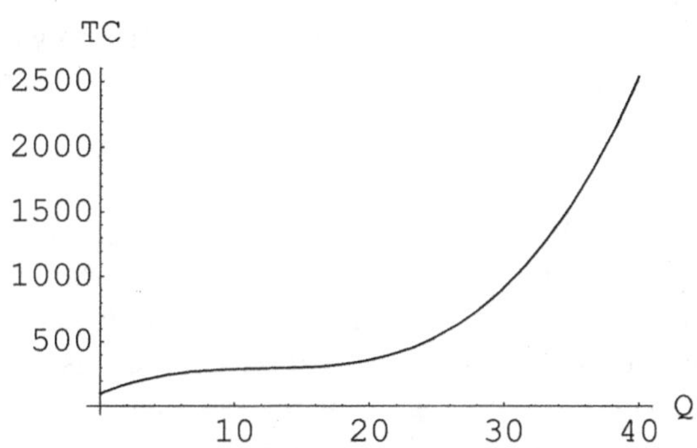

To find TC corresponding to 30 units of Q, substitute for Q in the equation to get TC = 910. Using Mathematica, TC is found as

```
In[27]:= TC /. Q -> 30
```

Output[27]= 910.

Example: Revenue Function

Demand for a product is given as Q = 100 -10P, where Q is quantity demanded and P is the price. Find the total revenue as a function of price and graph the function. Find the revenue corresponding to P = $4.

Solution: Total revenue (TR) is equal to price times quantity, TR = P.Q. Therefore, TR = 100P -10P^2 . This is a second degree polynomial with a line of symmetry P = $\frac{-b}{2a}$= 5 and a vertex (5, 250), and TR-intercept (0, 0). Using this information the function can be graphed. Using *Mathematica*, the graph is,

```
In[28]:= Clear[p, tr]
         Clear[p, tr]
         tr = 100 p - 10 p^2
         Plot[tr, {p, 0, 10}, AxesLabel -> {"P", "TR"}]
```

Output[28]= $100\,p - 10\,p^2$

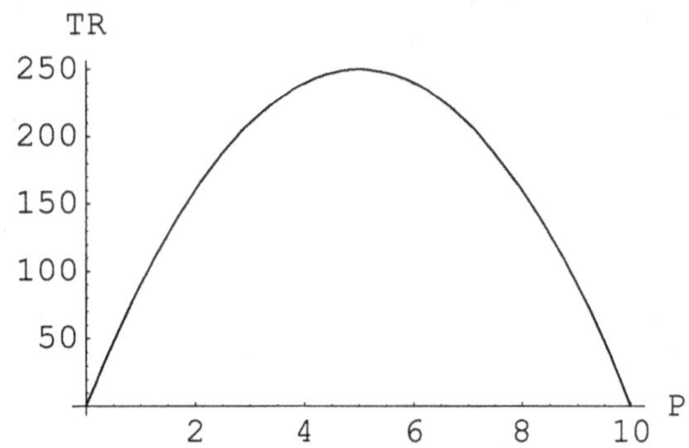

To find total revenue corresponding to P = 4, substitute for P in the TR equation to get TR = 240.

```
In[29]:= tr /. p -> 4
```

Output[29]= 240

<u>Rational functions:</u> Rational functions are derived by forming ratios of polynomial functions. Examples of rational functions are $y = f(x) = \frac{x-1}{x+1}$, $x \neq -1$, and $y = f(x) = \frac{x-1}{x^2-x-2}$, $x \neq -1$ & 2. There is not a single shape that describes the graphs of all rational functions. The shape of the graphs depends on the degree of the polynomial on the numerator and denominators, the zeroes of the polynomials in the denominator, and the asymptotes of the function.

The horizontal asymptotes of a function are the y-values of the function as x approaches to $\mp\infty$. They are derived by finding the limit of f(x) as $x \to \mp\infty$. The limits of the functions will be studied in chapter 6. The vertical asymptotes of a function are the x-values of the function as y approaches to $\mp\infty$. They are found by finding the zeroes of the polynomial in the denominator. Here, it helps to know that, in general, if the numerator and denominator of a rational function are of the same degree, as $x \to \mp\infty$, lim f(x) approaches to the ratio of the coefficients of xs that have the largest exponent in the numerator and the denominator. If the degree of the polynomial in the denominator is larger than the degree of the polynomial in the numerator, the function has y = 0 as its horizontal asymptote. If the degree of the polynomial in the numerator is larger than the degree of the polynomial in the denominator, the function may have a slant asymptote.

Example: Graph $y = f(x) = \frac{x-1}{2x+1}$.

Solution: The function has a vertical asymptote at $x = -\frac{1}{2}$ (this is derived by finding the zeroes of the denominator), and a horizontal asymptote at $y = \frac{1}{2}$ (this is found by finding $\text{Lim}\, f(x)_{x\to\infty}$, which is the ratio of the x-coefficients in the numerator and denominator). The function has an x-intercept of (1, 0) and y-intercept of (0, -1). These information are summarized in table below.

x	$-\infty$	$-\frac{1}{2}$	0	$\frac{1}{2}$	1	$+\infty$
f (x)	$\frac{1}{2}$	$+\infty \mid -\infty$	-1	$-\frac{1}{4}$	0	$\frac{1}{2}$

The table helps with graphing the function. Using *Mathematica*, the function can be graphed as,

```
In[30]:= Clear[x, y]
         y = (x - 1) / (2 x + 1);
         Plot[y, {x, -10, 10}, AxesLabel -> {"X", "Y"}]
```

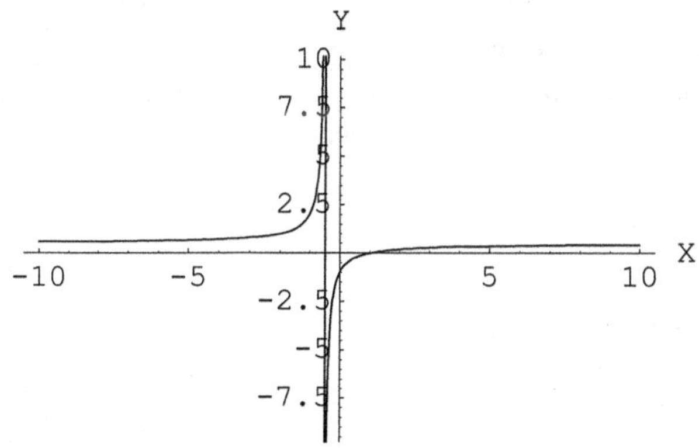

Example: Graph y = f(x) = $\frac{x+2}{x^2-5x+4}$.

Solution: The function has two vertical asymptote at x = 1 & 4 (these are derived by finding the zeroes of the denominator), and a horizontal asymptote at y = 0 (this is found by finding $\text{Lim}\, f(x)_{x\to\infty}$. Since the degree of the polynomial in the enumerator is larger than the degree of the polynomial in the numerator, as x→∓ ∞, y→0).

The function has an x-intercept of (-2, 0) and y-intercept of (0, $\frac{1}{2}$). These information are summarized in table below.

x	$-\infty$	-2	0	1	2	3	4	$+\infty$
f (x)	0	0	$\frac{1}{2}$	$+\infty \mid -\infty$	-2	-2.5	$-\infty \mid +\infty$	0

The table helps with graphing the function. Using *Mathematica*, the function is graphed as,

```
In[31]:= Clear[x, y]
         y = (x + 2) / (x^2 - 5 x + 4);
         Plot[y, {x, -8, 8}, AxesLabel -> {"X", "Y"}]
```

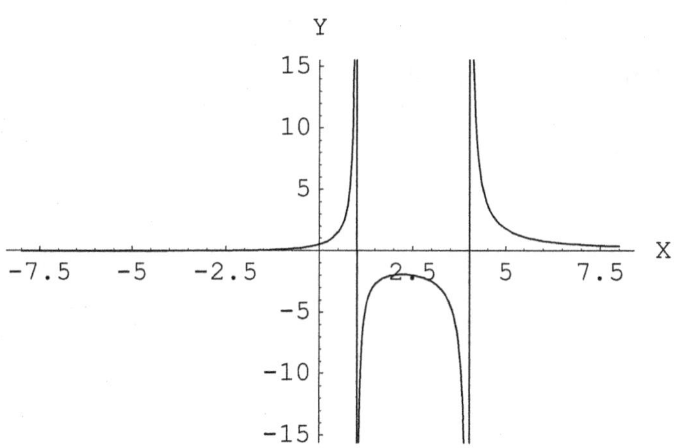

Do More Practice: Change the denominator of the function to x^2- 4x + 4. This give the denominator a multiple zero of x = 2. Graph the function and observe the changes.

2. Change the denominator of the function to x^2- 4x + 5. In this case the denominator has no zeroes. Graph the function and observe the changes.

3. Graph y = $\frac{x^3+1}{x^2+3x-4}$. This function has slant asymptote y = x - 3.

Example: Present Value of Constant Cash Flows

Suppose a constant stream of income at the amount of \$R is accrued at the end of each year for the next n years. The present value of this cash flow at a discount rate of r can be formulated as PV = $\sum_{t=1}^{n} \frac{R}{(1+r)^t}$. If the stream of income is accrued forever, the present value of the perpetuity equals to PV = $\frac{R}{r}$. That is, PV is an inverse function of r. With a numerical values of R = \$100, the relation between PV and can be graphed at the r range of 1% to 20% as shown in the *Mathematica* solution below.

```
In[32]:= Clear[R, r, PV]
         PV = 100/r ;
         Plot[PV, {r, .01, .2}, AxesLabel → {"r", "PV"}]
```

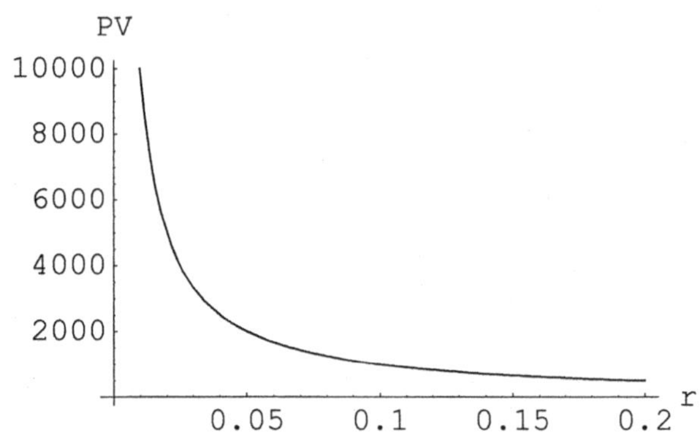

Example: Constant Elasticity Demand Function

Demand for a product is given as Q = $\frac{100}{P}$. Graph the demand function and find the price elasticity of demand.

Solution: The conventional demand shows P in the vertical axes and Q in the horizontal axes. Following this convention, the function can be written as P = $\frac{100}{Q}$. This function has a horizontal asymptote at Q = 0 and a vertical asymptote at the P = 0. The function is a decreasing function as Q increases. The plot of function is shown below.

```
In[33]:= Clear[p, Q]
         Plot[100 / Q, {Q, .1, 5}, AxesLabel -> {"Q", "P"}]
```

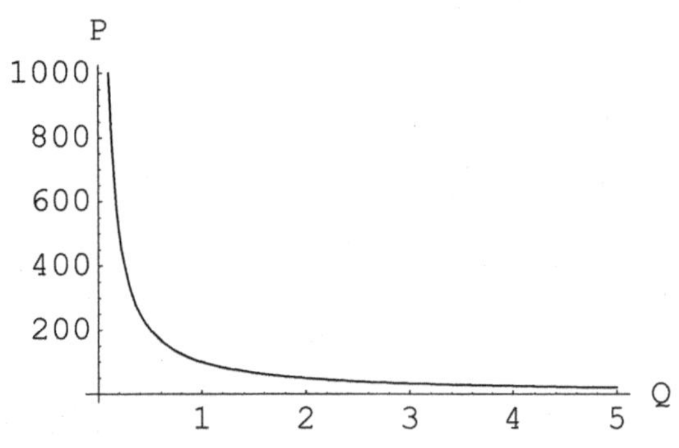

Note: To avoid division by zero, the domain of the function is defined from .1 to 5.

The demand function above has a property that price elasticity of demand, defined as $e_p = \frac{\Delta Q\%}{\Delta P\%}$ is the same everywhere along the demand curve, (e_P = 1). This is because $e_p = \frac{\Delta Q\%}{\Delta P\%} = \frac{\Delta Q}{\Delta P} . \frac{P}{Q}$, where $\frac{\Delta Q}{\Delta P}$ is the slope of the demand function. Since P.Q = 100, $\frac{\Delta Q}{\Delta P} = \frac{Q}{P}$ and the elasticity e_p = 1.

Explicit Algebraic Functions Algebraic functions are generated by the application of some of the algebraic operations on x. Examples are $y = f(x) = \sqrt{x^2 - 4}$ and $Y = f(x) = \frac{\sqrt{1+x}}{\sqrt{1+x^2}}$. To graph the algebraic functions, define the domain of the function, find the asymptotes if any exists, find the x and y intercepts and sketch the graph. The following examples show the graphs of the few explicit algebraic functions.

Example: Graph $y = f(x) = \sqrt{x^2 - 4}$.

Solution: The domain of the function is $x \leq -2$ and $x \geq 2$. For $-2 < x < 2$ the function is not defined. The function has x-intercepts at x = -2 & 2. Finding few more points helps to graph the function.

Using *Mathematica*, the graph of the function is,

```
In[34]:= Clear[x, y]
         y = Sqrt[x^2 - 4];
         Plot[y, {x, -5, 5}, AxesLabel -> {"X", "Y"}]
```

```
Plot::plnr : y is not a machine-size real number at x = -1.63777.

Plot::plnr : y is not a machine-size real number at x = -1.85858.

Plot::plnr : y is not a machine-size real number at x = -1.96775.

General::stop :
 Further output of Plot::plnr will be suppressed during this calculation.
```

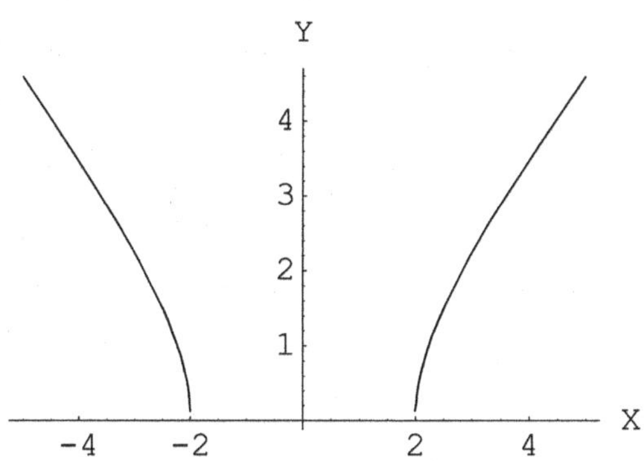

Example: Graph $y = f(x) = \frac{\sqrt{1+x}}{\sqrt{1+x^2}}$.

Solution: The domain of the function is $x \geq -1$. For $x < -1$ the function is not defined. The function has an x-intercept at x = - 1 and an asymptote at y = 0 . This is because the degree of the polynomial in the denominator is larger than the degree of the polynomial in the numerator. As $x \to \mp \infty$, $y \to 0$. Finding few more points helps to graph the function. Using *Mathematica*, the graph of the function is,

```
In[35]:= Clear[x, y]
         y = Sqrt[1 + x] / Sqrt[1 + x^2];
         Plot[y, {x, -1, 20}, AxesLabel -> {"X", "Y"}]
```

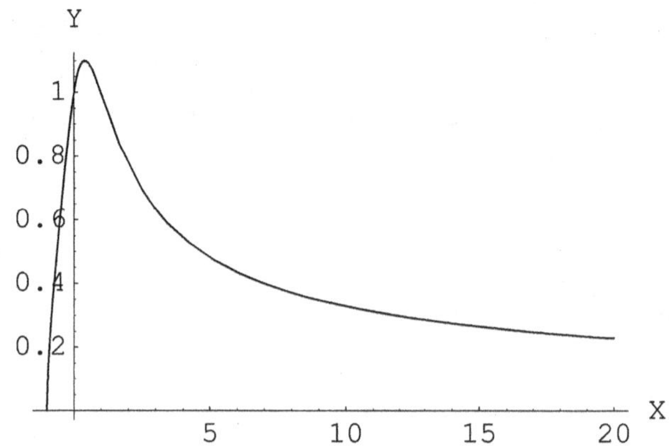

Exponential Functions: The general form of exponential functions is y = f(x) = b^x, where b is positive and x is a real number. This function has the property that the range of y is the set of positive real numbers. Exponential functions of this form are continuous and strictly increasing if b >1, and continuous and strictly decreasing if b < 1.

Example: Graph $y = 2^x$.

Solution: Since b > 1, the function is strictly increasing. As $x \to +\infty$, $y \to +\infty$. As $x \to -\infty$, $y \to 0$. Therefore, y = 0 is the horizontal asymptote. At x = 0, the y-intercept is y = 1. Using these information the function can be graphed. Using *Mathematica*, the graph of the function is,

```
In[36]:= Clear[x, y]
         y = 2^x;
         Plot[y, {x, -5, 5}, AxesLabel -> {"X", "Y"}]
```

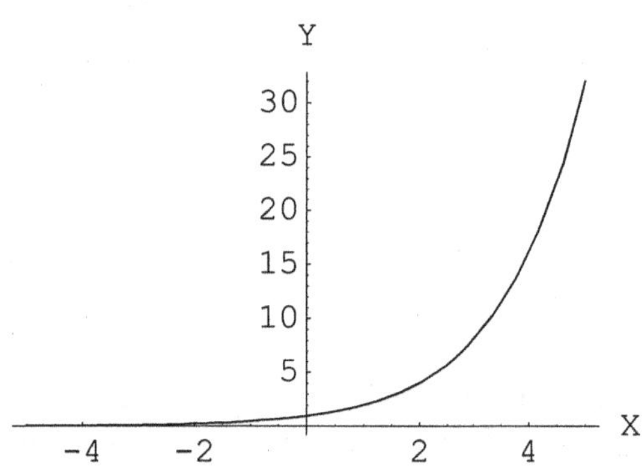

Example: Graph $y = (\frac{1}{2})^x$.

Solution: Since 0 < b < 1, the function is strictly decreasing. As $x \to +\infty$, $y \to 0$. Therefore, y = 0 is the horizontal asymptote. As $x \to -\infty$, $y \to +\infty$. At x = 0, the y-intercept is y = 1. Using these information the function can be graphed. Using *Mathematica*, the graph of the function is,

```
In[37]:= Clear[x, y]
         y = (1 / 2)^x;
         Plot[y, {x, -5, 5}, AxesLabel -> {"X", "Y"}]
```

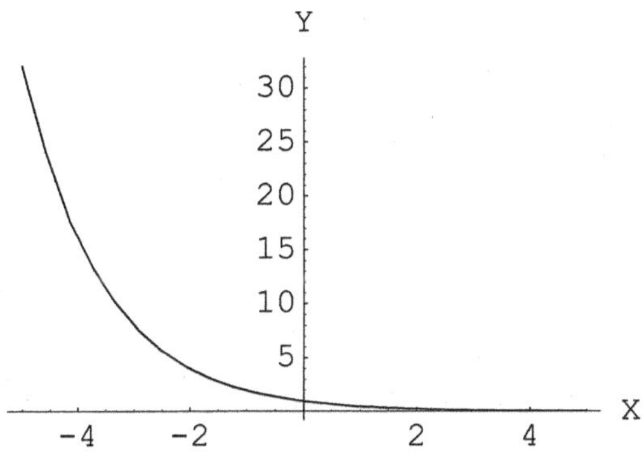

Example: Investment

Plot the time path of the values of \$100 invested at an annual rate of 5%. What is the value of the investment after 5 years?

Solution: Use the compounding formula $S = A(1 + r)^t$, where S is the compounded value after t years, A is the initial value, and r is the interest rate, to formulate the problem as $S = 100(1 + .05)^t$. Since $b = 1.05 > 0$, the exponential function is increasing function with a S intercept of (0, 100). The function is graphed below.

```
In[38]:= Clear[S, t]
         S = 100 (1 + .05)^t;
         Plot[S, {t, 0, 30}, AxesLabel -> {"t", "S"}]
```

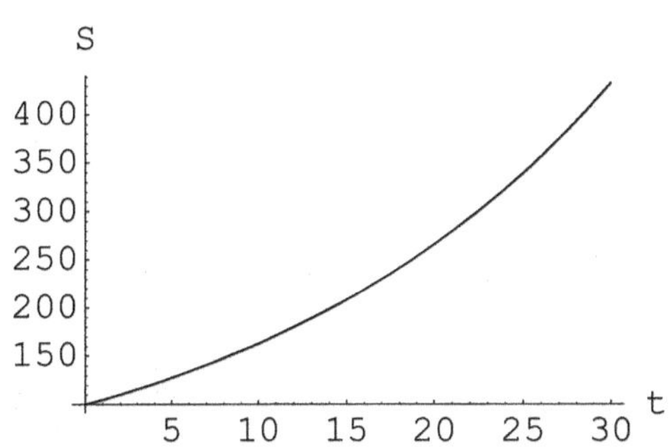

For t = 5, S = 127.63.

```
In[39]:= S /. t -> 5

Output[39]= 127.628
```

Example: Depreciation

The half-life of a machine with an initial value of \$120000 is 5 years. Find a model that shows the value of the machine after t years. Find the depreciated value of the machine after 20 years.

Solution: The half-life of five years means that every five years the machine loses half of its value. Therefore, the value of the machine is depreciating according to an exponential function model $V = 120000(\frac{1}{2})^{\frac{t}{5}}$. At this rate of depreciation, the value of the machine is \$60000 after 5 years, \$30000 after 10 years and so on. After 20 years, the value of the machinery is $V = 12000(\frac{1}{2})^4 = \$750$. Using *Mathematica*, the graph of the function is

```
In[40]:= Clear[V, t]
         V = 12000 (1 / 2) ^ (t / 5);
         Plot[V, {t, 0, 60}]
```

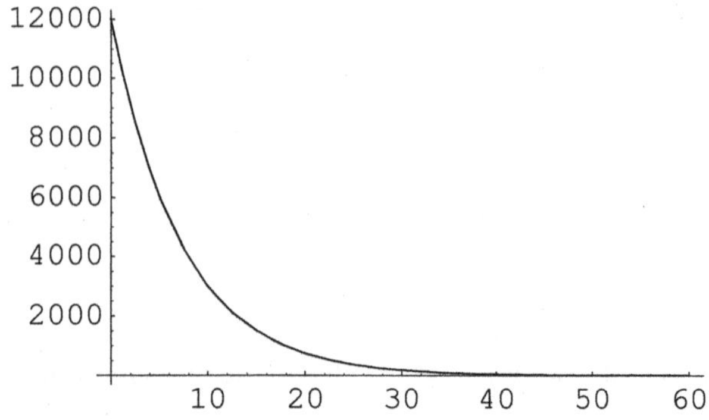

After 20 years, the depreciated value of the machine is $750.

```
In[41]:= V /. t → 20

Output[41]= 750
```

Logarithmic Functions Every exponential function of the form $y = f(x) = b^x$ can be transformed to $\log_b y = x$. The graph of logarithmic functions are similar to the exponential functions. In graphing logarithmic functions, you need to interchange only the x and y axis in the corresponding exponential functions. The following figures compares graphs of $y = e^x$ and $\ln y = x$.

```
In[42]:= Clear[x]
         Plot[E^x, {x, -2, 2}, AxesLabel -> {"X", "Y"}]
```

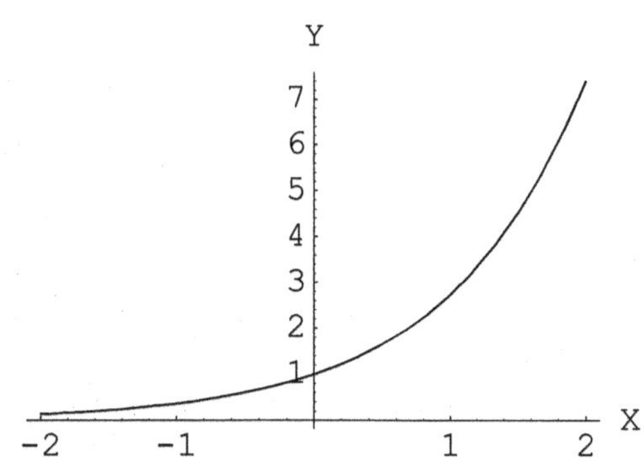


```
In[43]:= Clear[y]
         Plot[Log[y], {y, 0, 7}, AxesLabel -> {"y", "x"}]
```

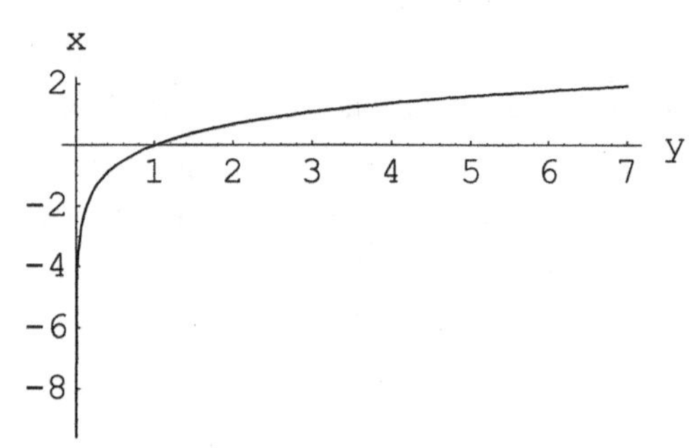

Example: Investment

Plot the time path of an initial investment of $100 deposited at a continuously compounded interest rate of 5%. What is the value of the investment after 5 years?

Solution: The compounded value of an initial investment of A at a rate of r after t years can be found using equation $y = Ae^{rt}$. Plot $y = 100e^{.05t}$ and find the value of y for t = 5. Note that, you may graph $y = 100e^{.05t}$ or $\ln y = \ln 100 + .05t$. The latter graph will provide you with a graph of lny. To find y at time t = 5, substitute for t in the equation to get $\ln y = \ln 100 + .25 = 4.86$. Take the anti log of 4.86 to find y = 128.40.

```
In[44]:= Clear[t, y]
         y = 100 E^(.05 t);
         Plot[y, {t, 0, 8}, AxesLabel -> {"t", "Y"}]
```

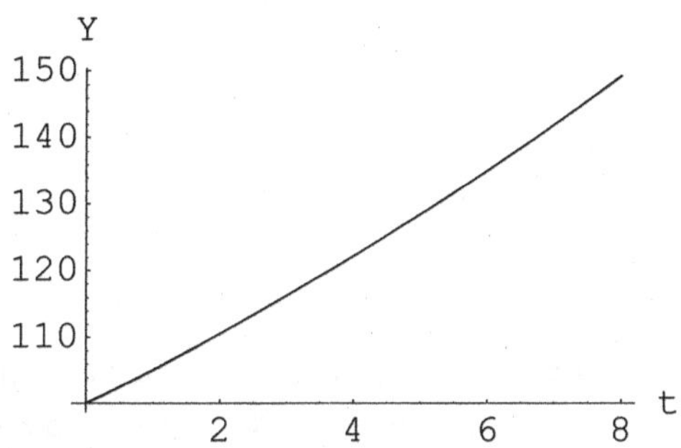

To find the value of y after 5 years substitute for t in the equation to get y = 129.40.

```
In[45]:= y /. t → 5
Output[45]= 128.403
```

Example: Sales

The sales (y) of a new product is estimated as $y = 28000e^{\frac{-4.62}{t}}$, where t is the number of years that the product has been in the market. Graph the sales function and find how many units will be sold after thirty, forty, and fifty years. Find the market saturation point.

Solution: To graph the function you may substitute few values for t and find the corresponding y values and plot the points. It is clear that as t approaches infinitely large numbers, $e^{\frac{-4.62}{t}}$ approaches to one and y approaches 28000 units. This is the **market saturation quantity**. To find y after 30, 40, and 50 years, either substitute for t in the equation above and find the corresponding y values or find $\ln y = \ln 28000 - \frac{4.62}{t}$ then substitute for the t-values of 30, 40, and 50. In this case, you have to take anti log to find the corresponding values of y.

Using *Mathematica*, the graph and the solution of the problem is as follows,

```
In[46]:= Clear[y, t]
         y = 28000 E^(-4.62 / t);
         Plot[y, {t, 1, 60}]
```

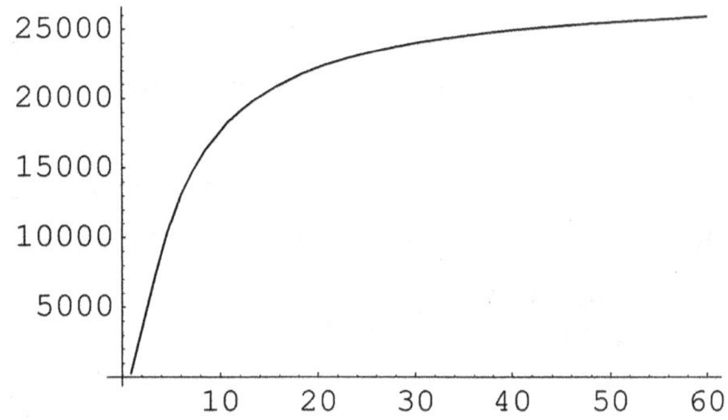

To find the value of y at thirty, forty, and fifty,

```
In[47]:= y /. t -> 30
         y /. t -> 40
         y /. t -> 50

Output[47]= 24003.6

            24945.8

            25528.7
```

the number of units sold after 30, 40, and 50 years are 24003, 24945, and 25528, respectively. To estimate the market saturation point, the graph of y shows that the S-shaped sales graph flattens around 25000 units of sale or after 40 years. In the extreme case, as $t \to \infty$, $y \to 28000$.

Graphing Exponential Functions with Negative Base: If the base of an exponential function is a negative number, the value of the function will oscillate as the exponent values changes from odd numbers to even numbers. Graphing the function follows the same rules of graphing exponential functions with positive base, except for the fact that the function here will not be a smooth function.

Example: Graph $y = (-2)^x$.

Solution: For values of x = 0, 1, 2, 3, 4, ..., the values of the corresponding values of y = 1, -2, 4, -8, 16, ... The graph of the function is,

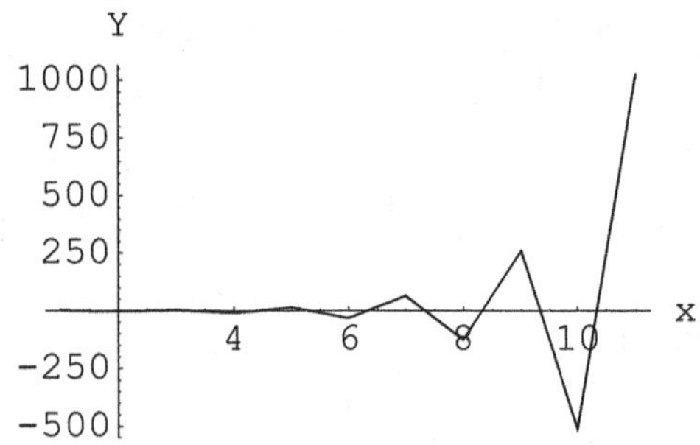

In *Mathematica*, use the command **Table** to generate the values for exponential function then use the ListPlot command to graph the function.

```
In[48]:= Clear[x, y, data]
         data = Table[(-2)^x, {x, 0, 10}];
         ListPlot[data, PlotJoined -> True, AxesLabel -> {"x", "Y"}]
```

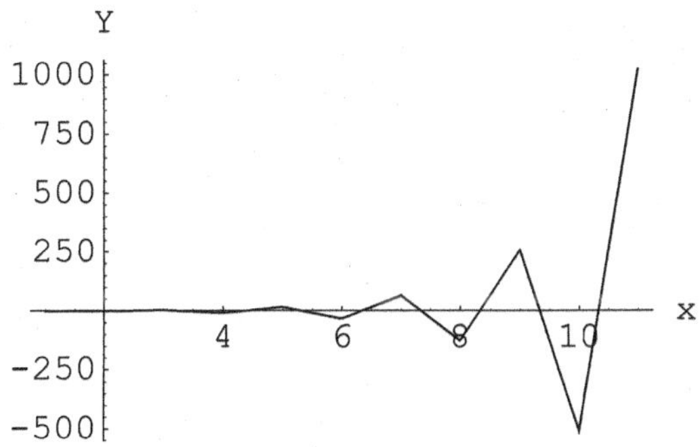

Example: Graph $y = (\frac{-1}{2})^x$.

Solution: Since the base of the exponential function is a negative number, the function will oscillate with the odd and even values for x. Find several values of y by substituting numbers for x and graph the function. Using *Mathematica*, make a table of values for y and ListPlot the data.

```
In[49]:= Clear[x, y, data]
         data = Table[(-.5)^x, {x, 0, 10}];
         ListPlot[data, PlotJoined -> True, AxesLabel ->{"x", "Y"}]
```

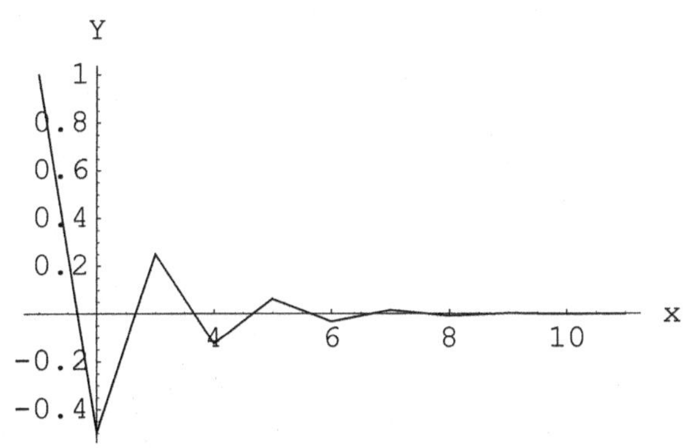

Compound Functions: A function that is specified by more than one expression is called compound function.

Example: Let $f(x) = \begin{cases} 2, & \text{if } x \leq 0 \\ 2 - x, & \text{if } x > 0 \end{cases}$, find f(-2), f(0), f(2) and graph the function.

Solution. Here the value of x determines which expression to be used. For x = -2 and x = 0, the values of f(-2) and f(0) are determined from the first expression and are equal to 2, respectively. For x = 2, the value of f(2) is determined from the second expression and is equal to 0. Below is the graph of the compound function.

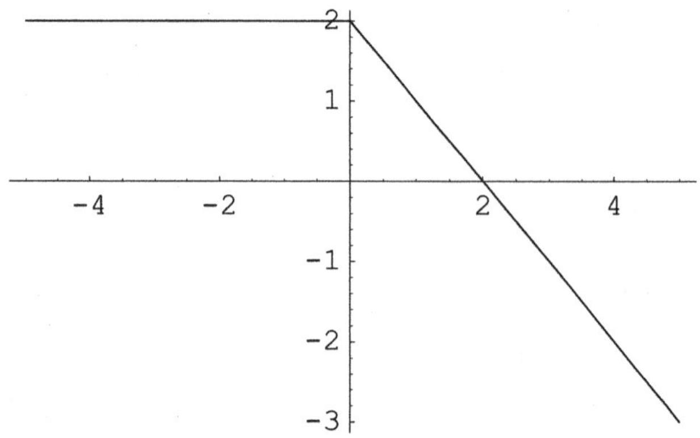

To graph the function using *Mathematica*,

```
In[50]:= Clear[x, p1, p2]
         p1 = Plot[2, {x, -5, 0}]
         p2 = Plot[2 - x, {x, 0, 5}]
         Show[p1, p2]
```

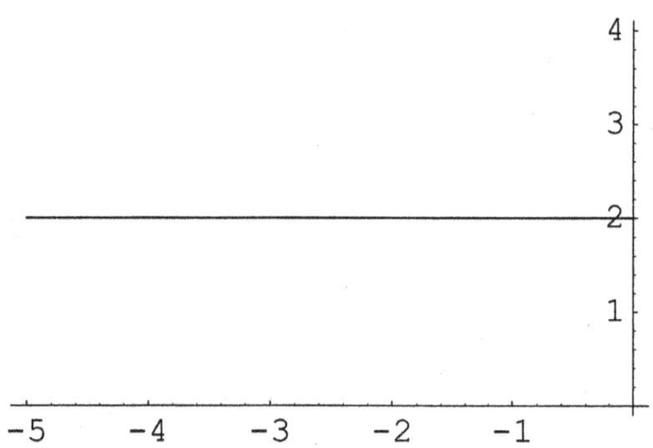

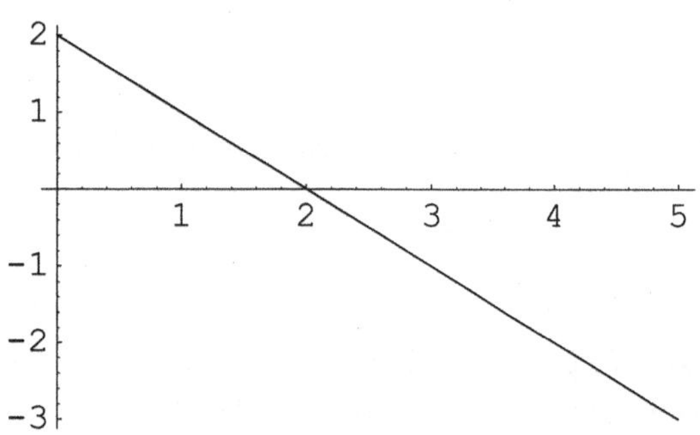

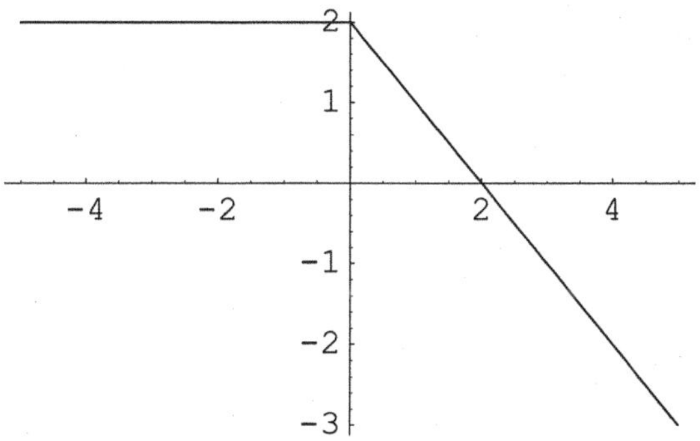

Example: Sales

A manufacturer of T-shirts has three rates for the wholesale price of its popular Halloween T-shirts. For quantities 0-100, the price is \$5 each. For quantities 101-320, the price follows p(q) = $5.(.999)^{(q-100)}$, where q is the quantity. Beyond 320 units, T-shirts are sold at the price \$4.00 each. Express the price quantity relations in the compound function form and graph the relation.

Solution: Here the price is a function of quantity sold. The compound function can be expressed as p(q) =

$$\begin{cases} 5, \text{ if} & 0 < q \le 100 \\ 5.(.999)^{(q-100)}, \text{ if } 101 & < q \le 320 \\ 4, \text{ if} & q > 320 \end{cases}$$

The graph of the price function is as:

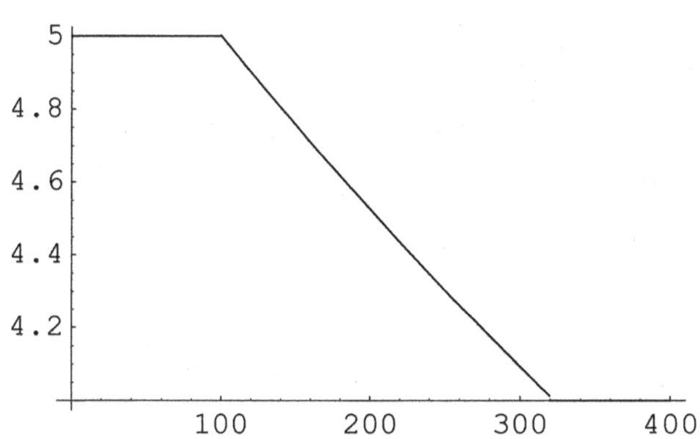


```
In[51]:= Clear[q, p1, p2, p3]
         p1 = Plot[5, {q, 0, 100}]
         p2 = Plot[5 (.999) ^ (q - 100), {q, 100, 320}]
         p3 = Plot[4, {q, 320, 400}]
         Show[p1, p2, p3]
```

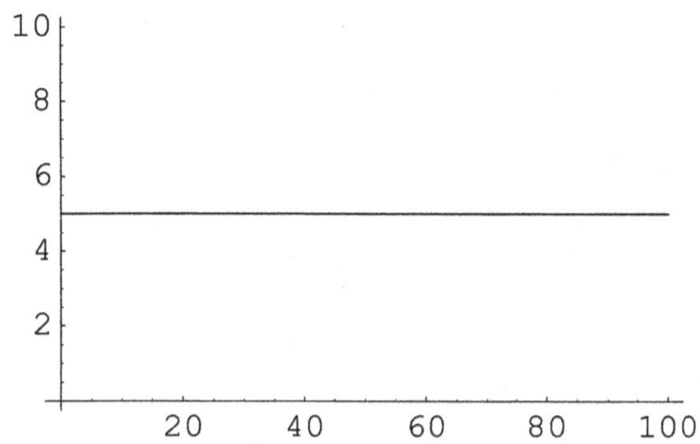

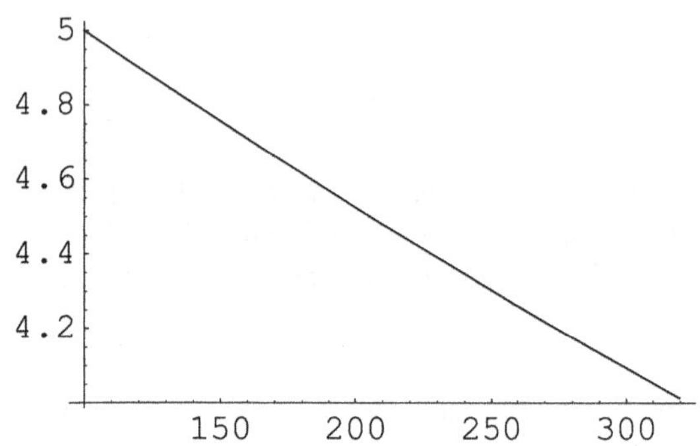

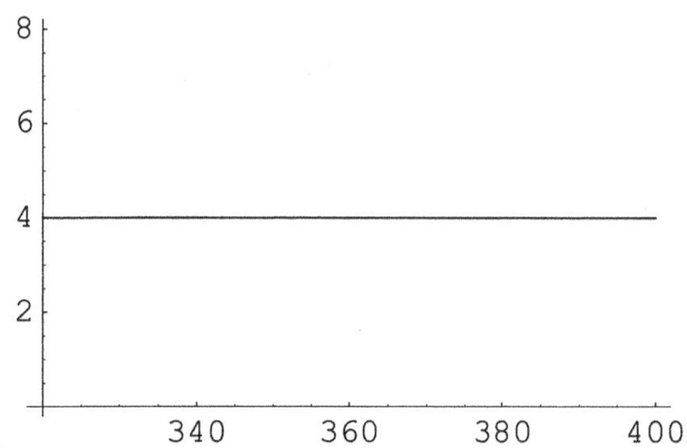

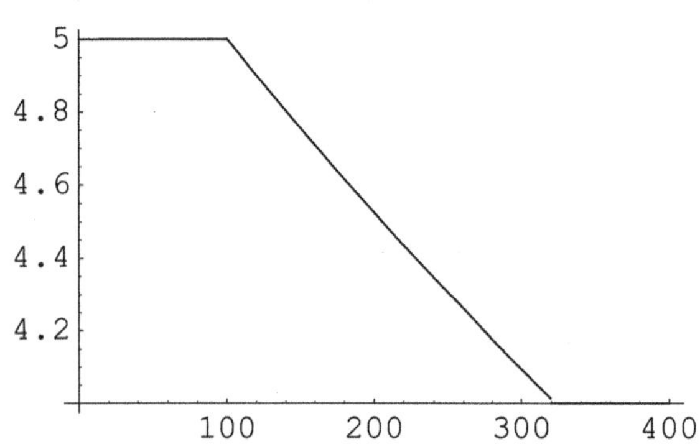

Periodic Functions: Many business and economic variables have patterns that repeat. A repeating pattern can best be represented by a periodic function. A function f is periodic if there exists a positive real number (c) such that $f(t + c) = f(t)$. The smallest repeating portion of the periodic function is called **cycle.** The reciprocal of the period ($\frac{1}{\text{period}}$) is called **frequency.** The distance between the highest or lowest point of the graph of a periodic function and its axis is called **amplitude**.

An example of periodic function is sine and cosine functions. A sine function, y = sin x has a period of 360°, frequency of $\frac{1}{360}$, and amplitude of 1.

Example: Graph y = sin x

Solution: The sine function has a values of zero at x = 0, π, & 2π, an amplitude of one at x = $\frac{\pi}{2}$& $\frac{3\pi}{2}$and a periodicity of 350°. Using these information the function can be graphed. Using *Mathematica*, the graph of the function is,

In[52]:=
```
Clear[x]
Plot[Sin[x], {x, 0, 2 Pi}, AxesLabel -> {"X", "Y"}]
```

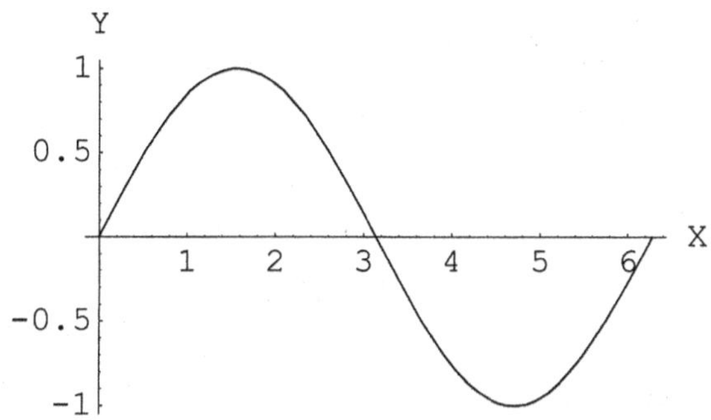

Do More Practice

1. Graph the function in the example for x = 0 to x = 4π.
2. Graph the function in the example for x = -2π to x = 2π.
3. Graph y = 2sinx for x = 0 to x = 2π and compare it to the graph of y = sinx.
4. Graph y = -sin x for x = 0 to x = 2π and compare it to the graph of y = sinx.
5. Graph y = sin2x for x = 0 to x = 2π and compare it to the graph of y = sinx.
6. Graph y = sin3x for x = 0 to x = 2π and compare it to the graph of y = sinx.
7. Graph y = cosx for x = 0 to 2π. Compare it to the graph of y = sinx.

Example: Graph y = tan(x)

Solution: The tan function has a value of zero at x = 0 & π, values of 1 and -1 at x = $\frac{\pi}{4}$ and x =$\frac{3\pi}{4}$, respectively, and an asymptote at x = $\frac{\pi}{2}$. The function has a periodicity of 180°.

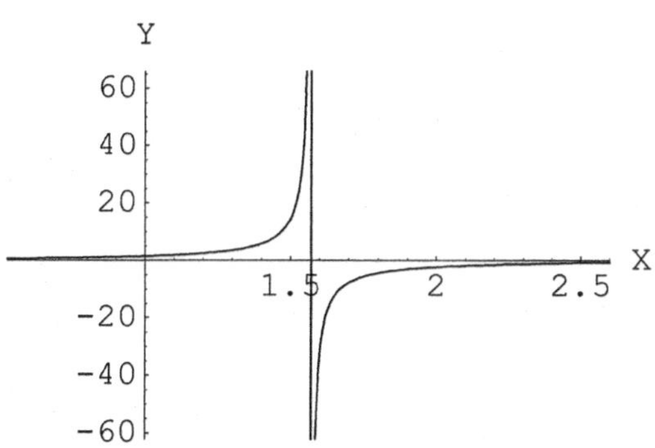

Using *Mathematica*, the function can be graphed as,

In[53]:=
```
Clear[x]
Plot[Tan[x], {x, 0, Pi}, AxesLabel -> {"X", "Y"}]
```

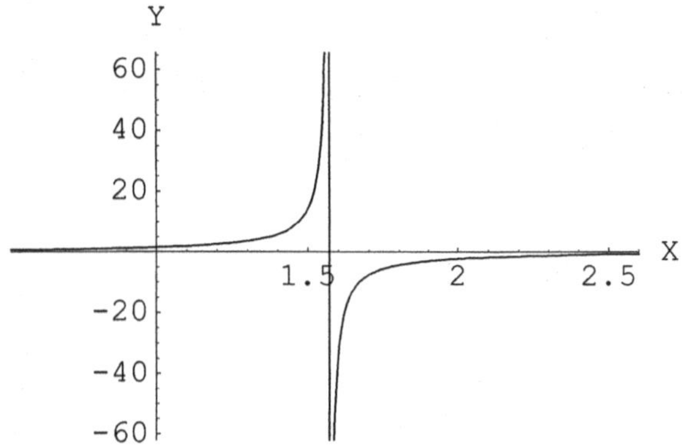

Do More Practice

1. Graph the function in the example for x = 0 to x = 2π.
2. Graph the function in the example for x = $-\pi$ to x = π.
3. Graph y = 2tanx for x = 0 to x = π and compare it to the graph of y = tanx.
4. Graph y = - tanx for x = 0 to x = π and compare it to the graph of y = tanx.
5. Graph y = tan2x for x = 0 to x = π and compare it to the graph of y = tanx.
6. Graph y = tan3x for x = 0 to x = π and compare it to the graph of y = tanx.
7. Graph y = cotx for x = 0 to π. Compare it to the graph of y = tanx.

Example: GDP Growth

Plot the time path of GDP for an economy which is growing at an average annual rate of 6% and experiences the standard business cycles of expansions and recessions with peaks and troughs of $\pm$ 25% average growth.

Solution: The GDP growth function (y) has a trend growth of 6% which can be expressed as y = .06t plus cycles with amplitudes of 25% over and below the trend. Thus, the overall function can be expressed as y = .06t + .25sint. By substituting values for t the function can be graphed. Using *Mathematica*, the graph of the function is,

```
In[54]:= Clear[t]
         Plot[.06 t + .25 Sin[t], {t, 0, 30}, AxesLabel -> {"t", "Y"}]
```

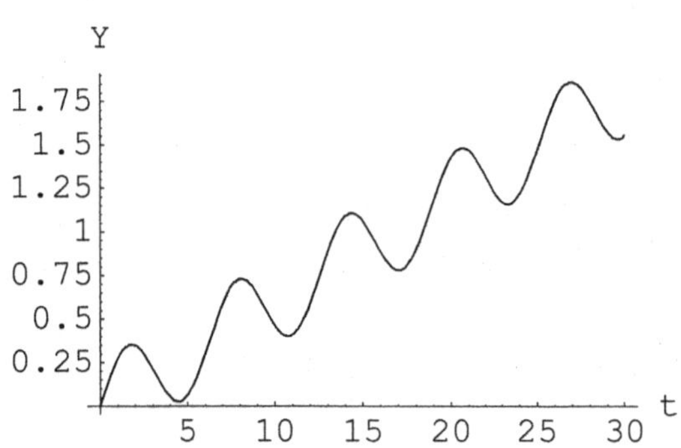

Problems

1- If y = f(x) = $2x^2$ - x + 1, find

a) f(0)   c) f(-1)   e) f(x + h)

b) f(1) d) f(a) f) f(x + h) - f(x)

2- If y = f(x) = $\frac{x^2-9}{x-3}$, find

a) f(0) c) f(3) e) f(x + h)

b) f(1) d) f(a) f) f(x + h) - f(h)

3- State the degree of each of the following polynomials

a) $3x^4 + 2x^2 - 5$

b) 2x + 5

c) 10

4- Find f(-1), f(0), f(1) for the following compound function, $f(x) = \begin{cases} 5, & \text{if } x \le 0 \\ 5 + x, & \text{if } x > 0 \end{cases}$. Graph the function

5- Find f(-5), f(-1), f(0), f(1), and f(5) for the following Compound function, $f(x) = \begin{cases} 0, \text{ if} & x \le 0 \\ .5x^2 - 3x, \text{ if} & 0 < x \le 1.5 \\ 1.5, \text{ if} & x > 3 \end{cases}$. Graph the function.

6- A movie theatre with a capacity of 350 seats has been selling tickets at a price of \$7.50 each. At this rate the theatre has been selling 150 tickets a day. A survey indicates that for each 50 cents rebate offered, the number of tickets sold will increase by 30 per day. Find the demand function of the movie theatre and graph it. Find the total revenue function and graph it. At what price the movie theatre will maximize its revenue? What is the price elasticity of demand at the revenue-maximizing price? (Note that, with linear demand function, revenue is maximized at the mid point of the demand line)

7- Price elasticity of demand for a product with a market price of P = \$10 is $\epsilon = -.5$. At this price, the firm is selling 6000 units of the good. Find the demand function of the firm. Find the total revenue function and graph it. At what price the firm will maximize its total revenue?

8- Graph the following functions,

a) $y = x^2 + 2$

b) $y = \frac{2x-1}{x+2}$

c) $y = x^3 - x + 1$

d) $y = 10 + x^2 - x^3$

e) $y = \sqrt{1+x}$

f) $y = (.1)^x$

g) $y = 3^x$

h) $y = (-1/2)^x$

i) $y = \log_{10} x$

j) $y = (-2)^x$

9- For the following total cost functions

$TC = .3q^2 + 3q + 30$

$TC = q^2 + 30q + 100$

$TC = q3 - 59q^2 + 1315q + 2000$

$TC = .1q3 - 2q^2 + 20q + 100$

where, TC is total cost and q is the output level, plot the total cost and variable cost functions. Plot average total cost, average variable cost, and average fixed cost functions.

10- For the following demand functions plot the demand functions. For each function, find total revenue as a function of price (p) and plot them.

$q = \frac{20}{P}$

$q = 20 - .5p$

$q = \frac{1000}{P^3}$

$p = 10 - .1q$

11- A firm depreciates its machine which was purchased at a price of $250000 at an annual rate of 15%. Plot the depreciation function and find the depreciated value of the machine after ten years. After how many years the machine will be worth half of its initial value?

12- The sales (S) of a new product is estimated as $S = 40000(1 - e^{kt})$, where t is the number of years that the product has been in the market. Find S as a function of t given that 10000 units are sold after 2 years. What is the market saturation quantity? Graph the sales function and find how many units will be sold after five years.

13- Graph the following functions. Find the cycle, frequency, and the amplitude of the functions.

a--$y = 4\sin x$	b- $y = \sin 4x$	c- $y = 4\sin x + 5$
d- $y = \sin 4x + 5$	e- $y = \sin(- 45^{o})$	f- $y = \sin(x + 45^{o})$
g- $y = .25\sin x - 1$	h- $y = .25\sin(x + 45^{o}) - 1$	i- $y = -\sin x$

14- Graph the functions in problem 13 replacing sine with cosine.

III. Multi-Variable Functions and Their Graphs: The Plot3D and ContourPlot Commands

Most real world business and economic problems involve more than one variable. For example, demand for goods depends on consumer income, price of substitute goods, price of complement goods and many other variables as well as the price of the product. In every production process, quantity output depends on labor input, capital input, technology, and other variables.

The analytic approach to multi-variable functions will be discussed later. Here, the three dimensional graphs of the functions and their contours will be used to study the relationship among the variables for certain economic relationships. The three dimensional graph of z = f(x, y) maps the set D (domain of the function) to set R (range of the function). A contour plots the relationship between x and y for a given value of z. Contour plots are useful in analyzing indifference curves in consumer theory or in studying isoquants and isocosts in theory of production.

Example: Production Function

In a simple linear production technology the relationship between output (Q) and labor input (L) and capital input (K) is given as Q = 6L + 4K. Graph the production function and the isoquants (contour) representing the trade off between L and K for different levels of output.

Solution: The production function shows the most efficient values of Q for different combinations of K and L. The contour plot shows the trade-off between labor input and capital input (isoquants) for the given levels of output. Here, the graph of the production function is a surface in L, K, and Q space. By substituting values for L and K and finding their corresponding Q values the output function can be graphed in three dimensions. To graph the contour plots, set Q to a fixed values, find K in terms of L and graph the relationship between K and L in K-L space.

```
In[55]:= Clear[L, K]
        Plot3D[6 L + 4 K, {L, 0, 10}, {K, 0, 10}, AxesLabel -> {"L", "K", "Q"} ,
         PlotLabel -> FontForm["Linear Production", {"Helvetica-Bold", 12}]]
```

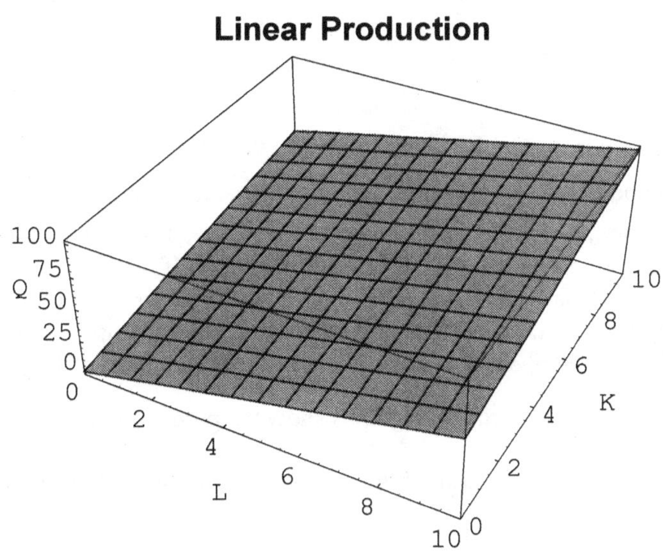

- SurfaceGraphics -

```
In[56]:=ContourPlot[6 L + 4 K, {L, 0, 10}, {K, 0, 10},
        AxesLabel -> {"L", "K"}, ContourShading -> False ,
         PlotLabel -> FontForm["Isoquants of Linear Production",
          {"Helvetica-Bold", 12}]]
```

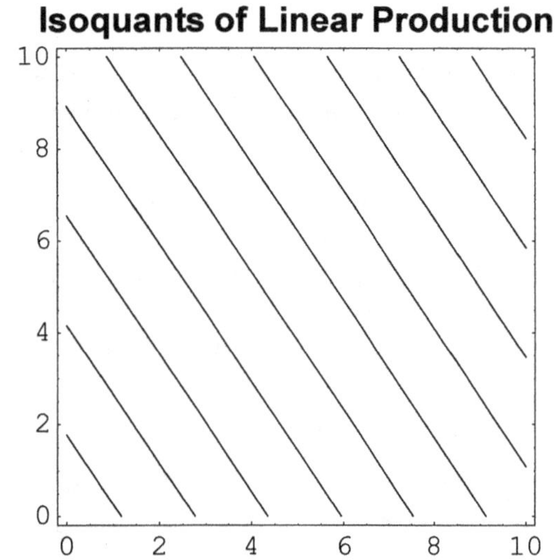


```
- ContourGraphics -
```

Example: Cobb-Douglas Production Function

One of the best known functions in the economic theory is the Cobb-Douglas function. Cobb-Douglas function has a general form of $\mathbf{Z = AX^{\alpha}Y^{\beta}}$. One very nice property of Cobb-Douglas function is that α and β are elasticities. That is, if the function represented a demand function, Q = $AP^{\alpha} y^{\beta}$, where Q is quantity, P is price, and y is income, α would be price elasticity of demand and β would be income elasticity of demand. If the function was a production function, Q = $AL^{\alpha} K^{\beta}$, where Q is output, L is labor input and K is capital input, α would be labor elasticity of output and β would be capital elasticity of output. The function has been successfully used in empirical studies of markets, production, and consumer theory . Suppose the relationship between output (Q), labor input (L), and capital input (K) is given as $Q = 10L^{.75}K^{.25}$. The following two graphs show the production function and its contour plot. The contour plots shows the trade-off between labor input and capital input (isoquants) at every fixed level of output.

```
Input[57] := Clear[Q, L, K]
Plot3D[10 L^(.75) K^(.25), {L, 0, 50},
 {K, 0, 50}, AxesLabel -> {"L", "K", "Q"},
 PlotLabel -> FontForm["Non-Linear Production", {"Helvetica-Bold", 12}]]
```

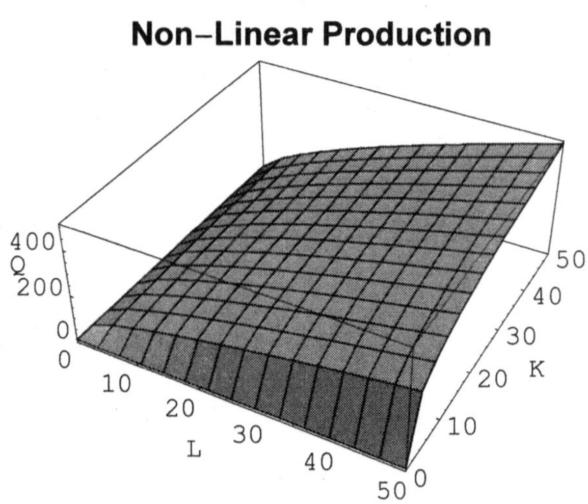

```
- SurfaceGraphics -
```

```
Input[58] := ContourPlot[10 L^(.75) K^(.25), {L, 0, 50}, {K, 0, 50},
  ContourShading -> False, AxesLabel -> {"L", "K"}, PlotLabel ->
   FontForm["Isoquants of Non-Linear Production", {"Helvetica-Bold", 12}]]
```

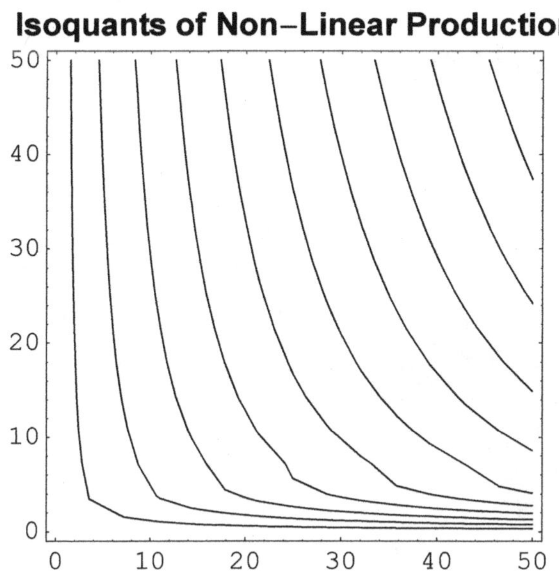


```
- ContourGraphics -
```

In the production function above, α and β adds up to one, .75 + .25 = 1. Such production function is called a **constant return to scale** production function. That is changing inputs by k proportion will change output by k proportion. However, if $\alpha + \beta > 1$, the function is called **increasing return to scale**. Increasing return to scale (or **economies of scale**) implies that changing inputs by k proportion will change the output by more than k proportion. If $\alpha + \beta < 1$, the function is **decreasing return to scale**. A decreasing return to scale (or **diseconomies of scale**) implies that changing inputs by k proportion will change output by less than k proportion.

Do More Practice:

1. Change The function to $Q = 10L^{1.2} L^{.8}$. Graph the function and its contours. Compare the graphs to ones in the example. Comment of the shape of the graphs and the scale of operation.

2. Change The function to $Q = 10L^{.5} L^{.3}$. Graph the function and its contours. Compare the graphs to ones in the example. Comment of the shape of the graphs and the scale of operation.

Example: Market

Demand for a product is given as $Q = 10\frac{y^2}{p^{.5}}$ where, Q is quantity, y is income, and p is the price of the product.

Graph the demand function and comment on the relationship between the variables. What are the price and income elasticities of demand?

Solution: It is clear from the graph that as income increases Q also increases and as price increases, Q decreases.

```
Input[59] := Clear[y, p]
Plot3D[(10 y^2) / (P^.5), {y, 1, 50}, {P, 1, 10},
 AxesLabel -> {"Y", "P", "Q"}, PlotLabel -> FontForm[
   "Demand as a Function of Income and Price", {"Helvetica-Bold", 12}]]
```

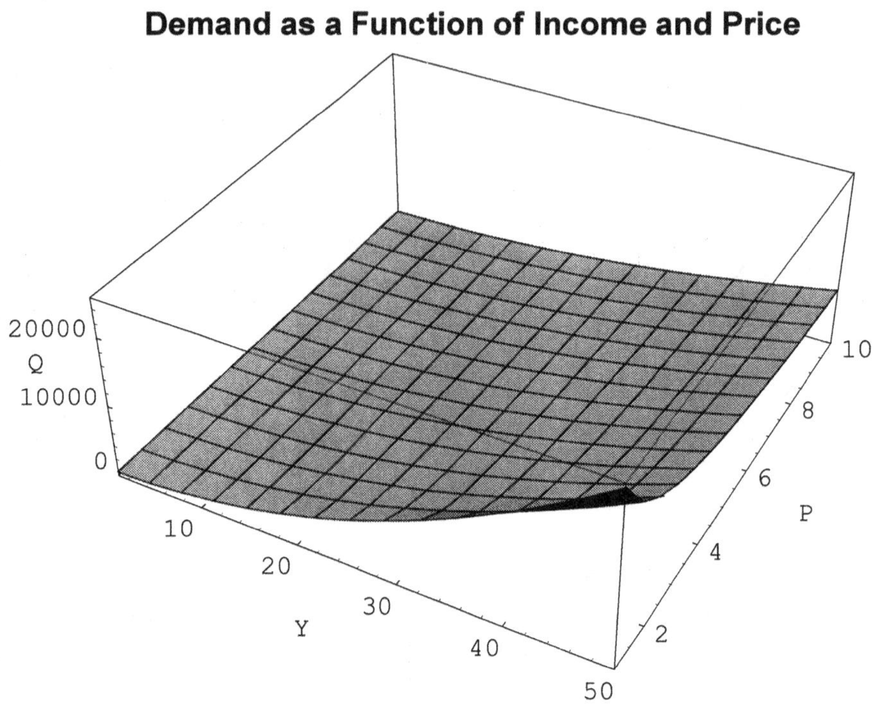

- SurfaceGraphics -

These relationships can be clearly observed by plotting the relationship between Q and Y for given levels of P and by plotting the relationship between Q and P for given levels by Y. To observe the relationship between y and Q, plot contours of Q-Y relations for fixed values of P. This is done by solving the demand equation above for P and plotting the contours of the resulting equation as,

```
Input[60]:= Clear[p, q, y, sol]
            sol = Solve[q == 10 y^2 / p^(.5), p]
```

Output[60]= $\{\{\text{Null}\ (p \to \frac{100.\, y^4}{q^2})\}\}$

```
Input[61]:= p[q_, y_] = p /. sol[[1]]
```

Output[61]= $\frac{100.\, y^4}{q^2}$

```
Input[62]:= ContourPlot[p[q, y], {y, 1, 50}, {q, 1, 20}, ContourShading -> False]
```

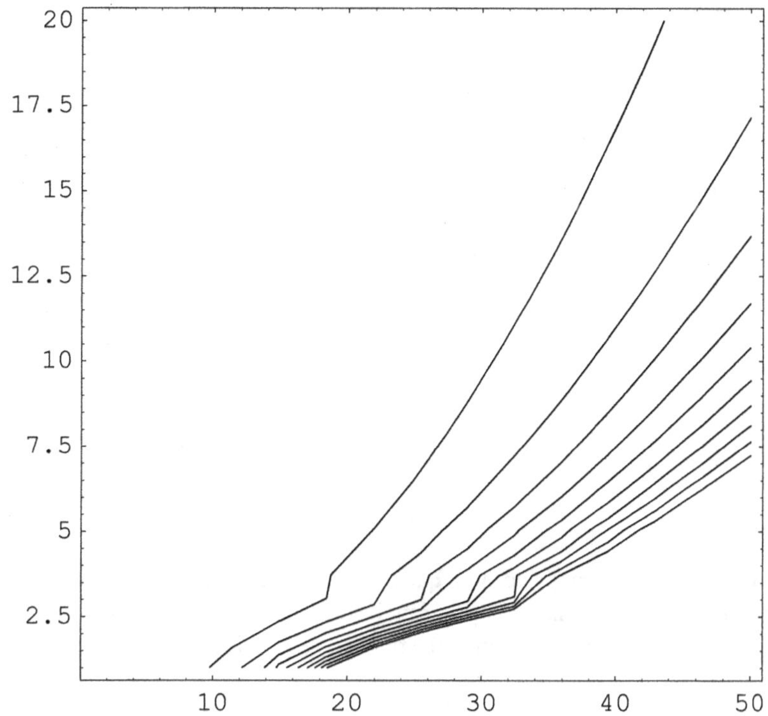

- ContourGraphics -

Obviously, the relationship between income and quantity demanded is a positive relation, as shown by the contours. However, what is interesting is that, at lower levels of income, the function is concave downward and at higher levels of income the function is concave upward. This implies that, given a fixed price, as income increases, Q increases at a decreasing rate up to a certain level of income. Beyond this level of income, as income increases, Q increases with an increasing rate (**note** that here, demand is income elastic, $e_y = 2$. Check this property for an income inelastic demand in the following "Do More Practice").

To observe the relationship between p and q, plot contours of Q-P relations for fixed values of y. This is done by solving the demand equation for y and plotting the contours of the resulting equation as,

Input[63]:= **Clear[p, q, y, sol]**
sol = Solve[q == 10 y^2 / p^(.5), y]

Output[63]= {{Null $(y \to -0.316228\, p^{0.25} \sqrt{q})$}, {Null $(y \to 0.316228\, p^{0.25} \sqrt{q})$}}

Input[64]:= **y[q_, p_] = y /. sol[[1]]**

Output[64]= $-0.316228\, p^{0.25} \sqrt{q}$

Input[65]:= **ContourPlot[y[q, p], {p, 1, 10}, {q, 1, 20}, ContourShading -> False]**

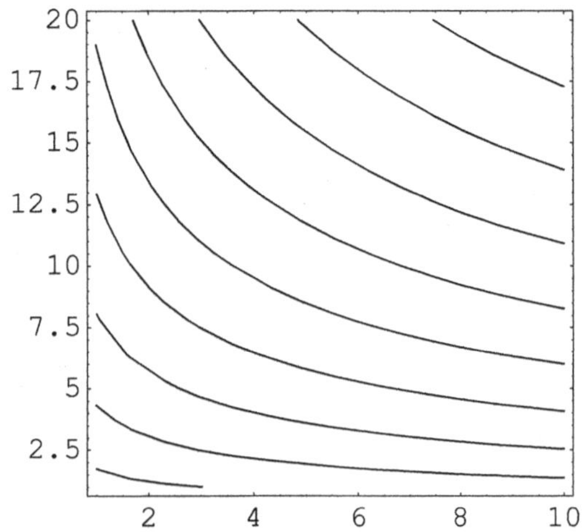

- ContourGraphics -

the contours above show the inverse relationship between price and quantity demanded for different levels of income (y). the price elasticity of demand is $e_p = .5$ and demand is price inelastic. **Note that,** in a Cobb-Douglas demand function of the form $Q = AP^{\alpha}Y^{\beta}$, α and β are price and income elasticities, respectively.

Do More Practice:

1. Change the demand function of the example above to $Q = 10\frac{y^{.7}}{p^{.5}}$. Graph the demand function and its Y-Q and Y-P contours. Comment on the behavior of Q with respect to y and P. What are the income and price elasticities and how do they affect the Y-Q and P-Q relationships?

2. Change the demand function of the example above to $Q = 10\frac{y^{.7}}{p^{2}}$. Graph the demand function and its Y-Q and Y-P contours. Comment on the behavior of Q with respect to y and P. What are the income and price elasticities and how do they affect the Y-Q and P-Q relationships?

Problems

1- Plot the following production functions and the contours of the functions (isoquants).

a) $Q = 5L + 2K$

b) $Q = 50K^2L^2 - K^3L^3$

c) $Q = .3L - .01L^2 + .4K^2 - .02K^3$

d) $Q = 100K^{.5}L^{.5}$

e) $Q = 40K^{1.2}L^{2.2}$

f) $Q = 10(2K^{-.5} + 3L^{-.5})^{(-\frac{8}{5})}$

2- Plot the following utility functions and the contours of the functions (indifference curves).

a) $U = \sqrt{xy}$

b) $U = \sqrt{x^2 + y^2}$

c) $U = (5x^{.5} + 4y^{.5})^2$

d) U = Log xy

IV. Parametric Functions: The **ParametricPlot** and **ParametricPlot3D** Commands

In the previous sections y coordinate of each point was defined as a function of x coordinate or x and other coordinates in case of multi-variable functions. In a parametric function x and y coordinates of each point are a function of some other variable. Suppose x and y are both a function of a third variable, say t. That is, x = g(t) and y = h(t). The functions relating x and y to t are called parametric functions and t is called a parameter. Equation relating y to x may be derived by solving x = g(t) for t and substituting it in y = h(t) to obtain y as a function of x, y = f(x). In business or economic problems, you may think of t as representing time, and the parametrization as describing the time path of the variables. The point (x(t), y(t)) describes a point on x, y coordinates reached at time t.

At times, parametric equations help with graphing complicated functions that are difficult to graph otherwise. For example, to graph $x = y^4 - 4y^2$,the function should be solved for y as functions of x and each function to be graphed independently. However, if we define y = t and write $x = t^4 - 4t^2$, the parametric functions can be graphed by substituting values for t and finding the corresponding values of x and y and graphing the function or by using the ParametricPlot function of calculators or computer programs. In general, if we are to graph an equation of the form x = f(y), we can use the parametric equations x = g(t), y = t and parametric plot of the equations.

Example: Graph $x = y^4 - 4y^2$.

Solution: Set y = t and $x = t^4 - 4t^2$. Find x and y for the given values of t and graph the x-y relations.

t	-2	-1	0	1	2	3	4
y	-2	-1	0	1	2	3	4
x	0	-3	0	-3	0	45	192

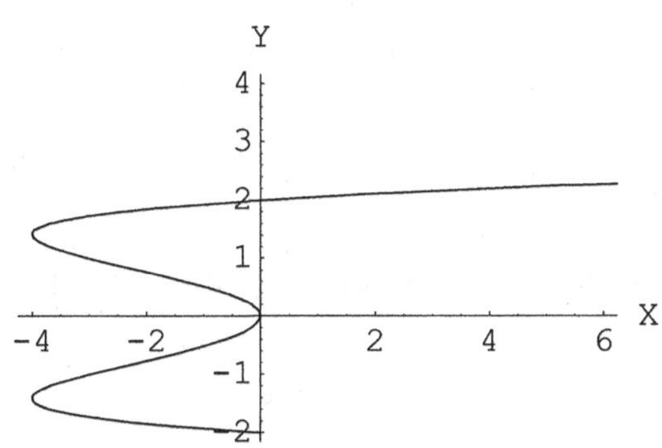

To graph the function using *Mathematica*, use the command ParametricPlot as follows.

```
Input[66]:= Clear[t]
            ParametricPlot[{t^4 - 4 t^2, t}, {t, -2, 10}, AxesLabel -> {"X", "Y"}]
```

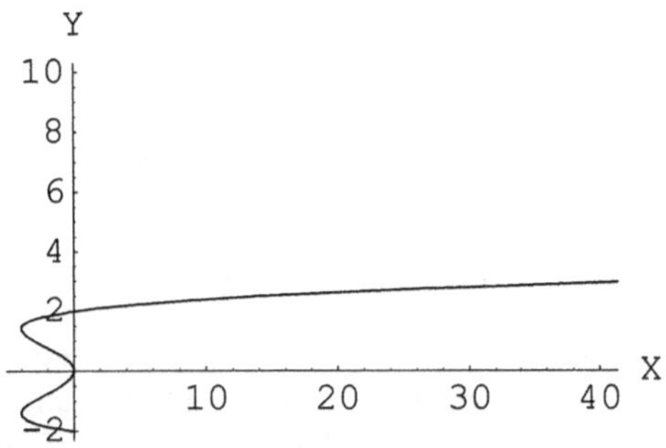

Example: Graph the parametric functions x = f(t) = t^3+ 2t, y = g(t) = t + 4 and identify the relationship between x and y.

Solution: Substitute for values for t and find the corresponding values of x and y and graph the x-y values on the x-y plane.

t	-3	-2	-1	0	1	2	3	4
y	1	2	3	4	5	6	7	6
x	-33	-12	-3	0	3	12	33	72

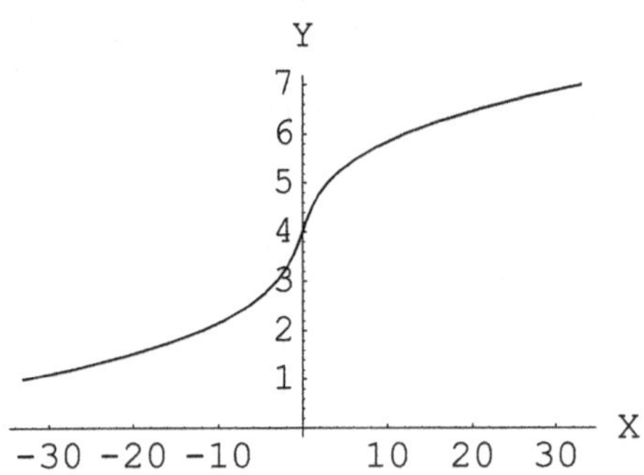

The relationship between x and y can be derived by finding t from y = t + 4 and substituting it in the equation of x. This will give x = $y^3 + 12y^2 + 14y + 72$.

Using Mathematica,the function can be graphed as

```
Input[67]:= Clear[t]
            ParametricPlot[{t^3 + 2 t, t + 4}, {t, -5, 5}, AxesLabel -> {"X", "Y"}]
```

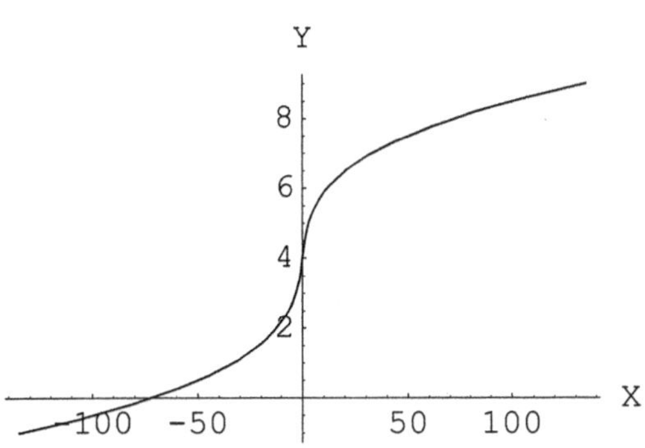

Note that, in parametric equations, each value of t determines a point (x, y) which can be plotted in x-y coordinate system. As t varies, the point (x, y) = (f(x), g(x)) varies and traces out a curve, called parametric curve.

Example: Production and Growth

Labor force in a cyclical farming sector of an economy is growing at a rate of 10% from an initial number of 10 million, $L = 10e^{.1t}$ The real production of the sector (RP) has a time path given by RP = 20t + 100sin . 9t. Plot the time path of L, RP, and the relationship between labor force and RP. Derive RP as a function of L.

Solution: The labor force and the real production equations are both expressed in terms of t. Substitute values for t and find the corresponding values of L and RP and graph the relationship between RP and L. To find RP in terms of L, using logarithmic transformation find t from the first equation , LogL = Log10 + .1t or t = 10(LogL - Log10). Substitute t in the RP equation to find RP in terms of L, RP = 200(LogL - Log10) + 1000sin . 9(LogL - Log10).

t	0	1	2	3	4	5	6	7	8	9
L	10	11.05	12.21	13.50	14.92	16.49	18.22	20.14	22.26	24.60
RP	0	90.33	137.38	102.74	35.75	2.25	42.72	141.68	239.36	276.99

The graph of the parametric curve is

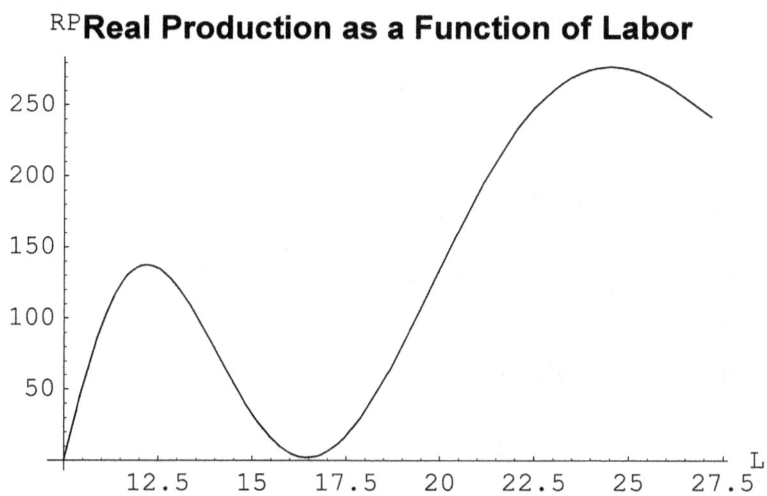

Using *Mathematica,* the following graphs show the graphs of the labor equation as a function of time, real production as a function of time, and the parametric plot, respectively.

```
Input[68] := Clear[t]
Plot[10 E^(.1 t), {t, 0, 10}, AxesLabel -> {"t", "L"},
 PlotLabel -> FontForm["Labor Growth Over Time", {"Helvetica-Bold", 12}]]
```

```
Input[69] := Clear[t]
Plot[20 t + 100 Sin[.9 t], {t, 0, 10}, AxesLabel -> {"t", "RP"},
  PlotLabel ->
   FontForm["Real Production Growth Over Time", {"Helvetica-Bold", 12}]]
```

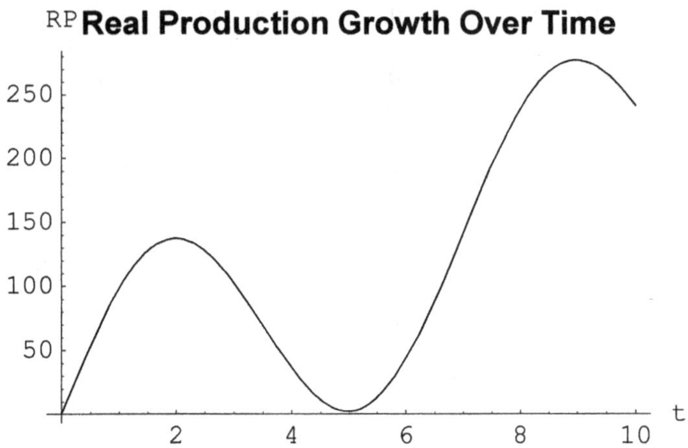


```
Input[70] := Clear[t]
ParametricPlot[{10 E^(.1 t), 20 t + 100 Sin[.9 t]}, {t, 0, 10},
    AxesLabel -> {"L", "RP"},
  PlotLabel -> FontForm["Real Production as a Function of Labor",
                 {"Helvetica-Bold", 12}]];
```

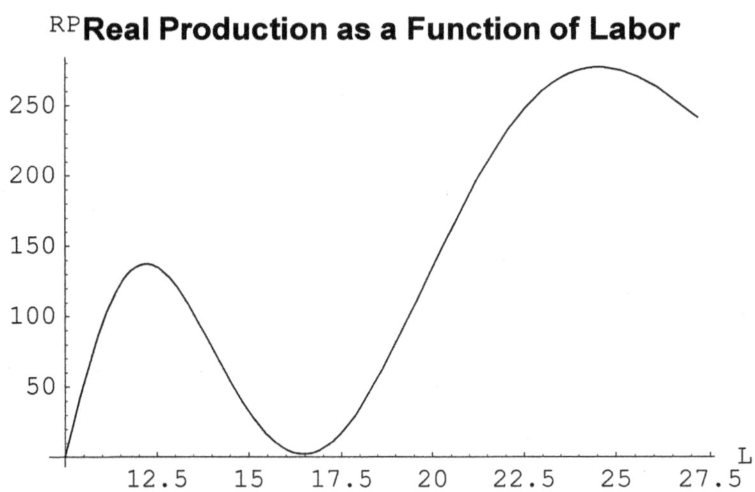

Example: Production Function

A very fast growing business starts with one unit of labor and one unit of capital and grows at an annual rate of 30% and 80% for labor and capital, respectively. The output (Q) of the business is cyclical and is growing as Q = t + sin(t). Graph the parametric equations.

Solution: Labor is growing as $L = e^{.3t}$, capital is growing as $K = e^{.8t}$, and $Q = t + \sin t$. Substitute values for t and find the corresponding L, K, and Q. The plot will be a three dimensional plot which is hard to graph. Using *Mathematica*, the parametric plot of the equations is shown below.

```
Input[71] := Clear[t]
ParametricPlot3D[ {E^(.3 t), E^(.8^t), t + Sin[t]}, {t, 0, 5},
    AxesLabel -> {"L", "K", "Q"},
 PlotLabel -> FontForm["Parametric Plot of Output as Function of Inputs",
   {"Helvetica-Bold", 12}]];
```

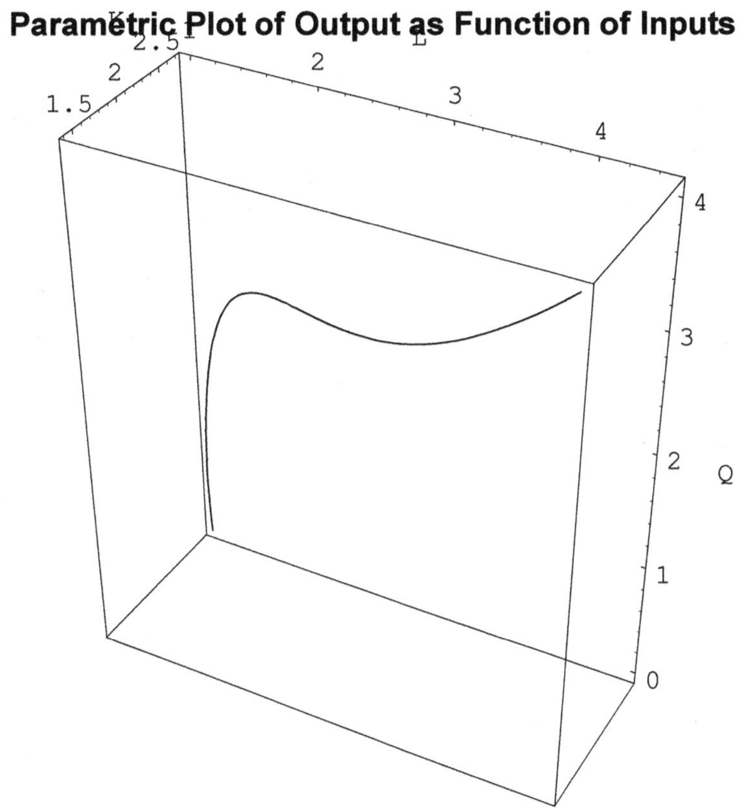

Problems

1. Graph the following equations.

a) $x = y^5 + y^3$

b) $x = y^3 - 3y^2$

c) $x = \sqrt{y^3 + 3y}$

d) $x = \sqrt[3]{t^2 - t}$

e) $x = \sqrt[3]{y^4 - 4y^2}$

2. Graph the following parametric equations.

a) $x = 10t^2 - 3t + 8$ and $y = 5t + 6$

b) $x = t^3 - 3t + 4$ and $y = 2t + 5$

c) $x = t^4 - 3t^3 + t$ and $y = -t + 8$

d) $x = \sqrt{t + 2} + 8$ and $y = t^2$

e) $x = .5^t$ and $y = 10 + t^2$

f) $x = 10e^t$ and $y = t - .5t^2$

g) $x = \sin t$ and $y = \sin 2t$

h) $x = \cos t$ and $y = t\sin t$

CHAPTER 3

FITTING EQUATIONS

A study of the relationship between two or more variables, frequently, involves collecting sets of data on the variables and finding how changes in one variable affects the values of the other variables. An initial investigation of the relationship between variables is usually done by plotting the scatter diagrams of sets of data. Although plotting scatter diagrams are highly recommended for initial investigation of the relationship between variables, they do not specify the form of the relationship between the variables and do not determine the strength of the relationship between the variables statistically. By finding the best fitting line or curve to sets of data, the relationship between the variables can be expressed in mathematical form and the strength of the relationship between variables can be evaluated statistically. The process of fitting a line or a curve to sets of data is called regression.

The most common procedure to fit a linear regression is the **least-squares technique**. The least-squares technique and regression analysis integrate differential calculus and statistics and provide a means of model building, estimation, statistical analysis and forecasting. The detailed techniques of model building, fitting curves and statistical analysis will not be discussed in this book. Here, we will briefly study the methods of fitting lines and curves using least-squares technique. The interested reader is referred to any standard Statistics or Econometrics text book.

TOOL BOX

Mathematica Commands	Function
data = {{1,3}, {4,7}, (5,3}, {9,4}}	Defines "data" as a list of 4 paired observations x, y.
N[Log[data]]	Transforms observations in "data" to natural log.
Fit[data, {1, x, x^2}, x]	Fits a best fitting parabola to pre-defined "data" table
Show[plot1, plot2]	Shows plots called plot1 and plot2 together
ListPlot[data]	Listplots the table called "data"
ListPlot3D [data]	3D Plot of a 3-dimensional list of data called "data"
PlotStyle	Sets plot options of color, thickness, and others
InterpolatingPolynomial	Fits a polynomial of degree n - 1 to n observations

I. Fitting Lines: The Fit & Show Commands

For every two points on a coordinate system we can fit a line that contains the points. Such a fit will be a perfect fit because the distances between the points and the fitted line, or the errors between the actual points and the fitted line, will be zero. In general, for every set of n data points we can fit a polynomial of order n -1 which passes through all the points. However, fitting such polynomial will limit degrees of freedom for any statistical test and will be of no practical value. In practice, the type of the function to be fitted to a set of data is decided by a combination of eyeballing (observing the shape of the relationship through the list plot of data), statistical tests of best fit, and by implications of relevant economic theory.

Example: Fit a line to points A(1, 3) and B(4, 2).

Solution: A line fitted to any two points is a perfect fit. Find the slope (m) of the line as the ratio of the rise to run, m = $\frac{2-3}{4-1} = \frac{-1}{3}$. Using y = mx + c, find the y-intercept as 3 = $\frac{-1}{3}$ + c, or c = $\frac{10}{3}$. The equation of the line is y = $\frac{-1}{3}$x + $\frac{10}{3}$. The points and the graph of the line is shown below

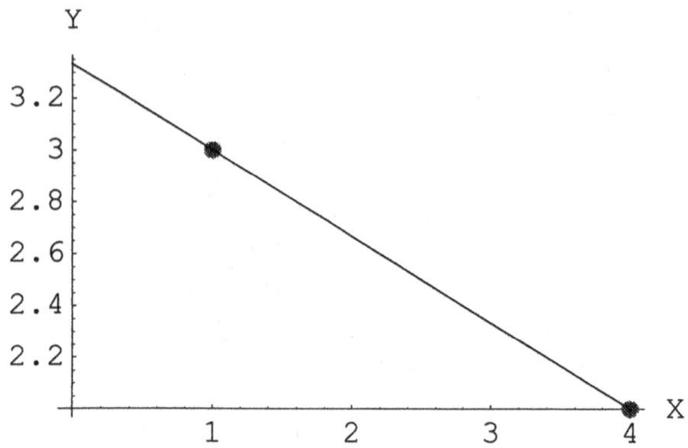

However, if the number of points is more than two and if the points are not all lying on the same line, any fitted line will not contain all the points. For every point, the distance between the fitted line and the point measures the error of the fit. The graph below shows five points {{1, 3}, {4, 2}, {3, 3}, {3, 2.5}, {4, 2.2}} and the best line fitted to the points. The

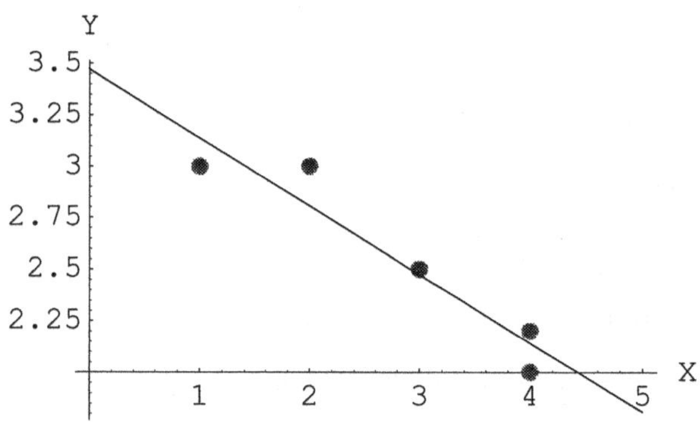

equation of the fitted line is y = 3.5 -. 33x. For every value of x, the table below shows the actual values of y, the estimated values of the y, and the errors of estimation, which are equal to actual minus estimated ys.

x	1	4	2	3	4
Actual **y**	3	2	3	2.5	2.2
Estimated **y**	3.17	2.18	2.84	2.51	2.18
Error	-.17	-.18	.16	-.01	.02

It is clear from the graph and the table that, if the fitted line is a good fit, the line will contain as many points or will pass through the points in such a way that the error distances from the estimated line will be minimum. In statistics and econometrics, the least-squares technique is concerned with fitting a line or a curve to the points in such a way that the sum of the squares of the error distances between the actual points and the fitted line or curve is minimized.

Let's assume a linear model, $Y_i = \alpha + \beta X_i + \epsilon_i$, where α and β are intercept and slope coefficients, respectively, and ε_i is the distance between actual point i and the estimated line. The least-squares technique estimates α and β in such a way that $\Sigma\epsilon_i^2$ is minimized. The least-square estimates of coefficients are $\hat{\beta} = \frac{\sum_{i=1}^{n} x_i y_i}{\sum_{i=1}^{n} x_i^2}$, where x_i and y_i are the deviations of X_i and Y_i from their corresponding means ($x_i = X_i - \bar{X}$ and $y_i = Y_i - \bar{Y}$) and $\hat{\alpha} = \bar{Y} - \hat{\beta}\bar{X}$. The

goodness of the fit is measured by **multiple correlation coefficient** (**R^2**), which utilizes $\sum\epsilon_i^2$ to measure the portion of the variations in y that is explained by X. That is, $R^2 = 1 - \frac{\sum\varepsilon_i^2}{\sum Y_i^2}$. The closer R^2 is to one, the better is the fit. The statistical significance of the estimated α and β coefficients are measured by t statistics.

In this chapter we will study the techniques of fitting lines and curves to data. The statistical tests of goodness of the fit and significance of coefficients will not be discussed here. The interested reader is referred to standard econometrics text books.

Example: Consumption Function and Marginal Propensity to Consume (MPC)

The following data represents annual disposable income and consumption of a country in millions of dollars. Fit the best fitting equation to data and estimate marginal propensity to consume.

Income (Y): 2300, 2360, 2395, 2407, 2389, 2430, 2395, 2413, 2430, 2467, 2480, 2490

Consumption (C): 2145, 2173, 2189, 2190, 2185, 2206, 2195, 2215, 2235, 2251, 2279, 2281

Solution: In any consumption model, consumption depends on income. Plot the data on the Y-C plane to study the functional form of the relationship between C and Y.

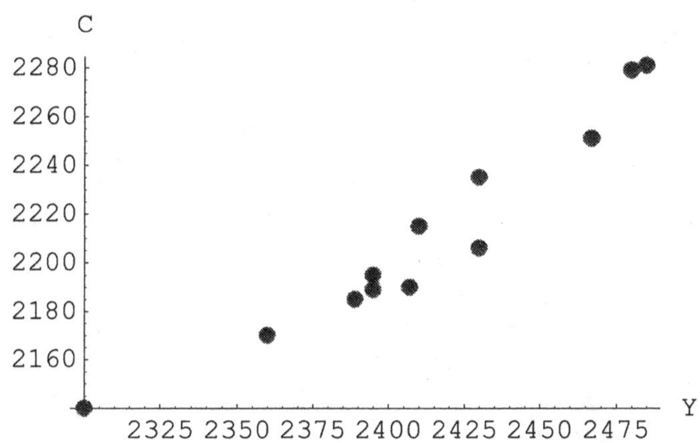

The eyeballing of the plot clearly shows that the relationship between Y and C is more of a linear relation. Therefore, we fit an equation of a line C = $\alpha + \beta$Y. The least square estimate of β is $\hat{\beta} = \frac{\sum_{i=1}^{n} c_i y_i}{\sum_{i=1}^{n} c_i^2}$, where c_i and y_i are deviations of C_i and Y_i from their corresponding means. The least-square estimate of α is $\hat{\alpha} = \bar{C} - \hat{\beta}\bar{Y}$. Using these two formulas, α and β can be estimated by finding the elements of $\hat{\alpha}$ and $\hat{\beta}$ as follows,

Observation	Y	C	$y = Y - \bar{Y}$	$c = C - \bar{C}$	y^2	c.y
1	2300	2145	-113	-67	12769	7571
2	2360	2173	-53	-39	2809	2067
3	2395	2189	-18	-23	324	414
4	2407	2190	-6	-22	36	132
5	2389	2185	-24	-27	576	648
6	2430	2206	17	-6	289	-102
7	2395	2195	-18	-17	324	306
8	2413	2215	0	3	0	0
9	2430	2235	17	23	289	391
10	2467	2251	54	39	2916	2106
11	2480	2279	67	67	4489	4489
12	2490	2281	77	69	5929	5313
Σ	**28956**	**26544**	0	0	**30750**	**23335**

the estimated β, $\hat{\beta} = \frac{23335}{30750} = .75886$, the estimated α, $\hat{\alpha} = 2212 - .75886(2413) = 380.87$, and the estimated equation is C = 380.87 + .76Y. The β coefficient of the consumption function is the marginal propensity to consume (MPC), which represents increase in consumption for every one dollar increase in income, $\beta = \frac{\Delta C}{\Delta Y} = .76$.

As you may notice, even for a two variable, simple regression model, finding the coefficients of a regression model needs large amount of numerical calculations. Fortunately, to day most scientific calculators have built-in programs that can run regression models. As well, computer programs such as TSP, SPSS, SAS, and others are available for such calculations. To solve the same problem using *Mathematica*, list data as pairs of observations then use **ListPlot** command to plot the data,

```
Input[1]:= Clear[data, linefit, p1, p2]
           data = {{2300, 2145}, {2360, 2173}, {2395, 2189},
             {2407, 2190}, {2389, 2185}, {2430, 2206}, {2395, 2195}, {2413, 2215},
             {2430, 2235}, {2467, 2251}, {2480, 2279}, {2490, 2281}} ;
           p1 = ListPlot[data, PlotStyle -> {RGBColor[1, 0, 0], PointSize[.03]},
             AxesLabel -> {"Y", "C"}];
```

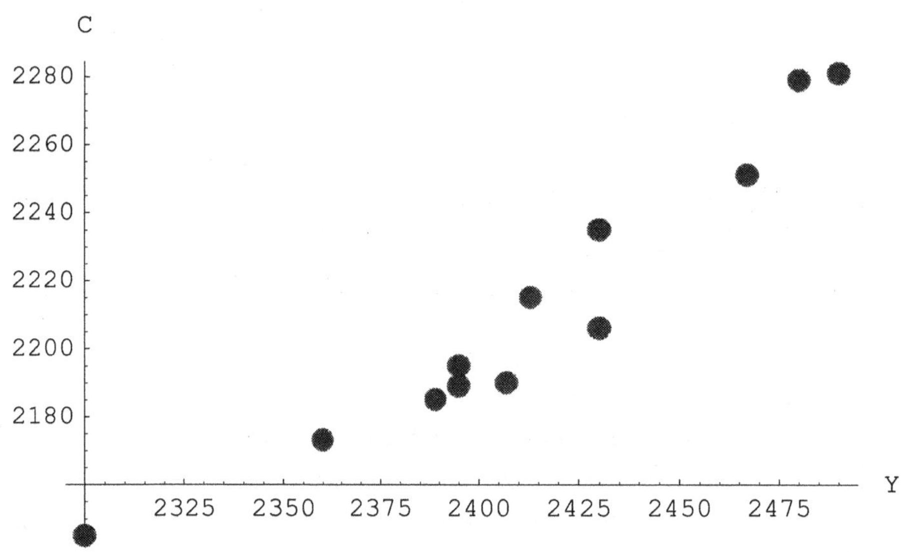

Although few observations are outliers, the eyeballing of the list plot clearly shows that the relationship between Y and C is more of a linear relation. Therefore, we fit an equation of a line C = $\alpha + \beta$Y. Use the **Fit** command to fit the line to list of data,

```
Input[2]:= linefit = Fit[data, {1, Y}, Y]
Output[2]= 380.867 + 0.758862 Y
```

The equation of the line is C = 380.87 + .76Y. The plot of the equation and the list plot of the data together is shown below

```
Input[3]:= p2 = Plot[linefit, {Y, 2200, 2500}];
```

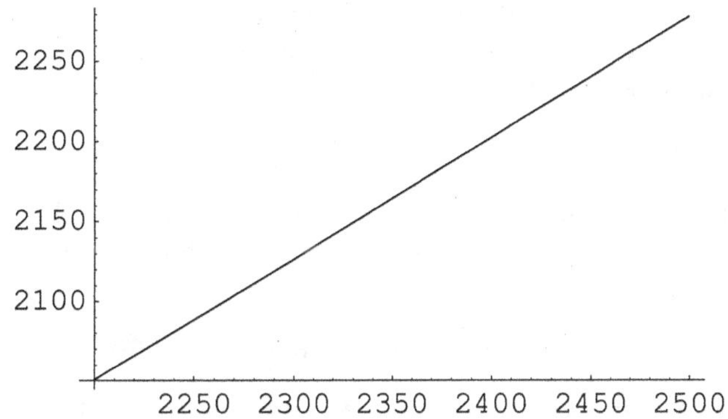

Input[4]:= Show[p1, p2]

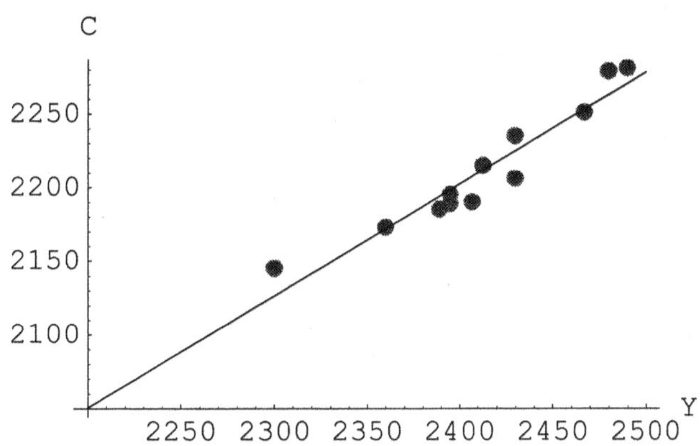

It is obvious that the least-squares estimate, C = 380.67 + .76Y, passes through several points but not all. Some of the points lie above the fitted line and others lie below the fitted line. But the estimated line best represents the points with the sum squared errors between the actual points and the fitted line being minimum.

To fit the best line or curve to a set of data take the following steps:

1- Identify the **independent and dependent variables**. A variable that determines or controls another variable is called independent variable. A variable that is controlled or depends on other variables is called dependent variable. In the example above, consumption (C) depends on income (Y) and not the other way around. Therefore, consumption is dependent variable and income is independent variable.

2- Plot the independent variable on the horizontal (x) axis and the dependent variable on the vertical (y) axis. If the data points appear to lie on a straight line, fit an equation of line. If data are scattered in a non-linear form, decide on one of the non linear functional forms discussed in chapter 2. If you choose a polynomial function, select the degree of the polynomial by eyeballing the plot or by studying the theoretical relationship between the variables. For example, production and cost functions, usually, follow a third degree polynomial. Total revenue function is usually a second degree polynomial. In general, economic theory is the best starting point for deciding the form of the relationship between the variables. Eyeballing of the plot gives an indication of the relationship for the specific set of data that you are working with. Finally, the functional form may be decided by statistical tests of best fit to the data set.

3- If the plot of the data, or the economic theory indicates no intercept for the line or the curve, constrain your fit to pass through the origin. (Note: In *Mathematica* command Fit[data, {1, x}, x}], the number "1" in the curved bracket refers to intercept. To fit a line that passes through the origin do not include "1"). Also, to fit polynomials of higher degree, include the polynomial in the curved bracket. For example, to fit a third-degree polynomial with no intercept, use the command Fit[data, {x, x^2, x^3}, x].

4- The fitted model may be used for studying the structural relationship between the variables or for forecasting. To forecast the future values of the dependent variable or its value at a certain point, an estimated value of the independent variables must be available.

Example: Demand and Price Elasticity

A businessperson collects the following data on the relationship between the price and the quantity of the product sold in the store. Find the relationship between the price and quantity and estimate the price elasticity of the demand at P = $7.50.

P: 6.00, 6.30, 6.50, 6.65, 6.75, 8.00, 8.10, 8.25, 7.10, 7.20, 6.90, 6.60, 8.15, 7.70, 8.20

Q: 255, 228, 212, 198, 185, 140, 135, 131, 178, 170, 193, 199, 142, 156, 139

Solution: Quantity demanded for a product (q) depends on the price of the product (p). Plot the relationship between p and q to decide on the functional form.

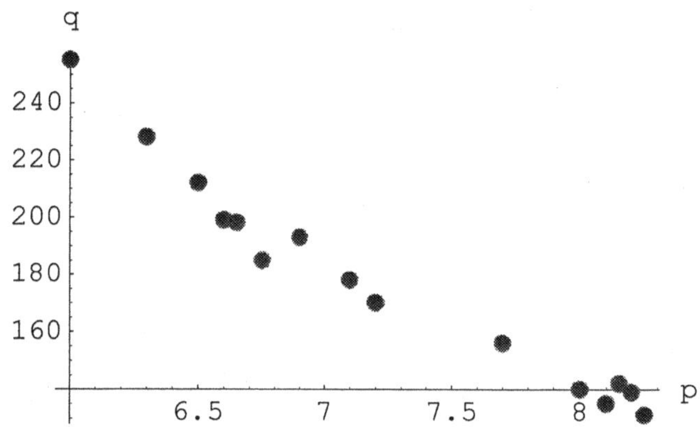

The plot of the relationship between p and q shows that a linear relationship may be appropriate way of presenting the relationship between p and q. As in the previous example, make a table to find the elements of the least-square estimators $\hat{\beta} = \frac{\sum_{i=1}^{n} c_i\, y_i}{\sum_{i=1}^{n} c_i^2}$ and $\hat{\alpha} = \overline{C} - \hat{\beta}\overline{Y}$ to estimate α and β in the equation $p = \alpha + \beta p + \epsilon$. Check your results with the estimated coefficients $\hat{\alpha} = 517.13$ and $\hat{\beta} = -47.01$. The demand function is q = 517.13 - 47.01p.

To find the price elasticity of demand at p = $7.50, find the corresponding q by substituting for p in the demand equation. This results in q = 164.55 units. The price elasticity of demand is defined as $\xi_p = \frac{\Delta q\,\%}{\Delta p\,\%} = \frac{\Delta q}{\Delta p} \cdot \frac{p}{q}$, where $\frac{\Delta q}{\Delta p}$ is the slope of the demand function, $\hat{\beta} = -47.01$. By substituting the values of p, q, and $\frac{\Delta q}{\Delta p}$ to ζ_p, the price elasticity of demand is estimated to be $\zeta_p = -2.15$.

To solve the problem using *Mathematica*,

```
Input[5] := Clear[p, q, data, demandfit]
data = {{6, 255}, {6.30, 228}, {6.50, 212}, {6.65, 198}, {6.75, 185},
   {8, 140}, {8.10, 135}, {8.25, 131}, {7.10, 178}, {7.20, 170},
   {6.9, 193}, {6.60, 199}, {8.15, 142}, {7.70, 156}, {8.20, 139}};
demandplot = ListPlot[data, PlotStyle → {RGBColor[1, 0, 0], PointSize[.03]},
  AxesLabel -> {"p", "q"}]
```

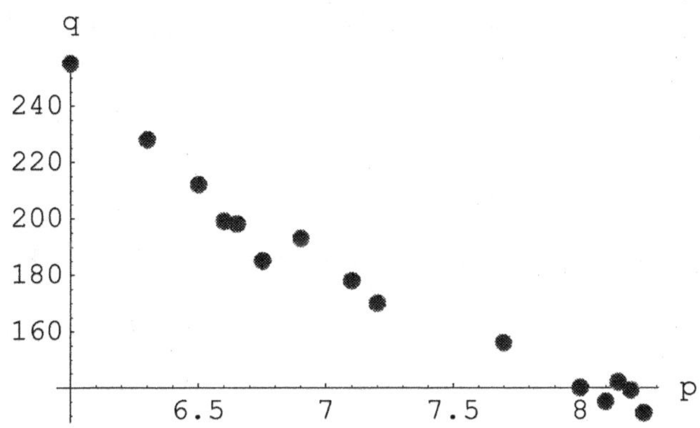

The eyeballing of the plot shows that the relationship between p and q is a linear relationship. To fit an equation of line to data, use the Fit command,

Input[6]:= `demandfit = Fit[data, {1, p}, p]`

Output[6]= `517.132 - 47.0109 p`

The demand equation is q = 517.13 - 47.10p. To graph the equation, use the Plot command as,

Input[7]:= `demandline = Plot[demandfit, {p, 5, 10}]`

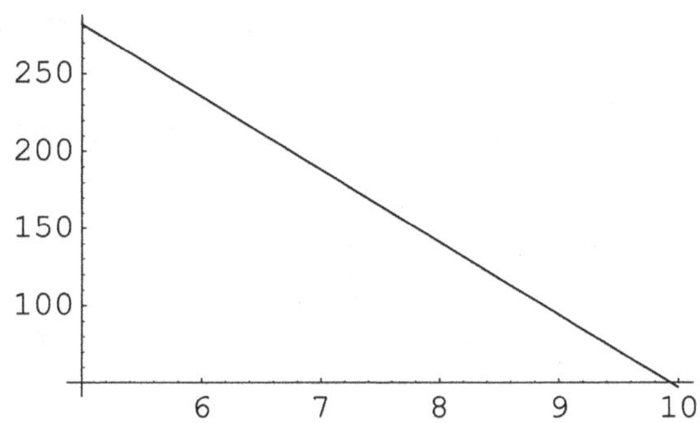

and to graph the list plot of dada and graph of the demand function together, use the Show command as,

Input[8]:= `Show[demandplot, demandline]`

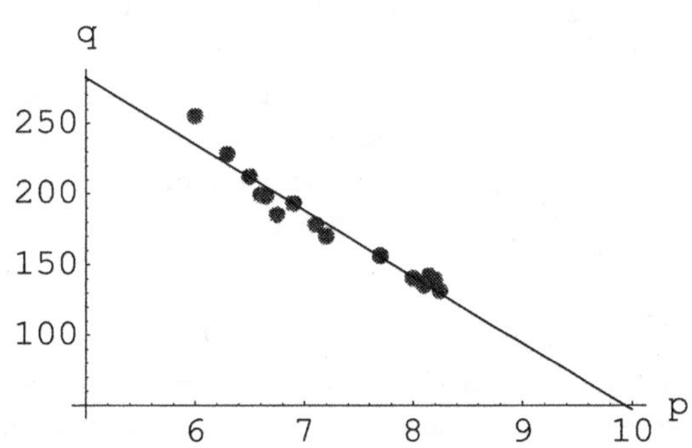

To find the price elasticity of demand at p = 7.50, first estimate q at this price using the estimated demand equation,

Input[9]:= `demandfit /. p → 7.50`

Output[9]= `164.55`

the estimated quantity corresponding to p = 7.50 is q = 164.55. By substituting for p, q, and $\frac{\Delta q}{\Delta p}$ in the elasticity formula $\xi_p = \frac{\Delta q\%}{\Delta p\%} = \frac{\Delta q}{\Delta p} \cdot \frac{p}{q}$, ζ_p is estimated to be ζ_p = -2.15.

Problems

1. Plot the following two points and fit a line to the points. Fine the errors between the points and the fitted lines.

a- (2, 6) and (-1, 4)

b- (0, 4) and (5, 8)

c- (1, 5), (3, 8), and (4, 9)

d- (0, 0), (-1, -1), (1, 2), (-2, -3), and (3, 5)

e- (3, 2), (4, 3), (5, 6), (9, 7), (2, 2), (6, 4), and (4, 4)

2. The following data are observations on the price (P) and the quantity demanded (Q) for a product.

P: 6, 7, 6.5, 6, 7.5, 7.3, 7, 7.8, 8, 8.2, 8.5, 7.5, 8, 7.3, 9, 9.5

Q: 67, 65, 67, 66, 67, 65, 66, 65, 64, 65, 65, 66, 65, 66, 60, 59

Plot the relationship between P and Q. Fit the best line representing the data. Estimate the quantity demanded for P = 10 and for P = 5.

3. The following data represents the relationship between labor input (L) and output (Q) in a business.

L: 10, 11, 9, 13, 16, 11, 13, 12, 11, 13, 9, 6, 17, 8, 14, 14, 13, 16, 18

Q: 28, 35, 24, 39, 44, 34, 38, 34, 35, 39, 28, 20, 50, 26, 42, 44, 40, 42, 52

Plot the relationship between L and Q. Fit the best line representing the data. Estimate Q for a labor input level of 20.

II. Fitting Curves

The consumption model may be the best example of the linear relationship between two economic variables. However, most economic relationships are non-linear, which require fitting curves rather than lines to data. Economic theory may have strong implication on the functional form of the variables under study. However, it is always advisable to plot the relationship between the variables and decide the functional form based on economic theory and what the plots represent. If there is a contrast between the economic theory and what your data presents, you may fit both models, analyze the differences in the fit, and explain why your data does not comply with economic theory.

The least-square method described in the previous section is applicable only to linear models. If the model is nonlinear, it should first be linearized by converting the variables, then the least square method can be applied to the converted model. In practice, the three types of non-linearity that happen are exponential models, polynomial models, and models with rational independent variable.

Log-Linearizing Exponential Models: To apply least squares to an exponential model of the form $Y = \alpha X^{\beta}\epsilon$, apply logarithmic transformation to the model as, $\ln Y = \ln\alpha + \beta \ln X + \ln\epsilon$. The new model is a log-linear model that can be written as $y = a + \beta x + e$, where y, a, x, and e are the natural logarithms of Y, α, X, and ϵ. The least square method can be applied to this model to estimate β and a. Note that since $a = \ln\alpha$, once a is estimated, the estimated α will be equal to $\hat{\alpha} = E^{\hat{a}}$, where E is the natural number, E = 2.718.

Example: Demand Function

The following data represents the relationship between price and quantity demanded for a product. Estimate the demand equation and find the price elasticity of demand.

P	5.50	5.00	6.30	6.80	5.80	6.2	6.5	5.1	7.2	7.5
Q	20.0	25.0	14.0	12.0	18.0	15.5	14.7	23.8	11.0	9.8

Solution: The plot of the data below shows a relationship that seems to be exponential function relationship between P and Q. Therefore, let's try a function of the form $Q = \alpha P^{\beta} \in$.

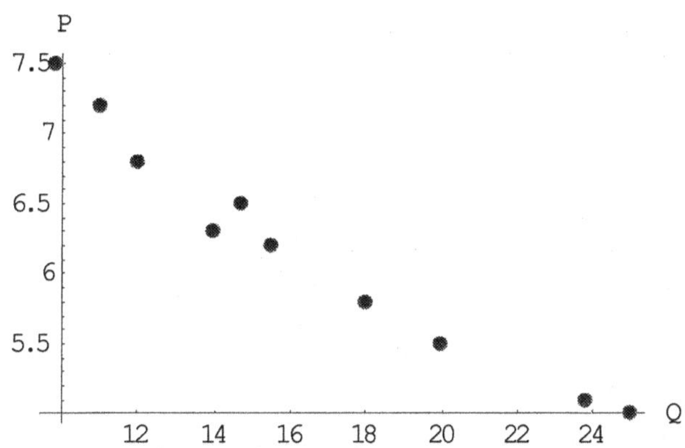

To fit a function, first log-linearize the function by applying the natural log transformation, Log Q = Log α + βLog P + Log $\in$. The log-linear equation can be written as LQ = a + βLP + e where, LQ, LP, a, and e are the natural logs of Q, P, α, and $\in$, respectively. Make a table of log of P and Q values and use the least square estimator formulas to estimate a and β as follows

observations	P	Q	LP	LQ	$lp = LP - \overline{LP}$	$lq = LQ - \overline{LQ}$	$lp * lq$	lp^2
1	5.5	20	1.70	3.00	-.1097	.2457	-.0269	.0120
2	5	25	1.61	3.22	-.2050	.4689	-.0961	.0420
3	6.3	14	1.84	2.64	.0262	-.1109	-.0029	.0007
4	6.8	12	1.92	2.48	.1025	-.2651	-.0272	.0105
5	5.8	18	1.76	2.89	-.0565	.1404	-.0079	.0032
6	6.2	15.5	1.82	2.74	.0101	-.0092	0	.0001
7	6.5	14.7	1.87	2.69	.0574	-.0622	-.0036	.0033
8	5.1	23.8	1.63	3.17	-.1852	.4197	-.0777	.0343
9	7.2	11	1.97	2.40	.1597	-.3521	-.0562	.0255
10	7.5	9.8	2.01	2.28	.2005	-.4676	-.0938	.0402
Σ	**61.9**	**163.8**	**18.13**	**27.51**	0	0	**-.3923**	**.1718**

the least-square estimators $\hat{\beta} = \frac{\sum_{i=1}^{n} lp_i \, lq_i}{\sum_{i=1}^{n} lp_i^2} = \frac{-.3923}{.1718} = -2.28$ and $\hat{a} = \overline{LQ} - \hat{\beta}\overline{LP} = 2.75 + 2.28(1.814) = 6.89$. Since $\hat{a}$ = Log($\hat{\alpha}$), to find $\hat{\alpha}$, take anti-log as, $\hat{\alpha} = E^{6.89} = 982.40$. Therefore, the demand equation can be written as $Q = 982.40 P^{-2.28}$.

To solve the same problem using *Mathematica*, first list the data and plot the relationship.

```
Input[10]:= Clear[P, Q, data, logdata, plot1, plot2, fit1]
           data = {{5.5, 20}, {5, 25}, {6.3, 14}, {6.8, 12}, {5.8, 18},
              {6.2, 15.5}, {6.5, 14.7}, {5.1, 23.8}, {7.2, 11}, {7.5, 9.8}};
```

```
plot1 = ListPlot[data,
  PlotStyle → {RGBColor[1, 0, 0], PointSize[.03]}, AxesLabel → {"P", "Q"}]
```

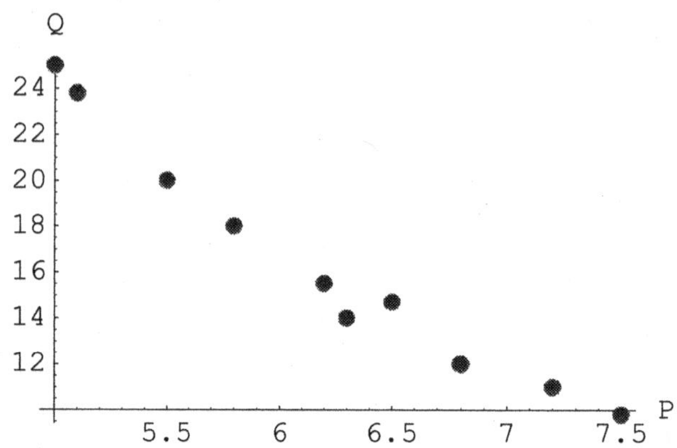

To find the logarithmic values of P and Q take the log of data and call it a name (here, logdata) and fit the equation of line to the transformed data

```
Input[11]:= logdata = N[Log[data]];
           fit1 = Fit[logdata, {1, x}, x]
```

```
Output[11]= 6.89498 - 2.28406 x
```

the fitted equation is LogQ = 6.89 - 2.28LogP. Since $\hat{\alpha}$ = Log($\hat{\alpha}$), to find $\hat{\alpha}$, take anti-log as, $\hat{\alpha} = E^{6.89} = 982.40$. Therefore, the demand equation can be written as Q = $982.40P^{-2.28}$. To plot the fitted equation, use the "Plot" command as,

```
Input[12]:= plot2 = Plot[982.40 P^(-2.28), {P, 5, 8}]
```

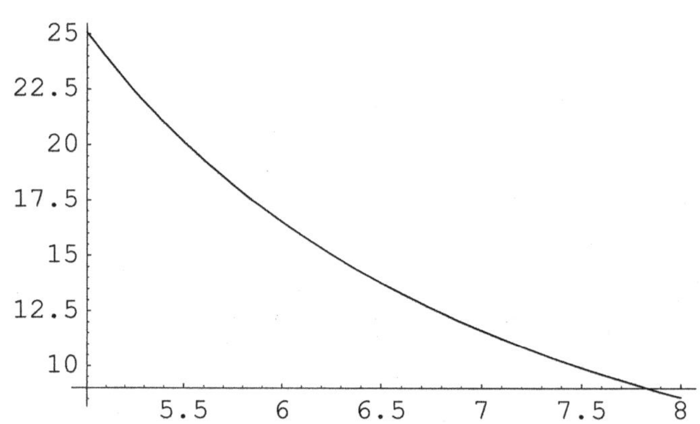

To show the actual data and the plot of the fitted equation together, use the Show command as follows

```
Input[13]:= Show[plot1, plot2]
```

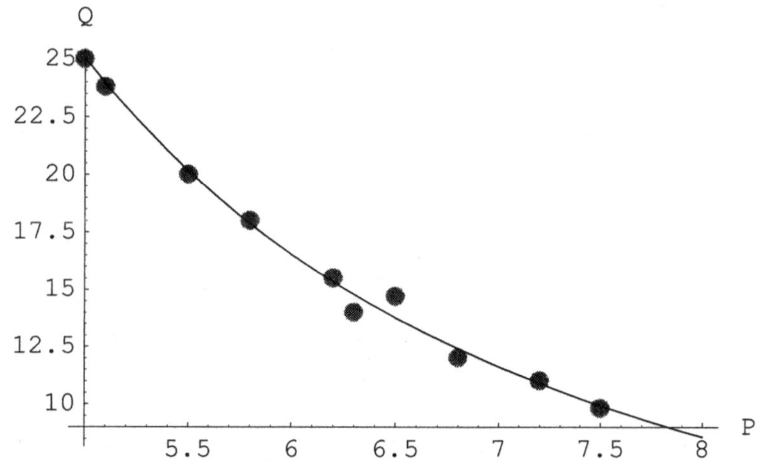

<u>Converting Polynomial Models:</u> To apply least squares to a polynomial model of the form $Y = \alpha + \beta X^n + \epsilon$, assume $Z = X^n$ and estimate the model $Y = \alpha + \beta Z + \epsilon$ as you would estimate a linear model.

Example: Production Function

A manufacturer has collected the following monthly data on the relationship between labor input and output in the operating range of the business for the past twelve month. Fit a best fitting model to the data.

L	23	21	22	26	28	25	23	20	22	24	25	29
Q	254	202	252	342	379	330	275	176	233	306	308	375

Solution: The plot of the relationship between L and Q below shows a quadratic from, $Q = \alpha + \beta L^2 + \in$.

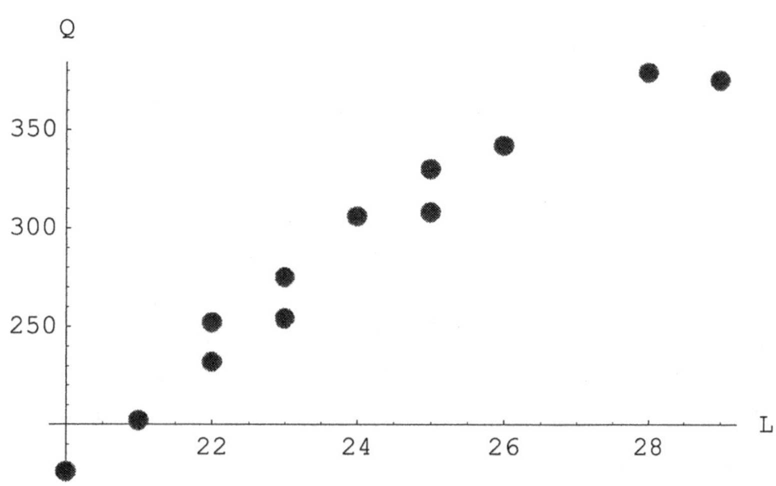

Let's assume that $z = L^2$. The model can be written as $Q = \alpha + \beta z + \in$ and can be estimated as a linear model. Make a table of data as follow

Observations	L	Q	Z	$z = Z - \bar{Z}$	$q = Q - \bar{Q}$	zq	z^2
1	23	254	529	-53.83	-32	1722.56	2897.67
2	21	202	441	-141.83	-84	11913.72	20115.75
3	22	252	484	-98.83	-34	3360.22	9767.37
4	26	342	676	93.16	56	5217.52	8680.65
5	28	379	784	201.16	93	18708.81	40469.37
6	25	330	625	42.16	44	1855.48	1778.31
7	23	275	529	-53.83	-11	592.13	2897.67
8	20	176	400	-182.83	-110	20111.30	33426.81
9	22	233	484	-98.83	-53	5237.99	9767.37
10	24	306	576	-6.83	20	-136.60	46.65
11	25	308	625	42.16	22	927.74	1778.31
12	29	375	841	258.17	89	22977.13	66651.75
Σ	**288**	**3432**	**6994**	0	0	**92488.00**	**198277.68**

the least-square estimators $\hat{\beta} = \frac{\sum_{i=1}^{n} z_i q_i}{\sum_{i=1}^{n} z_i^2} = \frac{92488.00}{198277.68} = .466$ and $\hat{\alpha} = \bar{Q} - \hat{\beta}\bar{Z} = 286 - .466457(582.833) = 14.13$. Therefore, the production function can be written as $Q = 14.41 + .466L^2$.

To solve the problem using *Mathematica*, use the following commands

Input[14]:=
```
Clear[L, Q, data, graph1, graph2, fit1]
data = {{23, 254}, {21, 202}, {22, 252}, {26, 342}, {28, 379}, {25, 330},
  {23, 275}, {20, 176}, {22, 233}, {24, 306}, {25, 308}, {29, 375}};
graph1 = ListPlot[data, PlotStyle → {RGBColor[1, 0, 0], PointSize[.03]},
  AxesLabel → {"L", "Q"}]
```

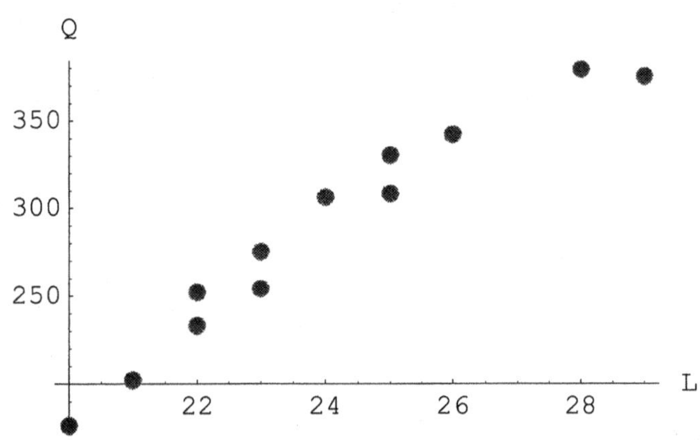

To fit a quadratic form $Q = \alpha + \beta L^2 + \epsilon$ to data, use the Fit command as,

Input[15]:= `fit1 = Fit[data, {1, L^2}, L]`

Output[15]= $14.1333 + 0.466457\, L^2$

to graph the fitted equation, use the Plot command

Input[16]:= `graph2 = Plot[fit1, {L, 20, 30}]`

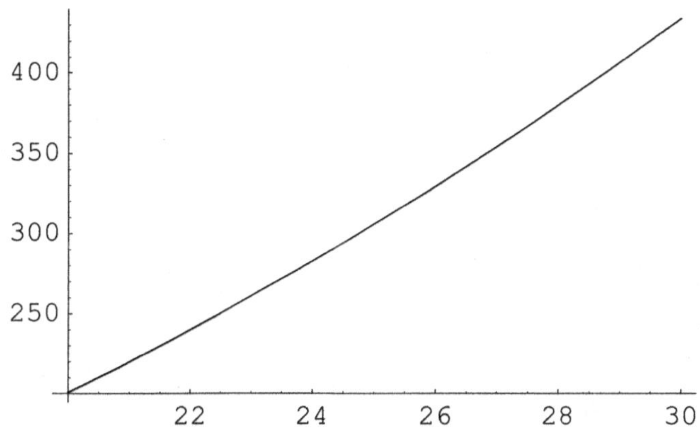

To show the actual data and the fitted curve together, use the Show command as,

Input[17]:= `Show[graph1, graph2]`

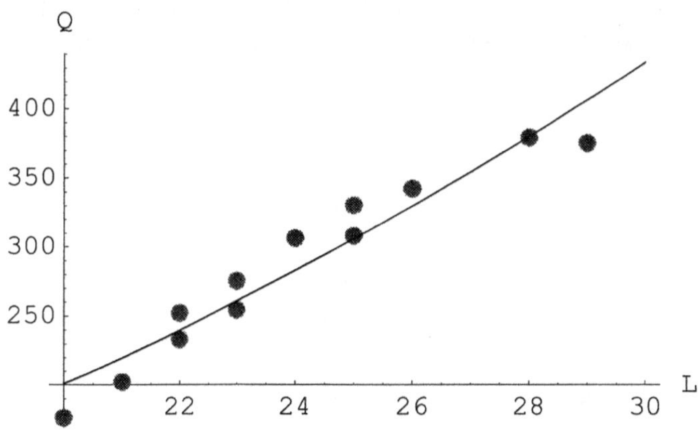

Converting Models Containing Rational Independent Variable: To apply least squares to a function of the form $Y = \alpha + \frac{\beta}{X} + \epsilon$, assume $Z = \frac{1}{X}$ and estimate the model $Y = \alpha + \beta Z + \epsilon$ as you would estimate a linear model.

Example: Inflation-Unemployment Trade-Off: Phillips Curve

In an empirical study of the relationship between inflation and unemployment for the 1861-1957 period, the British economist A.W. Phillips found an inverse relationship between inflation and unemployment. Later studies in US economy was supportive of such trade-off, at least in the short run. Such trade-off between inflation and unemployment was called Phillips Curve. Suppose the following ten annual observations on inflation (Y) and unemployment (X) has been collected. Estimate a best fitting equation to data and find the inflation rate if the unemployment rate were 3%.

X	3.0	4.5	4.0	3.8	2.5	2.2	3.8	4.5	5.5	6.2
Y	5.2	4.2	3.2	4.9	5.9	7.3	4.3	4.2	2.6	2.2

Solution: Plot the relationship between inflation and unemployment as follows

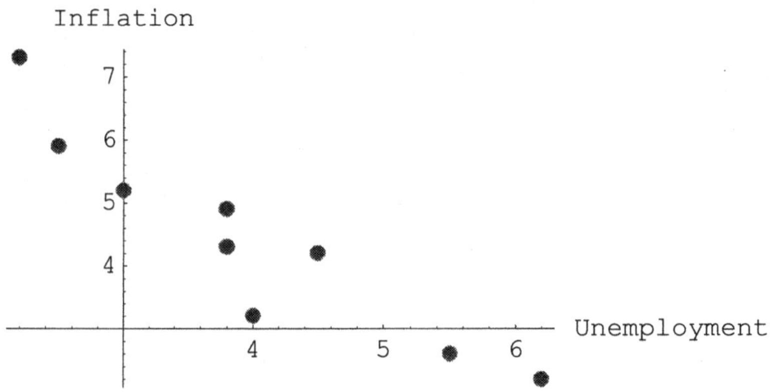

the plot of the relationship between the two variables shows an inverse relationship between inflation and unemployment. Let's assume a model of the Phillips curve as $Y = \alpha + \frac{\beta}{X} + \in$. To covert the model to a linear model, let's assume that $Z = \frac{1}{X}$. Therefore, the model can be written as $Y = \alpha + \beta Z + \in$. Make a table of data and find the least square estimates of α and β.

Observations	X	Y	$Z = \frac{1}{X}$	$y = Y - \bar{Y}$	$z = Z - \bar{Z}$	y.z	z^2
1	3.0	5.2	.33	.80	.06	.047	.0034
2	4.5	4.2	.22	-.20	-.05	.011	.0028
3	4.0	3.2	.25	-1.20	-.03	.030	.0006
4	3.8	4.9	.26	.50	-.01	-.006	.0001
5	2.5	5.9	.40	1.50	.12	.186	.0156
6	2.2	7.3	.46	2.90	.18	.521	.0322
7	3.8	4.3	.26	-.10	-.01	.001	.0002
8	4.5	4.2	.22	-.20	-.05	.011	.0028
9	5.5	2.6	.19	-1.80	-.09	.168	.0087
10	6.2	2.2	.16	-2.20	-.12	.250	.0129
Σ	**40.0**	**44.0**	**2.75**	**0**	**0**	**1.219**	**.0793**

the least-squares estimate of $\hat{\beta} = \frac{\sum_{i=1}^{n} z_i\, y_i}{\sum_{i=1}^{n} z_i^2} = \frac{1.219}{.0793} = 15.37$ and $\hat{\alpha} = \bar{Y} - \hat{\beta}\bar{Z} = 4.4 - 15.37(.275) = .173$. Therefore, the relationship between inflation and unemployment can be written as $Y = .173 + \frac{15.37}{X}$. To estimate the inflation rate corresponding to 3% rate of unemployment, substitute for X = 3, to get Y = 5.30.

To solve the problem using *Mathematica*, list the data and plot the relation between the two variables

```
Input[18] := Clear[data, graph1, graph2, fit1]
data = {{3, 5.2}, {4.5, 4.2}, {4, 3.2}, {3.8, 4.9}, {2.5, 5.9},
   {2.2, 7.3}, {3.8, 4.3}, {4.5, 4.2}, {5.5, 2.6}, {6.2, 2.2}};
graph1 = ListPlot[data, PlotStyle -> {RGBColor[1, 0, 0], PointSize[.03]},
  AxesLabel → {"Unemployment", "Inflation"}];
```

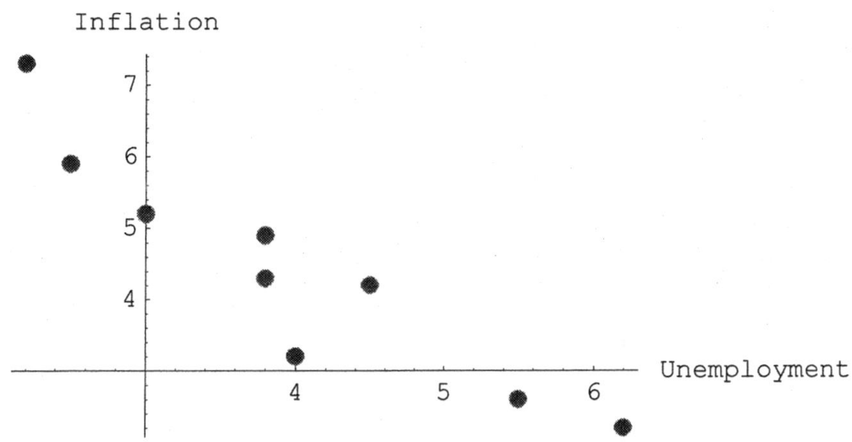

to fit the equation, use the fit command as,

Input[19}:= **fit1 = Fit[data, {1, 1 / x}, x]**

Output[19]= 0.172464 + $\frac{15.3631}{x}$

to graph the estimated equation use the Plot command,

Input[20]:= **graph2 = Plot[fit1, {x, 2, 7}, AxesLabel → {"Unemployment", "Inflation"}];**

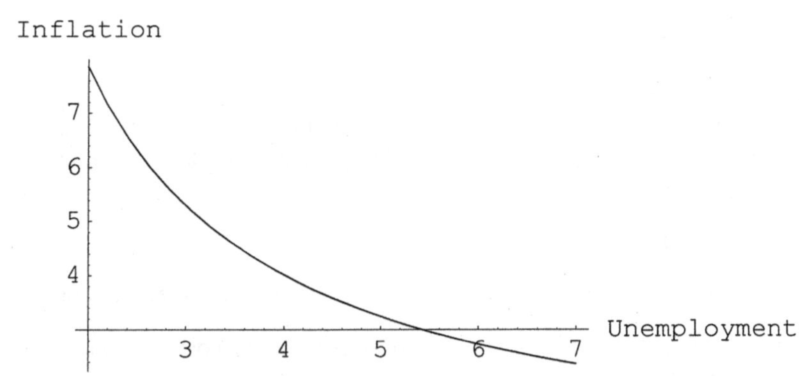

to show the actual data and the estimated equation together, use the show command as,

Input[21]:= **Show[graph1, graph2]**

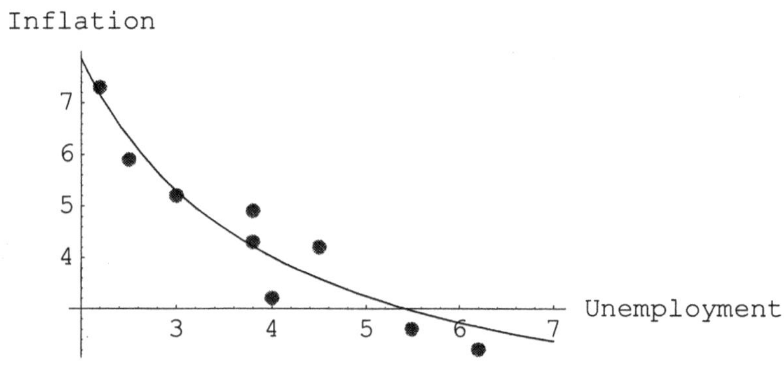

To estimate the output level at an input level of 9 units, evaluate the estimated equation as x→9.

Input[22]:= `fit1 /. x -> 3`

Output[22]= `5.2935`

More On Fitting Polynomials: In fitting polynomials to data, so far, we have assumed a function of the form Y = $\alpha + \beta X^n + \epsilon$. What if the polynomial functional form is Y = $\alpha + \beta_1 X + \beta_2 X^2 + ... + \epsilon$. Even though we have only one variable (X) as independent variable, the conversion of the model to a linear form changes the nature of the regression to a multi-variable regression model. That is, to transform a third degree polynomial Y = $\alpha + \beta_1 X + \beta_2 X^2 + \beta_3 X^3 + \epsilon$ to a linear form, assume Z = X^2 and W = X^3. The model can be written as Y = $\alpha + \beta_1 X + \beta_2 Z + \beta_3 W + \epsilon$. Estimating α and βs in this model is the subject of fitting planes and surfaces to data discussed in the next section. Fitting planes and surfaces to set of data follows the least squares techniques of estimating α and βs in such a way that the sum squared residuals between actual data and the estimated plane are minimized. However, the numerical calculations are very much involved and time consuming and using a computer software to estimate the coefficients becomes almost unavoidable.

Example: Cost Function

The cost function is a relationship between output and cost of producing output. Suppose the following data represents the relationship between the quantity of output (Q) and cost of producing output (cost)

Q	0	3	7	12	16	21	23	25	29	32	34
Cost	20	130	251	386	414	416	604	830	1082	1332	1702

Plot the relationship between cost and output and fit the equation representing the relationship.

Solution: Economic theory would argue that the relationship between cost and output is often a third-degree polynomial with an intercept. Due to division of labor and initial economies of scale, the total cost initially increases with a decreasing rate. Later, as the diminishing productivity catches up, total cost increases with an increasing rate. The list plot of data below supports such relationship between cost and output.

```
Input[23] :=  Clear[data, graph1, graph2, fit1]
data = {{0, 20}, {3, 130}, {7, 251}, { 12, 386}, {16, 414}, {21, 416},
   {23, 604}, {25, 830}, {29, 1082}, {32, 1332}, {34, 1702}};
graph1 = ListPlot[data, PlotStyle -> {RGBColor[1, 0, 0], PointSize[.03]},
  AxesLabel → {"Q", "Cost"}];
```

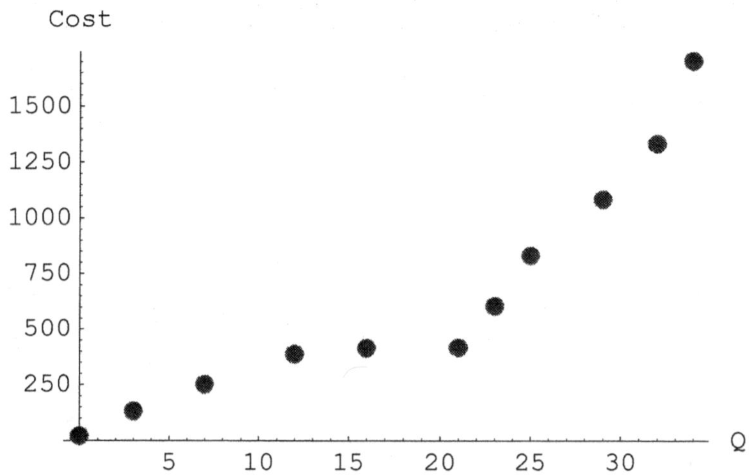

the best functional form representing this data is cost = $\alpha + \beta Q + \gamma Q^2 + \delta Q^3$. Using *Mathematica*, the coefficients can be estimated as

Input[24]:= **fit1 = Fit[data, {1, x, x^2, x^3}, x]**

Output[24]= $10.8771 + 55.4434\, x - 3.50889\, x^2 + 0.0977175\, x^3$

the estimated function is cost = $10.88 + 55.44Q - 3.51Q^2 + 0.10Q^3$. The list plot of actual data and the estimated function are graphed together below.

Input[25]:= **graph2 = Plot[fit1, {x, 0, 40}, AxesLabel → {"Q", "Cost"}];**

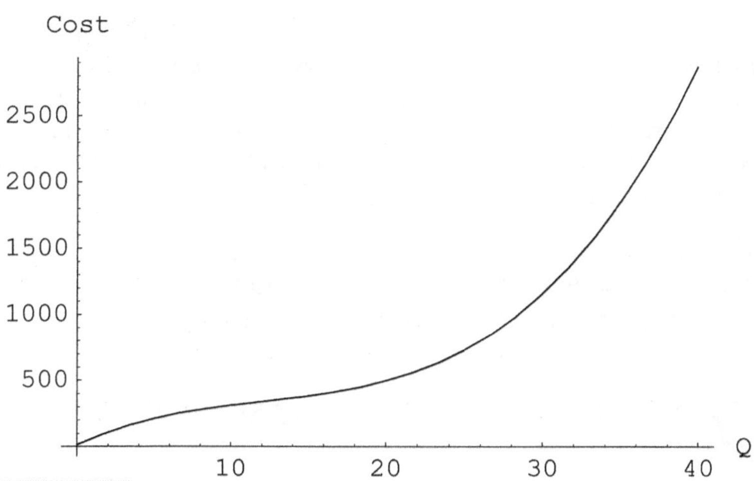

Input[26]:= **Show[graph1, graph2]**

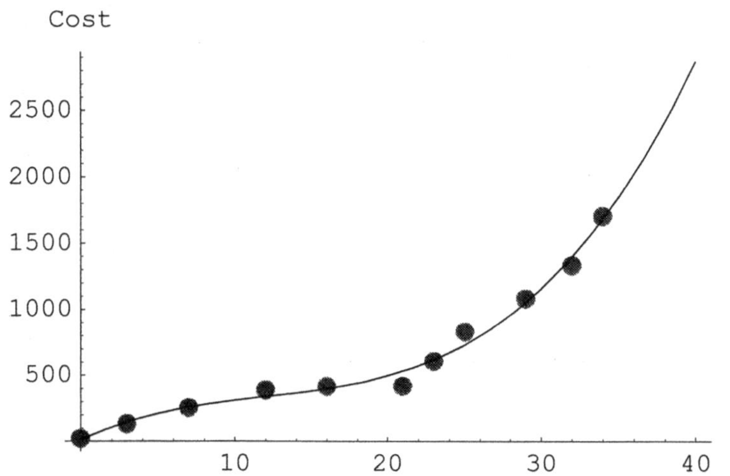

Example: Total Product, Arc Marginal Product, and Average Product Functions

The relationship between labor input (L) and output (Q) is given below,

L	3	5	8	10	14	16	19	22	25	26
Q	127	235	448	610	893	1012	1168	1318	1339	1335

find production function (TP) by fitting a curve to the data. Find arc marginal product (AMP) and average product (AP) of labor. ListPlot the AMP and AP data. Fit appropriate models to each data. Graph the total product, arc marginal product, and average product curves. Write your observations on the relationship between the AMP and AP curves.

Solution: Economic theory assumes a third-degree polynomial between labor input and output in the short-run. Initially, as labor input increase output increases with increasing rate due to efficiency obtained from division of labor. In the short-run, assuming capital input are fixed, beyond certain level of labor input (inflection point) diminishing return will catch up and output will increase with a decreasing rate. As labor input increases further beyond maximum output level, marginal product of labor turns negative and output will decrease. The graph below shows the listplot of data and the three phases of the production.

```
Input[27]:= Clear[data, graph1, graph2, fit1]
            data = {{3, 127}, {5, 235}, {8, 448}, {10, 610}, {14, 893},
               {16, 1012}, {19, 1168}, {22, 1318}, {25, 1339}, {26, 1335}};
```

```
graph1 = ListPlot[data, PlotStyle -> {RGBColor[1, 0, 0], PointSize[.03]},
  AxesLabel → {"Labor Input", "Output"}]
```

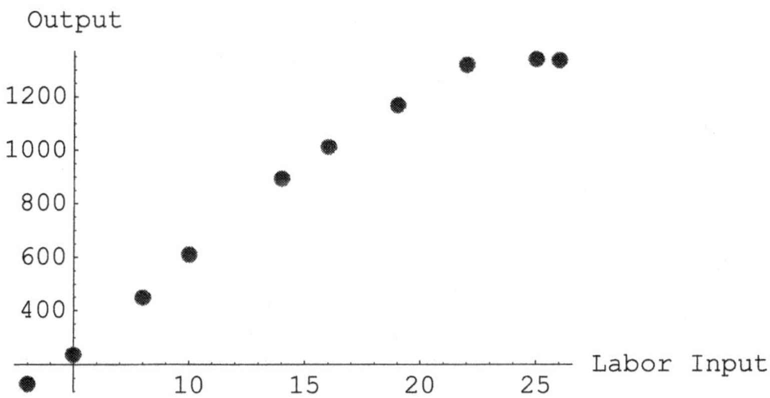

The listplot confirms that the mathematical from of the function can be expressed as $Q = \alpha + \beta L + \gamma L^2 + \lambda L^3 + \epsilon$. To fit the curve to data using *Mathematica*, use the following command,

```
Input[27]:= fit1 = Fit[data, {1, x, x^2, x^3}, x]
```

Output[27]= $-24.6047 + 39.7765\, x + 3.3551\, x^2 - 0.110395\, x^3$

the fitted equation is $Q = -24.60 + 39.78L + 3.36L^2 - 0.11L^3$ and the graph of the fitted equation is

```
Input[28]:= graph2 = Plot[fit1, {x, 0, 30},
        PlotLabel -> {"TP"}, AxesLabel → {"Labor Input", "Output"}]
```

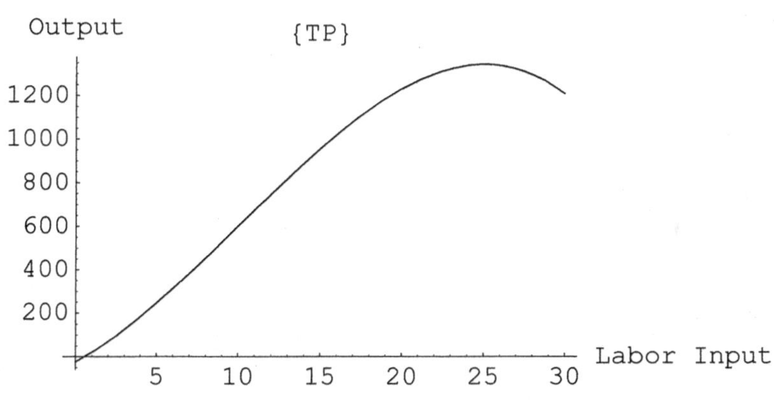

to graph the function and the listplot of data together use the following command

```
Input[29]:= Show[graph1, graph2]
```

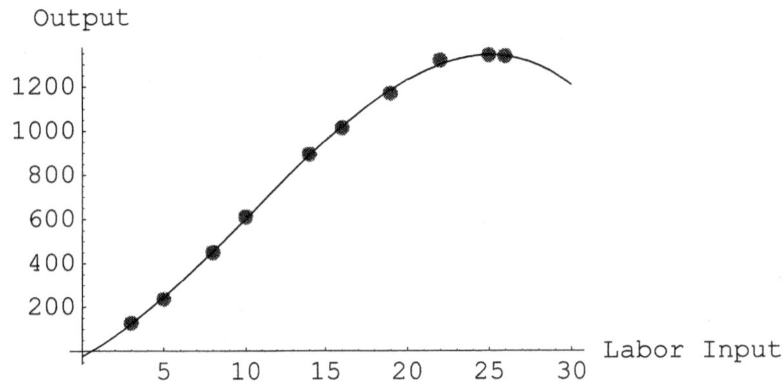

The arc marginal product of a variable input is defined as increase in total product per unit increase in variable input, assuming other inputs held constant. Here, the arc marginal product of labor is, AMP = $\frac{DQ}{DL}$, where DQ is the change in Q and DL is the change in labor input. Average product is total product per unit of input, AP = $\frac{Q}{L}$. Once AMP and AP are generated from the original data, listplot the data to decide on the proper functional forms

and fit the functions. (**Note:** For generating DQ, DL, and AMP through the program, see Appendix at the end of this chapter).

```
Input[30] :=  Clear[AMP, AP, fit2, fit3, graph3, graph4]
AMP = {54, 71, 81, 70.75, 59.5, 52, 50, 7, -4};
graph3 = ListPlot[AMP,  PlotStyle -> {RGBColor[1, 0, 0], PointSize[.03]},
  AxesLabel → {"Labor Input", "AMP"}]
```

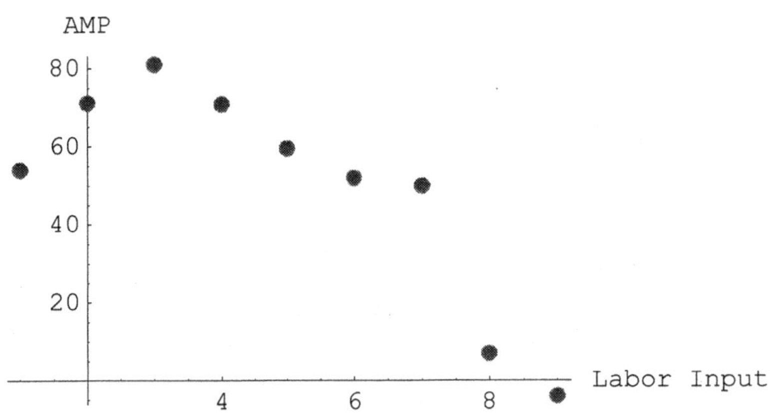

the listplot of AMP data shows that a parabola seems to be the appropriate form of the function

```
Input[31]:= fit2 = Fit[AMP, {1, x, x^2}, x]
```

Output[31]= $43.8929 + 17.3315\,x - 2.5744\,x^2$

the marginal product function is AMP = 43.89 + 17.33L - 2.57L^2, the graph of which is presented below,

Input[32]:=
```
graph4 = Plot[fit2, {x, 0, 10}, PlotLabel -> {"AMP"},
    PlotStyle -> {RGBColor[1, 0, 1]}, AxesLabel → {"Labor Input", "AMP"}];
```

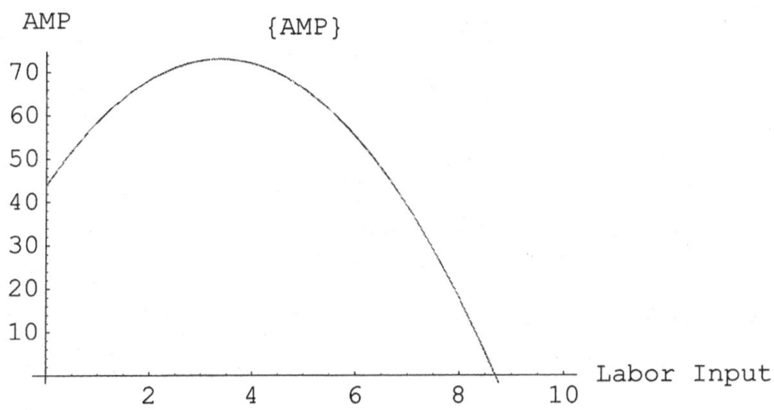

the following graph shows the estimated curve od AMP and the actual data together

Input[33]:=
```
Show[graph3, graph4]
```

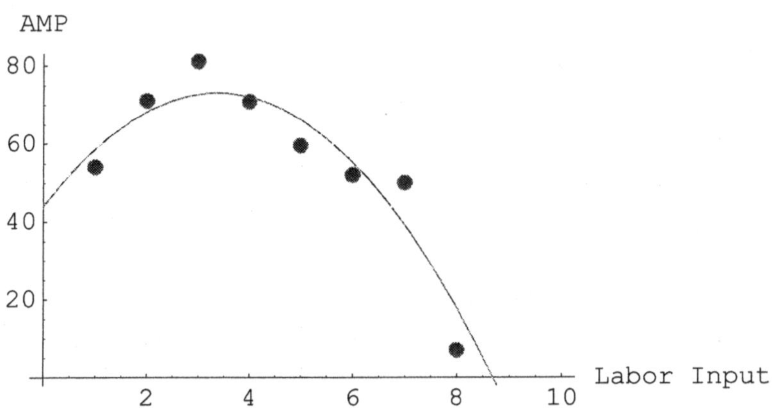

The average product of labor is AP = $\frac{Q}{L}$. Listplot the data to decide the appropriate form of the function.

Input[34]:=
```
AP = {42.33, 47, 56, 61, 63.79, 63.25, 61.47, 59.91, 53.56, 51.35};
ListPlot[AP, PlotStyle -> {RGBColor[1, 0, 0], PointSize[.03]}]
```

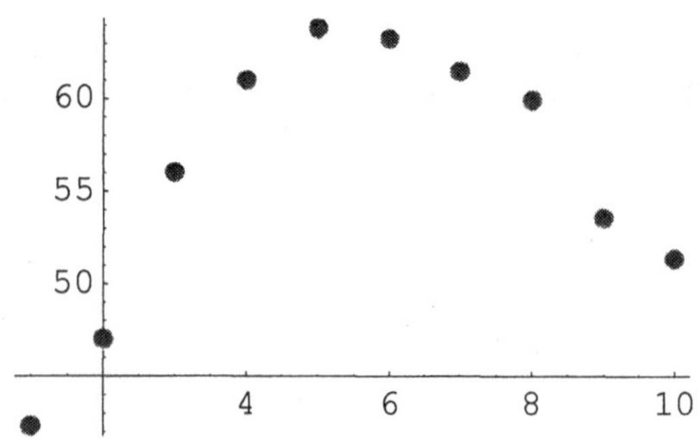

The AP plot also shows that a parabola is an appropriate form of the function to present the data. Therefore, fit a function of the form AP = $\alpha + \beta L + \gamma L^2 + \in$

Input[35]:= **fit3 = Fit[AP, {1, x, x^2}, x]**

Output[35]= 32.0253 + 10.4057 x - 0.864697 x^2

the AP function is AP = 32.03 + 10.41 L - 0.86 L^2,
the graph of which is presented below

Input[36]:= **graph5 = Plot[fit3, {x, 0, 10}, PlotLabel -> {"AP"}, PlotStyle -> {RGBColor[1, 1, 0]}];**

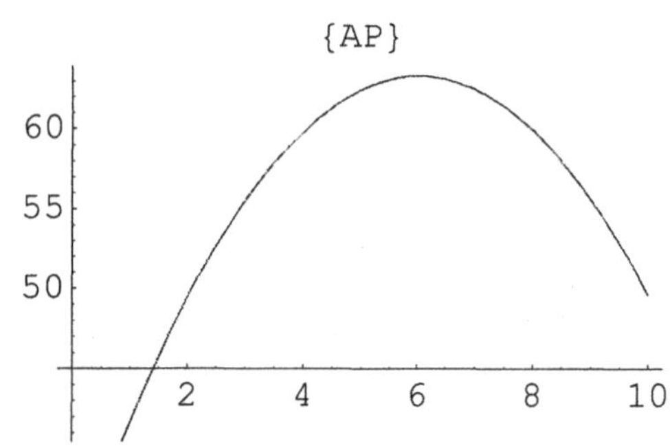

the arc marginal product and average product function are graphed together below

Input[37]:= **Show[graph4, graph5]**

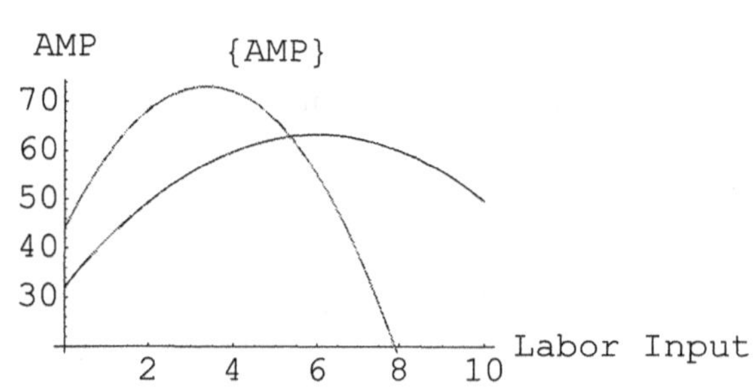

As the graphs of AMP and AP show as long as marginal product is higher than average product, average product increases. The two curve meet at the maximum AP, beyond which the average product of labor will be decreasing. To see the relationship between total product (TP), marginal product (AMP), and average product (AP), graph all three curve as follows

```
Input[38]:= Show[ graph2]
            Show[graph4, graph5]
```

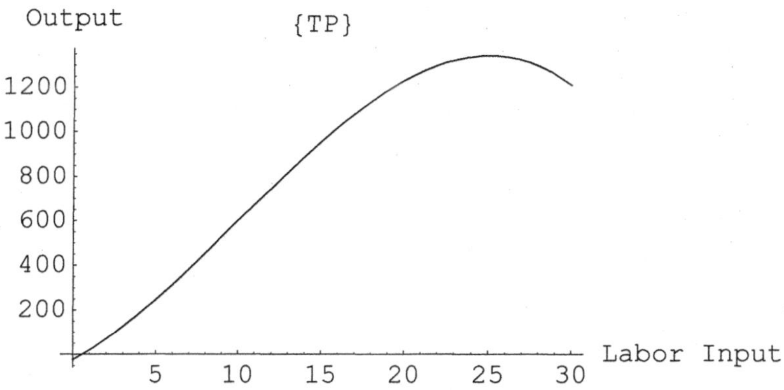

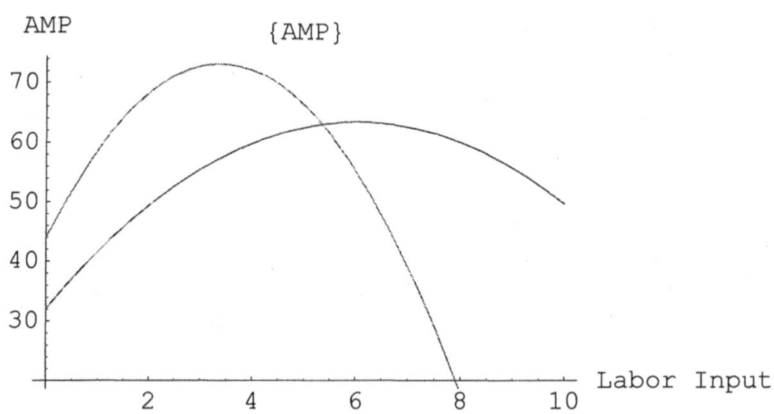

The graphs are similar to the standard TP, AP, and AMP curves in any economics text book. As long as Q increases with an increasing rate, AMP is rising. At the inflection point of the total output curve, AMP reaches maximum and beyond that point it decreases. As long as AMP is higher than AP, AP is rising. AMP passes through the maximum point of AP curve, beyond which both curves are declining. At the maximum output level AMP = 0.

Problems

1- The following data represents total cost (TC) of producing a good at different levels of output (Q).

Q	0	1	2	3	4	5	6	7	8	9	10
TC	65	111	140	163	176	194	228	273	334	418	514

a) Plot the relationship between TC and Q and decide on the degree of the polynomial.

b) Find marginal cost (MC) of production (MC is defined as increase in TC for one unit increase in Q).

c) Find average variable cost (AVC) of production (AVC is defined as $\frac{VC}{Q}$).

d) Find average total cost (ATC) of production (ATC is defined as $\frac{TC}{Q}$)

e) Plot MC, AVC and ATC.

f) Fit appropriate polynomials to TC, MC, AVC and ATC.

g) Show ATC, AVC, and MC together on one coordinate system and comment on the relationship between the three curves.

h) Find TC corresponding to 11 units of output.

2- The following data represents total cost (TC) of producing a good at different levels of output (Q).

Q	0	1	2	3	4	5	6	7	8	9	10
TC	100	116	128	135	144	158	184	228	295	386	519

a) Plot the relationship between TC and Q and decide on the degree of the polynomial.

b) Find marginal cost (MC) of production (MC is defined as increase in TC for one unit increase in Q).

c) Find average variable cost (AVC) of production (AVC is defined as $\frac{VC}{Q}$).

d) Find average total cost (ATC) of production (ATC is defined as $\frac{TC}{Q}$)

e) Plot MC, AVC and ATC.

f) Fit appropriate polynomials to TC, MC, AVC and ATC.

g) Show ATC, AVC, and MC together on one coordinate system and comment on the relationship between the three curves.

h) Find TC corresponding to 12 units of output.

3- The following data represents the relationship between labor input (L) and output (Q) of a manufacturing business.

L	0	1	2	3	4	5	6	7	8	9	10
Q	0	19	40	49	64	78	87	92	95	96	94

a) Plot the relationship between L and Q

b) Find arc marginal product (MP)of labor.

c) Find average product of labor (AP).

d) Fit appropriate functions to Q, MP, and AP.

e) Plot Q, MP, and AP.

f) Comment on the relationships between the three curves.

g) Find output level corresponding to 11 units of labor

4- The following data represents the relationship between labor input (L) and output (Q) of a manufacturing business.

L	0	1	2	3	4	5	6	7	8	9	10
Q	0	45	135	395	605	773	865	897	911	916	917

a) Plot the relationship between L and Q

b) Find arc marginal product (MP)of labor.

c) Find average product of labor (AP).

d) Fit appropriate functions to Q, MP, and AP.

e) Plot Q, MP, and AP.

f) Comment on the relationships between the three curves.

g) Find output level corresponding to 11 units of labor

III. Fitting Planes and Surfaces: The **ListPlot3D** Command

Every economic variable may be explained by a number of other variables. For example, consumption may depend on price level and interest rate, as well as income. Output in a manufacturing business may depend on capital and state of the technology, as well as labor input, and etc. Explaining a variable in terms of more than one variable requires building multi-variable models and is the subject of multiple regression analysis. A multi-variate relationship may have linear form or non-linear form. The form of the function may be decided by using statistical tests or by plotting the relationship between the dependent variable and independent variables and eyeballing the trends.

Fitting planes and surfaces follows the same principles of fitting line in the least squared method discussed in the previous section. The objective is to fit a plane to set of data in such a way that the sum squared errors between the actual data and the plane is minimized. In practice estimating coefficients of multiple regression models requires large amount of data manipulation and calculations. Therefore, the use of statistical packages such as E-Views, SAS, or SPSS becomes a necessity. Here, we will use *Mathematica* to fit some multiple regression models.

Example: Production Function

The relationship between labor input (L), capital input (K), and output (Q) is given as

L	0	2	3	5	6	9	12	14	15	15	15
K	0	5	6	6	6	7	10	10	12	15	18
Q	0	24	30	34	46	64	68	68	78	90	102

Plot the relationship between labor input and output.

Plot the relationship between labor input and output.

Plot the relationship between labor input, capital input, and output.

Fit a model explaining output as a function of inputs.

Find the output level corresponding to 6 units of labor and 6 units of capital

Solution: To plot the relationship between labor input and output, make a table of data (data1) consisting of labor input and output and list plot the data. (**Note:** For generating data1 through the *Mathematica* commands, see, 2.

Data Manipulation of a Table in Appendix at the end of this chapter)

```
Input[39] :  Clear[data, data1, data2, fit4]
data = {{0, 0, 0}, {2, 5, 24}, {3, 6, 30}, {5, 6, 34}, {6, 6, 46}, {9, 7, 64},
   {12, 10, 68}, {14, 10, 68}, {15, 12, 78}, {15, 15, 90}, {15, 18, 102}};
```

```
data1 = {{0, 0}, {2, 24}, {3, 30}, {5, 34}, {6, 46},
   {9, 64}, {12, 68}, {14, 68}, {15, 78}, {15, 90}, {15, 102}};

ListPlot[data1,
 PlotStyle -> {RGBColor[1, 0, 0], PointSize[.03]}, AxesLabel → {"L", "Q"}]
```

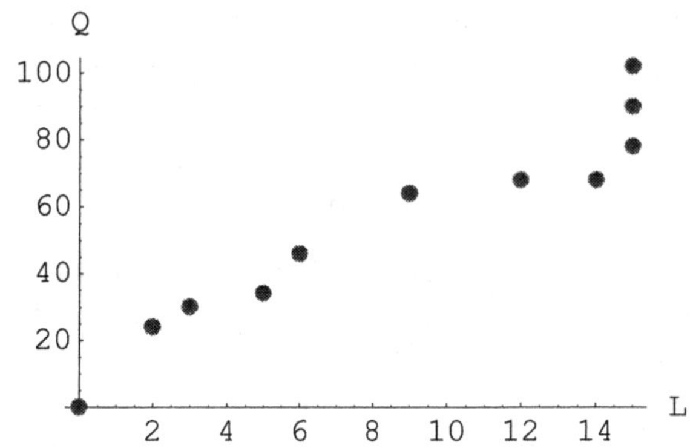

To plot the relationship between capital input and output, make a table of data (data2) consisting of capital input and output and list plot the data.

```
Input[40] :=  data2 = {{0, 0}, {5, 24}, {6, 30}, {6, 34}, {6, 46},
   {7, 64}, {10, 68}, {10, 68}, {12, 78}, {15, 90}, {18, 102}};

ListPlot[data2,
 PlotStyle -> {RGBColor[1, 0, 0], PointSize[.03]}, AxesLabel → {"K", "Q"}]
```

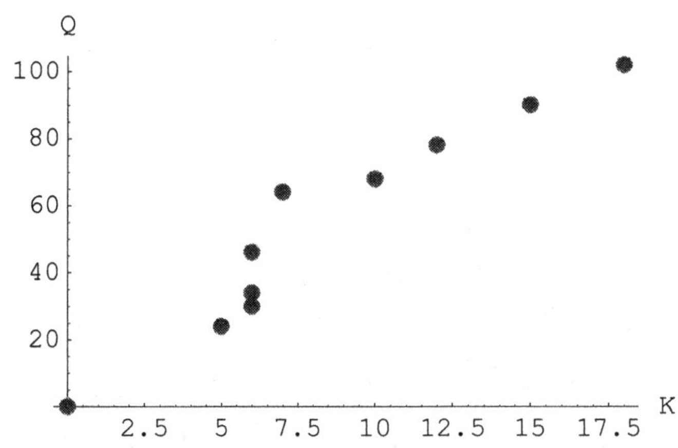

The two graphs above show the relationship between labor input and output and the relationship between capital input and output, respectively. Although the individual graphs show a nonlinear relationship between each input and output, the graph of the relationship between both inputs and output (below) shows that the relationship may be approximated by a linear surface.

```
Input[41]:= ListPlot3D[data, AxesLabel → {"L", "K", "Q"}]
```

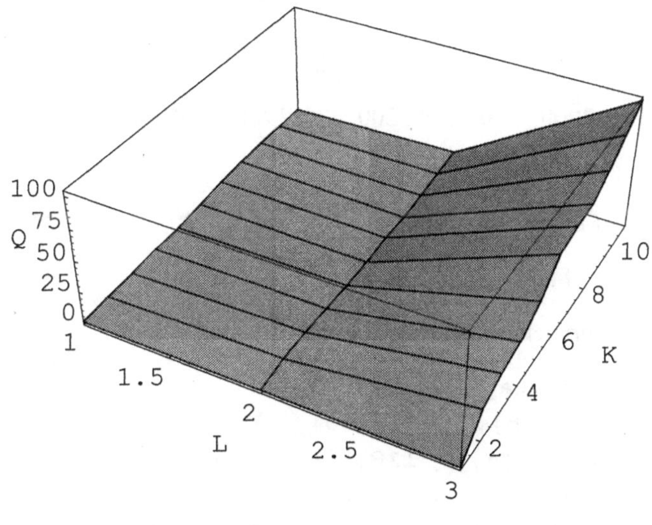

Assuming a linear surface, the production function can be estimated by fitting a model of the form Q = α L + βK. Using *Mathematica*, α and β can be estimated by using the "Fit" command as,

```
Input[42]:= fit4 = Fit[data,{x,y}, {x,y}]
```

```
Output[42]= 2.66566 x + 3.53405 y
```

the estimated model is Q = 2.67L + 3.53K, the graph of which is presented below.

```
Input[43]:= Plot3D[fit4, {x, 0, 6}, {y, 0, 6}]
```

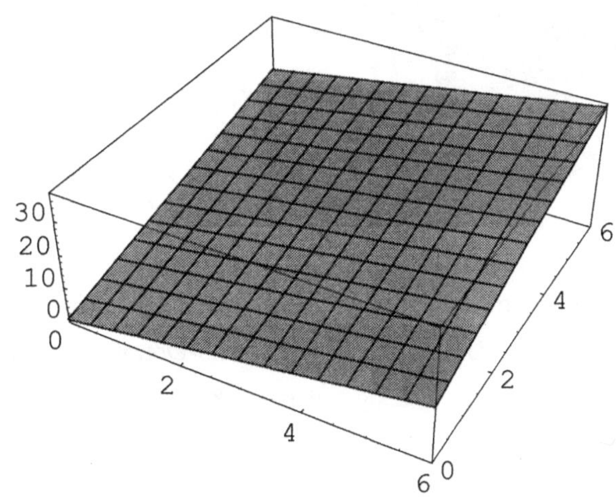

To evaluate the output corresponding to L = 6 and K = 6, evaluate the estimated function at this point as,

```
Input[44]:=   fit4/. {x -> 6, y -> 6}
```

```
Output[44]= 37.1982
```

the estimated output corresponding to 6 units of labor input and 6 units of capital input is 37 units.

Example: Cobb-Douglas Production Function

Data for labor input (L), capital input (K), and output (Q) is given as

L: 1, 2, 1, 3, 3, 2, 4, 4, 4, 4, 3, 3, 5, 5, 5, 5, 5, 4

K: 1, 1, 2, 1, 2, 3, 1, 2, 3, 4, 3, 4, 1, 2, 3, 4, 5, 5

Q: 10, 35, 40, 70, 290, 318, 118, 490, 1100, 2000, 670, 1200, 200, 730, 1700, 3000, 4800, 3000

Plot the relationship between labor input, capital input, and output.

Fit a model explaining output as a function of inputs.

Find the output level corresponding to 6 units of labor and 6 units of capital

Solution: List the data and plot the relationship between the variables.

Input[45]:=
```
Clear[data, Q, x, y, logdata]
data = {{1, 1, 10}, {2, 1, 35}, {1, 2, 40}, {3, 1, 70},
   {3, 2, 290}, {2, 3, 318}, {4, 1, 118}, {4, 2, 490}, {4, 3, 1100},
   {4, 4, 2000}, {3, 3, 670}, {3, 4, 1200}, {5, 1, 200}, {5, 2, 730},
   {5, 3, 1700}, {5, 4, 3000}, {5, 5, 4800}, {4, 5, 3000}};

ListPlot3D[data]
```

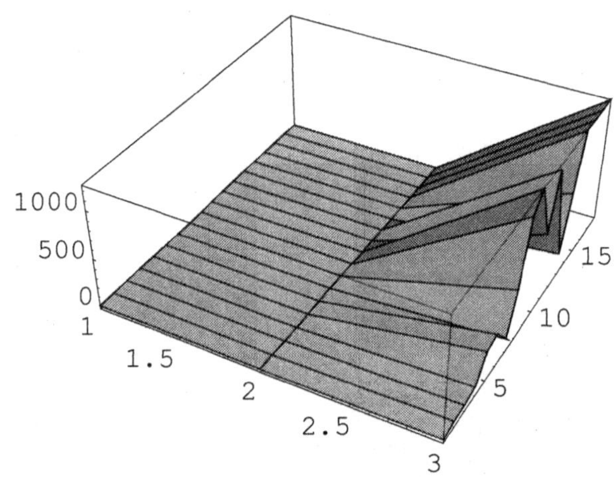

The list plot of the data clearly shows that the relationship between inputs and output is not a linear surface. Therefore, we may try a Cob-Douglas model, $Q = AL^{\alpha}K^{\beta}e^{\varepsilon}$. To fit a Cob-Douglas model, use logarithmic transformation to transform the function to Log Q = Log A + αLog L + βLog K + ε, then fit a log-linear model to the transformed data.

Input[46]:=
```
logdata = N[Log[data]];
Fit[logdata, {1, x, y}, {x, y}]
```

Output[46]= `Null (2.29063 + 1.81933 x + 2.00848 y)`

The loglinear model is LogQ = 2.29 + 1.82LogL + 2LogK. To write the model in its Cob-Douglas form, first find A. Since LogA = 2.29, A = $e^{2.29}$, where e is the natural number.

Input[47]:= `N[E^2.29063]`

Output[47]= `9.88116`

A = 9.88 and the model can be written in its cob-Douglas form as Q = $9.88L^{1.82}K^{2}$. To plot the function, use

Plot3D command as,

Input[48]:= `Plot3D[9.88 x^1.8 y^2.0085, {x, 0, 10}, {y, 0, 10}]`

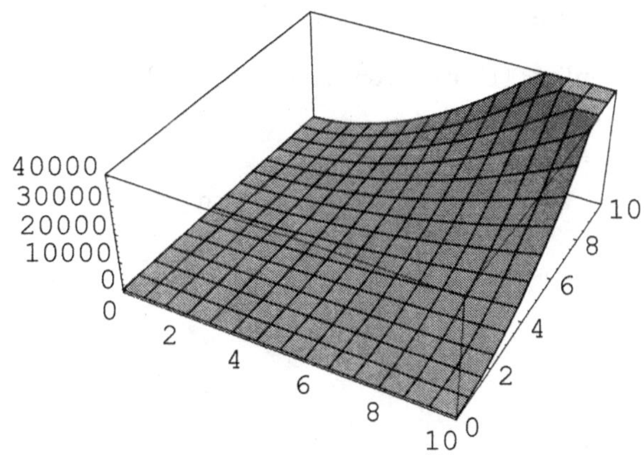

To find Q corresponding to L = 6 and K = 6, substitute for L and K in the production function

Input[49]:=
```
Q = 9.88 x^1.82 y^2;
Q /. {x -> 6, y -> 6}
```

Output[49]= 9274.59

the output corresponding to six units of labor and six units of capital is 9274.59 units.

Problems

1. Data for labor input (L), capital input (K), and output (Q) is given as

L: 1, 1, 2, 3, 3, 5, 5, 4, 6, 6, 7, 8, 9, 10

K: 1, 2, 2, 4, 3, 4, 5, 5, 7, 6, 8, 9, 8, 10

Q: 11, 14, 22, 38, 32, 50, 55, 52, 70, 66, 83, 91, 98, 110

Plot the relationship between labor input and output.

Plot the relationship between capital input and output.

Plot the relationship between labor input, capital input, and output.

Fit a model explaining output as a function of inputs.

Find the output level corresponding to 6 units of labor and 6 units of capital

2- Data for labor input (L), capital input (K), and output (Q) is given as

L	1	1	2	3	3	5	5	4	6	6	7	8	9	10
K	1	2	2	4	3	4	5	5	7	6	8	9	8	10
Q	2	7	40	340	200	1154	1730	1020	4895	3720	9000	15360	16460	31700

a) Plot the relationship between labor input and output.

b) Plot the relationship between capital input and output.

c) Plot the relationship between labor input, capital input, and output.

d) Fit a model explaining output as a function of inputs.

e) Find the output level corresponding to 11 units of labor and 11 units of capital.

3- A consumer expresses her total ordinal utility (TU) from consuming two goods x and y as follows.

x	1	0	1	1	2	3	3	5	5	4	6	6	7	8	9	10
y	0	1	1	2	2	4	3	4	5	5	7	6	8	9	8	10
U	0	0	1	1.5	2	3.5	3	4.5	5	4.5	6.5	6	7.5	8.5	8.5	10

a) Plot the relationship between x, y, and U.

b) Fit the best fitting function to the data

IV. Finding Exact Fits: The **InterpolatingPolynomial** Command

The least-squares method fits the best model of your specification to data. The estimated model will best generate the data, with the sum squares of errors between the actual points and the estimated model being minimum. However, the model will not exactly reproduce data. To reproduce data exactly, you can fit a polynomial that will pass through all the points. In general, for every set of n data points we can fit a polynomial of degree n - 1 which passes through all the points. The function "InterpolationgPolynomial" in *Mathematica* estimates a polynomial that fits the data exactly. Such polynomials may have a limited use, or no use, in applied economics but they may be useful in various algebraic algorithms.

Example: Fit an exact polynomial to the following four points, A(1,5), B(3,3), C(5,6), D(8,8), and plot the function.

Solution: For the four data points of this example a third degree polynomial will fit exactly and will pass through all the four points. List the data and use the *Mathematica* command InterpolatingPolynomial to find the function.

```
Input[50]:= Clear[data, interpol]
            data = {{1, 5}, {3, 3}, {5, 6}, {8, 8}};
            interpol = InterpolatingPolynomial[data, x]
            Plot[interpol, {x, 0, 10}]
```

Output[50]= $5 + (-1 + (\frac{5}{8} - \frac{19}{168}(-5 + x))(-3 + x))(-1 + x)$

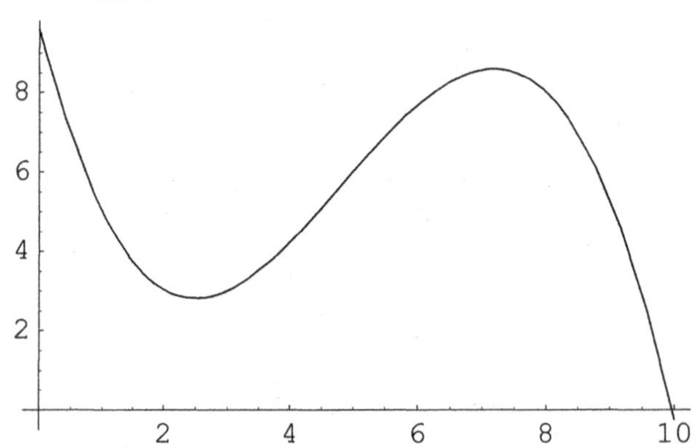

Problems

Fit an exact polynomial to the following points find the equation of the polynomial and graph it.

1. (2, 5) and (-1, 6)
2. (2,2), (4, -1), and (5, 0)
3. (0,0), (-1, -3), and (1, 3)
4. (0, 3), (4, 2), and (-1, 6)
5. (4,-1), (5, 3), (1,1), and (-2,0)
6. (1, 1), (3, -1), (0, 5), and (-6, 1)

Appendix: Data Manipulation

1. Generating DQ and **DX Using *Mathematica*:** The data set below is a paired observations on X and Q. To generate DQ = Q_t - Q_{t-1} and DX = X_t - X_{t-1}through the *Mathematica* commands, load the data manipulation package by using the command <<Statistics`DataManipulation` and list your data as follows,

```
Input[51]:= << Statistics`DataManipulation`
```

```
Clear[data, L, Q, MP, AP]
data = {{2, 13}, {4, 36}, {7, 72}, {8, 87},
   {11, 123}, {14, 138}, {16, 140}, {17, 139}, {20, 121}};
```

by using the command X = Column[data, 1] select the first column of the data set. This will give the observations on X. Qs are the second elements of each pair of observation. Find Q using the command Q = Column[data, 2].

```
Input[52]:= L = Column[data, 1]

Output[52]= {2, 4, 7, 8, 11, 14, 16, 17, 20}

Input[53]:= Q = Column[data, 2]

Output[53]= {13, 36, 72, 87, 123, 138, 140, 139, 121}
```

The data set has nine observations. Since lagging a variable drops one of the observations from the end point (the 9th observation here), Q_{t-1} is generated by the command Drop[Q, {9}]. Q_t is generated by taking observations two through nine using the command Take[Q, {2, 9}]. Therefore, DQ is the difference between the two generated sets. The same applies to DL. AMP is a table generated by taking the ratio of the two variables, $\frac{DQ}{DL}$.

```
Input[54]:= DQ = Take[Q, {2, 9}] - Drop[Q, {9}]

Output[54]= {23, 36, 15, 36, 15, 2, -1, -18}

Input[55]:= DL = Take[L, {2, 9}] - Drop[L, {9}]

Output[55]= {2, 3, 1, 3, 3, 2, 1, 3}

Input[56]:= A MP = Table[DQ / DL]

Output[56]= Set::write : Tag Times in A MP is Protected.
```

```
{23/2, 12, 15, 12, 5, 1, -1, -6}
```

AP is generated by a table of $\frac{Q}{L}$.

```
Input[57]:= ap = Table[Q / L]

Output[57]= {13/2, 9, 72/7, 87/8, 123/11, 69/7, 35/4, 139/17, 121/20}
```

2. Data Manipulation of a Table: Suppose you have a three-columns table of data with observations on X, Y, and Z and you would like to make new tables with paired observations on X and Y, X and Z, and Y and Z. To generate new tables, first load the data manipulation package <<Statistics`DataManipulation`, type your data set, and use the command Column[data, {1, 2}] for generating a table of X and Y observations, Column[data, {1, 3}] for generating a table of X and Z observations, and Column[data, {2, 3}] for generating a table consisting of observations on Y and Z.

```
Input[58]:= << Statistics`DataManipulation`

Clear[data, dataxy, dataxz, datayz]

data = {{0, 0, 0}, {2, 5, 24}, {3, 6, 30}, {5, 6, 34}, {6, 6, 46}, {9, 7, 64},
   {12, 10, 68}, {14, 10, 68}, {15, 12, 78}, {15, 15, 90}, {15, 18, 102}};

dataxy = Column[data, {1, 2}]

Output[58]= {{0, 0}, {2, 5}, {3, 6}, {5, 6}, {6, 6},
             {9, 7}, {12, 10}, {14, 10}, {15, 12}, {15, 15}, {15, 18}}

Input[59]:= datayz = Column[data, {2, 3}]

Output[59]= {{0, 0}, {5, 24}, {6, 30}, {6, 34}, {6, 46},
             {7, 64}, {10, 68}, {10, 68}, {12, 78}, {15, 90}, {18, 102}}

Input[60]:= dataxz = Column[data, {1, 3}]

Output[60]= {{0, 0}, {2, 24}, {3, 30}, {5, 34}, {6, 46},
             {9, 64}, {12, 68}, {14, 68}, {15, 78}, {15, 90}, {15, 102}}
```

CHAPTER 4

ELEMENTS OF MATRIX ALGEBRA

Matrix algebra began as the study of systems of linear equations. Today, it provides a basic framework for the study of modern algebra, calculus of several variables, and differential equations. Although algebraic operations on matrices may seem a bit abstract, they do serve a great purpose by simplifying the mathematical notations and operations of the underlying models of economic problems. Chapters 6 through 8 of this book employ matrix algebra in optimizing problems, where the function has two or more independent variables. Chapters 9 and 10 utilize the concept of rank and linear dependence as well as matrix algebra in solving linear programming problems and linear economic models. Chapter 11 makes an extensive use of concepts developed in this chapter and chapter 5, such as eigenvalues and eigenvectors, in solving systems of first order differential equations.

TOOL BOX

Mathematica Commands	Function
a = {5, 3, 2}	a is a 3x1 column vector or a 1x3 row vector.
B = {{3, 4, 5}, {4, 5,1}, {-1, 2,4}}	B is a 3x3 matrix with the given elements
MatrixForm[B]	Writes B in matrix form
MatrixForm [A.B]	Pre-multiplies B by A and writes the product in matrix form
Det[B]	Finds determinant of matrix B
Transpose[B]	Transposes matrix B
Inverse[B]	Inverses matrix B
Tr[B]	Finds trace of matrix B
IdentityMatrix[n]	Generates a nxn identity matrix
Dot [a , b]	Finds dot product a & b vectors of dimension 3 or higher
Cross [a, b]	Finds cross product a & b vectors

Note: To generate a matrix in *Mathematica,* take the following steps:

Click on "File", click on "Palettes", and click on "2 Basic Calculations" to open the basic calculation window. In the new window, click on "List and Matrices", click on "Creating Lists and Matrices", and click on the matrix or vector icon. To expand a matrix or a vector, click on one element of matrix or vector on the screen and push "Ctrl" and "Enter" keys to add a row, and push "Ctrl" and "," keys to add a column.

I. Algebra of Matrices: The **MatrixForm, Det, Transpose, Inverse,** and **Tr** Commands

Matrix: A matrix is a rectangular array of elements arranged in rows or columns as

$$\mathbf{A} = \begin{pmatrix} a11 & a12 & . & . & . & . & . & . & . & a1n \\ a21 & a22 & . & . & . & . & . & . & . & a2n \\ . & . & . & . & . & . & . & . & . & . \\ . & . & . & . & . & . & . & . & . & . \\ am1 & am2 & . & . & . & . & . & . & . & amn \end{pmatrix}$$

A has **mxn** elements arranged in m rows and n columns, and is said to be of **order m by n.** The elements in the ith row and j^{th} column is represented by a_{ij}. Matrices provide a powerful way of organizing and manipulating data. In many instances a large amounts of information can be summarized in matrix form. Matrices are usually denoted by bold capital characters.

Example: A car manufacturer is producing three models of cars, sedan (S), full-size (F), and Vans (V). Total production (TP) of the three models (in thousands) for the years 1994-1998 is summarized in a 5x3 matrix,

$$\mathbf{TP} = \begin{pmatrix} 45 & 32 & 28 \\ 52 & 38 & 31 \\ 66 & 41 & 39 \\ 72 & 42 & 45 \\ 78 & 45 & 52 \end{pmatrix} \quad (S \;\; F \;\; V)$$

, where the columns represent the models and the rows represent the years, with the first row representing 1994.

Vectors: A matrix of order 1xn is called a **row vector**. For example, **b** = (b1 b2 b3) is a 1x3 row vector. A matrix of order mx1 is a **column vector**. For example, $\mathbf{c} = \begin{pmatrix} c1 \\ c2 \\ c3 \end{pmatrix}$ is a 3x1 column vector. With these definitions, a matrix can also be viewed as a set of column vectors or row vectors. That is the matrix **A** above can be written as

$$\mathbf{A} = (\,a1 \;\; a2 \;\; . \; . \; . \;\; an\,), \text{ where } \mathbf{a1} = \begin{pmatrix} a11 \\ a21 \\ . \\ . \\ am1 \end{pmatrix}, \mathbf{a2} = \begin{pmatrix} a12 \\ a22 \\ . \\ . \\ am2 \end{pmatrix}, \text{ and so on.}$$

In physical science, the term vector is used to indicate a quantity (such as force and velocity) that has both magnitude and direction. In such cases, a vector is often represented by an arrow or directed line segment. The length of the arrow represents the magnitude and the tip of the arrow points in the direction of the vector. In business and economics, a vector may represent a collection of data for one variable over time (a table of time series data) or it may represent a cross section collection of data on a variable at one time. The k elements of a row or column vector **a** can be viewed as the coordinates of a point in a k-dimensional space. Vectors are usually denoted by bold lower case character.

Example: Price of the sedan cars (in thousand dollars) for 1994-1998 may be represented by a 1x5 vector **pt** = (9.2 9.6 1.2 1.8 11.2). This is an example of Time-series data on one variable presented in a vector form. The prices of the three models of cars for the 1994 may be represented by a 1x3 vector **pc** = (9.2 11.4 18.75). This is an example of a cross-section data presented in a vector form.

Algebra of Matrices: The algebra of matrices utilizes the following rules:

Equality of Matrices: Two matrices **A** and **B** are said to be equal when they are of the same order and all the corresponding elements of **A** and **B** are equal.

Example: Find x and y in a way that the two matrices $\mathbf{A} = \begin{pmatrix} 2 & -1 \\ 3 & 4 \end{pmatrix}$ and $\mathbf{B} = \begin{pmatrix} 2 & -1 \\ .5x & y-x \end{pmatrix}$ are equal.

Solution: For **A** and **B** to be equal .5x = 3 and y - x = 4 should be satisfied. Solving these two equations together gives x = 6 and y = 10.

Addition of Matrices: If **A** and **B** are of the same order, then **A** + **B** is obtained by adding the corresponding elements of **A** and **B**.

Example: Two matrices **A** and **B** are given below. Find **A** + **B**.

$\mathbf{A} = \begin{pmatrix} 3 & -1 \\ 0 & 10 \end{pmatrix}$, $\mathbf{B} = \begin{pmatrix} 2 & 4 \\ -7 & 5 \end{pmatrix}$

Solution: Adding up the corresponding elements of the matrices gives $\begin{pmatrix} 3+2 & -1+4 \\ 0-7 & 10+5 \end{pmatrix} = \begin{pmatrix} 5 & 3 \\ -7 & 15 \end{pmatrix}$. To solve the problem using *Mathematica*, define the matrices, add them, and write in matrix form using "MatrixForm" command

Input[1]:=
```
Clear[A, B]
A = (3 -1
     0 10) ; B = (2 4
                  -7 5);

MatrixForm[A + B]
```
Output[1]= $\begin{pmatrix} 5 & 3 \\ -7 & 15 \end{pmatrix}$

The Difference of Matrices: If **A** and **B** are of the same order, then **A** - **B** is obtained by subtracting the corresponding elements of **B** from **A**.

Example: For the two matrices **A** and **B** of the preceding example find **A** - **B**.

Solution: Subtracting the corresponding elements of the two matrices gives $\begin{pmatrix} 3-2 & -1-4 \\ 0+7 & 10-5 \end{pmatrix} = \begin{pmatrix} 1 & -5 \\ 7 & 5 \end{pmatrix}$.

Using *Mathematica*, the problem can be done as,

Input[2]:=
```
Clear[A, B]
A = (3 -1
     0 10) ; B = (2 4
                  -7 5);

MatrixForm[A - B]
```
Output[2]= $\begin{pmatrix} 1 & -5 \\ 7 & 5 \end{pmatrix}$

Scalar Multiplication: If k is a scalar, the scalar multiplication of **A** is obtained by multiplying each element of **A** by k.

Example: For the matrix $\mathbf{A} = \begin{pmatrix} 3 & -1 \\ 0 & 10 \end{pmatrix}$ find k**A**.

Solution: Multiply each element of the matrix **A** by k to get $\begin{pmatrix} 3k & -k \\ 0 & 10k \end{pmatrix}$. In *Mathematica,*

Input[3]:=
```
Clear[A]
A = (3 -1
     0 10);

MatrixForm[k A]
```
Output[3]= $\begin{pmatrix} 3\,k & -k \\ 0 & 10\,k \end{pmatrix}$

Example: Unit Cost

A manufacturer has two plant p1 and p2. The cost of inputs per unit of labor, capital, and material in each plant are shown in the following matrix, **C** = $\begin{pmatrix} 23 & 35 \\ 42 & 38 \\ 33 & 40 \end{pmatrix}$ (columns labelled (p1 p2)), where the columns are representing the plants 1 and 2 and the rows 1-3 are representing labor, capital, and material unit costs, respectively. It is expected that, the next year's unit costs will rise by 6%. Find the expected unit cost matrix for the next year.

Solution: Multiply all elements of the cost matrix by 1.06 to get $\begin{pmatrix} 24.15 & 36.75 \\ 44.1 & 39.9 \\ 34.65 & 42. \end{pmatrix}$. In *Mathematica,*

Input[4]:=
```
cost = (23 35
        42 38
        33 40);

MatrixForm [1.05 cost]
```
Output[4]= $\begin{pmatrix} 24.15 & 36.75 \\ 44.1 & 39.9 \\ 34.65 & 42. \end{pmatrix}$

Matrix Multiplication: If **A** is of order mxn and **B** is of order nxp, then the dot product of **A.B** is a matrix of order mxp, where the ij^{th} element is obtained by multiplying the elements of the ith row of the first matrix by the corresponding elements of the jth column of the second matrix and summing over all terms. For this to be possible, it is clear that the number of columns in the first matrix (n) should be equal to the number of rows in the second matrix (n). If that is the case, the matrices are said to be **conformable** for multiplication.

Example: Find the dot product of **A** = $\begin{pmatrix} a11 & a12 \\ a21 & a22 \end{pmatrix}$ and **B** = $\begin{pmatrix} b11 & b12 \\ b21 & b22 \end{pmatrix}$.

Solution: The product is a 2x2 matrix with elements as $\begin{pmatrix} a11.b11 + a12.b21 & a11.b12 + a12.b22 \\ a21.b11 + a22.b21 & a21.b12 + a22.b22 \end{pmatrix}$.

To do the same multiplication using *Mathematica*, **note that**, in *Mathematica* the dot product of matrices is done using "." rather than "*" or space. The latter symbols are reserved for a special multiplication which will be explained in the next section.

Input[5]:=
```
Clear[A, B]
A = (a11 a12
     a21 a22); B = (b11 b12
                    b21 b22);
MatrixForm[ A.B ]
```
Output[5]= $\begin{pmatrix} a11\,b11 + a12\,b21 & a11\,b12 + a12\,b22 \\ a21\,b11 + a22\,b21 & a21\,b12 + a22\,b22 \end{pmatrix}\begin{pmatrix} a11\,b11 + a12\,b21 & a11\,b12 + a12\,b22 \\ a21\,b11 + a22\,b21 & a21\,b12 + a22\,b22 \end{pmatrix}$

$$\begin{pmatrix} a11\,b11 + a12\,b21 & a11\,b12 + a12\,b22 \\ a21\,b11 + a22\,b21 & a21\,b12 + a22\,b22 \end{pmatrix}$$

Example: Find **A.B** for **A** and **B** given as $\mathbf{A} = \begin{pmatrix} 3 & -1 \\ 0 & 10 \end{pmatrix}$ $\mathbf{B} = \begin{pmatrix} 2 & 4 \\ -7 & 5 \end{pmatrix}$.

Solution: Let's call the product **C**, **C** = **A.B**. To find the c11 element of the product matrix (element in the first row first column), multiply the elements of the first row of **A** with the elements of the corresponding elements of the first column of **B** and add the products. That is, c11 = 3x2 + (-1)(-7) = 13. To find the c12 element of the product matrix (element in the first row second column), multiply elements of the first row of **A** with the corresponding elements of the second column of **B** and add the products. That is, c12 = 3x4 + (-1)(5) = 7. Continue to find the other elements. The resulting matrix is $\begin{pmatrix} 13 & 7 \\ -70 & 50 \end{pmatrix}$. Using *Mathematica*, the product is found as

Input[6]:=
```
Clear[A, B]
A = (3 -1; 0 10);   B = (2 4; -7 5);

MatrixForm[A.B]
```

Output[6]= $\begin{pmatrix} 13 & 7 \\ -70 & 50 \end{pmatrix}$

It is important to note that not all the laws of ordinary scalar algebra hold for matrix algebra. The following are some of the laws of matrix operations.

The Commutative law of addition holds. If **A** and **B** are of the same order, **A** + **B** = **B** + **A**.

The Associative law of addition holds. That is, (**A** + **B**) + **C** =**A** + (**B** + **C**).

The Commutative law of multiplication **does not hold** for product of matrices, except in the case of some special square matrices. That is, in general **A.B** $\neq$ **B.A**.

The Associative law of multiplication holds. That is, (**A.B**).**C** = **A**.(**B.C**).

The Distributive law holds. That is, **A.(B + C)** = **A.B** + **A.C**, and(**B** + **C**).**A** = **B.A** + **C.A**.

The Distributive law of scalar multiplication holds. That is,k(**A** + **B**) = k**A** + k**B**, and (v + k)**A** = v**A** + k**A**.

Example: Input Requirement

A manufacturer is producing three products x, y, and z. Each unit of x requires 5 units of labor and 1 unit of capital. Each unit of y requires 2 units of labor and 4 units of capital. Each unit of z requires 3 units of labor and 2 units of capital. Write the matrix of input requirement. If the manufacturer decides to produce 50 units of x, 20 units of y, and 80 units of z how many units of labor and capital are needed.

Solution: The matrix of input requirements per unit of output (**IR**) is, **IR** = $\begin{pmatrix} 5 & 2 & 3 \\ 1 & 4 & 2 \end{pmatrix}$. The first row of the matrix shows labor requirements per unit of outputs and the second column shows the capital requirements per unit of outputs. The output vector is **Q** = $\begin{pmatrix} 50 \\ 20 \\ 89 \end{pmatrix}$. Therefore, the total input requirement can be found by finding the dot product of the two matrices $\begin{pmatrix} 5 & 2 & 3 \\ 1 & 4 & 2 \end{pmatrix}.\begin{pmatrix} 50 \\ 20 \\ 89 \end{pmatrix} = \begin{pmatrix} 530 \\ 290 \end{pmatrix}$. That is, 530 units of labor and 290 units of capital. Using *Mathematica,* define the matrices and find the product.

Input[7]:=
```
Clear[IR, Q]
```

$$IR = \begin{pmatrix} 5 & 2 & 3 \\ 1 & 4 & 2 \end{pmatrix}; \qquad Q = \begin{pmatrix} 50 \\ 20 \\ 89 \end{pmatrix};$$

```
MatrixForm[IR.Q]
```

Output[7]= $\begin{pmatrix} 557 \\ 308 \end{pmatrix}$

Note that, **IR** is a 2x3 matrix and **Q** is a 3x1 vector. The matrices are conformable for multiplication and the product is a 2x1 vector.

Example: Total Revenue

A firm operating in three locations manufactures and sells four products (A, B, C, and D), which are sold at market prices $P_A = 8$, $P_B = 22$, P_C 13, and $P_D = 16$. Location 1 supplies 350 units of A and 598 units of D. Location 2 supplies 960 units of A, 1020 units of C, and 560 units of D. Location 3 supplies 2040 units of B and 975 units of D. Write the total revenue of the producer in matrix form and find total revenue for each location and the total revenue of the firm.

Input[8]:=
```
Clear[TR, P]
```

$$TR = \begin{pmatrix} 350 & 0 & 0 & 598 \\ 960 & 0 & 1020 & 560 \\ 0 & 2040 & 0 & 975 \end{pmatrix};$$

$$P = \begin{pmatrix} 8 \\ 22 \\ 13 \\ 16 \end{pmatrix};$$

```
MatrixForm[TR.P]
```

Output[8]= $\begin{pmatrix} 12368 \\ 29900 \\ 60480 \end{pmatrix}$

Total Revenue of the Firm = 12368 + 29900 + 60480 = 102748

Direct Multiplication of Matrices: There are many cases in which a column or a row of a matrix needs to be scaled up or down. In these cases direct or "*" multiplication is a useful tool. Direct product of two matrices is a matrix with the elements that are the product of the corresponding elements of the two matrices.

Example: Find the direct product of $\mathbf{A} = \begin{pmatrix} a1 & b1 \\ c1 & d1 \end{pmatrix}$ and $\mathbf{B} = \begin{pmatrix} a2 & b2 \\ c2 & d2 \end{pmatrix}$.

Solution: The direct product of the two matrices is $\begin{pmatrix} a1a2 & b1b2 \\ c1c2 & d1d2 \end{pmatrix}$.

Using *Mathematica*, direct product of the two matrices is,

Input[9]:= $\text{MatrixForm}\left[\begin{pmatrix} a1 & b1 \\ c1 & d1 \end{pmatrix} * \begin{pmatrix} a2 & b2 \\ c2 & d2 \end{pmatrix}\right]$

Output[9]= $\begin{pmatrix} a1\,a2 & b1\,b2 \\ c1\,c2 & d1\,d2 \end{pmatrix}$

Example: Find the direct product of the $\mathbf{A} = \begin{pmatrix} 3 & 1 \\ 7 & 2 \end{pmatrix}$ and $\mathbf{B} = \begin{pmatrix} 5 & 2 \\ 1 & -3 \end{pmatrix}$.

Solution: The direct product of the two matrices is a matrix with the elements equal to the product of the corresponding elements of the two matrices, $\begin{pmatrix} 15 & 2 \\ 7 & -6 \end{pmatrix}$. Using *Mathematica*,

Input[10]:= `MatrixForm[`$\left[\begin{pmatrix} 3 & 1 \\ 7 & 2 \end{pmatrix} * \begin{pmatrix} 5 & 2 \\ 1 & -3 \end{pmatrix}\right]$

Output[10]= $\begin{pmatrix} 15 & 2 \\ 7 & -6 \end{pmatrix}$

Example: Input-Output Models

The input-output table of a two-sector economy is given as,

To Industry:

From Industry	1	2	Consuption	Total
1	15	10	25	50
2	20	8	12	40

find the input-output coefficients.

Solution: The input-output model is a simplified descriptive model of the economy, describing the interrelationships within the producing sectors (here, industries 1 and 2) and the relationship between producing and consuming sectors of the economy. (Input-Output models are studied in Chapter 10). In these models, every industry in the economy buys variable inputs from other industries and sells variable inputs to other industries and final goods to the consuming sectors. The main objective of the input-output analysis is to determine the level of output that each industry should produce in order to satisfy the total demand for the product of the industry in the economy.

The inter-industry matrix (yellow in the table above) represents the quantity of goods sold by industry i to industry j. The last column in the table represents the total goods produced by the sectors. For practical purposes, it is customary to convert the inter-industry matrix of the table to input-output coefficients by dividing each element to corresponding total output. This can be done by finding the direct product of the inter-industry matrix and a conversion matrix. The conversion matrix here is a matrix which its rows are the inverse of the total outputs

$\begin{pmatrix} 15 & 10 \\ 20 & 8 \end{pmatrix} * \begin{pmatrix} \frac{1}{50} & \frac{1}{40} \\ \frac{1}{50} & \frac{1}{40} \end{pmatrix} = \begin{pmatrix} .30 & .25 \\ .40 & .20 \end{pmatrix}$. Using *Mathematica*,

Input[12]:= `N[MatrixForm[`$\left[\begin{pmatrix} 15 & 10 \\ 20 & 8 \end{pmatrix} * \begin{pmatrix} 1/50 & 1/40 \\ 1/50 & 1/40 \end{pmatrix}\right]\right]$

Output[12]= $\begin{pmatrix} 0.3 & 0.25 \\ 0.4 & 0.2 \end{pmatrix}$

Therefore, the input-output table with the input-output coefficients can be written as,

To Industry :

From Industry	1	2	Consuption	Total
1	.30	.25	25	50
2	.40	.20	12	40

Transposition: One operation on matrices that has no equivalent in scalar algebra is called transposition. The transpose of **A**, denoted **A'** or A^t, is obtained from **A** by interchanging rows and columns. That is, the first row of **A** becomes the first column of **A'**, the second row of **A** becomes the second column of **A'** and so on.

Example: Find the transpose of the following matrix, $\mathbf{A} = \begin{pmatrix} a11 & a12 \\ a21 & a22 \\ a31 & a32 \end{pmatrix}$.

Solution: Change the columns and rows of **A** to find **A'** as, $\mathbf{A'} = \begin{pmatrix} a11 & a21 & a31 \\ a12 & a22 & a32 \end{pmatrix}$. In *Mathematica*, use the command **Transpose**[A] as follows,

```
Input[13]:= Clear[A]
            A = (a11 a12
                 a21 a22
                 a31 a32);
            MatrixForm[Transpose[A]]

Output[13]= Null (a11 a21 a31
                  a12 a22 a32)
```

Example : Find the transpose of the following matrix, $\mathbf{A} = \begin{pmatrix} 3 & 2 \\ 0 & 1 \end{pmatrix}$.

Solution: The transpose is $\mathbf{A'} = \begin{pmatrix} 3 & 0 \\ 2 & 1 \end{pmatrix}$. Using *Mathematica*,

```
Input[14]:= Clear[A]
            A = (3 2
                 0 1);
            MatrixForm[Transpose[A]]
Output[14]= (3 0
             2 1)
```

Transposition is often used in matrix multiplication to make two matrices conformable for multiplication. For example if **A** is a matrix of order 3x2 and **B** is a matrix of 4x2, **B** can't be pre-multiplied by **A**. However, transposing **B** turns it to a 2x4 matrix and make it conformable for pre-multiplication by **A**. The product of **A.B'** will be a 3x4 matrix.

Example: Total Spending

A consumer selects a basket of goods containing three items with the prices (**P**) and quantities (**Q**) given as **P** = (2 4 11), **Q** = (5 27 88), respectively. Find the total spending (**TS**) on the goods.

Solution: **P** and **Q** are given as two 1x3 row vector. To find **P.Q**, **Q** should be transposed to a column vector of 3x1. That is, the total spending **TS** = **P.Q'** = $\begin{pmatrix} 2 & 4 & 11 \end{pmatrix} \begin{pmatrix} 5 \\ 27 \\ 88 \end{pmatrix}$ = 1086. **Note** that, the product of two vectors that generates a scalar number is called the **inner product** of the two vectors. The inner product and two other types of vector multiplications are discussed in the section titled "Products of Vectors" in this chapter.

Using *Mathematica*, the inner product of the two vectors is,

```
Input[15]:= Clear[Q, P, TS]
            Q = (2 4 11);    P = (5 27 88);
            TS = P.Transpose[Q]

Output[15]= {{1086}}
```

The theorems of the transposition of matrices are given below, without the proofs. Students are required to verify them using numerical values for **A**, **B**, and **C** **(Note: A' stands for the transpose of A).**

(A')' = A

(A + B)' = A' + B'

(A.B)' = B'.A'

(A.B.C)' = C'.B'.A'

(A.B.C)' = [(A.B).C]' = C'.(A.B)' = C'.B'.A'

Trace The Trace of a square matrix is defined as the sum of the elements in the principal diagonal, $\sum a_{ii}$. The trace of **A** is denoted by tr(**A**).

Example: Find the trace of **A** = $\begin{pmatrix} a11 & a12 & a13 \\ a21 & a22 & a23 \\ a31 & a32 & a33 \end{pmatrix}$.

Solution: Add the diagonal elements, tr(**A**) = a11 + a22 + a33.

In *Mathematica,* use the command **Tr**[A] to find the trace.

Input[16]:=
```
Clear[A]
A = (a11 a12 a13
     a21 a22 a23
     a31 a32 a33);
Tr[A]
```

Output[16]= a11 + a22 + a33

Alternatively, you can use the following command

Input[17]:= `Sum[A[[i, i]], {i, 3}]`

Output[17]= a11 + a22 + a33

Example: Find the trace of **A** = $\begin{pmatrix} 5 & -1 & 0 & 3 \\ 2 & 9 & 21 & 3 \\ -5 & 7 & 9 & 11 \\ 5 & 4 & 6 & 12 \end{pmatrix}$.

Solution: The sum of the elements in the principal diagonal is 35. Using *Mathematica,*

Input[18]:=
```
Clear[A]
A = ( 5 -1  0  3
      2  9 21  3
     -5  7  9 11
      5  4  6 12);
Tr[A]
```

Output[18]= 35

Alternatively, you can use the command

Input[19]:= `Sum[A[[i, i]], {i, 4}]`

Output[19]= 35

Inverse Matrix: Given matrix **A**, if a matrix $\mathbf{A}^{-1}$ exists such that $\mathbf{AA}^{-1} = \mathbf{A}^{-1}\mathbf{A} = \mathbf{I}$, then $\mathbf{A}^{-1}$ is called the inverse of **A**. Here, **I** is a **unit matrix** or **identity matrix.** A unit matrix or identity matrix is a square matrix of order nxn with "1" in the principal diagonal and zeros everywhere else. In order to establish the matrix $\mathbf{A}^{-1}$, we need a few definitions.

Determinant: Associated with any square matrix **A**, there is a scalar quantity called the determinant of **A**, denoted by Det[**A**] or |**A**|. This quantity is obtained by summing various products of the elements of **A**.

Example: Find the determinant of $\mathbf{A} = \begin{pmatrix} a11 & a12 \\ a21 & a22 \end{pmatrix}$.

Solution: Determinant of a 2x2 matrix is the difference between the product of the diagonal and off-diagonal elements. That is, Det[**A**] = a11a22 - a21a12.

In *Mathematica*, use the command **Det**[A] to find the determinant.

Input[20]:=
```
Clear[A]
A = (a11 a12
     a21 a22);
Det[A]
```

Output[20]= -a12 a21 + a11 a22

For higher order matrices, finding determinants require more calculation. There are several ways to find determinant of higher-order matrices. One way, called **column expansion** is as follows:

1- Take a_{11} element of the matrix **A**, delete the corresponding row and column, and form a new matrix ($\mathbf{A}_{11}$) with the remaining elements of the matrix. A determinant formed by deleting a row and a column corresponding to one element of a matrix is called **principal minor**.

2- Take a_{21} element of the matrix **A**, delete the corresponding row and column, and form a new matrix ($\mathbf{A}_{21}$) with the remaining elements of the matrix.

3- Continue this for all the elements of the matrix **A** in column one.

4- Det[**A**] = a_{11}Det[$\mathbf{A}_{11}$]- a_{21}Det[$\mathbf{A}_{21}$] + a_{31}Det[$\mathbf{A}_{31}$] ....

This was a column expansion. Note that, the sign of the a_{ij} coefficients alternates, beginning with positive. The sign of the a_{ij} coefficients follows the sign convention $(-1)^{i+j}$, where i+j is the sum of subscripts in a_{ij}. That is, for a_{11} the sign is $(-1)^2= +1$ and for a_{32} the sign is $(-1)^5 = -1$.

If you start with a_{11} and continue with selecting a_{ij}s in the first row rather than the first column and follow the steps above the method is called **row expansion**. Both row and column expansion will result in the same number for determinant. The question of which one to choose is answered by the magnitude of the numbers in the first row or first column. Obviously, numbers such as zero and one easier to use in numerical multiplication than numbers of larger magnitude.

Example: Find the determinant of $\mathbf{B} = \begin{pmatrix} 3 & 1 & 0 \\ 2 & 5 & 6 \\ 6 & 3 & 1 \end{pmatrix}$

Solution: Using the column expansion, the determinant can be found as: $\text{Det}[\mathbf{B}] = 3\begin{pmatrix} 5 & 6 \\ 3 & 1 \end{pmatrix} - 2\begin{pmatrix} 1 & 0 \\ 3 & 1 \end{pmatrix} + 6\begin{pmatrix} 1 & 0 \\ 5 & 6 \end{pmatrix}$ = 3(5x1 - 3x6) -2(1x1 - 3x0) + 6(1x6 - 5x0) = -5. However, the elements such as zero and one in the first row make it easier to find the determinant using the row expansion method. The row expansion would be: Det[**B**] = $3\begin{pmatrix} 5 & 6 \\ 3 & 1 \end{pmatrix} - 1\begin{pmatrix} 2 & 6 \\ 6 & 1 \end{pmatrix} + 0\begin{pmatrix} 2 & 5 \\ 6 & 3 \end{pmatrix}$ = 3(5x1 - 3x6) -1(2x1 - 6x6) = -5. In *Mathematica*, use the command "Det" to find the determinant,

Input[21]:=
```
Clear[B]
B = (3 1 0
     2 5 6
     6 3 1);
Det[B]
```

Output[21]= `-5`

In physical science the absolute value of the determinant of a 2x2 matrix **A** represents the area of the parallelogram formed by the columns of **A**. That is, for **A** = (***a*** ***b***) = $\begin{pmatrix} 4 & 3 \\ 1 & 2 \end{pmatrix}$ the area of the parallelogram formed by vectors **a** = $\begin{pmatrix} 4 \\ 1 \end{pmatrix}$ and **b** = $\begin{pmatrix} 3 \\ 2 \end{pmatrix}$ is (4)(2) - (1)(3) = 5. The column vectors of a 3x3 matrix form a solid that is a parallelotope. The determinant of the matrix will be the volume of this solid.

Properties of Determinants:

The following properties of determinants are given without proofs. Students are required to verify them by using numerical examples.

1- Det[**A'**] = Det [**A**]

2- Det[**A.B**] = Det[**A**].[**B**]

3- Interchanging any two columns (or rows) of **A** changes the sign of the determinant of **A**.

4- If every element of a row (or a column) of **A** is multiplied by a scalar k to give a new matrix **B**, then Det[**B**] = kDet[**A**].

5- If every element of an nth order matrix **A** is multiplied by k to give a new matrix **B**, then Det[**B**] = k^nDet[**A**].

6- A determinant is unaltered in value when to any row (column) a constant multiple of any other row (column) is added. This property can be used to cut the number of arithmetic operations for finding the determinant.

Co-factors: $c_{ij} = (-1)^{i+j}$ Det[A_{ij}]s are called co-factors (coefficients). Hence, c_{11} = Det[$\mathbf{A}_{11}$], c_{12} = -Det[$\mathbf{A}_{12}$], c_{13} = +Det[$\mathbf{A}_{13}$], and so on.

Inverse of a Matrix:

To find the inverse of a matrix **A**, take the following steps.

1. Find co-factors of **A** and form a new matrix so that its elements are corresponding co-factors of **A**.
2. Transpose the new matrix and call it adj **A**, the **adjoint of A**.
3. Divide all the elements of adj **A** by determinant of **A** to get $\mathbf{A}^{-1}$.

Note that, step 3 is possible only if Det[**A**] ≠ 0, in which case, the matrix **A** is said to be **nonsingular**. When Det[**A**] = 0, the matrix is called a **singular matrix**, and the inverse does not exist.

Example: Find the inverse of the **A** = $\begin{pmatrix} 3 & 1 & 0 \\ 2 & 5 & 4 \\ 6 & 3 & -1 \end{pmatrix}$.

Step1: Find the co-factors,

$c11 = Det\begin{pmatrix} 5 & 4 \\ 3 & -1 \end{pmatrix} = -17$ $c12 = -Det\begin{pmatrix} 2 & 4 \\ 6 & -1 \end{pmatrix} = 26$ $c13 = Det\begin{pmatrix} 2 & 5 \\ 6 & 3 \end{pmatrix} = -24$

$c21 = -Det\begin{pmatrix} 1 & 0 \\ 3 & -1 \end{pmatrix} = 1$ $c22 = Det\begin{pmatrix} 3 & 0 \\ 6 & -1 \end{pmatrix} = -3$ $c23 = -Det\begin{pmatrix} 3 & 1 \\ 6 & 3 \end{pmatrix} = -3$

$c31 = Det\begin{pmatrix} 1 & 0 \\ 5 & 4 \end{pmatrix} = 4$ $c32 = -Det\begin{pmatrix} 3 & 0 \\ 2 & 4 \end{pmatrix} = -12$ $c33 = Det\begin{pmatrix} 3 & 1 \\ 2 & 5 \end{pmatrix} = 13$

Step 2: adj **A** is,

$$\begin{pmatrix} -17 & 26 & -24 \\ 1 & -3 & -3 \\ 4 & -12 & 13 \end{pmatrix}$$

Step 3: Find Det[**A**] and divide the elements of Adj **A** by Det[**A**]. The determinant of **A** is, Det[**A**] = 3(-17) +2(1) + 6(4) = -25. Divide all the elements of adj **A** by -25 to get inverse of **A**.

$$\mathbf{A}^{-1} = \begin{pmatrix} .68 & -.04 & -.16 \\ -1.04 & .12 & .48 \\ .96 & .12 & -.52 \end{pmatrix}$$

Solving through *Mathematica,* you need to use the command **Inverse**[A] as follows,

Input[22]:=
```
Clear[A]
A = (3 1 0
     2 5 4
     6 3 -1);
MatrixForm[Inverse[A]]
```

Output[22]= $\begin{pmatrix} \frac{17}{25} & -\frac{1}{25} & -\frac{4}{25} \\ -\frac{26}{25} & \frac{3}{25} & \frac{12}{25} \\ \frac{24}{25} & \frac{3}{25} & -\frac{13}{25} \end{pmatrix}$

Inverse of a 2x2 Matrix: For a 2x2 matrix **A**, a short cut to find the inverse is to switch a11 with a22, change the sign of a12 and a21, and divide the elements of the new matrix (Adj **A**) by the determinant of **A**.

Example: Find the inverse of $\mathbf{A} = \begin{pmatrix} 2 & 3 \\ 1 & 4 \end{pmatrix}$.

Solution: Switch a11 with a22, change the sign of a12 and a21, and divide all the elements by Det[**A**]. That is, $A^{-1} = \frac{1}{5}\begin{pmatrix} 4 & -3 \\ -1 & 2 \end{pmatrix} = \begin{pmatrix} .8 & -.6 \\ -.2 & .4 \end{pmatrix}$. In *Mathematica,*

Input[23]:=
```
Clear[A]
A = (2 3
     1 4);
MatrixForm[Inverse[A]]
```

Output[23]= $\begin{pmatrix} \frac{4}{5} & -\frac{3}{5} \\ -\frac{1}{5} & \frac{2}{5} \end{pmatrix}$

Properties of Inverse Matrices: The following are properties of inverse matrices. Students are required to demonstrate them using numerical examples.

1- $(\mathbf{A.B})^{-1} = \mathbf{B}^{-1}.\mathbf{A}^{-1}$

2- $(\mathbf{A}')^{-1} = (\mathbf{A}^{-1})'$

3- $\text{Det}[\mathbf{A}^{-1}] = \dfrac{1}{\text{Det}[A]}$

Example: Application in Econometrics

In a regression model **y** = **Xb** + **e**, where **y** is a nx1 vector of dependent variable, **X** is a nxm matrix of independent variables, **b** is a mx1 vector of coefficients, and **e** is a nx1 vector of stochastic errors, **b** can be estimated as $\hat{\boldsymbol{b}}$ = $(\mathbf{X'.X})^{-1}\mathbf{X'.y}$. For the given information on a 3x10 matrix $\mathbf{X'} = \begin{pmatrix} 2 & 1 & 4 & 5 & 12 & 11 & 7 & 9 & 5 & 4 \\ 4 & 5 & 6 & 2 & 4 & 3 & 9 & 11 & 8 & 7 \\ 5 & 3 & 8 & 9 & 9 & 8 & 5 & 8 & 9 & 5 \end{pmatrix}$ and a 1x10 vector $\mathbf{y'} = (8\ \ 3\ \ 16\ \ 20\ \ 66\ \ 58\ \ 34\ \ 42\ \ 25\ \ 15)$ estimate the **b** coefficients.

Solution: $\mathbf{X'.X} = \begin{pmatrix} 482 & 358 & 458 \\ 358 & 421 & 401 \\ 458 & 401 & 519 \end{pmatrix}$, $(X'.X)^{-1} = \begin{pmatrix} 0.0129 & -0.0005 & -0.0110 \\ -0.0005 & 0.0090 & -0.0065 \\ -0.0110 & -0.0065 & 0.0167 \end{pmatrix}$, and $\mathbf{X'.y} = \begin{pmatrix} 2414 \\ 1694 \\ 2221 \end{pmatrix}$. Therefore $\mathbf{b} = \begin{pmatrix} 0.0129 & -0.0005 & -0.0110 \\ -0.0005 & 0.0090 & -0.0065 \\ -0.0110 & -0.0065 & 0.0167 \end{pmatrix} . \begin{pmatrix} 2414 \\ 1694 \\ 2221 \end{pmatrix} = \begin{pmatrix} 5.86 \\ -0.42 \\ -0.57 \end{pmatrix}$

In *Mathematica*, define the matrices and formulate the **b** as follows,

Input[24]:= `Clear[X, y, b]`

$$\mathbf{X} = \begin{pmatrix} 2 & 4 & 5 \\ 1 & 5 & 3 \\ 4 & 6 & 8 \\ 5 & 2 & 9 \\ 12 & 4 & 9 \\ 11 & 3 & 8 \\ 7 & 9 & 5 \\ 9 & 11 & 8 \\ 5 & 8 & 9 \\ 4 & 7 & 5 \end{pmatrix}; \quad y = \begin{pmatrix} 8 \\ 3 \\ 16 \\ 20 \\ 66 \\ 58 \\ 34 \\ 42 \\ 25 \\ 15 \end{pmatrix};$$

```
b = N[Inverse[Transpose[X].X].(Transpose[X].y)]
```

Output[24]= `{{5.85581}, {-0.415783}, {-0.566918}}`

Product of Vectors: Three types of vector products are **inner product**, **outer product**, and **cross product** of vectors.

A *Mathematica* Note: You can generate vectors using either the Palettes function of the *Mathematica* or by listing through the curved brackets { }. It is important to note that, vectors generated through curved brackets can be used more easily in vector operations that will follow in this chapter (commands such as Dot and Cross). If you produce vectors through the Palettes function of the *Mathematica*, vectors have to be conformable for matrix or vector operations. Otherwise *Mathematica* will return an error message. For this reason, vectors are generated through the brackets in the following sections.

Inner Product or Dot Product of Vectors: The dot product or the **inner product** of two vectors **a** = {a1, a2, a3} and **b** = {b1, b2, b3}is a scalar number found by multiplying corresponding elements of the two vectors and adding up. Here, the dot product of **a.b** = a1b1 + a2b2 + a3b3.

In *Mathematica,* you can use the `Dot` [a , b] command of the ***Mathematica*** to find the inner product of two vectors **a** and **b**. If you generate vectors using pallets function of the *Mathematica*, the two vectors should be conformable for inner product. That is, if **a** and **b** are both a 3x1 row vectors, to find their inner product you should multiply **a** by **b'** to get a scalar number.

Example: Total Revenue

The annual average price and the total quantity sold of a product is given as **p** = {28.50, 32.25, 33.15} and **q** = { 23400, 22150, 26300}, respectively. Find the total revenue (TR).

Solution: Total revenue is price time quantity or the inner product of price and quantity vectors, TR = **p.q'** = 28.50(23400) + 32.25(22150) + 33.15(26300) = 2,253,080.

Using *Mathematica,*

```
Input[25]:= Clear[p, q, TR]
            p = {28.50, 32.25, 33.15};
            q = {23400, 22150, 26300};
            TR = p.q
```

Output[25]= 2.25308×10^6

Alternatively,

```
Input[26]:= TR = Dot[p, q]
```

Output[26]= 2.25308×10^6

Note again that, if you are solving the problem using *Mathematica*, you should generate the vectors using the brackets "{ }" rather than the Pallets. Otherwise, you can't use the *Mathematica* command "Dot" or multiply the vectors using "." without making the vectors conformable for multiplication for dot product.

Outer Product of Vectors: The outer product of two vectors **a** and **b** is a matrix which its elements are the product of the corresponding elements of the two vectors. For example, the outer product of the 3x1 column vector **a** = $\begin{pmatrix} 3 \\ 1 \\ -2 \end{pmatrix}$ and the 1x4 row vector **b** = $\begin{pmatrix} 2 \\ 0 \\ 5 \\ 7 \end{pmatrix}$ is a 3x4 matrix that its elements are the product of the corresponding elements of the two vectors **a.b'** = $\begin{pmatrix} 3 \\ 1 \\ -2 \end{pmatrix}\begin{pmatrix} 2 & 0 & 5 & 7 \end{pmatrix} = \begin{pmatrix} 6 & 0 & 15 & 21 \\ 2 & 0 & 5 & 7 \\ -4 & 0 & -10 & -14 \end{pmatrix}$.

Using *Mathematica*, the outer product of the two vectors is,

```
Input[27]:= Clear[a, b]
```

$$a = \begin{pmatrix} 3 \\ 1 \\ -2 \end{pmatrix}; \quad b = \begin{pmatrix} 2 & 0 & 5 & 7 \end{pmatrix};$$

```
            MatrixForm[a.b]
```

Output[27]= $\begin{pmatrix} 6 & 0 & 15 & 21 \\ 2 & 0 & 5 & 7 \\ -4 & 0 & -10 & -14 \end{pmatrix}$

Cross Product of Vectors:

The cross product **u** × **v** of two vectors **u** and **v** is a vector. The cross product is defined if the vector **u** and **v** are three-dimensional vectors or higher. To find the cross product of two vectors **u** = {a1, b1, c1} and **v** ={a2, b2, c2}, write the two vectors in a matrix form, with the first row as a defining row with the i, j, k elements. Select the ith element, delete the corresponding row and column, and find the determinant of remaining matrix. This will be the first element of the cross product. Select the jth element, delete the corresponding row and column, and find the determinant of the remaining matrix. Change the sign of this number and make it the second element of the cross product. Continue with the process to find the third element of the cross product vector. Note that, the sign of the elements alternate, starting with no sign change for the first element.

Example: Find the cross product of **u** = {a1, b1, c1} and **v** = {a2, b2, c2}.

Solution: Write the two vectors in a matrix form as $\begin{pmatrix} i & j & k \\ a1 & b1 & c1 \\ a2 & b2 & c2 \end{pmatrix}$. Delete the column and row corresponding to the ith element to get $\begin{pmatrix} b1 & c1 \\ b2 & c2 \end{pmatrix}$. The determinant of this matrix is b1c2 - b2c1 which is the first element of the cross product **u** × **v**. Delete the row and column corresponding to the jth element to get $\begin{pmatrix} a1 & c1 \\ a2 & c2 \end{pmatrix}$. The determinant of this matrix is a1c2 - a2c1. The negative of this number is the second element of the cross product. Find the third element by deleting the row and column corresponding to the kth and finding the determinant. The cross product is **u** × **v** = { b1c2 – b2c1, a1c2 – a2c1, a1b2 – a2b1 }. **Note** that, the sign of the elements alternate, starting no sign change for the first element). Using *Mathematica*,

```
Input[28]:= Clear[u, v]
            u = {a1, b1, c1};
            v = {a2, b2, c2};
            MatrixForm[u×v]
```

Output[28]= $\begin{pmatrix} -b2\ c1 + b1\ c2 \\ a2\ c1 - a1\ c2 \\ -a2\ b1 + a1\ b2 \end{pmatrix}$

Not 1: The "×" in the cross product is not the alphabet character "x" of the keyboard. It is the "×" in the "3 BasicInput" of "Palettes" .

Note 2: For cross and dot multiplication, generate the vectors using curved bracket, Don't generate it using the Palettes. Alternatively, you can use the Cross command.

```
Input[29]:= MatrixForm[Cross[u, v]]
```

Output[29]= $\begin{pmatrix} -b2\ c1 + b1\ c2 \\ a2\ c1 - a1\ c2 \\ -a2\ b1 + a1\ b2 \end{pmatrix}$

Conversion of Vectors using direct Product:

The direct product or element by element product of two vectors produces a vector with elements which are the product of the corresponding elements of the two vectors. For example, the product of **u** = {a1, b1, c1} and **v** = {a2, b2, c2} is a vector **w** = **u*v** = {a1a2, b1b2, c1c2}.

Example: Find u*v for u = {2, 5, 12} and v = {3, -1, 10}

Solution: u*v = {6, -5, 120}. Using *Mathematica*, the direct product is,

```
Input[30]:= Clear[u, v]
            u = {2, 5, 12};
            v = {3, -1, 10};
            u * v

Output[30]= {6, -5, 120}
```

Problems

1- Given

$$A = \begin{pmatrix} 1 & 1 \\ 0 & -1 \end{pmatrix} B = \begin{pmatrix} 2 & -1 \\ 0 & 1 \end{pmatrix} \qquad C = \begin{pmatrix} 2 & 2 \\ 0 & -1 \end{pmatrix}$$

a- Find **A** + **C'** and **A'** - **C**.

b- Find **A.B'** and **C'. B'**.

c- Find $\mathbf{A'.C^{-1}A}$ and $\mathbf{B.C^{-1}.A}$.

d- Show that [(**A.B**)'.**C'**]' = **C.A.B**

2- Find the inverse of the following matrices,

$$\mathbf{A} = \begin{pmatrix} 1 & 0 \\ 3 & 2 \end{pmatrix} \qquad \mathbf{B} = \begin{pmatrix} 1 & 3 \\ 2 & 10 \end{pmatrix}$$

$$\mathbf{C} = \begin{pmatrix} 1 & 3 & 5 \\ 0 & 2 & 4 \\ 4 & 2 & 6 \end{pmatrix} \qquad \mathbf{D} = \begin{pmatrix} 1 & 0 & 1 \\ 0 & 1 & 1 \\ 0 & 0 & 1 \end{pmatrix}$$

3 Write the following system of equations

$$2x_1 + x_2 + 3x_3 = 8$$

$$x_1 + x_2 - x_3 = 2$$

$$3x_1 - 2x_2 + x_3 = 5$$

in the matrix form **A.X** = **C**, where **A** is the matrix of constant coefficients, $X' = [x_1 \ \ x_2 \ \ x_3]$, and **C** is the vector of constants **C'** = (8 2 5), and find $\mathbf{X = A^{-1} \cdot C}$.

II. Special Matrices: The IdentityMatrix Command

The following special matrices will be frequently referred to and used in solving problems.

Identity Matrix: The unit or identity matrix is a square matrix of order nxn with units in the principal diagonal and zeros everywhere else. The following matrix is a 4x4 identity matrix.

$$\mathbf{I}=\begin{pmatrix}1&0&0&0\\0&1&0&0\\0&0&1&0\\0&0&0&1\end{pmatrix}$$

It is easy to verify that **I.A** = **A.I** = **A**.

In *Mathematica* you can generate identity matrix using command IdentityMatrix[n], where n is the order of the identity matrix.

Example: Generate a 4x4 identity matrix.

Input[31]:= `MatrixForm[IdentityMatrix[4]]`

Output[31]= $\begin{pmatrix}1&0&0&0\\0&1&0&0\\0&0&1&0\\0&0&0&1\end{pmatrix}$

Scalar Matrix: Scalar matrices have a common scalar element in the principal diagonal and zeros everywhere else. The following matrix is a 4x4 scalar matrix with a common scalar element 7.

$$\begin{pmatrix}7&0&0&0\\0&7&0&0\\0&0&7&0\\0&0&0&7\end{pmatrix}$$

Scalar multiplication of a matrix **A** is equal to scalar matrix multiplication of that matrix. That is, k**A** = (k**I**)**A** = **A**(k**I**) = **A**k.

Example: Using *Mathematica*, create a 3x3 scalar matrix with a scalar unit 5.

Input[32]:= `MatrixForm[5 IdentityMatrix[3]]`

Output[32]= $\begin{pmatrix}5&0&0\\0&5&0\\0&0&5\end{pmatrix}$

Diagonal Matrix: A diagonal matrix has scalar elements, not necessarily equal, in the principal diagonal and zeros in the off-diagonal positions. The following matrix is an example of a 4x4 diagonal matrix.

$$\begin{pmatrix}5&0&0&0\\0&-2&0&0\\0&0&1&0\\0&0&0&6\end{pmatrix}$$

Symmetric Matrix: If **A** = **A'**, then **A** is called a symmetric matrix. It is obvious that **A** should be a square matrix. For example, if

$$\mathbf{A}=\begin{pmatrix}5&3&8\\3&7&0\\8&0&4\end{pmatrix} \qquad \mathbf{A}'=\begin{pmatrix}5&3&8\\3&7&0\\8&0&4\end{pmatrix} \qquad \text{and } \mathbf{A}=\mathbf{A}'$$

Orthogonal Matrix: A matrix **A** is said to be orthogonal if Det[**A**] $\neq 0$ and **A'A** = **I**.

Example: Show that $\mathbf{A} = \begin{pmatrix} \frac{\sqrt{2}}{2} & -\frac{\sqrt{2}}{2} \\ \frac{\sqrt{2}}{2} & \frac{\sqrt{2}}{2} \end{pmatrix}$ is an orthogonal matrix.

Solution: Det[**A**] = 1≠ 0 and **A'.A** = $\begin{pmatrix} 1 & 0 \\ 0 & 1 \end{pmatrix}$. Therefore, **A** is an orthogonal matrix.

Input[33]:=
```
Clear[A]
A = ( √2/2   -√2/2
      √2/2    √2/2 );
MatrixForm[A.Transpose[A]]
```

Output[33]= $\begin{pmatrix} 1 & 0 \\ 0 & 1 \end{pmatrix}$

Orthogonal Vectors: Two vectors **u** and **v** of the same order are called **orthogonal vectors** or **perpendicular vectors** if **u'.v** = 0.

Example: Show that **u** = $\begin{pmatrix} 2 \\ -3 \\ 1 \end{pmatrix}$ and **v** = $\begin{pmatrix} 1 \\ 2 \\ 4 \end{pmatrix}$ are orthogonal vectors.

Solution: u'v = $\begin{pmatrix} 2 & -3 & 1 \end{pmatrix}\begin{pmatrix} 1 \\ 2 \\ 4 \end{pmatrix}$ = 0. Using *Mathematica*,

Input[34]:=
```
Clear[u, v]
u = ( 2        v = ( 1
     -3 );           2 );
      1 )            4 )

Transpose[u].v
```

Output[34]= {{0}}

Idempotent Matrix: The matrix **A** is an idempotent matrix if **A.A** = **A**. That is, an idempotent matrix is one that upon dot multiplication by itself remains unchanged.

Example: Prove that $\mathbf{A} = \begin{pmatrix} \frac{2}{3} & \frac{1}{3} & \frac{1}{3} \\ \frac{1}{3} & \frac{2}{3} & -\frac{1}{3} \\ \frac{1}{3} & -\frac{1}{3} & \frac{2}{3} \end{pmatrix}$ is an Idempotent matrix.

Solution: **A.A** = **A**, thus **A** is an idempotent matrix. Using *Mathematica*,

Input[35]:= MatrixForm[$\begin{pmatrix} \frac{2}{3} & \frac{1}{3} & \frac{1}{3} \\ \frac{1}{3} & \frac{2}{3} & -\frac{1}{3} \\ \frac{1}{3} & -\frac{1}{3} & \frac{2}{3} \end{pmatrix} . \begin{pmatrix} \frac{2}{3} & \frac{1}{3} & \frac{1}{3} \\ \frac{1}{3} & \frac{2}{3} & -\frac{1}{3} \\ \frac{1}{3} & -\frac{1}{3} & \frac{2}{3} \end{pmatrix}$]

Output[35]= $\begin{pmatrix} \frac{2}{3} & \frac{1}{3} & \frac{1}{3} \\ \frac{1}{3} & \frac{2}{3} & -\frac{1}{3} \\ \frac{1}{3} & -\frac{1}{3} & \frac{2}{3} \end{pmatrix}$

Idempotent Matrices and Statistics: A very useful device for data manipulation in statistics is a vector **i** which contains a column of 1s. This vector can be operated on vectors of the same order, say **x**, to find the sum, mean, and deviations from mean. The sum of the elements of **x** is $\sum_{i=1}^{n} x_i = x_1 + x_2 + \ldots + x_n =$ **i'x**. Since $\sum_{i=1}^{n} x_i =$

$n\bar{x}$, $\bar{x} = \frac{1}{n}\mathbf{i'x}$. The set of deviations from mean, then, is $\mathbf{x} - \mathbf{i}\frac{1}{n}\,\mathbf{i'x} = .$ Since $\frac{1}{n} = (\boldsymbol{i}'\,\boldsymbol{i})^{-1}$, thus $\mathbf{x} - \mathbf{i}\frac{1}{n}\,\mathbf{i'x} = \mathbf{x} - \mathbf{i}(\boldsymbol{i}'\,\boldsymbol{i})^{-1}\mathbf{i'x} = (\mathbf{I} - \mathbf{i}(\boldsymbol{i}'\,\boldsymbol{i})^{-1}\mathbf{i'})\mathbf{x}$, where **I** is a unit matrix. Denoting $(\mathbf{I} - \mathbf{i}(\boldsymbol{i}'\,\boldsymbol{i})^{-1}\mathbf{i'})$ by $\boldsymbol{M^o}$, $\boldsymbol{M^o} = (\mathbf{I} - \mathbf{i}(\boldsymbol{i}'\,\boldsymbol{i})^{-1}\mathbf{i'})$ is called the **fundamental matrix** in statistics and is used in transforming data to deviation from their mean.

Example: Variance

Prove that the variance of a variable **x**, containing n observations and defined as Var(**x**) = $\sum_{i=1}^{n}(x_i - \bar{x})^2$ can be written as $\mathbf{x'}\boldsymbol{M^o}\mathbf{x}$, where $\boldsymbol{M^o} = (\mathbf{I} - \mathbf{i}(\boldsymbol{i}'\,\boldsymbol{i})^{-1}\mathbf{i'})$.

Solution: The set of deviations from mean can be written as $\boldsymbol{M^o}\mathbf{x}$. The variance, which is the sum square of deviations from mean, can be written in matrix as $(\boldsymbol{M^o}\mathbf{x})'(\boldsymbol{M^o}\mathbf{x}) = \mathbf{x'}\boldsymbol{M^{o'}}\boldsymbol{M^o}\mathbf{x}$. Since M^o is an idempotent matrix, $\boldsymbol{M^{o'}}\boldsymbol{M^o} = \boldsymbol{M^o}$, and Var(x) = $\mathbf{x'}\boldsymbol{M^o}\mathbf{x}$.

Problems

1. Show that $\mathbf{X}.(\mathbf{X'}.\mathbf{X})^{-1}.\mathbf{X'}$ and $\mathbf{I} - \mathbf{X}.(\mathbf{X'}.\mathbf{X})^{-1}.\mathbf{X'}$ are idempotent matrices.

2. Show that the following matrix is an orthogonal matrix.

$$\begin{pmatrix} \frac{\sqrt{3}}{3} & -\frac{\sqrt{2}}{2} & \frac{\sqrt{6}}{6} \\ \frac{\sqrt{3}}{3} & \frac{\sqrt{2}}{2} & \frac{\sqrt{6}}{6} \\ \frac{\sqrt{3}}{3} & 0 & -\frac{\sqrt{6}}{6} \end{pmatrix}$$

3. Find vectors that are orthogonal to the following vectors

   a) $\mathbf{u'} = (1 \;\; 0 \;\; -1)$

   b) $\mathbf{u'} = (3 \;\; -2 \;\; 5)$

4. Using the fundamental matrix $\boldsymbol{M^o} = (\mathbf{I} - \mathbf{i}(\boldsymbol{i}'\,\boldsymbol{i})^{-1}\mathbf{i'})$,

   a) Prove that M^o is an idempotent matrix.

   b) Prove that sum deviations from mean of any variable equals zero.

5. Prove that the covariance between two variables **x** and **y**, defined as Cov(x, y) = $\sum_{i=1}^{n}(x_i - \bar{x})(y_i - \bar{y})$ can be written as $\mathbf{x'}\boldsymbol{M^o}\mathbf{y}$, where $\boldsymbol{M^o} = (\mathbf{I} - \mathbf{i}(\boldsymbol{i}'\,\boldsymbol{i})^{-1}\mathbf{i'})$.

CHAPTER 5

SPECIAL TOPICS IN MATRIX ALGEBRA

This chapter extends the algebra matrices to some more advanced topics in linear algebra that have great deal of applications in statistics, econometrics, and solutions of dynamic models that apply difference and differential equations. The main topics covered in this chapter are rank and linear dependence, solution of system of equations, eigenvalues and eigenvectors, and quadratic forms.

TOOL BOX

Mathematica Commands	Function
Solve[$\begin{pmatrix} a & b \\ c & d \end{pmatrix}.\begin{pmatrix} x \\ y \end{pmatrix}==\begin{pmatrix} e \\ f \end{pmatrix}$, {x, y}]	Solves the system of simultaneous equations for x & y.
Simplify[expr]	Simplifies the "expr" in the argument
Eigenvalues[A]	Gives Eigenvalues of matrix A
Eigenvectors[A]	Gives eigenvectors of A
Eigensystem[A]	Gives eigenvalues and the corresponding eigenvectors of A

I. Rank and Linear Dependence

Consider the set of homogenous equations **Ax** = **0**, where **A** is a square matrix of mxm coefficients, and **x** is a vector of mx1 unknowns. If **A** is nonsingular, Det[**A**] ≠ 0, the only solution to **Ax** = **0** will be a trivial solution of **x** = **0**. The m columns of **A** then are said to be linearly independent, the rank of **A** is m, r(**A**) = m, and **A** is said to be of full rank. It follows that the m rows of **A** are also linearly independent.

Example: Solve $\begin{pmatrix} 1 & 3 & -1 \\ 0 & 1 & 2 \\ 4 & 1 & 2 \end{pmatrix}\begin{pmatrix} x1 \\ x2 \\ x3 \end{pmatrix} = \begin{pmatrix} 0 \\ 0 \\ 0 \end{pmatrix}$ for x1, x2 , and x3.

Solution: Solving the three equations simultaneously results in x1 = x2 = x3 = 0. This implies that the coefficients matrix (**A**) is non-singular and the **A** matrix has a full-rank, r(**A**) = 3. Using *Mathematica*,

```
In[1]:= Clear[A, x, x1, x2, x3]
        A = (1  3  -1
             0  1   2
             4  1   2);        x = (x1
                                     x2
                                     x3);
        Det[A]

Out[1]= 28

In[2]:= Solve[(1  3  -1      (x1        (0
               0  1   2   .   x2    ==   0  , {x1, x2, x3}]
               4  1   2)      x3)        0)

Out[2]= {{x1 → 0, x2 → 0, x3 → 0}}
```

In this example, Det[**A**] = 28≠ 0 and the only solution to the system of equations is $x_1 = x_2 = x_3 = 0$. The columns (and the rows) of **A** are linearly independent and **A** has the full rank of r(**A**) = 3. However, if Det**A**] = 0, then **Ax** = **0** would have a nontrivial solution and the m columns (or rows) of **A** would be linearly dependent and the rank of **A** would be less than 3, r(**A**) < 3.

Example: Given that $\mathbf{A} = \begin{pmatrix} 2 & 3 & 4 \\ -1 & 0 & 1 \\ 1 & 1 & 1 \end{pmatrix}$, find the solution of **Ax** = **0**

Solution: Solving the equations simultaneously results in x1 = x3 and x2 = -2x3. The determinant of matrix **A** is zero and **A** is a singular matrix, with a rank less than 3. The equations provide a set of solutions for xs not all equal to zero. For example, for $x_3 = 1$, x1 = 1 and x2 = -2. Also, any other scalar multiple of these values will be a solution to the equation. Furthermore, the inspection of the matrix shows that any pair of vectors is linearly independent. Hence, the rank of matrix **A** is two. Using *Mathematica,*

```
In[2]:= Clear[A, x, x1, x2, x3]
        A = ( 2  3  4
             -1  0  1
              1  1  1 );        x = ( x1
                                      x2
                                      x3 );

        Det[A]

Out[2]= 0

In[3]:= Solve[A.x == 0, {x1, x2, x3}]

Solve::svars : Equations may not give solutions for all "solve" variables.

Out[3]= {{x1 → x3, x2 → -2 x3}}
```

In general, the rank of a matrix is defined as the maximum number of linearly independent columns or rows.

II. Solution of System of Equations: The Solve[$\begin{pmatrix} a & b \\ c & d \end{pmatrix}.\begin{pmatrix} x \\ y \end{pmatrix}==\begin{pmatrix} e \\ f \end{pmatrix}$, {x, y}] Command

If **A** is a nonsingular matrix, the set of equations **Ax** = **h** has the solution **x** = A^{-1}**h**. This solution yields an easy way of finding the **x** vector, called **Cramer's Rule**. Every x_i is found as the ratio of two determinants, the numerator of which is the determinant of **A** matrix in which the ith column is replaced with the **h** vector, and the denominator is Det[**A**].

Example: Find the solution of the system of equations $\begin{pmatrix} 2 & 3 & -1 \\ 1 & 2 & 1 \\ 1 & -5 & 3 \end{pmatrix}\begin{pmatrix} x1 \\ x2 \\ x3 \end{pmatrix} = \begin{pmatrix} -2 \\ 3 \\ 16 \end{pmatrix}$.

Solution: To find x_1, replace the first column of **A** with **h** and find the determinant of the new matrix. x_1 is found by dividing this determinant by Det[**A**]. That is, $x1 = \frac{\text{Det}[\begin{pmatrix} -2 & 3 & -1 \\ 3 & 2 & 1 \\ 16 & -5 & 3 \end{pmatrix}]}{\text{DEt}[\begin{pmatrix} 2 & 3 & -1 \\ 1 & 2 & 1 \\ 1 & -5 & 3 \end{pmatrix}]} = \frac{46}{23} = 2$. To find x_2, replace the second column of **A** with **h** and find the determinant of the new matrix. x_2 is found by dividing this determinant by

Det[**A**]. That is, x2 = $\frac{\text{Det}\left[\begin{pmatrix} 2 & -2 & -1 \\ 1 & 3 & 1 \\ 1 & 16 & 3 \end{pmatrix}\right]}{\text{Det}\left[\begin{pmatrix} 2 & 3 & -1 \\ 1 & 2 & 1 \\ 1 & -5 & 3 \end{pmatrix}\right]} = \frac{-23}{23}$ = -1. To find x_3, replace the third column of **A** with **h** and find the determinant of the new matrix. x_3 is found by dividing this determinant by Det[**A**]. That is, x3 = $\frac{\text{Det}\left[\begin{pmatrix} 2 & 3 & -2 \\ 1 & 2 & 3 \\ 1 & -5 & 16 \end{pmatrix}\right]}{\text{Det}\left[\begin{pmatrix} 2 & 3 & -1 \\ 1 & 2 & 1 \\ 1 & -5 & 3 \end{pmatrix}\right]}$ = $\frac{69}{23} = 3$.

Cramer's rule provides a general rule for solving system of equations and is useful in derivation of some economic concepts that deals with the solution of system of equations (in Chapter 10 we use the rule to derive the Slutsky equation.)

To solve the same system of equations using *Mathematica,* use the Solve command as below,

```
In[4]:= Clear[A, B, x]
        A = (2  3  -1
             1  2   1
             1 -5   3);   x = (x1
                                 x2
                                 x3);   B = (-2
                                              3
                                              16);

        Solve[A.x == B, {x1, x2, x3}]

Out[4]= {{x1 → 2, x2 → -1, x3 → 3}}
```

An alternative way of solving the same system of equations is to use the Solve by listing the equations as follows

```
In[5]:=
Solve[{2 x1 + 3 x2 - x3 == -2, x1 + 2 x2 + x3 == 3, x1 - 5 x2 + 3 x3 == 16}, {x1, x2, x3}]

Out[5]= {{x1 → 2, x2 → -1, x3 → 3}}
```

Example: Aggregate Demand (AD) Aggregate Supply (AS) Model

In a simple AD-AS model of the economy, AD and AS are expressed as functions of price level (P) and interest rate (r): AD = 100 -.2P -.8r and AS = 18 + .45P + 1.2r. Solve the system for the equilibrium condition in terms of r and find the equilibrium output and price levels for r = 5% and r = 10%.

Solution: The system can be written as $\begin{pmatrix} \text{AD} \\ \text{AS} \end{pmatrix} = \begin{pmatrix} -.2 & -.8 \\ .45 & 1.2 \end{pmatrix}\begin{pmatrix} P \\ r \end{pmatrix} + \begin{pmatrix} 100 \\ 18 \end{pmatrix}$. Since at the equilibrium point AD = AS, set AD = AS and solve for P in terms of r. That is, P = 126.15 - 3.08r. Substitute for P in the AD equation to find AD in terms of r only. That is, AD = 74.77 - .18r. Using these two equations, AD and P can be found for r = 5% and r = 10%. For r = 5%, the equilibrium output (AD) is 74.76 and the equilibrium price level is 126. For r = 10%, the equilibrium output is 74.75 and the equilibrium price level is 125.85. Using *Mathematica,*

```
In[6]:= Clear[AD, AS, p, r]
        sol = Solve[(AD
                     AS) == (-.2  -.8
                             .45  1.2).(p
                                        r) + (100
                                              18), {AD, p}]

Out[6]= {{AD → 100. - 0.444444 (-18. + AS - 1.2 r) - 0.8 r,
          p → 2.22222 (-18. + AS - 1.2 r)}}
```

In[7]:= `Simplify[sol]`

Out[7]= `{{AD → 108. - 0.444444 AS - 0.266667 r, p → -40. + 2.22222 AS - 2.66667 r}}`

In[8]:= `sol /. r → .05`

Out[8]= `{{AD → 99.96 - 0.444444 (-18.06 + AS), p → 2.22222 (-18.06 + AS)}}`

In[9]:= `sol /. r → .10`

Out[9]= `{{AD → 99.92 - 0.444444 (-18.12 + AS), p → 2.22222 (-18.12 + AS)}}`

Do More Practice: Solve the preceding problem for r = 15% and 20%.

2- Make a list of paired observations on r and p for the r values of 5%, 10%, 15%, and 20%. List plot the relationship between r and p. Comment on the slope of the plot. What does it imply on the relationship between interest rate and inflation?

3- Make a list of paired observations on r and AD for the r values of 5%, 10%, 15%, and 20%. List plot the relationship between r and AD. Comment on the slope of the plot. What does it imply on the relationship between interest rate and output (AD)?

Problems

1. Find the rank of the following matrices

$$\mathbf{A} = \begin{pmatrix} 1 & 0 & 1 \\ 0 & 2 & 3 \\ -1 & 0 & 2 \end{pmatrix} \qquad \mathbf{B} = \begin{pmatrix} 1 & 2 & 1 \\ 3 & 0 & 5 \\ 2 & 10 & 0 \end{pmatrix}$$

$$\mathbf{C} = \begin{pmatrix} 1 & 0 & 2 \\ 0 & 2 & 1 \\ 2 & 2 & -1 \end{pmatrix} \qquad \mathbf{D} = \begin{pmatrix} 2 & 2 & -6 \\ 0 & 1 & 3 \\ 1 & 3 & 3 \end{pmatrix}$$

$$\mathbf{E} = \begin{pmatrix} 1 & 65 & -3 \\ -2 & -13 & 6 \\ 1 & 65 & -3 \end{pmatrix} \qquad \mathbf{F} = \begin{pmatrix} 1 & 0 & 0 \\ 0 & 1 & 0 \\ 0 & 0 & 1 \end{pmatrix}$$

2. Solve the following systems of equations

a) $2x + 3y = 8$

$-x + 2y = 3$

$2x + 3y + z = 8$

b) $x - y + 2z = -1$

$x + 2y - 3z = 5$

$x - 2y + z = 1$

c) $2x + 3y - z = 3$

$x + y + 2z = 6$

$x + y + 2z = 5$

d) $2x - y + z = 2$

$x + 4y + 5z = 11$

3. Demand and supply for a product are given as Q_d = 125 - 2P + .03y and Q_s = 5 + P, respectively. Q_d is the quantity demanded, which is expressed in terms of the price of the product (P) and income (Y). Q_s is the quantity supplied.

a) Solve the market model for the equilibrium quantity and price in terms of income.

b) Find the quantity demanded at the income levels of 500, 1000, 1500, and 2000.

c)Make a list plot of the relationship between income and quantity for the income levels mentioned in (b).

d) Comment on the relationship between income and quantity demanded.

e) Find the income elasticity of demand between income level of 1000 and 1500. (Note: income elasticity of demand is define as the percent change in quantity demanded over percent change in income, $\xi_y = \frac{\Delta Q\%}{\Delta Y\%}$).

III. Eigenvalues and Eigenvectors: The Eigenvalues[A] & Eigenvectors[A] Commands

The eigenvalue problem is defined as that of finding values of a scalar k such that **Ax** = kx, where **A** is some nxn matrix, k is the eigenvalue of **A**, and **x** , (**x** ≠ **0**), is the eigenvector associated with k. The eigenvalue problem is the basic tool for diagonalizing matrices, finding solutions of differential equations, and solving many other problems in calculus and matrix algebra.

Equation **Ax** = kx can be written as (**A** - k**I**)**x** = **0** and has a nontrivial solution if (**A** - kx) is singular, or Det[**A** - k**I**] = 0. This equation yields a polynomial in the unknown k, which may be solved for k. The solution of this equation will render the eigenvalue(s) k and the corresponding eigenvector(s) **x**. Alternative names of eigenvalues and eigenvectors are **characteristic roots** and **characteristic vectors** and **latent roots** and **latent vectors**.

Example: Find the eigenvalues and the eigenvectors of **A** $=\begin{pmatrix} 1 & 1 \\ 4 & 1 \end{pmatrix}$.

Solution: Equation (**A** - k**I**)**x** $= \left(\begin{pmatrix} 1 & 1 \\ 4 & 1 \end{pmatrix} - \begin{pmatrix} k & 0 \\ 0 & k \end{pmatrix}\right)\begin{pmatrix} x1 \\ x2 \end{pmatrix} = \begin{pmatrix} 0 \\ 0 \end{pmatrix}$ will have nontrivial solution if the determinant of $\left|\begin{pmatrix} 1 & 1 \\ 4 & 1 \end{pmatrix} - \begin{pmatrix} k & 0 \\ 0 & k \end{pmatrix}\right| = 0$. This will render a polynomial, $(1 - k)^2 - 4 = 0$, the solution set of which is $k_1 = -1$ and $k_2 = 3$. To find the eigenvectors corresponding to these two eigenvalues, substitute for the value of k in the equation (**A** - k**I**)**x** $= \left(\begin{pmatrix} 1 & 1 \\ 4 & 1 \end{pmatrix} - \begin{pmatrix} k & 0 \\ 0 & k \end{pmatrix}\right)\begin{pmatrix} x1 \\ x2 \end{pmatrix} = \begin{pmatrix} 0 \\ 0 \end{pmatrix}$ and solve for x1 and x2. For $k_1 = -1$, the equation yields $\begin{pmatrix} 2 & 1 \\ 4 & 2 \end{pmatrix}\begin{pmatrix} x1 \\ x2 \end{pmatrix} = \begin{pmatrix} 0 \\ 0 \end{pmatrix}$, the solution of which is $2x_1 + x_2 = 0$. For any arbitrary value of x_1 there is a value for x_2. For example, if $x_1 = -1$, then $x_2 = 2$. Therefore, the eigenvector corresponding to k1 = -1 is **x1** $= \begin{pmatrix} 1 \\ -2 \end{pmatrix}$, or any scalar multiple of **x**1.

For k2 = 3, the equation (**A** - k**I**)**x** $= \left(\begin{pmatrix} 1 & 1 \\ 4 & 1 \end{pmatrix} - \begin{pmatrix} k & 0 \\ 0 & k \end{pmatrix}\right)\begin{pmatrix} x1 \\ x2 \end{pmatrix} = \begin{pmatrix} 0 \\ 0 \end{pmatrix}$ yields $\begin{pmatrix} -2 & 1 \\ 4 & -2 \end{pmatrix}\begin{pmatrix} x1 \\ x2 \end{pmatrix} = \begin{pmatrix} 0 \\ 0 \end{pmatrix}$, the solution of which is $2x_1 - x_2 = 0$. For an arbitrary value of x1, say $x_1 = 1$, $x_2 = 2$. Therefore, the eigenvector corresponding to $k_2 = 3$ is $\mathbf{x}_2 = \begin{pmatrix} 1 \\ 2 \end{pmatrix}$, or any scalar multiple of $\mathbf{x}_2$. **Note** that, the elements of eigenvector are not independent of each other and one element is always selected arbitrarily. Therefore, if **x** satisfies equation $(\mathbf{A} - k\mathbf{I})\mathbf{x} = \mathbf{0}$ for given value of k, so does c**x**, where c is any constant.

In *Mathematica,* to find the eigenvalues and eigenvectors use the commands **Eigenvalues**[A] and **Eigenvectors**[A] to find the eigenvalues and eigenvectors, respectively, or use the command **Eigensystem**[A] to find the eigenvalues and Eigenvectors together, with one command. For the same example above the solution using *Mathematica* is,

```
In[10]:= Clear[A]
         A = (1  1
              4  1);
         Eigenvalues[A]

Out[10]= {-1, 3}

In[11]:= Eigenvectors[A]

Out[11]= {{-1, 2}, {1, 2}}
```

Alternatively, you may use the command "Eigensystem" to get the eigenvalues and eigenvectors.

```
In[12]:= Eigensystem[A]

Out[12]= {{-1, 3}, {{-1, 2}, {1, 2}}}{{-1, 3}, {{-1, 2}, {1, 2}}}
```

Example: Find the eigenvalues and the eigenvectors of $\mathbf{B} = \begin{pmatrix} 0 & 1 \\ 1 & 0 \end{pmatrix}$.

Solution: Follow the steps in the preceding example by forming (**B** - k**I**)**x** and setting Det[**B** - k**I**] = 0. This will yield $\left(\begin{pmatrix} 0 & 1 \\ 1 & 0 \end{pmatrix} - \begin{pmatrix} k & 0 \\ 0 & k \end{pmatrix}\right)\begin{pmatrix} x1 \\ x2 \end{pmatrix} = \begin{pmatrix} 0 \\ 0 \end{pmatrix}$ or $k^2 - 1 = 0$, the solution of which is $k_1 = 1$ and $k_2 = -1$. These are the eigenvalues of **B**. To find the eigenvectors corresponding to k_1 and k_2, form (**B** - k**I**)**x** = **0**. For $k_1 = 1$, this yields $\begin{pmatrix} -1 & 1 \\ 1 & -1 \end{pmatrix}\begin{pmatrix} x1 \\ x2 \end{pmatrix} = \begin{pmatrix} 0 \\ 0 \end{pmatrix}$ or x1 = x2. For an arbitrary value of x1 = 1, x2 = 1. Therefore, the eigenvector corresponding to $k_1 = 1$ is $\begin{pmatrix} 1 \\ 1 \end{pmatrix}$, or any constant c multiple of this vector, $c\begin{pmatrix} 1 \\ 1 \end{pmatrix}$. For $k_2 = -1$, (**B** - k**I**)**x** = **0** yields $\begin{pmatrix} 1 & 1 \\ 1 & 1 \end{pmatrix}\begin{pmatrix} x1 \\ x2 \end{pmatrix} = \begin{pmatrix} 0 \\ 0 \end{pmatrix}$ or x1 = -x2. For an arbitrary value of x1 = 1, x2 = -1. Therefore, the eigenvector corresponding to $k_2 = -1$ is $\begin{pmatrix} 1 \\ -1 \end{pmatrix}$, or any constant d multiple of this vector, $d\begin{pmatrix} 1 \\ -1 \end{pmatrix}$. The eigensystem of the **B** matrix is {{1, -1}, {{1, 1}, {1, -1}}, where the first bracket ({1, -1}) represents eigenvalues and the remaining two brackets are the corresponding eigenvectors. Using *Mathematica*, the eigensystem is,

```
In[13]:= Clear[B]
         B = (0  1
              1  0);
         Eigensystem[B]

Out[13]= {{-1, 1}, {{-1, 1}, {1, 1}}}
```

Symmetric Matrices and Orthogonal Eigenvectors If the matrix **A** is a symmetric matrix, the corresponding eigenvectors will be orthogonal. Two vectors, **x** and **y** are said to be orthogonal if their dot product is zero, **x.y** = **y.x** = 0, and **x.x** ≠ 0, **y.y** ≠ 0.

Example: Show that the eigenvectors of the symmetric matrix **A** = $\begin{pmatrix} 1 & 1 \\ 1 & -1 \end{pmatrix}$ are orthogonal.

Solution: Following the steps of the previous examples, the eigenvalues of the matrix are found to be $\sqrt{2}$ and -$\sqrt{2}$. The corresponding eigenvectors are **x1** = $\begin{pmatrix} 1-\sqrt{2} \\ 1 \end{pmatrix}$, and **x2** = $\begin{pmatrix} 1+\sqrt{2} \\ 1 \end{pmatrix}$, respectively. Since **x1'.x2** = **x2'x1** = **0, x1'x1** = 4 - 2$\sqrt{2}$, and 4 + 2$\sqrt{2}$, the two eigenvectors are orthogonal. Using *Mathematica*, the problem can be solved as,

In[14]:= `Eigenvectors[` $\begin{pmatrix} 1 & 1 \\ 1 & -1 \end{pmatrix}$ `]`

Out[14]= `{{1 - √2, 1}, {1 + √2, 1}}`

The matrix has two eigenvectors, **x** = $\{1-\sqrt{2}, 1\}$ and **y** = $\{1+\sqrt{2}, 1\}$. The dot product of the two vectors is

In[15]:=
```
Clear[x1, x2]
x1 = {1 - Sqrt[2], 1};  x2 = {1 + Sqrt[2], 1};
N[Dot[x1, x2]]
N[Dot[x1, x1]]
N[Dot[x2, x2]]
```

Out[15]= -2.22045×10^{-16}

1.17157

6.82843

Since **x1'.x2** = 0, **x1'x1** ≠ 0, and **x2'.x2** ≠ 0, **x1** and **x2** are orthogonal.

Example: Find the eigenvalues and the eigenvectors of the symmetric matrix **A** = $\begin{pmatrix} 1 & 2 \\ 2 & 4 \end{pmatrix}$ and verify that the eigenvectors are orthogonal.

Solution: Form the eigenvalue equation (**A** - **kI**)**x** = $\left(\begin{pmatrix} 1 & 2 \\ 2 & 4 \end{pmatrix} - \begin{pmatrix} k & 0 \\ 0 & k \end{pmatrix}\right)\begin{pmatrix} x1 \\ x2 \end{pmatrix} = \begin{pmatrix} 0 \\ 0 \end{pmatrix}$ and solve for ks. This equation will have nontrivial solution if the determinant of $\left|\begin{pmatrix} 1 & 2 \\ 2 & 4 \end{pmatrix} - \begin{pmatrix} k & 0 \\ 0 & k \end{pmatrix}\right| = 0$. The determinant renders the polynomial k(k - 5) = 0, the solution set of which is $k_1 = 0$ and $k_2 = 5$. To find the eigenvectors corresponding to these two eigenvalues, substitute for the value of ks in the equation (**A** - **kI**)**x** = $\left(\begin{pmatrix} 1 & 2 \\ 2 & 4 \end{pmatrix} - \begin{pmatrix} k & 0 \\ 0 & k \end{pmatrix}\right)\begin{pmatrix} x1 \\ x2 \end{pmatrix} = \begin{pmatrix} 0 \\ 0 \end{pmatrix}$ and solve for x1 and x2. For $k_1 = 0$, the equation yields $\begin{pmatrix} 1 & 2 \\ 2 & 4 \end{pmatrix}\begin{pmatrix} x1 \\ x2 \end{pmatrix} = \begin{pmatrix} 0 \\ 0 \end{pmatrix}$, the solution of which is $x_1 + 2x_2 = 0$. For an arbitrary value of $x_2 = 1$, x1 = -2. Therefore, the eigenvector corresponding to k1 = 0 is **x1** = $\begin{pmatrix} -2 \\ 1 \end{pmatrix}$, or any scalar multiple of **x1**. For k2 = 5, the equation (**A** - **kI**)**x** = $\left(\begin{pmatrix} 1 & 2 \\ 2 & 4 \end{pmatrix} - \begin{pmatrix} 5 & 0 \\ 0 & 5 \end{pmatrix}\right)\begin{pmatrix} x1 \\ x2 \end{pmatrix} = \begin{pmatrix} 0 \\ 0 \end{pmatrix}$ yields $\begin{pmatrix} -4 & 2 \\ 2 & -1 \end{pmatrix}\begin{pmatrix} x1 \\ x2 \end{pmatrix} = \begin{pmatrix} 0 \\ 0 \end{pmatrix}$, the solution of which is $2x_1 - x_2 = 0$. For an arbitrary value of x1, say $x_1 = 1$, $x_2 = 2$. Therefore, the eigenvector corresponding to $k_2 = 5$ is $\mathbf{x_2} = \begin{pmatrix} 1 \\ 2 \end{pmatrix}$, or any scalar multiple of $\mathbf{x_2}$.

To test for the orthogonality of **x1** and **x2**, fine **x1'.x2**, **x1'.x1**, and **x2'.x2**. Since **x1'.x2** = **x2'.x1** = 0 and **x1'.x1** =

x2'.x2 = 5 ≠ 0, the two eigenvectors are orthogonal to each other. To solve the same problem using *Mathematica*,

```
In[16]:= Clear[A, x1, x2]
         A = (1 2
              2 4);

         Eigensystem[(1 2
                      2 4)]

Out[16]= {{0, 5}, {{-2, 1}, {1, 2}}}
```

The matrix has two eigenvectors **x1** = $\begin{pmatrix} -2 \\ 1 \end{pmatrix}$ and **x2** = $\begin{pmatrix} 1 \\ 2 \end{pmatrix}$. The two vectors **x1** and **x2** are orthogonal if **x1'.x2** = **x2'.x1** = 0, **x1'.x1** ≠ 0, and **x2'.x2** ≠ 0. The following operations show that the two vectors are orthogonal.

```
In[17]:= x1 = {-2, 1}; x2 = {1, 2};
         Dot[x1, x2]
         Dot[x2, x1]
         Dot[x1, x1]
         Dot[x2, x2]

Out[17]= 0

         0

         5

         5
```

Normalized Eigenvectors : We may normalize eigenvectors of a matrix by setting their length at unity. The length of a vector **x** = {x1, x2} is L = $\sqrt{x1^2 + x2^2}$. To normalize the vector **x,** set L = 1 and solve for x1 and x2.

Example Normalize the eigenvectors of the matrix **A** = $\begin{pmatrix} 1 & 2 \\ 2 & 4 \end{pmatrix}$ of the preceding example.

Solution: The two eigenvectors of the **A** matrix are **x1**= {-2, 1} and **x2**= {1, 2}. Let us denote the elements of **x1** as x_1 and x_2. To normalize $\mathbf{x}_1$, set its length to unity. That is, $x_1^2 + x_2^2 = 1$. But, x_1 and x_2 are not independent of each other and the relation $x_1 = -2x_2$ always holds between the two (this is the relation that was found in solving the equation of eigenvalues for eigenvectors). Therefore, we can find x_1 and x_2 by solving the two equations $x_1 = -2x_2$ and $2x_2^2 + x_2^2 = 1$ for x_1 and x_2. This gives two solutions for x_2 which are $x_2 = \frac{1}{\sqrt{5}}$ and $\frac{-1}{\sqrt{5}}$. The corresponding solutions for x_1 are $x_1 = \frac{-2}{\sqrt{5}}$ and $\frac{2}{\sqrt{5}}$, respectively. Therefore, the The two sets of normalized vectors corresponding to **x1** = {-2, 1} are $\{\frac{-2}{\sqrt{5}}, \frac{1}{\sqrt{5}}\}$ and $\{\frac{2}{\sqrt{5}}, \frac{-1}{\sqrt{5}}\}$.

The same procedure follows for the **x2** = {1, 2}. Denoting the elements of the vector **x2** by x1 and x2, set its length to unity, $x1^2 + x2^2 = 1$. Also, x1 and x2 are not independent of each other and that x2 = 2x1 always holds between the two elements of the vector. Therefore, solve the two equations together to find x1 and x2. The two set of normalized vectors corresponding to **x2** = {1, 2} are $\{\frac{-1}{\sqrt{5}}, \frac{-2}{\sqrt{5}}\}$ and $\{\frac{1}{\sqrt{5}}, \frac{2}{\sqrt{5}}\}$.

Mathematica my help with solving the system equations for finding the normalized vectors. Use the following commands to solve the equations. To find the normalized vectors corresponding to **x1**= {-2, 1},

```
In[18]:= Clear[x1, x2]
          Solve[{Sqrt[x1^2 + x2^2] == 1, x1 == -2 x2}, {x1, x2}]
```

```
Solve::ifun : Inverse functions are
   being used by Solve, so some solutions may not be found.
```

Out[18]= $\{\{x1 \to -\frac{2}{\sqrt{5}}, x2 \to \frac{1}{\sqrt{5}}\}, \{x1 \to \frac{2}{\sqrt{5}}, x2 \to -\frac{1}{\sqrt{5}}\}\}$

To find the normalized vectors corresponding to **x2**= {1, 2},

```
In[19]:= Clear[x1, x2]
         Solve[{Sqrt[x1^2 + x2^2] == 1, x2 == 2 x1}, {x1, x2}]
```

```
Solve::ifun : Inverse functions are
   being used by Solve, so some solutions may not be found.
```

Out[19]= $\{\{x1 \to -\frac{1}{\sqrt{5}}, x2 \to -\frac{2}{\sqrt{5}}\}, \{x1 \to \frac{1}{\sqrt{5}}, x2 \to \frac{2}{\sqrt{5}}\}\}$

Note: Normalized vectors of a symmetric matrix have the property that they are **orthonormal.** Two vectors **x** and **y** are orthonormal if **x'.y** = **y'.x** = 0 and **x'.x** = **y'.y** = 1

Example: Verify that the normalized vectors of the preceding example are orthonormal.

Solution: Take one of the normalized vectors corresponding to **x**1, say $\{\frac{-2}{\sqrt{5}}, \frac{1}{\sqrt{5}}\}$ and call it **xn**. Take one of the normalized vectors corresponding to **x**2, say $\{\frac{-1}{\sqrt{5}}, \frac{-2}{\sqrt{5}}\}$, and call it **yn**. If **xn** and **yn** are orthonormal, the following relations should hold between the two **xn'.yn** = **yn'.xn** = 0 and **xn'.xn** = **yn'.yn** = 1. The dot products of the vectors confirm that these relations in fact hold between **xn** and **yn**. These relationships can also be checked using *Mathematica* as follows.

```
In[20]:= x = {-2 / Sqrt[5], 1 / Sqrt[5]};   y = {-1 / Sqrt[5], -2 / Sqrt[5]};
         N[Dot[x, y]]
         N[Dot[y, x]]
         N[Dot[x, x]]
         N[Dot[y, y]]

Out[20]= 0.

         0.

         1.

         1.
```

The same can be proved for the second set of normalized vectors $\{\frac{2}{\sqrt{5}}, \frac{-1}{\sqrt{5}}\}$ and $\{\frac{1}{\sqrt{5}}, \frac{2}{\sqrt{5}}\}$.

Diagonalizing Symmetric Matrices

In general a symmetric matrix **A** of order nxn has n eigenvalues, possibly not all distinct. Corresponding to these roots are a set of n orthonormal eigenvectors. A matrix **X**, formed by using these orthonormal eigenvectors as its columns, will have the property that **X'.X** = I. The product **X'.A.X** produces a diagonal matrix with eigenvalues of **A** displayed on its principal diagonal. Therefore, any symmetric matrix **A** can be diagonalized by pre and post multiplying it by a matrix with columns that are normalized eigenvectors of matrix **A**.

Example: Find the matrix **X** that will diagonalize **A** = $\begin{pmatrix} 1 & 4 \\ 4 & 1 \end{pmatrix}$. Find the diagonalized matrix.

Solution: The matrix has two distinct eigenvalues -3 and 5. Corresponding to these eigenvalues are two eigenvectors **x** = {-1, 1} and **y** = {1, 1}, respectively. To normalize the eigenvectors, set the length of each vector to unity.

For **x** = {-1, 1} this means that the sum square of the elements is equal to one, $x^2{}_1 + x^2{}_2 = 1$. Also, since x1 = -x2 always hold between x1 and z2. To find x1 and x2, solve the two equations together. The two set of solutions for **x** are $\{\frac{-1}{\sqrt{2}}, \frac{1}{\sqrt{2}}\}$ and $\{\frac{1}{\sqrt{2}}, \frac{-1}{\sqrt{2}}\}$.

To find the solutions for **y**, follow the same procedure and solve for y_1 and y_2 using $y^2{}_1 + y^2{}_2 = 1$ and $y_1 = y_2$. The two set of solutions for **y** are $\{\frac{-1}{\sqrt{2}}, \frac{-1}{\sqrt{2}}\}$ and $\{\frac{1}{\sqrt{2}}, \frac{1}{\sqrt{2}}\}$. The orthonormal matrices formed by the normalized eigenvectors are, $\begin{pmatrix} \frac{-1}{\sqrt{2}} & \frac{-1}{\sqrt{2}} \\ \frac{1}{\sqrt{2}} & \frac{-1}{\sqrt{2}} \end{pmatrix}$ and $\begin{pmatrix} \frac{1}{\sqrt{2}} & \frac{1}{\sqrt{2}} \\ \frac{-1}{\sqrt{2}} & \frac{1}{\sqrt{2}} \end{pmatrix}$. These are the **X** matrices, each with the properties that **X'.X** = **I** and **X'.A.X** is a diagonal matrix with the eigenvalues in the principal diagonal and zeroes everywhere else, $\begin{pmatrix} -3 & 0 \\ 0 & 5 \end{pmatrix}$. These can be confirmed by multiplying the relevant matrices as in the equations **X'.X** = **I** and **X'.A.X**.

Mathematica can be use to facilitate the calculation involved as follows: To find the eigenvalues and the eigenvectors, use the command "EigenSystems" as follows,

```
In[21]:= Eigensystem[(1 4
                      4 1)]

Out[21]= {{-3, 5}, {{-1, 1}, {1, 1}}}
```

To solve the system of equations for finding the normalized eigenvectors for **x** use the following commands,

```
In[22]:= Clear[x1, x2]
         Solve[{x1^2 + x2^2 == 1, x1 == -x2}, {x1, x2}]

Out[22]= {{x1 → -1/√2, x2 → 1/√2}, {x1 → 1/√2, x2 → -1/√2}}
```

the two set of solutions for **x** are $\{\frac{-1}{\sqrt{2}}, \frac{1}{\sqrt{2}}\}$ and $\{\frac{1}{\sqrt{2}}, \frac{-1}{\sqrt{2}}\}$.

To solve the system of equations for finding the normalized eigenvectors of **y,** use the following commands,

```
In[23]:= Clear[v1, v2]
         Solve[{y1^2 + y2^2 == 1, y1 == y2}, {y1, y2}]

Out[23]= {{y1 → -1/√2, y2 → -1/√2}, {y1 → 1/√2, y2 → 1/√2}}
```

The two set of solutions for **y** are $\{\frac{-1}{\sqrt{2}}, \frac{-1}{\sqrt{2}}\}$ and $\{\frac{1}{\sqrt{2}}, \frac{1}{\sqrt{2}}\}$. To test the properties **X'.X** = **I** and **X'.A.X** is a diagonal matrix with the eigenvalues in the principal diagonal and zeroes everywhere else use the following commands.

```
In[24]:= MatrixForm[Transpose[(-1/√2  -1/√2
                               1/√2  -1/√2)].(-1/√2  -1/√2
                                               1/√2  -1/√2)]

Out[24]= (1 0
          0 1)

In[25]:= diag = Transpose[(-1/√2  -1/√2
                            1/√2  -1/√2)].(1 4
                                           4 1).(-1/√2  -1/√2
                                                  1/√2  -1/√2);
         MatrixForm[N[diag]]

Out[25]= (-3. 0.
          0.  5.)
```

Problems

1 Find the eigenvalues and the eigenvectors of the following matrices

a) $\mathbf{A} = \begin{pmatrix} 1 & 1 \\ 1 & 3 \end{pmatrix}$ b) $\mathbf{A} = \begin{pmatrix} 1 & 3 \\ 3 & 1 \end{pmatrix}$

c) $\mathbf{A} = \begin{pmatrix} 0 & 0 & 1 \\ 0 & 0 & 1 \\ 1 & 1 & 0 \end{pmatrix}$ d) $\mathbf{A} = \begin{pmatrix} 2 & 1 & 0 \\ 0 & 1 & 0 \\ 1 & 0 & 1 \end{pmatrix}$

2. Show that if A is a singular matrix, $k = 0$ is an eigenvalue, and conversely.

3. Show that if k is an eigenvalue of A, 1/k is an eigenvalue of A^{-1}.

4. Diagonalize the following matrices

a) $\mathbf{A} = \begin{pmatrix} 2 & 1 \\ 2 & 2 \end{pmatrix}$ b) $\mathbf{A} = \begin{pmatrix} 9 & 3 \\ 3 & 1 \end{pmatrix}$ c) $\mathbf{A} = \begin{pmatrix} 3 & 0 & 3 \\ 0 & 3 & 4 \\ 3 & 4 & 3 \end{pmatrix}$

IV. Quadratic Forms and Positive Definite Matrices

As well as their application in statistics, quadratic forms and tests of positive definiteness of matrices are frequently used in deciding the critical points (maxima, minima, inflection points) of multi-variate functions. If **x** denotes a nx1 vector and **A** is a nxn symmetric matrix, then **x'Ax** is a scalar and is called a quadratic form. For example, if $\mathbf{x} = \begin{pmatrix} x1 \\ x2 \end{pmatrix}$ and $\mathbf{A} = \begin{pmatrix} a11 & a12 \\ a21 & a22 \end{pmatrix}$ hen, $\mathbf{x'Ax} = a11x1^2 + (a12 + a21)x1x2 + a22x2^2$.

```
In[26]:= Clear[x1, x2]
         Transpose[(x1
                    x2)].(a11 a12
                          a21 a22).(x1
                                    x2)

Out[26]= {{x1 (a11 x1 + a21 x2) + x2 (a12 x1 + a22 x2)}}
```

The restriction that matrix **A** be symmetric is not a serious one, because any non-symmetric matrix can be transformed to symmetric one by averaging the off-diagonal elements. For example, the quadratic forms of the non-symmetric matrix $\mathbf{A} = \begin{pmatrix} 1 & 3 \\ 5 & 2 \end{pmatrix}$ and the symmetric matrix $\mathbf{A^*} = \begin{pmatrix} 1 & 4 \\ 4 & 2 \end{pmatrix}$ are the same. However, **A*** is derived from **A** by averaging the off-diagonal elements. To test this, let's assume that $\mathbf{x} = \begin{pmatrix} 1 \\ 1 \end{pmatrix}$ and find **x'Ax** = **x'A*x** = 11. Using *Mathematica,*

```
In[27]:= Transpose[(1
                    1)].(1 3
                         5 2).(1
                               1)

Out[27]= {{11}}

In[28]:= Transpose[(1
                    1)].(1 4
                         4 2).(1
                               1)

Out[28]= {{11}}
```

both products result in the same value, 11.

A quadratic form is said to be **positive definite** and the matrix **A** is said to be positive definite if and only if **X'AX** > 0 for all **X** ≠ **0**. A quadratic form and the associated matrix are said to be **positive semi-definite** if **X'AX** ≥ 0 for all **X** ≠ **0**.

Another way of testing for positive definiteness of quadratic forms is by finding the eigenvalues of their associated matrices. A matrix **A** is said to be positive definite if and only if all its eigenvalues are positive. It can also be proved that a positive definite matrix has a determinant that is positive.

Example: Show that the quadratic form $x^2 + 2xy + 2y^2$ is positive definite.

Solution: The corresponding matrix of the coefficients is $\begin{pmatrix} 1 & 0 \\ 2 & 2 \end{pmatrix}$. The eigenvalues of this matrix are 1 and 2. Since both eigenvalues are positive, the quadratic form and its associated matrix are positive definite. Using *Mathematica*,

In[29]:= `Eigenvalues[` $\begin{pmatrix} 1 & 0 \\ 2 & 2 \end{pmatrix}$ `]`

Out[129]= `{1, 2}`

Note that, to form the matrix associated with the quadratic form $ax^2 + bxy + cy^2$ the coefficients of the squared variables (x^2 and y^2) form the diagonal elements and the coefficients of the cross product (xy) form the off-diagonal elements.

Example: Test for the positive definiteness of the quadratic form $x^2 + 3y^2 - z^2 + 5xy - 2xz + 5yz$.

Solution: Write the quadratic form in a matrix product form and find the eigenvalues of the corresponding matrix. That is,

$x^2 + 3y^2 + z^2 + 5xy - 2xz + 7yz = (x \;\; y \;\; z)\begin{pmatrix} 1 & 5 & -2 \\ 0 & 3 & 7 \\ 0 & 0 & 1 \end{pmatrix}\begin{pmatrix} x \\ y \\ z \end{pmatrix}$. The eigenvalues of the associated matrix is 1, 1, and 3.

Since all the eigenvalues are positive, the matrix and its corresponding quadratic form are positive definite. To find the eigenvalues of the matrix, you can use *Mathematica* as,

In[30]:= `Eigenvalues[` $\begin{pmatrix} 1 & 5 & -2 \\ 0 & 3 & 7 \\ 0 & 0 & 1 \end{pmatrix}$ `]`

Out[30]= `{1, 1, 3}`

Problems

1. Write the corresponding coefficient matrices of the following quadratic forms and transform the matrices into symmetric matrices

   a) $x^2 + 3xy + y^2$

   b) $3x^2 + 2xy + y^2$

   c) $3x^2 + 2zy + y^2 + 2z^2$

2. Show that the following quadratic forms are positive definite:

   a) $3x^2 + 2xy + 3y^2$

   b) $2x^2 + xy + 3y^2$

   c) $x^2 + 4xy + 2y^2 + 3z^2$

   d) $3x^2 + xy + 2y^2$

CHAPTER 6

DIFFERENTIAL CALCULUS

Many business and economic problem solving and decisions are based on the cost and benefit analysis or more specifically on the marginal cost and marginal benefit analysis. For example, to maximize profit, a business in a perfectly competitive market sets marginal cost of the product equal to the price of the product; a producer in a monopolistic market sets marginal cost equal to marginal revenue. The concept of "**marginal**" in economics is nothing but the concept of the "**derivative**" in mathematics. Marginal cost and marginal revenue are the derivatives of the total cost and the total revenue functions with respect to output. The mathematical concept of the derivative is also of extreme importance for optimizing problems. Maximizing revenue, maximizing profit, or minimizing cost are examples of optimizing problems that can be solved using derivatives. Differential calculus is the branch of mathematics which deals with the calculation of the derivatives of functions and with the application of these derivatives. The derivative is the most fundamental concept studied in this chapter. However, the concepts of limit and continuity, which have direct bearing on differentiability and derivatives of the functions, will be studied first.

TOOL BOX

***Mathematica* Commands**	**Function**
`Limit[f(x), x -> n]`	Finds limit of f(x) as x approaches n
`D [f(x), x]`	Finds derivative of f(x) with respect to x
`D [f(x), {x, n}]`	Finds the nth order derivative of f(x) with respect to x
`f'[x]`	Finds derivative of f(x) if it is defined in "f[x_]: = " format
`f''[x]`	Finds second derivative of f(x) if it is defined in "f[x_]: = " format
`f'[n]`	Finds derivative of f(x) and evaluates it at x = n
`Dt[f(x, y), x]`	Find total derivative of f(x, y) with respect to x

I. Limit and Continuity: The `Limit` Command

Limit: Let the function f(x) be defined on the set of real numbers and let x_o be a real number. If f(x) approaches the real number L when x approaches x_o, we say that f(x) has the limit L at x_o, and we write $f(x) \underset{x \to x_o}{\to} L$, or

$\lim_{x \to x_o} f(x) = L$.

Example: Find the limit of $f(x) = x^2 + 1$ as x approaches 3.

Solution: When x approaches 3, $x^2 + 1$ approaches 10. Thus, f(x) has the limit 10 at 3.

The limit of a function f(x) as x approaches x_o can be found using *Mathematica*. To find the limit of a function with *Mathematica*, use the command `Limit[f(x), x -> n]` as follows:

```
In[1]:= Clear[x]
         Limit[x^2 + 1, x -> 3]

Out[1]= 10 Null
```

Example: Find the limit of f(x) = $\frac{x+1}{x}$ as x approaches 1 and ∞.

Solution: As x approaches 1, f(x) approaches 2 and has the limit 2 at x = 1. As x approaches an infinitely large number, ∞, f(x) is undefined. However, if we write f(x) as f(x) = 1 + $\frac{1}{x}$, as x → ∞, $\frac{1}{x}$ → 0 and the limit of f(x) approaches 1. To solve the problem using *Mathematica*,

```
In[1]:= Clear[x]
        Limit[(x + 1) / x, x -> 1]

Out[1]= 2

In[2]:= Limit[(x + 1) / x, x -> Infinity]

Out[2]= 1
```

Example: Find the limit of f(x) = $\frac{4x}{x}$ as x approaches 0.

Solution: Here, f(x) = 4 if x ≠ 0. For x = 0, f(x) is not defined. However, as x approaches zero, f(x) approaches the limit 4.

This example shows that, in many occasions the value of f(x) at x_0 has nothing to do with the limit of f(x) as x approaches to x_0. f(x) may equal L at x_0, may be different from L at x_0, or may be undefined at x_0. The next example reemphasizes this point.

Example: Find the limit of f(x) = $\frac{\sin x}{x}$ as x approaches zero.

Solution: At x = 0, f(x) is not defined. However, it can be shown (by direct substitution) that as x approaches zero, f(x) approaches 1. To solve the problem using *Mathematica*,

```
In[3]:= Clear[x]
        Limit[Sin[x] / x, x -> 0]

Out[3]= 1
```

Right-Side Limit Left-Side Limit: A function f(x) approaches a limit L as x approaches x_0 if and only if its **right-side limit,** denoted by $\lim_{x\to x_0^+} f(x)$, is equal to its **left-side limit**, denoted by $\lim_{x\to x_0^-} f(x)$. That is, whether x approach x_0 from the right side of x_0 (x values that are larger than x_0), or the left side of x_0 (x values that are smaller than x_0), f(x_0) will be the same. The following graph illustrates this point for f(x) = $x^2 + 2$. In the graph of y = f(x) = $x^2 + 2$ below, the right- side limit equals to left-side limit at x_0 = 1 and the function has limit at this point.

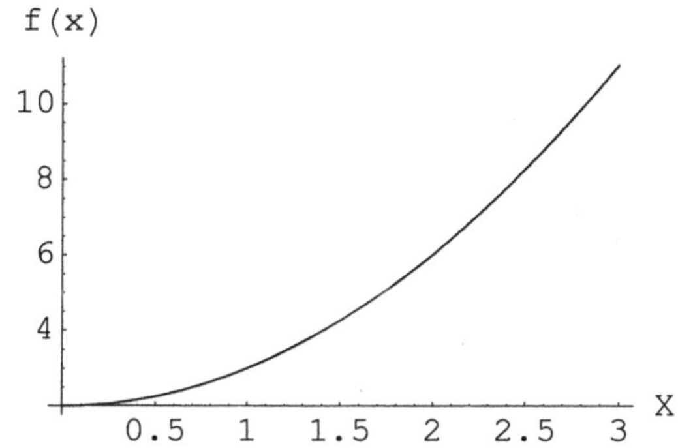

However, for the function $y = \frac{x}{x-1}$, the right-side limit is different from the left-side limit at $x_0 = 1$. As x approaches $x_0 = 1$ from the x values of $x < 1$, f(x) approaches $-\infty$. As x approaches $x_0 = 1$ from the x values of $x > x_0$, f(x) approaches $+\infty$. Therefore, the function does not have limit at $x_0 = 1$.

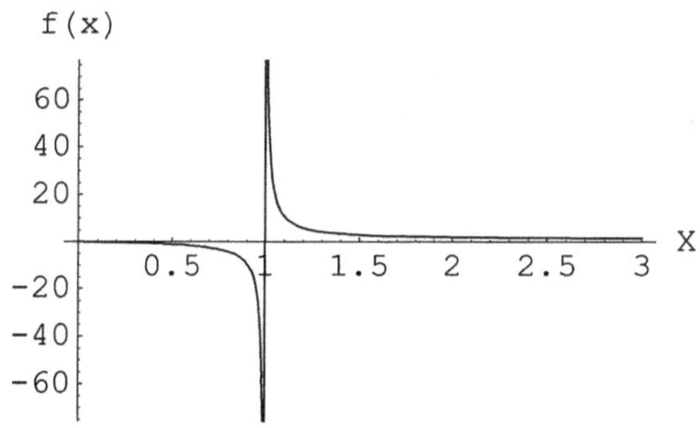

The geometric interpretation of the limit, as presented by the two graphs above, are of great importance for the concept of continuity and differentiability which will be studied in the next sections.

Example: Sales of New Products

Sales of new products are usually estimated using the function $y = c\, e^{\frac{k}{t}}$, where t is the number of years that the product has been in the market and c and k are unknown parameters. Find c and k and express y as a function of t for a product that has sold 6000 units after 3 years and that the market saturation quantity for the product is 28000 units. Graph the sales function and find how many units will be sold after ten years.

Solution: The market saturation quantity is given to be 28000 units. This implies that, as $t \to \infty$, $y \to 28000$. In other words, the $\lim_{t\to\infty} y = 28000$. To find the limit of y as $t \to \infty$, substitute for t in the equation of y. Since $\frac{k}{t} \to 0$ as $t \to \infty$, $e^{\frac{k}{t}} \to 1$ and the limit of y as $t \to \infty$ is y = c. This gives c = 28000. Also, For t = 3, y = 6000, this gives $6000 = c\, e^{\frac{k}{3}}$. By substituting for c and solving the latter equation for k, k = -4.62. Therefore, the sales equation is $y = 28000e^{\frac{-4.62}{t}}$. To find the estimated sales after 10 years, set t = 10 in the equation to find y = 17640 units.

The sales of new products, generally, have a standard "S" shaped curve. The sales increases with increasing rate at the early years of introduction to market. As time passes, sales increases with a decreasing rate, then it flattens out as the market saturates. To graph the function, either find few points of the function by substituting values for t and finding the corresponding values of y and graphing them or by using the "Plot" command of the *Mathematica*. To solve the problem using *Mathematica*, formulate the problem in *Mathematica* and solve the equations as follow,

```
In[4]:= Clear[y, c, k, t]
        y = c E^(k/t);
        N[Solve[{Limit[y, t -> Infinity] == 28000, 6000 == c E^(k/3)}, {c, k}]]
```

```
Solve::ifun : Inverse functions are
   being used by Solve, so some solutions may not be found.
```

Out[4]= `{{Null² (c → 28000.), Null² (k → -4.62134)}}`

the function is $y = 28000\, e^{\frac{-4.62}{t}}$. To graph the function, use the Plot command as,

In[5]:=
```
Clear[y, t]
y = 28000 E^(-4.62/t);
Plot[y, {t, 1, 30}, AxesLabel -> {"t", "Y"}]
```

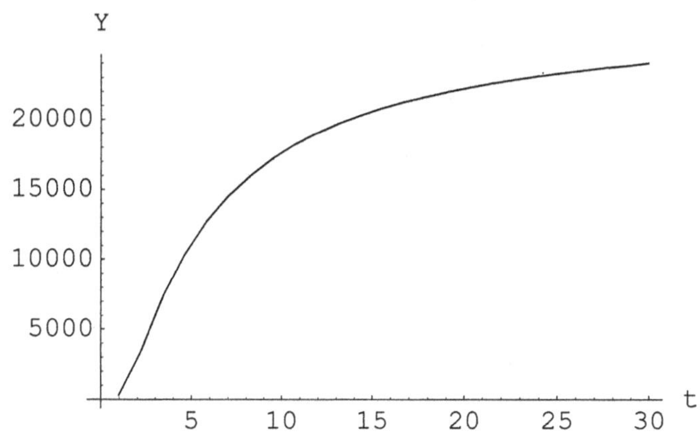

To find the value of y at t = 10,

In[6]:= `y /. t -> 10`

Out[6]= `17640.617640.6`

<u>Continuity:</u> A function $f(x)$ is said to be continuous at $x = x_o$ if and only if $f(x_o)$ and the $\lim_{x \to x_o} f(x)$ exist and $\lim_{x \to x_o} f(x) = f(x_o)$.

Note that, in the discussion of limits, we are concerned only with the limiting values of $f(x)$ as x approaches x_o. That is, $f(x_o)$ may or may not exist, and $\lim_{x \to x_o} f(x)$ may or may not be equal to $f(x_o)$. For continuity, however, a function is said to be continuous if the three conditions below are satisfied:

1- $f(x_o)$ exist

2- $\lim_{x \to x_o} f(x)$ exists.

3- $\lim_{x \to x_o} f(x) = f(x_o)$

If the three conditions hold for all the values of x in the domain of function, the function is called **everywhere continuous**. Otherwise, the function is called **piecewise continuous**.

Example: Verify continuity of $f(x) = x^2 + x + 1$ at x = 1.

Solution: Find f(1) and $\lim_{x \to 1} f(x)$. f(1) = 3 and $\lim_{x \to 1} f(x) = 3$. Since both f(x) and $\lim_{x \to 1} f(x)$ exist and are equal to each other, the function satisfies all the three conditions for continuity and f(x) is continuous at x = 1. The plot of the function, below, also shows the continuity at x = 1. To check the conditions using *Mathematica,* use the following commands

```
In[7]:= Clear[x]
        f[x_] := x^2 + x + 1
        f[1]
Out[7]= 3
In[8]:= Limit[x^2 + x + 1, x -> 1]
Out[8]= 3
In[9]:= Plot[x^2 + x + 1, {x, 0, 3}, AxesLabel → {"X", "Y"}]
```

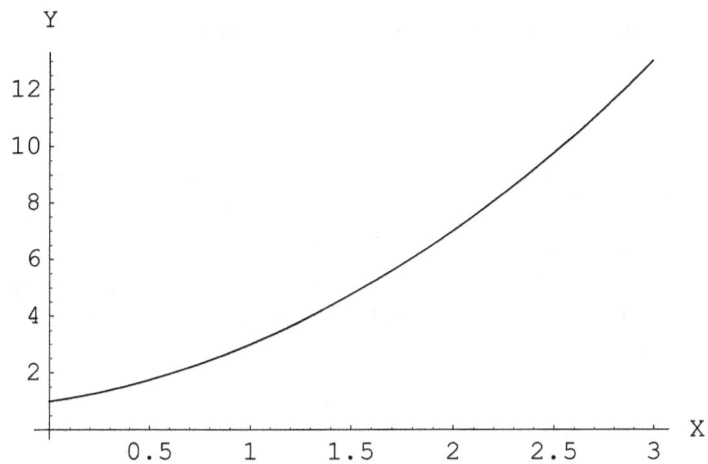

Example: Verify the continuity of $f(x) = \frac{X+1}{X-1}$, at x = 1.

Solution: f(-1) does not exist and $\lim_{x \to 1} f(x) = \infty$. Therefore, f(x) is not continuous at x = 1. The plot of the function, below, also shows discontinuity at x = 1. Using *Mathematica,*

```
In[10]:= Clear[x]
         f[x_] := (x + 1) / (x - 1)
         f[1]
Out[10]= Power::infy : Infinite expression 1/0 encountered.

         ComplexInfinity
In[11]:= Limit[(x + 1) / (x - 1), x -> 1]
Out[11]= ∞
In[12]:= Plot[(x + 1) / (x - 1), {x, 0, 3}, AxesLabel → {"X", "Y"}]
```

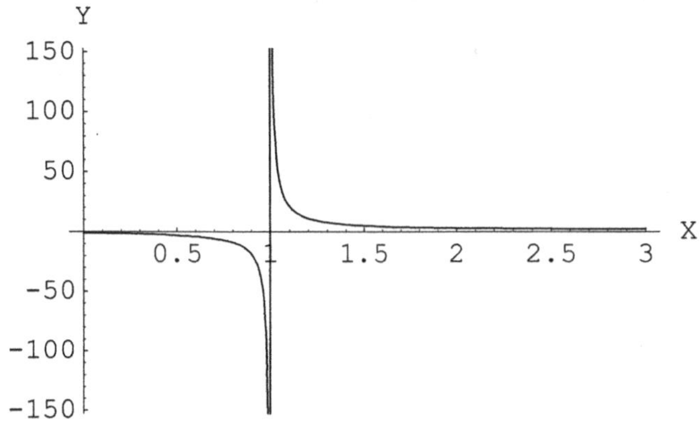

Note that, under the definition of continuity, functions that have kinks are continuous. The next example shows the continuity of a kinked function.

Example: Verify the continuity of $f(x) = |x - 1| + 2$ at x = 1.

Solution: Find f(1) and $\lim_{x \to 1} f(x)$. Since f(1) and $\lim_{x \to 1} f(x)$ exist and are equal to each other, f(x) is continuous at x = 1. **Note** that, as x→1 from the x-values x < 1, the left-side limit exists and is equal to 2. When x→1 from the x-values x > 1, the right-side limit exists and is equal to 2. Since the right-side limit equals to the left-side limit, f(x) has a limit as x→1 and the limit is equal to 2. Also f(1) = 2. Therefore, the function is continuous at x = 1. The following graph shows the graph of the function and confirms its continuity, even though the function is a kinked function. Using *Mathematica* to test for continuity of the function and plot the function.

```
In[13]:= Clear[x]
         Plot[Abs[x - 1] + 2, {x, -10, 10}]
```

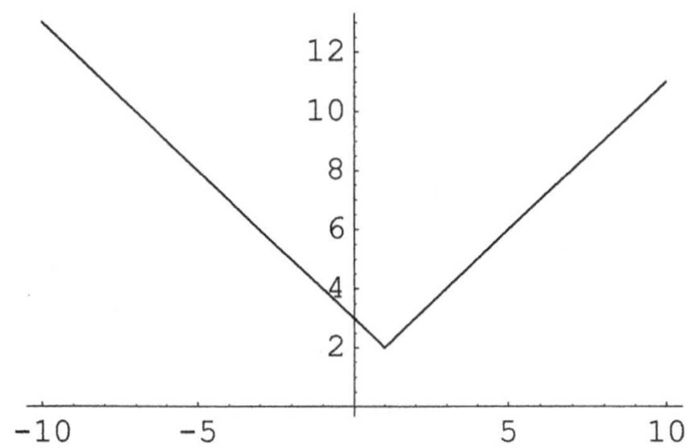


```
In[14]:= f[x_] := Abs[x - 1] + 2
         f[1]

Out[14]= 2 Null

In[15]:= Limit[Abs[x - 1] + 2, x -> 1]

Out[15]= 2

In[16]:= Plot[Abs[x - 1] + 2, {x, 0, 2}, AxesLabel → {"X", "Y"}]
```

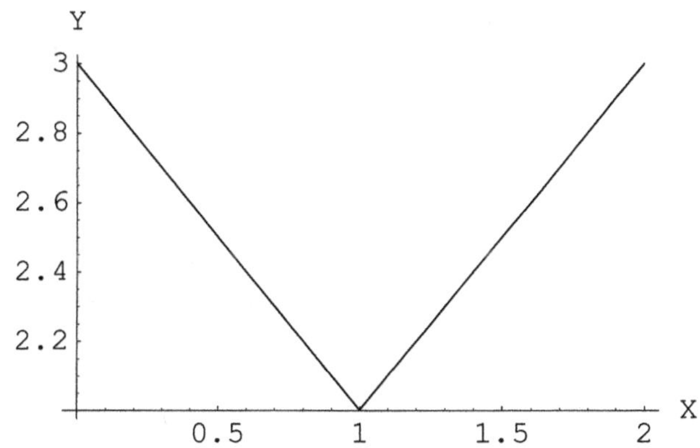

Example: Verify the continuity of $y = f(x) = \begin{cases} 2 & \text{if } x \le 0 \\ x^2 & \text{if } x > 0 \end{cases}$ at $x_o = 0$.

Solution: The graph of the function, below, shows that as x approaches zero from the left side, $\lim_{x\to 0^-} f(x) = 2$. However, as x approaches zero from the right side, $\lim_{x\to 0^+} f(x) \to 0$. Since the left-side limit is not equal to the right-side limit as x approaches zero the function is not continuous at x = 0.

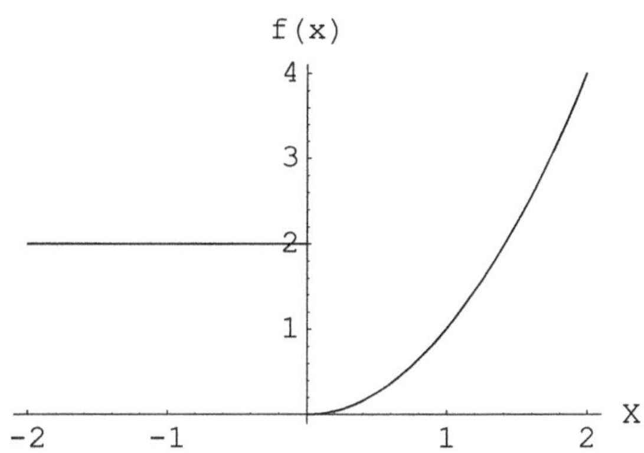

Problems

1. Find the limit of the following sequence of numbers

   a) 1, 2, 3, 4, ...

   b) 10, 9, 8, 7, ...

   c) $\frac{1}{2}, \frac{2}{3}, \frac{3}{4}, \frac{4}{5}, \ldots\ldots$

   d) $\frac{1}{2}, \frac{1}{3}, \frac{1}{4}, \frac{1}{5}, \ldots\ldots$

2. For an economy with an aggregate consumption function of C = 500 + 0.7Y, the effect of a one dollar increase in autonomous spending can be listed as:

Time	1	2	3	4	. . .
Increase in Income	1	0.7	0.7^2	0.7^3	. . .

   a) What is the effect on income as $t \to \infty$?

   b) What is the total effect on income?

3. Find the limit of the following functions as x approaches the given values.

a) $\lim_{x\to 1} f(x) = x^2 - 1$

b) $\lim_{x\to -2} f(x) = \frac{2x+4}{2x-4}$

c) $\lim_{x\to 3} f(x) = \frac{x+3}{x-3}$

d) $\lim_{x\to 1} f(x) = x(1+x)$

e) $\lim_{x\to 1} f(x) = \frac{x^3-1}{x-1}$

f) $\lim_{x\to 0} f(x) = \frac{\text{Cos}\,x}{\text{Sinx}}$

4. Verify the continuity of the following functions at the given points.

a) $f(x) = x + 3;$ $x = 1$

b) $f(x) = x^2 - 1;$ $x = 1$

c) $f(x) = \frac{1}{x} 1/x;$ $x = 0$

d) $f(x) = \frac{1}{(x-1)^2};$ $x = 1$

e) $f(x) = \frac{1-x}{1+x};$ $x = 1$

f) $f(x) = |x - 3| + 1;$ $x = 3$

g) $f(x) = |x - 2|;$ $x = 2$

h) $f(x) = \{\begin{matrix} 2 & \text{if } x \le 0 \\ x^2+2 & \text{if } x > 0 \end{matrix};$ $x = 0$

i) $f(x) = \{\begin{matrix} 1-x & \text{if } x < 1 \\ -1+x & \text{if } x \ge 1 \end{matrix};$ $x = 1$

j) $f(x) = \{\begin{matrix} x^2+1 & \text{if } x \le 1 \\ 1 & \text{if } x > 1 \end{matrix};$ $x = 1$

II. Derivatives and Differentiability: The D [f(x), x] Command

Derivative of a Function: The derivative is the most fundamental concept in differential calculus. It is concerned with the instantaneous rate of change of a function f(x) with respect to x at a point, say x_0. Let the function f given by $y = f(x)$ be continuous in the open interval $a < x < b$ which contains the point x_0. If we represent a small increment in x_0 by Δx and the corresponding increment in y by Δy, $\Delta y = f(x_0 + \Delta x) - f(x_0)$, the following quotient, $\frac{\Delta y}{\Delta x} = \frac{f(x_0+\Delta x) - f(x_0)}{\Delta x}$ is called a **difference quotient**. The difference quotient shows the average rate of change of $f(x)$ with respect to x, over the interval Δx. Regardless of how small we make Δx, we can not use the difference quotient to compute the instantaneous rate of change of y with respect to x at the point x_0. The difficulty is that at x_0, we have no interval (Δx) over which we could take an average. However, if $\frac{\Delta y}{\Delta x}$ approaches a limit as Δx approaches zero, it is reasonable to define the instantaneous rate of change of y with respect to x at $x = x_0$ as the limit of the difference quotient. Therefore, the instantaneous rate of change of y with respect to x at $x = x_0$, which may be denoted by $f'(x_0)$, is defined as,

$$f'(x_0) = \lim_{\Delta x \to 0} \frac{\Delta y}{\Delta x} = \lim_{\Delta x \to 0} \frac{f(x_0 + \Delta x) - f(x_0)}{\Delta x}$$

The expression $f'(x_0)$ is called the derivative of $f(x)$ with respect to x at $x = x_0$. If the $\lim_{\Delta x \to 0} \frac{\Delta y}{\Delta x}$ exists at this point, we say that $f(x)$ is differentiable at x_0 or has a derivative at x_0.

Example: Find the derivative of $y = f(x) = x^2 + 2x$ at $x = 2$.

Solution : The difference quotient of the function is $\frac{\Delta y}{\Delta x} = \frac{f(x_0 + \Delta x) - f(x_0)}{\Delta x} = \frac{(2 + \Delta x)^2 + 2(2 + \Delta x) - 8}{\Delta x} = \frac{\Delta x(\Delta x + 6)}{\Delta x} = \Delta x + 6$. The limit of the difference quotient (at $x_0 = 2$) as $\Delta x \to 0$ is $f' = \lim_{\Delta x \to 0} \frac{\Delta y}{\Delta x} = 6$.

To follow the steps using *Mathematica*, define the function, then find the limit of the difference quotient as $\Delta x \to 0$.

```
In[17]:= Clear[x]
         f[x_] := x^2 + 2 x
         Limit[(f[2 + Δx] - f[2]) / Δx, Δx -> 0]

Out[17]= 6
```

Example: Find $f'(3)$ for the function f defined by $y = f(x) = \frac{1}{x}$.

Solution: Find the limit of the difference quotient $\frac{\Delta y}{\Delta x} = \frac{f(3 + \Delta x) - f(3)}{\Delta x}$ as $\Delta x \to 0$. The difference quotient is, $\frac{\Delta y}{\Delta x} = \frac{\frac{1}{3+\Delta x} - \frac{1}{3}}{\Delta x} = \frac{\frac{-\Delta x}{3(3+\Delta x)}}{\Delta x} = \frac{-1}{3(3+\Delta x)}$. At $x_0 = 3$, $\lim_{\Delta x \to 0} \frac{\Delta y}{\Delta x} = f'(3) = \frac{-1}{9}$. Using *Mathematica*,

```
In[18]:= Clear[x]
         f[x_] := 1 / x
         Limit[(f[3 + Δx] - f[3]) / Δx, Δx -> 0]

Out[18]= - 1/9
```

The instantaneous rate of change expression, $f'(x_0)$ above, may be defined for the set of points at which f is differentiable, rather than the point $x = x_0$. Such a function may be denoted by y', $f'(y)$, $\frac{dy}{dx}$, or f_x and is defined as, $y' = f'(x) = \frac{dy}{dx} = f_x = \lim_{dx \to 0} \frac{dy}{dx} = \lim_{dx \to 0} \frac{f(x + dx) - f(x)}{dx}$.

Example: Find the derivative of $y = f(x) = 3x^2 - 2x + 5$.

Solution: $y' = f'(x) = \lim_{dx \to 0} \frac{f(x + dx) - f(x)}{dx} = \lim_{dx \to 0} \frac{3(x + dx)^2 - 2(x + dx) + 5 - (3x^2 - 2x + 5)}{dx} = \lim_{dx \to 0} \frac{dx(3\,dx + 6x - 2)}{dx} = 6x - 2$.

Using *Mathematica*, The solution is derived as,

```
In[19]:= Clear[x]
         f[x_] := 3 x^2 - 2 x + 5
         Limit[(f[x + dx] - f[x]) / dx, dx -> 0]

Out[19]= Null^2 (-2 + 6 x)
```

Example: Find the derivative of $y = f(x) = \frac{1}{1-x^2}$.

Solution: $y' = f'(x) = \lim_{dx \to 0} \frac{f(x+dx) - f(x)}{dx} = \lim_{dx \to 0} = \frac{\frac{1}{1-(x+dx)^2} - \frac{1}{1-x^2}}{dx} = \lim_{dx \to 0} \frac{\frac{1-x^2-1+(x+dx)^2}{(1-x^2)(1-(x+dx)^2)}}{dx} = \lim_{dx \to 0} \frac{\frac{dx(2x+dx)}{(1-x^2)(1-(x+dx)^2)}}{dx} = \frac{2x}{(1-x^2)^2}$.

Using *Mathematica*,

In[20]:=
```
Clear[x]
f[x_] := 1 / (1 - x^2)
Limit[(f[x + dx] - f[x]) / dx, dx → 0]
```

Out[20]= $\frac{2x}{(-1+x^2)^2}$

Example: Average Cost and Marginal Cost:

The total cost of producing q units of a product is given as TC = 4500 + 9q + $.06q^2$. Find the average rate of change of TC with respect to q when the output level is changed from: (i) 100 to 110 units, (ii) from 100 to 105 units. Find the instantaneous rate of change of TC with respect to q when q = 100. (**Note:** the instantaneous rate of change of TC with respect to q at q_o, or the derivative of TC with respect to q at q_o, is the same as the marginal cost at q_o)

Solution: The average rate of change is found using the difference quotient formula. When q changes from 100 to 110 units, Δq = 10 units. Therefore, we can write:

$$\frac{\Delta TC}{\Delta q} = \frac{f(100+\Delta q) - f(100)}{\Delta q} = \frac{4500 + 9\,(100+\Delta q) + .06\,(100+\Delta q)^2 - 6000}{\Delta q} = \frac{\Delta q(21 + .06\,\Delta q)}{\Delta q} = 21 + .06\Delta q$$

for Δq = 10, $\frac{\Delta TC}{\Delta q}$ = 21.6. For a change in q from 100 units to 105 units Δq = 5 and $\frac{\Delta TC}{\Delta q}$ = 21.3. To find the instantaneous rate of change of TC with respect to q, find TC ' (100) as follows:

$$TC'(100) = \frac{dTC}{dq} = \lim_{dq \to 0} \frac{f(100 + dq) - f(100)}{dq} = \lim_{dq \to 0} \frac{4500 + 9\,(100 + dq) + .06\,(100 + dq)^2 - 6000}{dq} = \lim_{dq \to 0} \frac{dq(21 + .06\,dq)}{dq} = 21.$$

You may formulate the average rate of change and the instantaneous rate of change into *Mathematica* and do the calculations using *Mathematica.* To find the average rate of change using *Mathematica*,

In[21]:=
```
Clear[TC, q, Δq]
f[q_] := 400 + 9 q + .06 q^2
Limit[(f[100 + Δq] - f[100]) / Δq, Δq → 10]
```

Out[21]= $\frac{70}{40325967}$

In[22]:=
```
Limit[(f[100 + Δq] - f[100]) / Δq, Δq → 5]
```

Out[22]= $\frac{205}{110228976}$

To find the instantaneous rate of change using *Mathematica*,

In[23]:=
```
Limit[(f[100 + Δq] - f[100]) / Δq, Δq → 0]
```

Out[23]= $\frac{200}{99980001}$

Do More Practice: Find the average rate of change of TC with respect to q as q changes from 100 to 102.

2- Find the average rate of change of TC with respect to q as q changes from 100 to 101.

3- What is the relationship between the average rate of change and the instantaneous rate of change in TC as Δq approach to zero.

Differentiability: A function f is differentiable at $x = x_0$ if $f'(x_0)$ exists. The function f is differentiable in an open interval (a, b) if it is differentiable at every point in that interval.

Continuity and differentiability are two desirable properties of functions. If a function f is differentiable at x_0, then f is continuous at x_0. However, continuity does not imply differentiability. That is, a function f may be continuous at point x_0, but not differentiable at that point. What makes a continuous function not differentiable at a point is the existence of a corner point or kink at that point.

Example: Prove that $f(x) = |x|$ is continuous at x = 0 but is not differentiable at this point.

Solution: $f(x)$ is continuous at $x = x_0$ if $f(x_0)$ exists and $\lim_{x \to x_0} f(x) = f(x_0)$. The function $f(x) = |x|$ is continuous at x = 0 because $\lim_{x \to 0^-} f(x) = \lim_{x \to 0^+} f(x) = f(0)$. However, the function is not differentiable at x = 0. This is because $f(x) = |x|$ has a kink at x = 0. For x > 0, |x| = x. If we choose Δx a positive number (right-side of x = 0) we have:

$$f'(x) = \lim_{\Delta x \to 0} \frac{|x + \Delta x| - |x|}{\Delta x} = \lim_{\Delta x \to 0} \frac{(x + \Delta x) - x}{\Delta x} = \lim_{\Delta x \to 0} \frac{\Delta x}{\Delta x} = 1$$

This implies that |x| is differentiable for any x > 0. Similarly, for x < 0, |x| = -x and Δx can be chosen such that $|x + \Delta x| = -(x + \Delta x)$. Therefore, for x < 0 we have:

$$f'(x) = \lim_{\Delta x \to 0} \frac{|x + \Delta x| - |x|}{\Delta x} = \lim_{\Delta x \to 0} \frac{-(x + \Delta x) - (-x)}{\Delta x} = \lim_{\Delta x \to 0} \frac{-\Delta x}{\Delta x} = -1$$

This implies that |x| is differentiable for any x < 0. For x = 0, we have:

$$f'(0) = \lim_{\Delta x \to 0} \frac{f(0 + \Delta x) - f(0)}{\Delta x} = \lim_{\Delta x \to 0} \frac{|0 + \Delta x| - |0|}{\Delta x} = \lim_{\Delta x \to 0} \frac{|\Delta x|}{\Delta x}$$

$f'(0) = \lim_{\Delta x \to 0} \frac{|\Delta x|}{\Delta x}$ does not yield a single value. The right-side limit of $f'(0) = \lim_{\Delta x \to 0^+} \frac{|\Delta x|}{\Delta x} = \frac{\Delta x}{\Delta x} = +1$. The let-side limit of $f'(0) = \lim_{\Delta x \to 0^-} \frac{|\Delta x|}{\Delta x} = \frac{-\Delta x}{\Delta x} = -1$. Since the right-side limit is different from the left-side limit, $f'(0)$ does not exits and $f(x) = |x|$ is not differentiable at x = 0.

This can be observed through the graph of $f(x) = |x|$ below. The function f is continuous at x = 0 but it has a kink at this point.

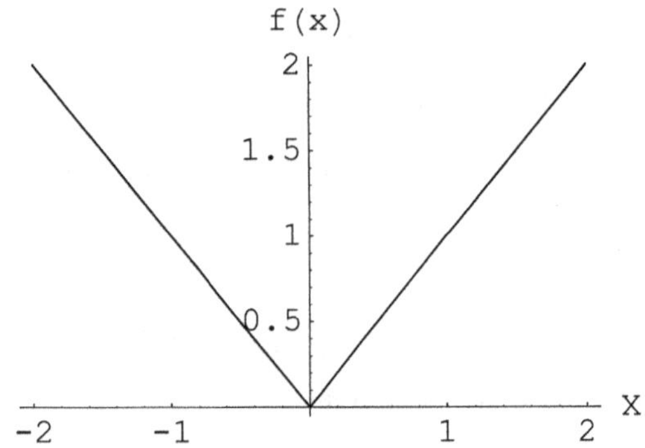

Similarly, the graph of $f'(x) = \begin{cases} -1 & \text{if } x < 0 \\ +1 & \text{if } x > 0 \end{cases}$ is shown below. As the graph shows $\lim_{x \to 0^-} f(x) \neq \lim_{x \to 0^+} f(x)$, thus $f'(0)$ does not exist.

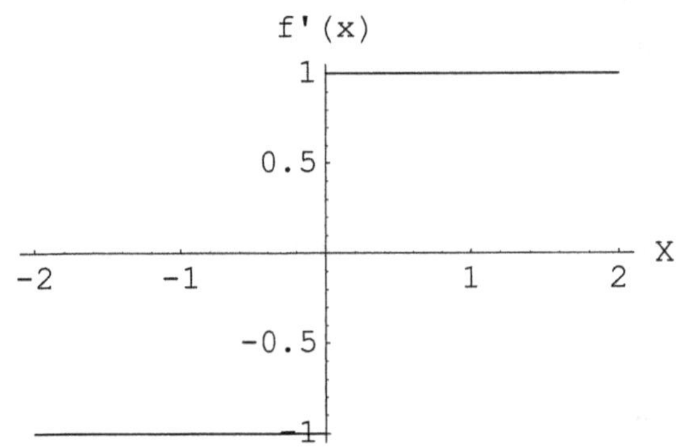

Differentiation Formulas: The basic method of finding the derivative of a function is to apply the definition of derivative. However, learning the following formulas enables us to compute derivatives much more rapidly. The proof of the formulas are not given here. Students are required to practice deriving at least the first three formulas using the definition of the derivative. In *Mathematica*, the command to find the derivatives of f(x) with respect to x is D [f(x), x].

Formula 1- If y = f(x) = c (constant), y' = 0.

Example: Find y' if y = 10.

Solution: Since y equals a constant number, y' = 0. Using *Mathematica*, the solution is,

```
In[24]:= Clear[x]
         D[10, x]

Out[24]= 0
```

Formula 2- If y = f(x) = cx^m, y' = cmx^{m-1}.

Example: Find derivative of y = $10x^3$.

Solution: The solution is $30x^2$. Using *Mathematica*, the solution is,

```
In[25]:= Clear[x]
         D[10 x^3, x]

Out[25]= 30 x^2
```

Formula 3- Let u be a differentiable function of x and $y = u^m$, then $y' = \frac{d(u^m)}{dx} = mu'u^{m-1}$, where $u' = \frac{du}{dx}$.

Example: Find y', if $y = f(x) = (4x^3 + 3x + 1)^5$.

Solution: Let $u = 4x^3 + 3x + 1$, then $\frac{du}{dx} = 12x^2 + 3$ and $y = u^5$. Therefore, $y' = \frac{dy}{dx} = 5(12x^2 + 3)(4x^3 + 3x + 1)^4$. Using *Mathematica*, the solution is,

```
In[26]:= Clear[x]
         D[(4 x^3 + 3 x + 1)^5, x]
```

Out[26]= $5\,(3 + 12\,x^2)\,(1 + 3\,x + 4\,x^3)^4$

Formula 4- If u and v are differentiable functions of x and y = u.v, then $y' = \frac{d(uv)}{dx} = u'v + v'u$, where $u' = \frac{du}{dx}$ and $v' = \frac{dv}{dx}$.

Example: Find y' if $y = f(x) = (5x^3 + 5x^2 + 1)(3x^3 + 2x - 1)$.

Solution: Let $u = (5x^3 + 5x^2 + 1)$ and $v = (3x^3 + 2x - 1)$. Hence, y = u.v, $u' = \frac{du}{dx} = 15x^2 + 10x$, $v' = \frac{dv}{dx} = 9x^2 + 2$, and

$y' = \frac{dy}{dx} = (15x^2 + 10x)(5x^3 + 5x^2 + 1) + (9x^2 + 2)(3x^3 + 2x - 1)$. Using *Mathematica*, the solution is,

```
In[27]:= Clear[x]
         D[(5 x^3 + 5 x^2 + 1) (3 x^3 + 2 x - 1), x]
```

Out[27]= $(10\,x + 15\,x^2)\,(-1 + 2\,x + 3\,x^3) + (2 + 9\,x^2)\,(1 + 5\,x^2 + 5\,x^3)$

Formula 5- If u and v are differentiable functions of x and $y = \frac{u}{v}$, then $y' = \frac{d(\frac{u}{v})}{dx} = \frac{u'v - v'u}{v^2}$, where $u' = \frac{du}{dx}$ and $v' = \frac{dv}{dx}$.

Example: If $y = f(x) = \frac{5x^2+3}{x^3-1}$, find $\frac{dy}{dx}$.

Solution: Let $u = 5x^2 + 3$ and $v = x^3 - 1$. Therefore, $y = \frac{u}{v}$, $u' = 10x$, $v' = 3x^2$, and $\frac{dy}{dx} = \frac{5x^4 + 9x^2 + 10x}{(x^3-1)^2}$. Using *Mathematica*, the solution is,

```
In[28]:= Clear[x]
         D[(5 x^2 + 3) / (x^3 - 1), x]
```

Out[28]= $-\frac{3\,x^2\,(3+5\,x^2)}{(-1+x^3)^2} + \frac{10\,x}{-1+x^3}$

Formula 6- If u is a differentiable function of x and y = lnu, then $y' = \frac{dy}{dx} = \frac{u'}{u}$, where $u' = \frac{du}{dx}$.

Example: If $y = f(x) = 10 + \ln(x^3 - 1)$, find y'.

Solution: Let $u = x^3 - 1$. Hence, $u' = 3x^2$, y = f(x) = 10 + lnu, and $y' = \frac{dy}{dx} = \frac{u'}{u} = \frac{3x^2}{x^3-1}$. Using *Mathematica*, the solution is,

```
In[29]:= Clear[x]
         D[Log[E, x^3 - 1] + 10, x]
```

Out[29]= $\frac{3\,x^2}{-1+x^3}$

Note: If the logarithm has a base other than e, first transform it to a natural logarithm, then differentiate. To transform a log to natural log, note that $\log_b^x = \log_e^x \log_b^e$.

Example: If y = f(x) = $\log_2^{x^2}$, find y'.

Solution: First transform the log to the base e. That is, y =$\log_e^{x^2} \log_2^e$. Now the derivative is, y' = $\frac{dy}{dx} = \frac{2}{x} \log_2^e = \frac{2}{x}\log_2^e$. Using *Mathematica*, the solution is,

```
In[30]:= Clear[x]
         D[Log[2, x^2], x]
```

Out[30]= $\frac{2}{x \, \text{Log}[2]}$

Formula 7- If u is a differentiable function of x and y = e^u, then y' = u'e^u, where u' = $\frac{du}{dx}$.

Example: If y = f(x) = e^{x^2}, find y'.

Solution: Let u = x^2. Therefore, y = e^u and y' = $\frac{dy}{dx}$ = u' e^u where u' = 2x. Using *Mathematica*, the solution is,

```
In[31]:= Clear[x]
         D[E^(x^2), x]
```

Out[31]= $2 e^{x^2} x$

Derivative and Slope of a Function: For a linear function y = mx + c, the slope of the line (m) represents its steepness. For the two arbitrary points, P(x_1, y_1) and Q(x_2, y_2) on a line segment, the slope of the line represents the ratio of rise to run $\left(\frac{\Delta y}{\Delta x}\right)$ and is measured by m = tan ⊖ = $\frac{y2-y1}{x2-x1} = \frac{\Delta y}{\Delta x}$, where ⊖ is the angle between the line and the x-axis. Obviously for the equation of a line, the slope m is the same everywhere along the line. However, for a non-linear function, y = *f*(x), the slope of the curve at a point P(xo, yo), denoted by mp, is the slope of the tangent to the curve at that point. In other words, the slope of *f(x)* at a point x = x_o is the instantaneous rate of change of the function at this point or $f'(x_o) = \lim_{\Delta x \to 0} \frac{\Delta y}{\Delta x}$. Therefore, to find the slope of a function at a given point, find the derivative of the function and evaluate it at that point.

Example: Find the slope of y = f(x) = x^2+ x - 5 at x = 2.

Solution: The slope of the function at x = 2 is f '(2). The derivative of the function is f '(x) = 2x + 1. Evaluated at x = 2, f '(2) = 5.

Using *Mathematica*, The slope of the function at x = 2 is calculated as,

```
In[32]:= Clear[x]
         D[x^2 + x - 5, x] /. x → 2
```

Out[32]= 5

Example: Demand Function

Demand for a product is given as Q = $\frac{100}{P^{.8}}$. Find the slope of the demand function at P = 12.

Solution: Find the derivative of Q with respect to P and evaluate it at P = 12. The derivative is Q' = $\frac{dQ}{dP} = -\frac{80}{P^{1.8}}$. Evaluated at P = 12, $\frac{dQ}{dP}$ = -. 91. (Note: Conventional market theory assumes P as the vertical axis and Q as the horizontal axis for demand and supply functions. With this assumption, the slope of the demand curve would be $\frac{dP}{dQ} = \frac{-1}{.91}$.)

To solve the problem using *Mathematica*,

```
In[33]:= Clear[Q, p]
         Q = 100 / p^(.8);
         slope = D[Q, p] /. p -> 12
Out[33]= -0.913195
```

Example: Price Elasticity of Demand

Demand for a product is given as $Q = \frac{100}{P^2}$. Find the price elasticity of demand at P = 2.

Solution: At P = 2, Q = 25. The price elasticity of demand is defined as $\epsilon_p = \frac{dQ}{dP} \cdot \frac{P}{Q}$ where, $\frac{dQ}{dP}$ is the slope of the demand curve or derivative of Q with respect to P. Therefore, find $\frac{dQ}{dP}$ and evaluate it at P = 2 then multiply it by $\frac{P}{Q}$.

```
In[34]:= Clear[Q, p]
         Q = 100 / (p^2);
         D[Q, p] * (p / Q) /. p -> 2
Out[34]= -2
```

Example: Production Function and Marginal product of Labor

For a single-input production function, $Q = 100L^{.8}$ find marginal product of labor.

Solution: Marginal product of labor (MPL) is the derivative of output with respect to labor, $\frac{dQ}{dL}$. Therefore, MPL is $\frac{dQ}{dL} = 80L^{-.2}$. Using *Mathematica*,

```
In[35]:= Clear[Q, L]
         Q = 100 (L^.8);
         D[Q, L]
```

Out[35]= $\frac{80.}{L^{0.2}}$

Example: Perfectly Competitive market

The long-run cost function of a firm in a perfectly competitive market is given as $TC = 26Q - 3.5Q^2 + \frac{1}{3}Q^3$. Find the long-run equilibrium output and price.

Solution: In a perfectly competitive market profit maximizing quantity is decided by setting marginal cost (MC) equal to price (P). Also, in the long-run equilibrium, prices are expected to fall to the lowest point of average total cost (ATC), where only a normal profit is left for the businesses. Since the marginal cost curve always passes through the minimum average cost point (see the graphs below), in the long-run, P = MC = ATC. Therefore, find the long-run equilibrium quantity and price by setting MC = ATC and solving the resulting equation. MC is the derivative of TC with respect to Q, $MC = 26 - 7Q + Q^2$ and $ATC = \frac{TC}{Q} = 26 - 3.5Q + \frac{1}{3}Q^2$. Set MC = ATC and solve for Q. This yields Q = 5.25. Since at any optimum solution P = MC, evaluate MC at Q = 5.25 to find P = 16.81.

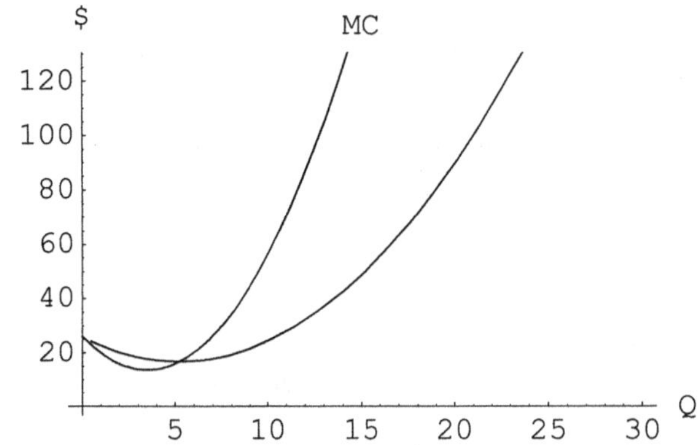

To solve the problem using *Mathematica*,

```
In[36]:= Clear[Q, tc, atc, mc, sol]
         tc = 26 Q - 3.5 Q^2 + (1 / 3) Q^3;
         atc = tc / Q;
         mc = D[tc, Q];
         sol = N[Solve[mc == atc, Q]]

Out[36]= {{Q → 0.}, {Q → 5.25}}

In[37]:= mc /. sol

Out[37]= {26., 16.8125}
```

The long-run equilibrium price is P = 16.81 with a quantity of Q = 5.25. To see the graphs of ATC and MC, plot them together as follows

```
In[38]:= atcplot =
           Plot[atc, {Q, 0.5, 30}, PlotLabel → "ATC", AxesLabel -> {"Q", "$"}];
         mcplot = Plot[mc, {Q, 0, 15}, PlotLabel → "MC", AxesLabel -> {"Q", "$"}];
         Show[mcplot, atcplot]
```

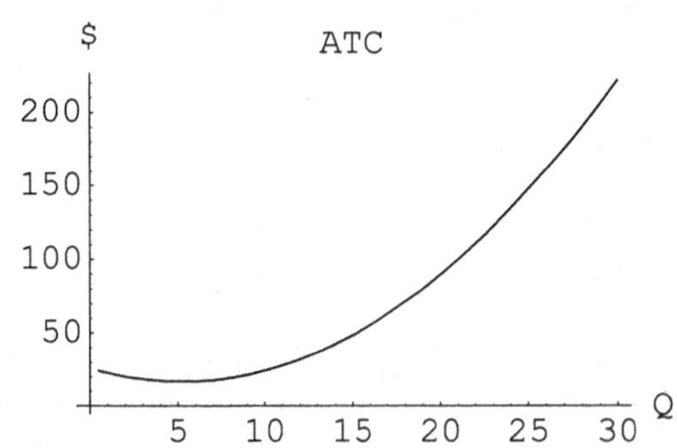

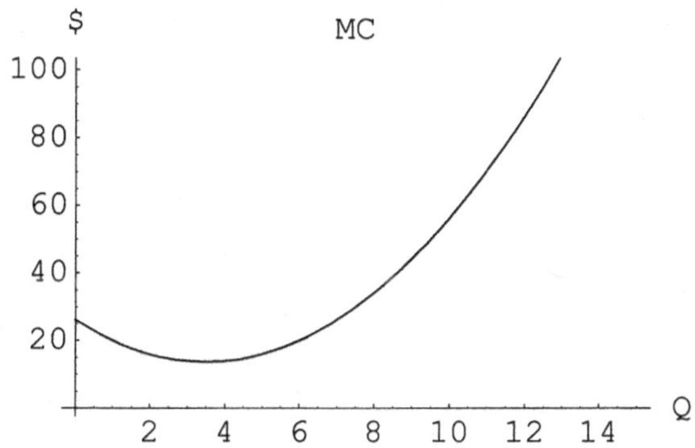

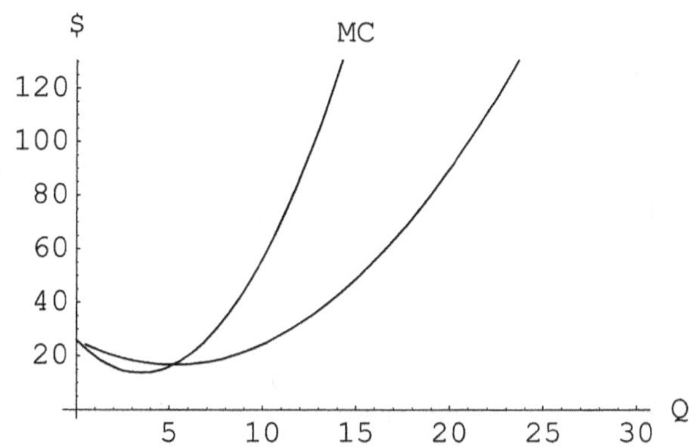

Problems

1 For the following functions find the average rate of change of f(x) with respect to x for the given change in x from x_1 to x_2.

a) $f(x) = 5;$ $x_1 = 10,\ x_2 = 15$

b) $f(x) = 2x + 3;$ $x_1 = 4,\ x_2 = 6$

c) $f(x) = x^2 - 4x + 5;$ $x_1 = 4,\ x_2 = 8$

d) $f(x) = \frac{6}{x};$ $x_1 = 3,\ x_2 = 5$

e) $f(x) = \frac{8}{2x-1};$ $x_1 = 0,\ x_2 = 1$

f) $f(x) = \frac{3}{x^2 - x + 1};$ $x_1 = 1,\ x_2 = 3$

2 For the functions (a) - (f) of the problem 1find the instantaneous rate of change of the function at the corresponding x_1 values.

3. Find the derivative of the following functions:

a) $y = f(x) = x^3 + 4x^2 - 2x + 8$

b) $y = f(x) = (x^4 + 3x^3 - 5x)^5$

c) $y = f(x) = \frac{(x^3 - 1)^4}{x^2}$

d) $y = f(x) = \sqrt{x^3 + 2x}$

e) $y = f(x) = \ln(x^3 + x)^2$

f) $y = f(x) = \ln(\frac{10}{x})$

g) $y = f(x) = \log10(x^2 - 1)$

h) $y = f(x) = e^{10x}$

i) $y = f(x) = 10^{5x}$

4. Find the slope of the following functions at the given points.

a) $y = f(x) = 2x + 10$; P(1, 12)

b) $y = f(x) = x^3 - 3x + 2$; P(1, 0)

c) $y = \frac{1+x}{1-x}$; P(0, 1)

d) $y = e^{2x}$; P(0, 1)

e) $y = \frac{x}{x^2+1}$; P(-1, $-\frac{1}{2}$)

f) $y = \log x$; P(10, 1)

5. Market demand and supply for a perfectly competitive industry is given as Qd = 7000 - 80P, Qs = 3000 + 120P. A firm in this market has the following total cost (TC) function $TC = 10 + 26q - 3.5q^2 + \frac{1}{3}q^3$.

a- What is the equilibrium price and quantity in this market?

b- Find the profit maximizing level of the output for the firm. What is the total profit? (Hint: For a firm in a perfectly competitive market the profit-maximizing quantity is decided by setting marginal cost (MC) equal to market price)

c- Is this firm in its long-run equilibrium?

6- Suppose the market in question 5 above is monopolized with a monopolist that produces the same good with a total cost of, $TC = 1685000 + 22.239q - Q^2 + .0003Q^3$.

a- What will be the level of output in the market? (Hint: A monopolist profit-maximizing quantity is decided by setting marginal cost (MC) equal to marginal revenue (MR))

f- At what price will the monopolist sell the product?

g- What will be the profit of the monopolist?

7. Find price elasticity of demand for the following demand functions. (Hint: Price elasticity of demand, ϵ_p, is defined as $\epsilon_p = \frac{dq}{dp} \cdot \frac{p}{q}$).

a) $q = 100 - p2$

b) $q = \frac{100}{P^3}$

8. A monopolist sells its product in two different markets, A and B, with differing price elasticities of demand, ϵ_A and ϵ_B. Prove that the profit maximizing condition requires that: $\frac{P_A}{P_B} = \frac{1-\frac{1}{\epsilon_B}}{1-\frac{1}{\epsilon_A}}$ where, P_A and P_B are market prices at A and B.

9. Prove that if demand for a product is elastic, an increase in price will decrease total revenue of the sellers and a decrease in price will increase total revenue.

10. Solve problem 7 replacing consumer spending for total revenue.

11. Total output of a firm as the function of labor input is given as Q = 100 + 20L + $10L^2$ - $.2L^3$. Find the average rate of change of Q with respect to L as L changes from: (i) 20 to 30, (ii) 20 to 25. Find the instantaneous rate of change of Q with respect to L at L = 20.

12. The following relation, C = a + blnY, a representation of the Engel curve, shows the relationship between the expenditure (C), and income (Y) of a consumer. Find the income elasticity. (Note:Income elasticity of demand is defined as $\in_Y = \frac{dC}{dY} \cdot \frac{Y}{C}$.

13. Let's assume that a researcher estimates the Engel curve of the problem 11 using an equation of the form lnC = a + blnY. What will be the income elasticity in this case.

14. Find the short-run labor elasticity of output for the following production function, $Q = 100L^{.75}K^{.25}$, assuming that K = 2401 units. (Note: Elasticity of output with respect to labor input is defined as $\epsilon_L = \frac{dQ}{dL} \cdot \frac{L}{Q}$.

III. Higher Order Derivatives: The D [f(x), {x, n}] Command

If f is a differentiable function, then its derivative f' is also a function that may have a derivative. Given y = f(x), $\frac{dy}{dx}$ defines the first derivative of y with respect to x. If we differentiate $\frac{dy}{dx}$ with respect to x, we obtain the second derivative of y with respect to x. The second derivative is denoted by $f''(x)$ or y'' or $\frac{d^2 y}{dx^2}$, or by f_{XX}. Derivatives of order higher than two are defined similarly.

Example: Find the first, second, and third derivatives of $y = f(x) = (x^3 - 1)^2$

Solution: The first derivative is $\frac{dy}{dx} = y' = f'(x) = f'_X = 6x^2(x^3 - 1)$, the second derivative is $\frac{d^2 y}{dx^2} = y'' = f''(x) = f'_{XX} = 30x^4 - 12x$, and the third derivative is $\frac{d^3 y}{dx^3} = y''' = f'''(x) = f'_{XXX} = 120x^3 - 12$.

The *Mathematica* command for the nth derivative of f(x) is **D** [f(x), {x, n}]. This command is used below to find the 1st, 2nd, and 3rd derivatives of the function given in the example.

```
In[39]:= Clear[y, x]
         y = (x^3 - 1)^2;
         D[y, {x, 1}]
Out[39]= 6 x² (-1 + x³)
In[40]:= D[y, {x, 2}]
Out[40]= 18 x⁴ + 12 x (-1 + x³)
In[41]:= D[y, {x, 3}]
Out[41]= 108 x³ + 12 (-1 + x³)
```

Example: Find the first, second, and third derivative of $y = f(x) = \frac{1}{x}$.

Solution: The first derivative is $\frac{dy}{dx} = y' = f'(x) = f'_X = -\frac{1}{x^2}$, The second derivative is $\frac{d^2 y}{dx^2} = y'' = f''(x) = f'_{XX} = \frac{2}{x^3}$, and the third derivative is $\frac{d^3 y}{dx^3} = y''' = f'''(x) = f'_{XXX} = \frac{-6}{x^4}$. Using ***Mathematica,***

```
In[42]:= Clear[y, x]
         y = 1/x;
         D[y, {x, 1}]
```

Out[42]= $-\frac{1}{x^2}$

```
In[43]:= D[y, {x, 2}]
```

Out[43]= $\frac{2}{x^3}$

```
In[44]:= D[y, {x, 3}]
```

Out[44]= $-\frac{6}{x^4}$

Since the first derivative of a function represents rate of change, the second derivative represents rate of change of a rate of change. In other words, the first derivative of a function is the slope function. The second derivative is the slope of the slope function.

In physical science, the second derivative represents acceleration. That is, if y = f(t) represents the position of an object that moves in a straight line, the first derivative f '(t) represents the velocity of the object as a function of time, and f "(t) represents acceleration. In business and economics, the second derivative represents scale of a change. For example, in the production process, if output is increasing, it may be increasing with an increasing rate or decreasing rate. The sign of the second derivative will show whether the increase is at an increasing rate or a decreasing rate. The cases that may occur are as follows:

Case1: f '$(x_o) > 0$ and f "$(x_o) > 0$. This implies that, at x_o, f is increasing at an increasing rate.

Example: Total Product, Marginal Product, and Economies of Scale

The relation between labor input (L) and output (Q) is given as Q = q(L) = $100L^{1.5}$. Find the first derivative and the second derivative of the function at $L_o = 25$ and comment on the scale effect at this point.

Solution: The first derivative of the function with respect to labor input (L) is q '(L) = $150L^{.5}$. This is the marginal product function. The second derivative of Q is q "(L) = $75L^{-.5}$. This function represents the slope of the marginal product function. The first derivative of q at L_o= 25 is q '(25) = 750 > 0. The second derivative evaluated at L_o = 25 is q "(25) = 15 > 0. Therefore, at L_o = 25, q(L) is increasing at an increasing rate. That is, at L_o = 25 not only the marginal product of labor is positive, the marginal product of labor will increase as extra units of labor are employed. This means that at L_o = 25production process is **increasing return to scale** or the function exhibits **economies of scale**. The *Mathematica* graph below is the graph of Q = $100L^{1.5}$ which is increasing and convex upward. To solve the problem using *Mathematica*,

```
In[45]:= Clear[Q, L]
         Q = 100 L^1.5;
         D[Q, L] /. L -> 25

Out[45]= 750.

In[46]:= D[Q, {L, 2}] /. L → 25

Out[46]= 15.

In[47]:= Plot[Q, {L, 0, 30}, AxesLabel -> {"L", "Q"}]
```

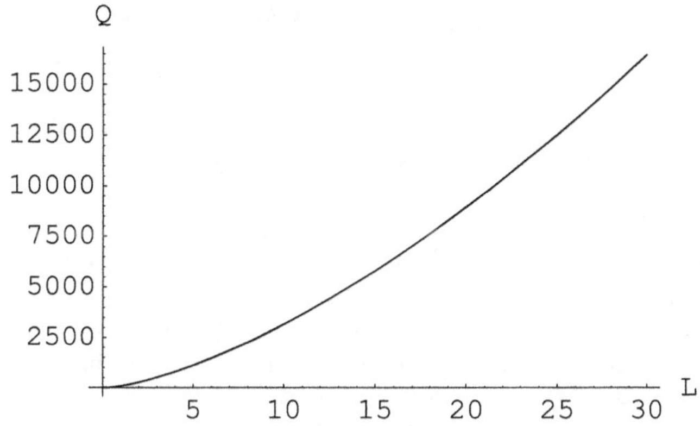

Case2: f '(x_o) > 0 and f "(x_o) < 0. This implies that, at x_o, f is increasing at a decreasing rate.

Example: Total Product, Marginal Product, and Diseconomies of Scale

The relation between labor input (L) and output (Q) is given as Q = q(L) = $100L^{.5}$. Find the first derivative and the second derivative of the function at $L_o = 25$ and comment on the scale effect at this point.

Solution: The first derivative of the function is q '(L) = $50L^{-.5}$. This is the marginal product function. The second derivative is q"(L) = -$25L^{-1.5}$. The first derivative of Q at L_o= 25 is q '(25) = 10 > 0. The second derivative evaluated at L_o = 25 is q "(25) = - .2 < 0. Therefore, at L_o = 25, q(L) is increasing at a decreasing rate. That is, the function exhibits **decreasing return to scale** or **diseconomies of scale**. The graph of the function, below, shows that the function is increasing and convex downward. The *Mathematica* solution to the problem is given below,

```
In[48]:= Clear[Q, L]
          Q = 100 L^.5;

         D[Q, L] /. L -> 25

Out[48]= 10.

In[49]:= D[Q, {L, 2}] /. L → 25

Out[49]= -0.2

In[50]:= Plot[Q, {L, 0, 30}, AxesLabel -> {"L", "Q"}]
```

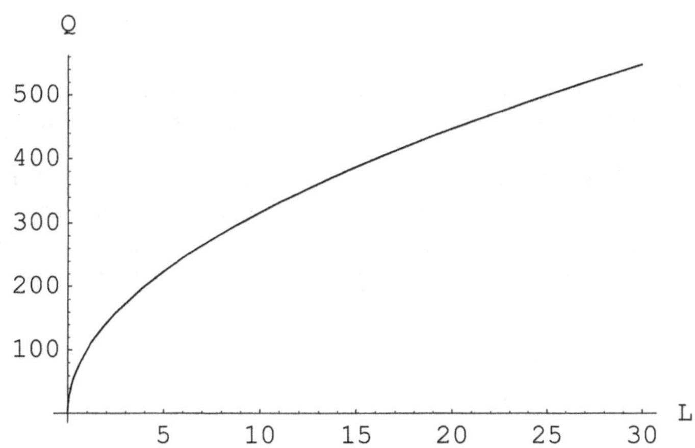

Case3: f '$(x_o) < 0$ and f "$(x_o) > 0$. This implies that, at x_o f is decreasing at an increasing rate.

Example: Average Variable Cost

Average variable cost of a firm (AVC) as a function of its output (Q) is given as AVC(Q) = f(Q) = 180 - 4.8Q + .08Q^2. Find the slope of the AVC curve at $Q_o = 10$. How does AVC change as output increase at the neighborhood of $Q_o = 10$.

Solution: The first derivative of the function gives the slope of the AVC. That is, f '(Q) = -4.8 + .16Q. At Q_o = 10, f '(10) = -3.2 < 0. This implies that AVC is a decreasing function of Q at $Q_o = 10$. The second derivative of AVC is f ' '(Q) = .16 > 0. That is, as Q increases at the neighborhood of $Q_o = 10$, AVC decreases with an increasing rate. As the graph of the function shows, at the neighborhood of $Q_o = 10$, the function is decreasing with an increasing rate (convex upward). The *Mathematica* solution to the problem is given below,

```
In[51]:= Clear[AVC, Q]
         AVC = 180 - 4.8 Q + .08 Q^2 ;
         D[AVC, Q] /. Q -> 10

Out[51]= -3.2

In[52]:= D[AVC, {Q, 2}] /. Q → 10

Out[52]= 0.16

In[53]:= Plot[AVC, {Q, 0, 50}, AxesLabel -> {"Q", "$"}]
```

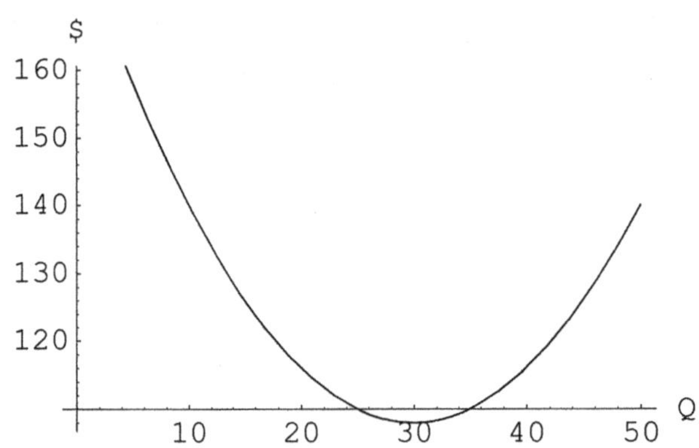

Case4: f '$(x_o) < 0$ and f "$(x_o) < 0$. This implies that, at x_o, f is decreasing at a decreasing rate.

Example: Average Product

Average product (AP) of a firm as a function of labor input (L) is given as AP(L) = f(L) = 200 + 5L -.3L^2. Find the slope of the AP at $L_o = 20$. How does AP change as labor input increase from $L_o = 20$.

Solution: The first derivative of the AP function is f '(L) = 5 - .6L. At $L_o = 20$, f '(20) = -7 < 0. That is, AP has a negative slope at $L_o = 20$. The second derivative of AP is f "(L) = -.6 < 0. That is as L increases from $L_o = 20$, AP deceases at a decreasing rate. The graph of the function, below, shows this fact. As L increases from $L_o = 20$, AP decreases with a decreasing rate (convex downward). The *Mathematica* solution to the problem is given below,

```
In[54]:= Clear[AP, L]
         AP = 2005 + 5 L - .3 L^2 ;
         D[AP, L] /. L -> 20

Out[54]= -7.

In[55]:= D[AP, {L, 2}] /. L → 20
```

Out[55]= -0.6

In[56]:=
```
Plot[AP, {L, 10, 30}, AxesLabel -> {"L", "AP"}]
```

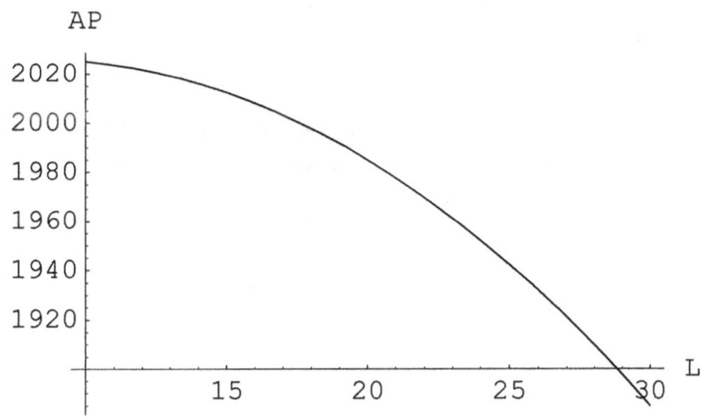

Example: Three Phases of Production Function

One of the conventional models of the production function is a third degree polynomial function that shows all three phases of production. Phase one is the starting or the low levels of input. In this phase, as input increases output increases with an increasing rate due to division of labor. Phase two is the middle phase, where the law of diminishing return sets in. In this phase, as input increases, output increases with a decreasing rate. The third phase is the output levels beyond the maximum level of output. In this phase as input increases, output decreases. The following graph is the presentation of such model given the relationship between labor input (L) and output (Q) expressed as Q = f(L) = 900L + $75L^2$ - L^3.

The following commands in *Mathematica* graph the function and finds the derivatives at different levels of label input that corresponds to three phases of production.

In[57]:=
```
Clear[Q, L]
Q = 900 L + 75 L^2 - 2 L^3;
Plot[Q, {L, 0, 35}, AxesLabel -> {"L", "Q"}]
```

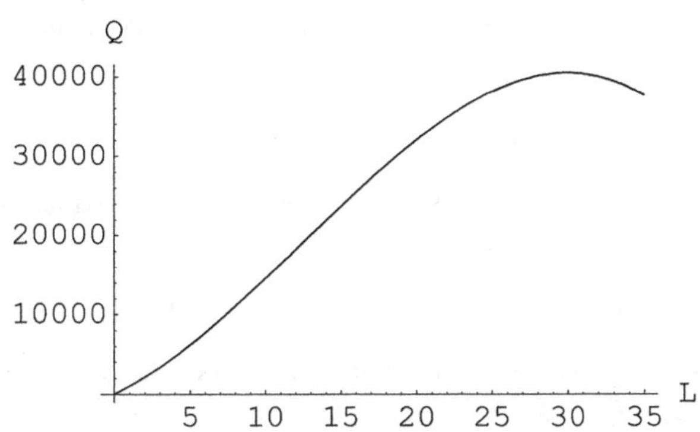

Phase one starts from labor input of zero to 12. At any point in the first phase, where output increases with increasing rate, the first derivative (the slope of the curve) and the second derivative (slope of the slope curve) has to be positive. We can test this for an input level of L_o = 10. The first derivative and second derivatives are f '(L) = 900 + 150L -3L^2 f "(L) = 150 - 6L, respectively. A L_o = 10, f '(10) = 1800 > 0 and f"(10) = 30 > 0. Since both f ' and f " are positive, at this point the function is increasing with an increasing rate. The *Mathematica* solution is given in the next page.

```
In[58]:= D[Q, L] /. L→10
Out[58]= 1800
In[59]:= D[Q, {L, 2}] /. L→10
Out[59]= 30
```

In the second phase, the first derivative (slope of the curve) is positive but the second derivative is negative. We can test this for an input level of $L_o = 20$. At this point f '(20) = 1500 > 0 and f "(20) < 0. The function is increasing with a decreasing rate. The mathematica solution is given below,

```
In[60]:= D[Q, L] /. L→20
Out[60]= 1500
In[61]:= D[Q, {L, 2}] /. L→20
Out[61]= -90
```

In the third phase, the first derivative (slope of the curve) is negative. We can test this for an input level of $L_o = 35$. At this point f'(35) = -1200 < 0. The *Mathematica* calculation is given below.

```
In[62]:= D[Q, L] /. L -> 35
Out[62]= -1200
```

Case 5: f '(x) = 0 and or f "(x) = 0. This will gives the critical points (maxima, minima, inflection point) of the function. This subject will be studied in the next chapter.

Taylor Series Expansion: Given a functional relation y = f(x), Taylor series expansion is a polynomial approximation to f(x) at a given point x = x_o. That is, $f(x) = f(x_o) + \frac{f'(x_o)(x-x_o)}{1} + \frac{f''(x_o)(x-x_o)^2}{2} + \frac{f'''(x_o)(x-x_o)^3}{6} + \ldots\ldots + R_n$, where, R_n is the remainder term. In general, Taylor series can be expressed as, $f(x) \approx f(x_o) + \sum_{i=1}^{n} \frac{1}{i!} \frac{d^i f(x_o)(x-x_o)^i}{dx^i}$, where $i!$ is the factorial of i, as n! = n(n-1)(n-2)........ 3.2.1. The choice of i is arbitrary. The higher is i, the more accurate is the approximation. The two most common and frequently used approximations are linear and quadratic approximations. The linear approximation can be expressed as, $f(x) \approx f(x_o) + \frac{f'(x_o)(x-x_o)}{1}$, or $f(x) \approx f(x_o) + f'(x_o)(x-x_o) \approx f'(x_o)x + [f(x_o) - x_o f'(x_o)]$, or f(x) = mx + b. The quadratic approximation is, $f(x) \approx f(x_o) + \frac{f'(x_o)(x-x_o)}{1} + \frac{f''(x_o)(x-x_o)^2}{2}$.

Example: Find a linear approximation to $y = x^2 - 4x + 5$, at $x_o = 1$.

Solution: f(x) ≈ f(1) + f '(1)(x-1). Here, f '(x) = 2x - 4, f '(x_o) = -2, and the linear approximation of f(x) ≈ 2-2(x-1) = 4 - 2x.

Example: Find linear and quadratic approximations for $y = f(x) = \sqrt{x-1}$ at $x_o = 5$.

Solution: $f'(x) = \frac{(x-1)^{-1/2}}{2}$, and $f''(x) = -\frac{(x-1)^{-3/2}}{4}$. Therefore, the linear approximation at x = 5 is $f(x) \approx 2 + \frac{1}{4}(x-5)$, and the quadratic approximation is $f(x) \approx f(5) + f'(5)(x-5) + \frac{1}{2}f''(5)(x-5) \approx 2 + \frac{1}{4}(x-5) - \frac{1}{4\sqrt[3]{16}}(x-5)^2$.

Problems

1. Find f '(x), f "(x), and f "'(x) for the following functions:

   a) $f(x) = 2x^3 - 3x^2 + 5x - 8$

   b) $f(x) = 3x^4 - 3x^{-2} - 5$

c) $f(x) = (2x - 3)^5(\frac{1}{x} - 3x^2)^{-4}$

d) $f(x) = \ln\frac{x+1}{x^2+1} + e^{2x}$

e) $f(x) = AL^{\alpha}$

2. For the cost function $TC = 20 + 50\,q - 3.75\,q^2 + 0.10\,q^3$ show that TC increases with a decreasing rate at the output level of Q = 10 and TC increases with an increasing rate at the output levels of Q = 20.

3. The total cost of repaying a home loan at an interest rate of r % per year is C = f(r).

a) What is the meaning of the derivative f '(r)?

b) What does the statement f '(10) = 800 mean?

c) Is f '(r) always positive or does it change in sign?

4. Find linear and quadratic approximations for the following functions at the given points.

a) $y = f(x) = \ln x;\quad x_o = 1$

b) $y = f(x) = \frac{1}{x};\quad x_o = 1$

c) $y = x^3 - 1;\quad x_o = 0$

d) $y = \sqrt{x^2 - 1};\ x_o = 1$, and $x_o = 2$.

IV. Some Helpful Rules: The `Dt` Command

There are many cases in which a function does not yield itself to a direct application of derivative formulas discussed in the earlier sections. The following rules help to differentiate such functions.

Chain Rule: Let f be given by y = f(z) and g by z = g(x). One way to obtain $\frac{dy}{dx}$ is to substitute for z in y = f(z), find a relation between y and x, and differentiate y with respect to x. However, it is often convenient to apply the following rule, which is known as the chain rule: $\frac{dy}{dx} = \frac{dy}{dz} \cdot \frac{dz}{dx}$.

Example: Let f be given by $y = f(z) = 2z^3 + 5z$ and g by $z = g(x) = (x^3 - 1)^2$. Find $\frac{dy}{dx}$.

Solution: Using the chain rule, $\frac{dy}{dx} = \frac{dy}{dz} \cdot \frac{dz}{dx} = (6z^2 + 5)[6x^2(x^3 - 1)]$. You may substitute for z to find $\frac{dy}{dx}$ in terms of x, $\frac{dy}{dx} = (6(x^3 - 1)^4 + 5)[6x^2(x^3 - 1)] = 6\,x^2\,(-1 + x^3)\,(5 + 6\,(-1 + x^3)^4)$.

In *Mathematica*, as long as the functions are defined, the command `D` [f(x), x] can be used to find the derivative of a function with respect to the variable of interest. For the same example as above, the *Mathematica* solution is:

```
In[63]:= Clear[x, y, z]
         y = 2 z^3 + 5 z;
         z = (x^3 - 1)^2;
         D[y, x]

Out[63]= 30 x^2 (-1 + x^3) + 36 x^2 (-1 + x^3)^5
```

The commands `Simplify[` *expr* `]` or `Factor[` *expr* `]` may be used to simplify or factor the solution as follows

```
In[64]:= Simplify[%]
Out[64]= 6 x^2 (-1 + x^3) (5 + 6 (-1 + x^3)^4)
```

Note that *Mathematica* substitutes for the value of z and differentiates y with respect to x.

Example: Let f be given by $y = f(z) = z^2 + 1$, g given by $z = g(x) = \ln x^2$, and h given by $x = h(t) = t$. Find $\frac{dy}{dt}$.

Solution: $\frac{dy}{dt} = \frac{dy}{dz}\cdot\frac{dz}{dx}\cdot\frac{dx}{dt} = 2z.\frac{2}{x}.1 = 4\frac{z}{x} = 4\ln\frac{x^2}{x} = 4\frac{\ln t^2}{t}$. Using ***Mathematica,***

```
In[65]:= Clear[x, y, z, t]
         y = z^2 + 1;
         z = Log[E, x^2];
         x = t;
         D[y, t]
Out[65]= 4 Log[t^2]/t
```

Implicit Differentiation: Given an implicit function F(x, y) = 0, the chain rule can be applied to find the derivative of the function without writing y as an explicit function of x.

Example: Let F(x, y) = 2x + 3y = 0. Find $\frac{dy}{dx}$.

Solution: $\frac{d}{dx}F(x, y) = 2 + 3\frac{dy}{dx} = 0$. Therefore, $\frac{dy}{dx} = -\frac{2}{3}$.

In ***Mathematica,*** Use the command **Dt**[f(x, y), x] to find the total derivative of the function and solve for Dt[y, x]. In the solution, Dt[y, x] is the same as $\frac{dy}{dx}$. For the example above the *Mathematica* solution is,

```
In[66]:= Clear[x, y]
         Dt[2 x + 3 y, x]
         Solve[% == 0, Dt[y, x]]
Out[66]= 2 + 3 Dt[y, x]
         {{Dt[y, x] -> -2/3}}
```

Example: Let $F(x, y) = 9x^2 + 4y^2 - 81 = 0$. Find $\frac{dy}{dx}$.

Solution: $\frac{d}{dx}F(x, y) = 18x + 8y\frac{dy}{dx} = 0$. Therefore, $\frac{dy}{dx} = -\frac{9x}{4y}$. In ***Mathematica,***

```
In[67]:= Clear[x, y]
         Dt[9 x^2 + 4 y^2 - 81, x];
         Solve[% == 0, Dt[y, x]]
Out[67]= {{Dt[y, x] -> -9 x/(4 y)}}
```

Example: Production Possibilities Frontier

Production possibilities frontier (PPF) shows the maximum amount of goods and services that a firm or a country can produce if all the available resources are used efficiently. Suppose that in a simple two-goods model of an economy PPF is represented by $x^2 + 9y^2 = 81$, where x and y are quantities of the two goods, food and fun, respectively. Find slope function of the PPF and prove that PPF is a decreasing function and is concave downward (that is, as more of x is produced the opportunity cost of producing it increases).

Solution: Differentiate the function and solve it for $\frac{dy}{dx}$. That is $2x + 18y\frac{dy}{dx} = 0$, or $\frac{dy}{dx} = \frac{-x}{9y}$. The slope of the PPF is $m = \frac{dy}{dx} = \frac{-x}{9y}$. Since, x and y are positive numbers, $m < 0$. To prove that the function is concave downward, we have to prove that the slope of the slope function or the second derivative of the PPF is always negative. Differentiate m with respect to x, $m' = \frac{d^2 y}{dx^2} = \frac{-9y + 9x\frac{dy}{dx}}{81y^2}$. Substitute for $\frac{dy}{dx}$ to get $m' = \frac{-9y^2 - x^2}{81y^3} = \frac{-81}{81y^3} = \frac{-1}{y^3} < 0$. That is, the PPF slopes downward and is concave downward.

To solve the problem and to graph it using *Mathematica,* use the following commands,

```
In[68]:= Clear[x, y]
         sol = Solve[Dt[x^2 + 9 y^2 - 81, x] == 0, Dt[y, x]]
         m = sol;
```

Out[68]= {{Dt[y, x] → $-\frac{x}{9y}$}}

The derivative of m with respect to x evaluated for $\frac{dy}{dx}$ is,

```
In[69]:= Dt[m, x] /. sol
```

Out[1]= {{{Dt[y, {x, 2}] → $-\frac{x^2}{81y^3} - \frac{1}{9y}$}}}

To observe the shape of the PPF , graph the function using the ContourPlot command as,

```
I n [ 6 9 ] : =
ContourPlot[x^2 + 9 y^2 - 81, {x, 0, 9}, {y, 0, 3}, ContourShading -> False]
```

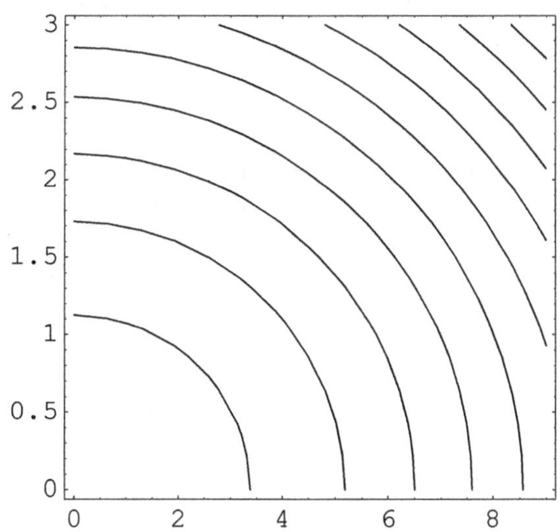

- ContourGraphics -

Differential of a Function: So far, we have interpreted $\frac{dy}{dx}$ as a single symbol and not as a quotient of dy and dx. We now assign separate meaning to dy and dx. We define the differential of a function f defined by $y = f(x)$ as $dy = f'(x)dx$. In this definition, dy, the differential of f, depends not only on x but also on dx. That is, to find a particular value of dy it is necessary to assign a value for x and a value for dx.

Example: Find the differential of $y = f(x) = x^2 - 2x$ when $x = 2$ and $dx = 1$.

Solution: $dy = (2x - 2)dx$. By substituting for the values of x and dx, $dy = 2$.

In *Mathematica,* use the command **Dt**[f(x, y)] to find the differential of the function. **Note** that, the command is used without defining the variable for which differentiation is done. Then solve for Dt[y], which is the same as dy, or solve for Dt[x], which is the same as dx.

Note that, although dx and dy are interpreted as separate variables, the ratio $\frac{dy}{dx}$ is still equal to f'(x).

```
In[70]:= Clear[x, y]
         sol1 = Dt[y - x^2 + 2 x];
         sol2 = Solve[sol1 == 0, Dt[y]];
         sol2 /. {x -> 2, Dt[x] -> 1}

Out[70]= {{Dt[y] → 2}}
```

The concept of differential of a function provides a convenient way of finding derivatives. This is specially true, when a function is in implicit form or the independent variable of a function is a function of other variables. The following examples illustrate derivatives of such functions.

Example: Find $\frac{dy}{dx}$ if $x^2 + y^2 - 9 = 0$.

Solution: Differential of the function is 2xdx + 2ydy = 0. Therefore, $\frac{dy}{dx} = -\frac{x}{y}$.

In ***Mathematica,*** the command "Dt[f(x, y), x]" finds differential of the function with respect to x. The command "Solve" then can be used to find Dt[y, x], which is the derivative of y with respect to x.

```
In[71]:= Clear[x, y]
         Dt[x^2 + y^2 - 9, x];
         Solve[% == 0, Dt[y, x]]

Out[71]= {{Dt[y, x] → -x/y}}
```

Example: Find $\frac{dy}{dt}$ if $y = f(x) = x^2 + 2x + 3$ and $x = g(t) = t^2 + 1$.

Solution: dy = f'(x)dx = 2xdx + 2dx and dx = g'(t)dt = 2tdt. Therefore, dy = 4xtdt + 4tdt, and $\frac{dy}{dt} = 4xt + 4t = 4t(x + 1) = 4t(t^2 + 2)$. In *Mathematica,* define both functions and use the command "Dt[y, t]" to find the derivative of y with respect to t.

```
In[72]:= Clear[x, y, t]
         y = x^2 + 2 x + 3;
         x = t^2 + 1;
         D[y, t]
         Simplify[%]

Out[72]= 4 t + 4 t (1 + t^2)

         4 t (2 + t^2)
```

The symbol "d" can be interpreted as the differential operator for the function to which it is applied. The operator follows the rules that apply to derivative rules (Differential Formulas). Therefore, if u and v are differentiable functions of x, we can write,

1- dc = 0, if c is constant.

2- d(u + v) = du + dv

3- $d(u^m) = mu^{m-1}du$

4- d(uv) = udv + vdu

5- $d(\frac{u}{v}) = \frac{vdu - udv}{v^2}$

6- $d(\ln u) = \frac{du}{u}$

7- $d(e^u) = e^u du$

Example: If $y = f(x) = (x^3 - 1)^4(x^2 + x + 10)^3$, find dy.

Solution: Let $u = (x^3 - 1)^4$ and let $v = (x^2 + x + 10)^3$. Using the formulas 3 and 4 above, dy = udv + vdu, where $du = 12x^2(x^3 - 1)^3dx$, and $dv = 3(2xdx + dx)(x^2 + x + 10)^2 = 3(2x + 1)(x^2 + x + 10)^2dx$. Hence, $dy=(x^3 - 1)^4[3(2x + 1)(x^2 + x + 10)^2]dx + (x^2 + x + 10)^3[12x^2(x^3 - 1)^3]dx$. In ***Mathematica,***

In[73]:=

```
Clear[x, y]
Dt[y - (x^3 - 1)^4 (x^2 + x + 10)^3];
Solve[% == 0, Dt[y]]
Simplify[%]
```

Out[73]=

```
{{Dt[y] → 12 x^2 (10 + x + x^2)^3 (-1 + x^3)^3 Dt[x] + 3 (10 + x + x^2)^2 (-1 + x^3)^4 (Dt[x] + 2 x Dt[x])}}

{{Dt[y] → 3 (10 + x + x^2)^2 (-1 + x^3)^3 (-1 - 2 x + 40 x^2 + 5 x^3 + 6 x^4) Dt[x]}}
```

Problems

1 Use the following functions to find $\frac{dy}{dx}$:

a) $y = f(z) = z^2 - z$; $z = g(x) = x^2 + 2x$

b) $y = f(z) = \ln(z^2 - z)$; $z = g(x) = \ln(x^2 + 2x)$

c) $y = f(z) = \frac{1}{z+1}$; $z = g(x) = \frac{x}{x^2+1}$

d) $y = f(z) = (3z^3 + 2z^2 + 10)^4$; $z = g(w) = (w^4 - 20)^2$; and $w = 5x^3 - 3x$

2 For the following functions, find $\frac{dy}{dx}$ by implicit differentiation:

a) $y = 2x^3 - 2y^{-2}$ b) $x^2 + xy + y^2 = 5$

c) $y = \ln \frac{x+1}{x^2+y}$ d) $\frac{1}{x} + \frac{1}{y} = 3$

3 Find dy for the following functions:

a) $y = f(x) = x^2 + 3x + 12$

b) $y = f(x) = \frac{x^2+1}{x+1}$

c) $y = f(x) = (2x + 1)^4(x^2 + x + 1)^3$

d) $y = f(x) = \ln (x^2 - 1)$

e) $y = f(x) = e^{4x}$

4. For the following demand functions find the slope of the demand curve and the price elasticity of demand at the given prices.

a) $Q = 100 - 2P$; $P = 8$

b) $Q = \frac{100}{P}$; P = 2

c) $Q = \frac{100}{P^3}$ P = 2

d) Q = 100 - lnP P = 10

e) Q = 100 - 3lnP P = 10

V. Differentiation of Functions of Two or More variables: The **Dt**[f(x, y), x] Command

A real-valued function of two real variables, defined by the expression z = f(x, y), is a mapping of set of ordered pairs (x, y) of real numbers to a set of f(x, y) of real numbers. The pairs (x, y) are the domain of the function and the set f(x, y) is the range of the function.

The definition of a real-valued function of *n* variables, defined by the expression u = f(x, y, z, ...), is the same as that of a function of two variables except that the domain of the function consists of ordered variables of more than two (x, y, z, ...) rather than ordered pairs (x, y).

Partial Derivatives: Let $y = f(x_1, x_2, ..., x_n)$ define a function of *n* variables. If all the n variables except x_j are fixed, y becomes a function of the one variable x_j. The derivative of y with respect to x_j is called **partial derivative** of y with respect to x_j and is denoted by $\frac{\partial y}{\partial x_j}$ or fx_j.

Example: Let $z = f(x, y) = x^2y^2 - xy + y + 5$. Find partial derivatives of z with respect to x and y.

Solution: Partial derivative of z with respect to x is $f_x = \frac{\partial z}{\partial x} = 2xy^2 - y$ and partial derivative of z with respect to y is $f_y = \frac{\partial z}{\partial y} = 2x^2y - x + 1$.

In *Mathematica,* use the derivative command "**D**". Use the variable for which the partial derivative is derived as the argument of the command.

```
In[74]:= Clear[x, y, z]
         z = x^2 y^2 - x y + y + 5;
         D[z, x]
Out[74]= -y + 2 x y^2
```

The partial derivative of z with respect to x is $\frac{\partial z}{\partial x} = -y + 2xy^2$. To find the partial derivative of z with respect to y, use the command D[z, y] as below,

```
In[75]:= D[z, y]
Out[75]= 1 - x + 2 x^2 y
```

the partial derivative of z with respect to y is $\frac{\partial z}{\partial y} = 1 - x + 2x^2y$.

Example: Cobb-Douglas Production Function

Consider a Cob-Douglas production function, $Q = AL^aK^b$. The partial derivative of Q with respect to L, (Q_L), is the marginal product of labor. This can be derived by holding the level of capital fixed and differentiating with respect to L. The partial derivative of Q with respect to K, (Q_K), is the marginal product of capital. This can be derived by holding the level of labor fixed and differentiating with respect to K In *Mathematica,*

In[76]:=
```
Clear[Q, L, K]
Q = A L^a K^b;
QL = D[Q, L]
```

Out[76]= $a\,A\,K^b\,L^{-1+a}$

In[77]:=
```
Qk = D[Q, K]
```

Out[77]= $A\,b\,K^{-1+b}\,L^a$

If the function is an implicit function of several variables, to find the partial derivatives, implicitly differentiate with respect to the variables and solve for the partial derivatives of interest.

Example: Let $x^2z^2 + y^2z^2 - 2xy - 2zy + 2z = 0$. Find $\frac{\partial z}{\partial x}$ and $\frac{\partial z}{\partial y}$.

Solution: To find $\frac{\partial z}{\partial x}$, let y be fixed and implicitly differentiate with respect to x. This gives, $2xz^2 + 2\frac{\partial z}{\partial x}zx^2 + 2y^2z\frac{\partial z}{\partial x} - 2y - 2y\frac{\partial z}{\partial x} = 0$. Solving for $\frac{\partial z}{\partial x}$, results in $\frac{\partial z}{\partial x} = \frac{y - xz^2}{zx^2 + zy^2 - y + 1}$. To find $\frac{\partial z}{\partial y}$, let x be fixed and implicitly differentiate with respect to y. This gives, $2x^2z\frac{\partial z}{\partial y} + 2yz^2 + 2y^2z\frac{\partial z}{\partial y} - 2x - 2z - 2y\frac{\partial z}{\partial y} + 2\frac{\partial z}{\partial y} = 0$. Solve for $\frac{\partial z}{\partial y}$ to get $\frac{\partial z}{\partial y} = \frac{x + z - y}{x^2z + y^2z - y + 1}$.

In *Mathematica,* use the implicit differentiation command "Dt" to differentiate the function with respect to the variable of interest, say x. Since the y variable is assumed to be fixed, set the partial derivatives of the z with respect to y equal to zero and solve for $\frac{\partial z}{\partial x}$. To solve the problem using *Mathematica*,

In[78]:=
```
Clear[x, y, z]
Dt[x^2 z^2 + y^2 z^2 - 2 x y - 2 z y + 2 z, x];
Solve[% == 0, Dt[z, x]] /. Dt[y, x] -> 0
```

Out[78]= $\left\{\left\{\text{Dt}[z, x] \to -\frac{-y+x\,z^2}{1-y+x^2\,z+y^2\,z}\right\}\right\}$

To find $\frac{\partial z}{\partial y}$, differentiate the function with respect to the y variable. Since the x variable is assumed to be fixed, set the partial derivatives of the z with respect to x equal to zero and solve for $\frac{\partial z}{\partial y}$.

In[79]:=
```
Clear[x, y, z]
Dt[x^2 z^2 + y^2 z^2 - 2 x y - 2 z y + 2 z, y];
Solve[% == 0, Dt[z, y]] /. Dt[x, y] -> 0
```

Out[79]= $\left\{\left\{\text{Dt}[z, y] \to -\frac{-x-z+y\,z^2}{1-y+x^2\,z+y^2\,z}\right\}\right\}$

Taylor Series Expansion of Multivariate Functions: The Taylor series expansion, discussed earlier in this chapter, can be applied to multivariate function also. For a multivariate function $y = f(\mathbf{x}) = f(x_1, x_2, \ldots\ldots x_n)$, the gradient vector $f'(\mathbf{x})$ is, $f'(\mathbf{x}) = \frac{df(x)}{dx} = \begin{pmatrix} \frac{\partial y}{\partial x1} \\ \frac{\partial y}{\partial x2} \\ . \\ \frac{\partial y}{\partial xn} \end{pmatrix} = \begin{pmatrix} f_1 \\ f_2 \\ . \\ f_n \end{pmatrix}$, and the second derivatives results in the Hessian matrix which is a square, symmetric matrix with the elements as, $\begin{pmatrix} \frac{\partial^2 y}{\partial x1\,\partial x1} & \frac{\partial^2 y}{\partial x1\,\partial x2} & . & . & \frac{\partial^2 y}{\partial x1\,\partial xn} \\ . & . & . & . & . \\ \frac{\partial^2 y}{\partial xn\,\partial x1} & \frac{\partial^2 y}{\partial xn\,\partial x2} & . & . & \frac{\partial^2 y}{\partial xn\,\partial xn} \end{pmatrix} = \begin{pmatrix} f_{11} & f_{12} & . & . & f_{1n} \\ f_{21} & f_{22} & . & . & f_{2n} \\ . & . & . & . & . \\ f_{n1} & f_{n2} & . & . & f_{nn} \end{pmatrix}$.

The first-order or linear Taylor series approximation of the function is $y \approx f(\boldsymbol{x_o}) + [\frac{df(x_o)}{dx_o}]'(\mathbf{x} - \boldsymbol{x_o}) = f(\boldsymbol{x_o}) + f'(x_o)(\mathbf{x} - \boldsymbol{x_o})$ or $y \approx [f(\boldsymbol{x_o}) - f'(\boldsymbol{x_o})\boldsymbol{x_o}] + f'(x_o)\mathbf{x} = \beta_o + \beta_1'\mathbf{x}$.

The second-order or quadratic approximation is $y \approx f(\boldsymbol{x_o}) + f'(\mathbf{x} - \boldsymbol{x_o}) + \frac{1}{2}(\mathbf{x} - \boldsymbol{x_o})'H_o(\mathbf{x} - \boldsymbol{x_o})$. Multiplication and collection of the terms results in $y \approx \beta_o + \beta_1'\mathbf{x} + \frac{1}{2}\mathbf{x}'H_o\mathbf{x}$ where, $\beta_o = f(\boldsymbol{x_o}) - f'(x_o)\boldsymbol{x_o} + \frac{1}{2}\boldsymbol{x_o}'H_o\boldsymbol{x_o}$ and $\beta_1 = -f'(\boldsymbol{x_o}) + H_o\boldsymbol{x_o}$.

Example: Find linear approximation for $z = f(x, y) = x^2 - xy + y^2$ at $x_o = y_o = 1$.

Solution: The gradient vector is $f'(x) = \begin{pmatrix} 2x - y \\ -x + 2y \end{pmatrix}$. Evaluated at the given point, $f'(x_o) = \begin{pmatrix} 1 \\ 1 \end{pmatrix}$, and $f(x_o) = 1$. The linear approximation of the function is $z \approx f(\boldsymbol{x_o}) + f'(\boldsymbol{x_o})(\mathbf{x} - \boldsymbol{x_o}) \approx 1 + (1 \;\; 1)\begin{pmatrix} x - 1 \\ y - 1 \end{pmatrix} = x + y + 1$.

Example: Find the linear and quadratic approximation for $z = f(x + y)$ at $x_o = 0$, $y_o = 1$.

Solution: The linear approximation of $z \approx f(\boldsymbol{x_o}) + f'(\boldsymbol{x_o})(\mathbf{x} - \boldsymbol{x_o})$. $f(\boldsymbol{x_o}) = \ln(0 + 1) = 0$, $f'(x) = \begin{pmatrix} \frac{1}{x+y} \\ \frac{1}{x+y} \end{pmatrix}$, and $f'(\boldsymbol{x_o}) = \begin{pmatrix} 1 \\ 1 \end{pmatrix}$. Therefore, $z \approx 0 + (1 \;\; 1)\begin{pmatrix} x \\ y - 1 \end{pmatrix} \approx x + y - 1$. For the quadratic approximation, we need to find $H = \begin{pmatrix} \frac{-1}{(x+y)^2} & \frac{-1}{(x+y)^2} \\ \frac{-1}{(x+y)^2} & \frac{-1}{(x+y)^2} \end{pmatrix}$. Evaluated at the given point, $x_o = 0$, $y_o = 1$, $H_o = \begin{pmatrix} -1 & -1 \\ -1 & -1 \end{pmatrix}$.

The quadratic approximation of $z \approx f(\boldsymbol{x_o}) + f'(\boldsymbol{x_o})(\mathbf{x} - \boldsymbol{x_o}) + (\mathbf{x} - \boldsymbol{x_o})'H_o(\mathbf{x} - \boldsymbol{x_o}) = 0 + (1 \;\; 1)\begin{pmatrix} x \\ y - 1 \end{pmatrix} + \frac{1}{2}(x \;\; y - 1)\begin{pmatrix} -1 & -1 \\ -1 & -1 \end{pmatrix}\begin{pmatrix} x \\ y - 1 \end{pmatrix} = -\frac{1}{2}(x + y)^2 + 2(x + y) - \frac{3}{2}$.

Example: Find the linear and quadratic approximation of z $f(x, y) = e^{xy}$ at $x_o = 0$, $y_o = 1$.

Solution: $f(\boldsymbol{x_o}) = 1$, $f'(x) = \begin{pmatrix} ye^{xy} \\ xe^{xy} \end{pmatrix}$, $f'(\boldsymbol{x_o}) = \begin{pmatrix} 1 \\ 0 \end{pmatrix}$, $\begin{pmatrix} y^2 e^{xy} & e^{xy} + xye^{xy} \\ e^{xy} + xye^{xy} & x^2 e^{xy} \end{pmatrix}$, and $H_o = \begin{pmatrix} 1 & 1 \\ 1 & 0 \end{pmatrix}$. Therefore, the linear approximation of $z \approx 1 + (1 \;\; 0)\begin{pmatrix} x \\ y - 1 \end{pmatrix} = 1 + x$ and the quadratic approximation of $z \approx 1 + (1 \;\; 0)\begin{pmatrix} x \\ y - 1 \end{pmatrix} + \frac{1}{2}(x \;\; y - 1)\begin{pmatrix} 1 & 1 \\ 1 & 0 \end{pmatrix}\begin{pmatrix} x \\ y - 1 \end{pmatrix} = \frac{1}{2}x^2 + xy + 1$.

Problems:

1. Find partial derivatives of the following functions with respect to x and y:

a) $z = x^2 + xy + y^2$ b) $z = e^{xy}$

c) $z = \frac{2x^5}{y^8}$ d) $z = 5x^{1.2}y^{.6}$

e) $Z = \ln \frac{x+y}{y^2}$ f) $z = 10(1 + xy)^2 - x^2y$

2. The relationship between labor input (L), capital input (K) and output (Q) is given as $Q = 100 L^{.7}K^{.3}$. Find marginal product of labor and marginal product of capital. (Hint: marginal product of labor is partial derivative of Q with respect to L. Marginal product of capital is partial derivative of Q with respect to K)

3. A consumer's utility (U) for two goods X and Y is a function of quantities of X and Y consumed. If the utility function is given as U = ln xy, find marginal utility of x and marginal utility of y. (Hint: Marginal utility of x is $\frac{\partial U}{\partial x}$ and marginal utility of y is $\frac{\partial U}{\partial y}$)

4. Demand for a product is estimated as $Q = 25P^{-.8}Y^{1.2}$ where, Q is the quantity demanded, P is the price, and y is the income. Find the price and income elasticities of the demand.

5. Find the labor and capital elasticities of output for the production function $Q = 100L^{.75}K^{.25}$. (Note: Elasticity of output with respect to labor input is defined as $\epsilon_L = \frac{dQ}{dL} \cdot \frac{L}{Q}$.

6. Find linear and quadratic approximations for the following functions at the given points.

a) $z = f(x, y) = \ln xy$; $x_o = 1, y_o = 1$.

b) $z = f(x, y) = x^3 - xy + y^3$; $x_o = 1, y_o = 1$.

c) $z = f(x, y) = e^x + e^y$; $x_o = 1, y_o = 1$.

VI. Special Topics in Differential Calculus

Jacobian Determinant: If two or more functions share a set of common variables, the partial derivatives of functions arranged into a determinant in a prescribed order is called the **Jacobian** determinant and is denoted byDet[J]. More formally, if we have *n* differentiable functions in *n* variables:

$$y_1 = f_1(x_1, x_2, \ldots\ldots, x_n)$$

$$y_2 = f_2(x_1, x_2, \ldots\ldots, x_n)$$

$$\ldots \quad \ldots\ldots\ldots$$

$$y_n = f_n(x_1, x_2, \ldots\ldots, x_n)$$

the Jacobian determinant is calculated as,

$$\text{Det}[J]| = \frac{\partial(y1, y2, \ldots\ldots yn)}{\partial(x1, x2, \ldots\ldots, xn)} = \begin{pmatrix} \frac{\partial y1}{\partial x1} & \frac{\partial y1}{\partial x2} & \cdot\ \cdot\ \cdot\ \cdot & \frac{\partial y1}{\partial xn} \\ \cdot & \cdot & \cdot\ \cdot\ \cdot\ \cdot & \cdot \\ \cdot & \cdot & \cdot\ \cdot\ \cdot\ \cdot & \cdot \\ \frac{\partial yn}{\partial x1} & \frac{\partial yn}{\partial x2} & \cdot\ \cdot\ \cdot\ \cdot & \frac{\partial yn}{\partial xn} \end{pmatrix}$$

The Jacobian is used for testing the existence of linear or nonlinear dependence among functions. If Det[J] = 0, the functions are said to be dependent.

Example: Verify the independence of the following functions:

$$y_1 = 2x_1 + 3x_2$$

$$y_2 = x_1 + 2x_2$$

Solution: The Jacobian determinant is $\text{Det}[J] = \frac{\partial(y1, y2)}{\partial(x1, x2)} = \text{Det}\begin{pmatrix} 2 & 3 \\ 1 & 2 \end{pmatrix} = 1$. Since Det[J] $\neq$ 0, the equations are independent.

To use *Mathematica*, define the functions and formulate the Jacobian determinant as defined above and find the determinant,

```
In[80]:= Clear[x1, x2, y1, y2]
         y1 = 2 x1 + 3 x2;
         y2 = x1 + 2 x2;
         Det[( D[y1, x1]  D[y1, x2] )]
             ( D[y2, x1]  D[y2, x2] )

Out[80]= 0
```

Example: Verify the independence of the following functions:

$y_1 = x_1^2 - x_1 x_2$

$y_2 = x_1^2 + x_2^2$

Solution: The Jacobian determinant is Det[J] = $\frac{\partial(y1, y2)}{\partial(x1, x2)} = \text{Det}\begin{pmatrix} 2\,x1 - x2 & -x1 \\ 2\,x1 & 2\,x2 \end{pmatrix}$ = $4x_1x_2$ $-2x2^2 + 2x1^2$. Since Det[J] $\neq 0$, the equations are independent. Using *Mathematica*, the solution is,

```
In[81]:= Clear[x1, x2, y1, y2]
         y1 = x1^2 - x1 x2;
         y2 = x1^2 + x2^2;
         Det[( D[y1, x1]  D[y1, x2] )]
             ( D[y2, x1]  D[y2, x2] )
```

Out[81]= $2\,x1^2 + 4\,x1\,x2 - 2\,x2^2$

Example: Verify the independence of the following equations:

$y_1 = x_1^2 + 2x_1x_2 + x_3^2$

$y_2 = x_1 + x_2 + x_3$

$y_3 = 2x_1x_3 + 2x_2x_3 + x_2^2$

The Jacobian determinant is Det[J] = $\frac{\partial(y1, y2, y3)}{\partial(x1, x2, x3)} = \text{Det}\begin{pmatrix} 2\,x1 + 2\,x2 & 2\,x2 & 2\,x3 \\ 1 & 1 & 1 \\ 2\,x3 & 2\,x2 + 2\,x3 & 2\,x1 + 2\,x2 \end{pmatrix}$. If evaluated, this determinant equals zero. Therefore, the equations are not independent from each other. This can be verified by *Mathematica* as,

```
In[82]:= Clear[x1, x2, x3, y1, y2, y3]
         y1 = x1^2 + 2 x1 x2 + x3^2;
         y2 = x1 + x2 + x3;
         y3 = 2 x1 x3 + 2 x2 x3 + x2^2;

             ( D[y1, x1]  D[y1, x2]  D[y1, x3] )
         Det[( D[y2, x1]  D[y2, x2]  D[y2, x3] )]
             ( D[y3, x1]  D[y3, x2]  D[y3, x3] )

Out[82]= 0
```

Since Det[J] = 0, the three equations are not independent. In fact, an inspection of equations shows that $y_3 = y_2^2 - y_1$.

Partial Derivatives of Higher Order and the Hessian Determinant: Given the function f defined by z = f(x, y), the partial derivatives of f with respect to x and y are functions of x and y. Continued partial differentiation of the function yield partial derivatives of higher order. Denoting the partial derivative of z = f(x, y) with respect to x as $f_x(x, y) = \frac{\partial z}{\partial x}$ and the partial derivative of z with respect to y as $f_y(x, y) = \frac{\partial z}{\partial y}$, the following notations are generally employed for the second-order partial derivatives:

Partial derivative of f_x (x, y) with respect to x is donated by, $\frac{\partial}{\partial x}\frac{\partial f}{\partial x} = \frac{\partial^2 f}{\partial x^2} = f_{xx}$.

Partial derivative of f_x (x, y) with respect to y is donated by, $\frac{\partial}{\partial x}\frac{\partial f}{\partial y} = \frac{\partial^2 f}{\partial x \partial y} = f_{xy}$.

Partial derivative of f_y (x, y) with respect to x is donated by, $\frac{\partial}{\partial y}\frac{\partial f}{\partial x} = \frac{\partial^2 f}{\partial y \partial x} = f_{yx}$.

Partial derivative of f_{yy} (x, y) with respect to x is donated by, $\frac{\partial}{\partial y}\frac{\partial f}{\partial y} = \frac{\partial^2 f}{\partial y^2} = f_{yy}$.

A determinant whose elements are the second-order partial derivatives is called a **Hessian** determinant and is denoted by |H| or Det[H]. For the function f, defined by z = f(x, y), the Hessian determinant is,

$$|H| = \text{Det}[H] = \begin{pmatrix} f_{xx} & f_{xy} \\ f_{yx} & f_{yy} \end{pmatrix}$$

Example: If z = f(x, y) = x^3 - xy + y^3, find the second-order partial derivative of the function and evaluate the Hessian determinant at P(1, 2, 7).

Solution: $f_x = \frac{\partial f}{\partial x} = 3x^2 - y$, $f_y = \frac{\partial f}{\partial y} = -x + 3y^2$, $f_{xx} = \frac{\partial^2 f}{\partial x^2} = 6x$, $f_{xy} = \frac{\partial^2 f}{\partial x \partial y} = -1$, $f_{yx} = \frac{\partial^2 f}{\partial y \partial x} = -1$, and $f_{yy} = \frac{\partial^2 f}{\partial y^2} = 6y$. The Hessian determinant is Det[H] = $\begin{pmatrix} 6x & -1 \\ -1 & 6y \end{pmatrix}$. Evaluated at P(1, 2, 7), Det[H] = $\begin{pmatrix} 6 & -1 \\ -1 & 12 \end{pmatrix}$ = 71. To evaluate Det[H]through *Mathematica*,

```
In[83]:= Clear[x, y, z]
         z = x^3 - x y + y^3;
         Det[( D[D[z, x], x]  D[D[z, x], y]
               D[D[z, y], x]  D[D[z, y], y] )] /. {x -> 1, y -> 2}

Out[83]= 71
```

Note that, by what is called the Young's Theorem $f_{xy} = f_{yx}$, for every x and y. Young's Theorem assures that the order of differentiation is immaterial in cross derivatives.

Differential of a Function of Two or More Variables: Let z = f(x, y) define a differentiable function of x and y. The differential of the function is defined as $dz = f_x dx + f_y dy$, where dx and dy are arbitrary real numbers.

Example: Given $z = f(x, y) = x^2 y + y^2 - x$, find dz.

Solution: $dz = 2xydx + x^2 dy + 2ydy - dx$. Using *Mathematica,* define the function and use the "D" command without an argument.

```
In[84]:= Clear[x, y, z]
         z = x^2 y + y^2 - x;
         Dt[z]

Out[84]= -Dt[x] + 2 x y Dt[x] + x² Dt[y] + 2 y Dt[y]
```

Example: Indifference Curve and Marginal Rate of Substitution

Given the utility function, U = U(x, y), find the marginal rate of substitution (MRS) of x for y.

Solution: Marginal rate of substitution of x for y (MRS) is the trade off rate between x and y on indifference curve in such a way that the consumer's utility stays unchanged. That is, MRS is the slope of the indifference curve at a point on indifference curve. Differential of U yields $dU = U_x dx + U_y dy$ where, U_x and U_y are the marginal utilities of x and y, respectively. For a given level of utility, (dU = 0). Therefore, $\frac{dy}{dx} = -\frac{U_x}{Uy}$ = MRS.

Example: MRS in a Cobb-Douglas Utility Function

For the Cob-Douglas utility function $U = Ax^{\alpha} y^{\beta}$ find marginal rate of substitution (MRS).

Solution: Marginal Rate of substitution of x for y is defined as $\frac{dy}{dx} = -\frac{U_x}{Uy}$, where U_x and U_y are marginal utilities of x and y, respectively. For the given Cob-Douglas utility function, $U_x = A\alpha x^{\alpha-1} y^{\beta}$ and $U_y = A\beta x^{\alpha} y^{\beta-1}$. Therefore, MRS = $-\frac{\alpha y}{\beta x}$. Using *Mathematica,*

```
In[85]:= Clear[x, y]
         U = A x^α y^β;
         Ux = D[U, x];
         Uy = D[U, y];
         MRS = -Ux / Uy

Out[85]= -(y α)/(x β)
```

Example: IS-LM model of Economy

A very standard and powerful tool of analyzing the effect of monetary and fiscal policy on output (y) and interest rate (r) is a simple model called IS-LM model. IS is downward sloping curve representing all the combination of y and r that the goods market is in equilibrium. LM is an upward sloping curve representing all the combination of y and r that the money market is in equilibrium. Suppose that the IS model of an economy is expressed by y = D(y, r) where D(y, r) is the aggregate demand as a function of income (y) and interest rate (r). Find the slope of the IS curve and prove that the slope is negative.

Solution: Differential of the function yields $D_y dy + D_r dr = dy$, where the slope $\frac{dr}{dy}$ is, $\frac{dr}{dy} = \frac{1 - D_y}{Dr}$. D_y is the derivative of aggregate demand with respect to income, which is marginal propensity to consume (MPC). Since 0 < MPC < 1, then $1 - D_y > 0$. D_r is the derivative of the aggregate demand with respect to interest rate, which conventionally is assumed to be negative, $D_r < 0$. Therefore, $\frac{1 - D_y}{D_r} < 0$.

Example: Find the second differential of $z = f(x, y)$.

Solution: The first differential of z yields $dz = f_x dx + f_y dy$. The second differential is a measure of the change in dz,

$$d(dz) = d^2z = \frac{\partial}{\partial x}(f_x dx + f_y dy)dx + \frac{\partial}{\partial y}(f_x dx + f_y dy)dy$$

$$d(dz) = d^2z = (f_{xx}dx + f_{yx}dy)dx + (f_{xy}dx + f_{yy}dy)dy = f_{xx}dx^2 + 2f_{xy}dxdy + f_{yy}dy^2$$

Or, in the matrix notation $d(dz) = d^2z = (\,dx \quad dy\,)\begin{pmatrix} f_{xx} & f_{xy} \\ f_{yx} & f_{yy} \end{pmatrix}\begin{pmatrix} dx \\ dy \end{pmatrix}$.

Total Derivative: The differential of the function $z = f(x, y)$ is $dz = f_x dx + f_y dy$. Dividing both sides of this equation by dx yields $\frac{dz}{dx} = f_x + f_y \frac{dy}{dx}$. Since the ratio $\frac{dz}{dx}$ may be interpreted as the derivative, we can regard it as a measure of the instantaneous rate of change of z with respect to x. The expression $\frac{dz}{dx} = f_x + f_y \frac{dy}{dx}$ is called **total derivative**, and the process of finding it is called total differentiation.

Example: Find total derivative $\frac{dz}{dx}$, given that $z = f(x, y) = 2x^3 - 5xy + y^2$ and $y = (x - 1)^2$.

Solution: $dz = f_x dx + f_y dy$, and $\frac{dz}{dx} = f_x + f_y \frac{dy}{dx} = 6x^2 - 5y + 2(-5x + 2y)(x - 1)$. Using *Mathematica*,

```
In[86]:= Clear[x, y, z]
         z = 2 x^3 - 5 x y + y^2;
         y = (x - 1)^2;
         D[z, x];
         Simplify[%]
```

Out[86]= $-9 + 32\,x - 21\,x^2 + 4\,x^3$

Example: Output Growth

Suppose that the real output of an economy (Q) is a function of labor (L) and capital (K), $Q = f(L, K)$. If the labor and capital are growing over time at the rates given by $L = g(t)$ and $K = h(t)$ respectively, find the rate of change of Q over time.

Solution: $dQ = f_K dK + f_L dL$, where $dL = g_t dt$ and $dK = h_t dt$. Thus, $\frac{dQ}{dt} = f_K h_t + f_L g_t$, where f_L and f_K are marginal products of labor and capital, respectively.

Differential Calculus in Matrix Notation: The rules for differentiation of vectors and matrices are given below. Students are required to verify them using simple numerical examples.

Rule 1. $\frac{\partial a'x}{\partial x} = \mathbf{a}$, where **a** and **x** are each nx1 vectors and the operation of differentiation is done with respect to the elements of the **x** vector. Therefore, partial derivatives of **a'x** with respect to the scalar **xi** gives *n* elements of the vector **a**.

Example: Find $\frac{\partial a' x}{\partial x}$, where $\mathbf{a}' = (5 \;\; 3 \;\; -1)$ and $\mathbf{x}' = (x1 \;\; x2 \;\; x3)$.

Solution: $\frac{\partial a' x}{\partial x1} = 5$, $\frac{\partial a' x}{\partial x2} = 3$, and $\frac{\partial a' x}{\partial x3} = -1$. Therefore, $\frac{\partial a' x}{\partial x} = (5 \;\; 3 \;\; -1)$.

Rule 2. $\frac{\partial \mathbf{x'Ax}}{\partial x} = 2\mathbf{Ax}$ or $\frac{\partial \mathbf{x'Ax}}{\partial x} = 2\mathbf{x'A}$, where **x** is a nx1 vector and **A** is a nxn symmetric matrix. The choice between the 2**Ax** and 2**A'x** is usually determined by the context in which differentiation takes place, and conformity of the resulting matrix with other matrices in equations.

Example: Find $\frac{\partial \mathbf{x'Ax}}{\partial x}$, where $\mathbf{x} = (x1 \;\; x2 \;\; x3)$ and $\mathbf{A} = \begin{pmatrix} 1 & -1 & 0 \\ -1 & 5 & 3 \\ 0 & 3 & 1 \end{pmatrix}$

Solution: $\mathbf{x'Ax} = x_1^2 - 2x_1x_2 + 6x_2x_3 + 5x_2^2 + x_3^2$. $\frac{\partial \mathbf{x'Ax}}{\partial x1} = 2x_1 - 2x_2$, $\frac{\partial \mathbf{x'Ax}}{\partial x2} = -2x_1 + 10x_2 + 6x_3$, and $\frac{\partial \mathbf{x'Ax}}{\partial x3} = 6x_2 + 2x_3$. Therefore, $\frac{\partial \mathbf{x'Ax}}{\partial x} = 2\begin{pmatrix} 1 & -1 & 0 \\ -1 & 5 & 3 \\ 0 & 3 & 1 \end{pmatrix}\begin{pmatrix} x1 \\ x2 \\ x3 \end{pmatrix}$.

Rule 3. $\mathbf{y} = f(\mathbf{x})$, where **y** is a nx1 vector each element of which is a function of m elements of vector **x.** That is, $yi = f_i(x_1, x_2 \ldots\ldots, x_m)$, $i = 1, \ldots., n$.

$$\frac{\partial y}{\partial x} = \begin{pmatrix} \frac{\partial y1}{\partial x1} & \frac{\partial y2}{\partial x1} & \cdots & \frac{\partial yn}{\partial x1} \\ \frac{\partial y1}{\partial x2} & \frac{\partial y2}{\partial x2} & \cdots & \frac{\partial yn}{\partial x2} \\ \cdot & \cdot & \cdots & \cdot \\ \cdot & \cdot & \cdots & \cdot \\ \cdot & \cdot & \cdots & \cdot \\ \frac{\partial y1}{\partial xm} & \frac{\partial y2}{\partial xm} & \cdots & \frac{\partial yn}{\partial xm} \end{pmatrix}$$

Each y_i is differentiated partially with respect to each x_j giving an mxn matrix of partial derivatives.

Example: Find $\frac{\partial y}{\partial x}$ for the following system of equations

$y1 = 2x_1 + 7x_2 + 2x_3$

$y2 = 3x_1 + 3x_2 + x_3$

$y3 = 2x_1 + 6x_2 + 3x_3$

$y4 = 4x_1 + 2x_2 + x_3$

$y5 = 6x_1 + x_2$

$$\frac{\partial y}{\partial x} = \begin{pmatrix} 2 & 3 & 2 & 4 & 6 \\ 7 & 3 & 6 & 2 & 1 \\ 2 & 1 & 3 & 1 & 0 \end{pmatrix}$$

$\frac{\partial y}{\partial x}$ can also be calculated using *Mathematica* as,

```
In[87]:= Clear[x1, x2, x3, y1, y2, y3, y4, y5]
         y1 = 2 x1 + 7 x2 + 2 x3;
         y2 = 3 x1 + 3 x2 + x3;
         y3 = 2 x1 + 6 x2 + 3 x3;
         y4 = 4 x1 + 2 x2 + x3;
         y5 = 6 x1 + x2;
         MatrixForm[( D[y1, x1]  D[y2, x1]  D[y3, x1]  D[y4, x1]  D[y5, x1]
                      D[y1, x2]  D[y2, x2]  D[y3, x2]  D[y4, x2]  D[y5, x2]
                      D[y1, x3]  D[y2, x3]  D[y3, x3]  D[y4, x3]  D[y5, x3] )]
```

$$\text{Out[87]}= \begin{pmatrix} 2 & 3 & 2 & 4 & 6 \\ 7 & 3 & 6 & 2 & 1 \\ 2 & 1 & 3 & 1 & 0 \end{pmatrix}$$

Problems

1. Given the utility function $U = U(x, y)$ and the budget constraint $P_X X + P_y Y = M$, find the equilibrium condition for the consumer. (Hint: At the equilibrium point, the budget constraint is tangent to the indifference curve. Therefore, at the equilibrium point, the slope of the budget constraint should be equal to the slope of the indifference curve).

2. Find the first-order and second-order differentials of the Cob-Douglas production function $Q = AL^a K^b$

3. In the IS-LM model of the economy LM represent all the combination of output (y) and interest rate (r) where the money market is in equilibrium. Suppose that, the LM model of the economy is expressed as $\frac{M_s}{p} = L(y, r)$, where $\frac{M_s}{P}$ is the real money supply and L(Y, r) is the demand for real balances. Find the slope of the LM curve and prove that slope of the LM curve is positive. (Hint: Demand for money is positively related with income and negatively related with interest rate.)

4. The production function of an economy is given as $Q = Q(L, K) = 20L^{.75}K^{.25}$, where the labor input (L) and the capital input (K) are growing at the rates $L = L(t) = .2t$, and $K = k(t) = .3t$. Find $\frac{dQ}{dt}$.

5. Find the input elasticities for the Cob-Douglas production function $Q = AL^a K^b$. (Hint: Elasticity of output with respect to labor (L) is defined as $\zeta_L = \frac{\partial Q}{\partial L}\frac{L}{Q}$.

6. For the production function, $Q = AK^{.25}L^{.75}$

a) Prove that the function is homogenous of degree one.(Hint: Prove that an *m* multiple of inputs will results in *m* multiple of output.)

b) Find the marginal product of labor, Q_L, and the marginal product of capital, Q_K.

c) Prove that if labor and capital are paid according to the value of their marginal products, total revenue will be exhausted, and the profit will be zero (Euler's Theorem). (Hint: The value of marginal product of labor or marginal revenue product of labor is equal to the marginal product of labor times the price of the product.)

7. For the following constant elasticity of substitution (CES) production function, $Q = A(aK^{-r} + bL^{-r})^{\frac{-1}{r}}$

a) Prove that the production function is homogenous of degree one.

b) Find the marginal product of capital and the marginal product of labor.

c) Prove that, if labor and capital are paid according to the value of their marginal products, total revenue will be exhausted and the profit will be zero.

8. For the following production functions find the elasticity of technical substitution of labor (L) for capital (K). (Hint: elasticity of technical substitution of labor for capital is defined as the ratio of percent change in $\frac{L}{K}$ to percent change in $\frac{r}{w}$, where w and r are wage and rental cost of capital, respectively).

a- $Q = AL^aK^b$

b- $Q = 100\, L^{1/2}K^{1/2}$

9. Find $\frac{\partial a'x}{\partial x}$, where $\mathbf{a}' = (3 \quad 1 \quad -1)$ and $\mathbf{x}' = (x1 \quad x2 \quad x3)$.

10. Find $\frac{\partial \mathbf{x}'\mathbf{A}\mathbf{x}}{\partial x}$, where $\mathbf{x}' = (x1 \quad x2 \quad x3)$ and $\mathbf{A} = \begin{pmatrix} 0 & 2 & 3 \\ 2 & 1 & -1 \\ -1 & 0 & 1 \end{pmatrix}$

11. For the function $\mathbf{Y} = f(\mathbf{X}) = \mathbf{XB} + \mathbf{u}$ where, $\mathbf{Y}$ is a nx1 vector, $\mathbf{x}$ is a nxk matrix, $\mathbf{B}$ is a kx1 vector, and $\mathbf{u}$ is a nx1 vector,

a- Find the first derivative of **u'u** with respect to **B.**

b- Find the second derivative of **u'u** with respect to **B.**

CHAPTER 7

UNCONSTRAINED OPTIMIZATION

Most of the optimizing problems in economics are constrained optimizing problems. In a constrained optimizing problem the economic unit optimizes an objective function given that certain constraints are to be satisfied. For example, in the theory of production, the optimizing problem may be formulated as maximizing output for a given level of cost, or minimizing cost of production for a certain level of output. However, there are many occasions in which the problem may be that of finding the optimum values of an objective function with no constraint attached. Such problems are unconstrained optimizing problems. Examples of such problems are maximizing total revenue, maximizing profit, or minimizing cost. Problems of this nature are simply problems of finding the critical values of a function in differential calculus.

Finding the critical values of a function is one of the best known and commonly applied areas of differential calculus. In this chapter the definitions and techniques of the preceding chapter will be utilized in finding the critical points of functions and solving some unconstrained optimizing problems in economics.

TOOL BOX

Mathematica Commands	Function
`Solve[D[y, x] == 0, x]`	Sets $\frac{dy}{dx} = 0$ and solves it for x
`D[y, x]/. x-> a`	Evaluates $\frac{dy}{dx}$ at x = a.
`D[y, {x, 2}]/. x->a`	Evaluates $\frac{dy^2}{d^2x}$ at x = a.
`Solve[{D[z,x] == 0, D[z,y]==0, {x,y}]`	Sets $\frac{\partial z}{\partial x} = 0$, $\frac{\partial z}{\partial y} = 0$ and solves them for x and y
`D[D[z, x], y]`	Finds $f_{xy} = \frac{\partial z^2}{\partial x \partial y}$ of the function z = f(x, y).
`D[D[z, y], x]`	Finds $f_{yx} = \frac{\partial z^2}{\partial y \partial x}$ of the function z = f(x, y).
$\begin{pmatrix} D[D[z, x], x] & D[D[z, x], y] \\ D[D[x, y], x] & D[D[z, y], y] \end{pmatrix}$	Hessian matrix, $H = \begin{pmatrix} f_{xx} & f_{xy} \\ f_{yx} & f_{yy} \end{pmatrix}$ of z = f(x, y).
$\begin{pmatrix} D[D[z, x], x] & D[D[z, x], y] \\ D[D[x, y], x] & D[D[z, y], y] \end{pmatrix}$`/.{x->a, y->b}`	Hessian evaluated at x = a and y = b.

I. Critical Points of a Function: The `Solve[D[y, x] == 0, x]` Command

The critical points of a function are its **maxima**, **minima**, and **inflection** points. A function f is said to have a **relative maxima** at x = x_o if $f(x_o)$ is the **absolute maximum** of f in some interval in the neighborhood of x_o which contains x_o. Similarly, f is said to have a **relative minimum** at x = x_1 if $f(x_1)$ is the **absolute minimum** of f in some interval in the neighborhood of x_1 which contains x_1. A maximum or a minimum of a function, whether relative or absolute, is called an **extremum**. The following figure illustrates the critical points the function y = f(x) = $x^3 - 3x^2 + 15$. Based on the definition of the critical points, point (0, 15) is a relative maxima. However, in the interval -1 < x < 1, this point is also an absolute maximum. Point (2, 11) is a relative minima. However, in the interval 1 < x < 3, this point is also an absolute minimum. Point (1, 13) is an inflection point.

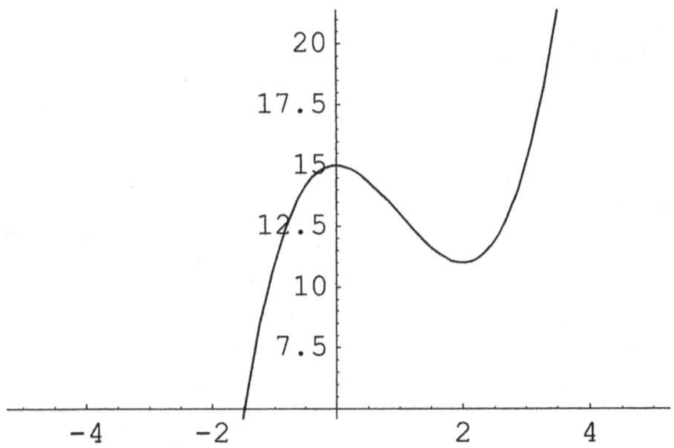

The points (0, 15) and (2, 11) are also called **interior extrema** versus **exterior extrema**, which are usually located at the end points of a graph.

<u>Derivatives and the Critical Points:</u> For a continuous function defined by y = f(x), the point (x_o, f(x_o)) is an interior extremum if x_o is in the domain of the function and f '(x_o) = 0. Therefore, to locate interior maxima or minima of a function we need only those values of x for which f '(x) = 0. The exterior maxima or minima can be tested for by examining the end points of an interval.

Example: Find the extrema of f defined by y = f(x) = $x^3 - 2x^2 + x + 1$ in the interval 0 < x < 2.

Solution: f '(x) = $3x^2 - 4x + 1 = 0$. The solution of the quadratic form yields, $x_1 = 1$ and $x_2 = \frac{1}{3}$. Therefore, the points (1, 1) and ($\frac{1}{3}$, $\frac{31}{27}$) are the interior extrema. The endpoints of the function in the domain 0 < x < 2 are (0, 1) and (2, 3). Thus, these two points are exterior extrema. To observe the points graphically, plot the function using the critical points and x and y intercepts as,

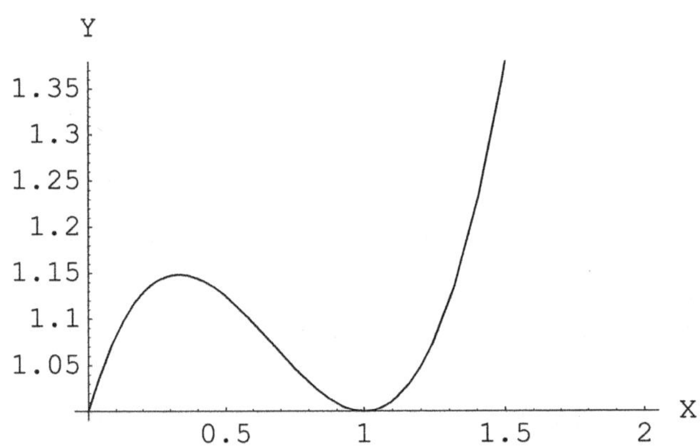

To solve the problem using *Mathematica*,

```
In[1]:= Clear[x, y, sol]
        y = x^3 - 2 x^2 + x + 1 ;
        sol = Solve[D[y, x] == 0, x]
```

Out[1]= {{x → $\frac{1}{3}$}, {x → 1}}

```
In[2]:= y /. sol
```

Out[2]= {$\frac{31}{27}$, 1}

To observe the points graphically, plot the function as,

In[3]:=
```
Plot[y, {x, 0, 2}, AxesLabel -> {"X", "Y"}]
```

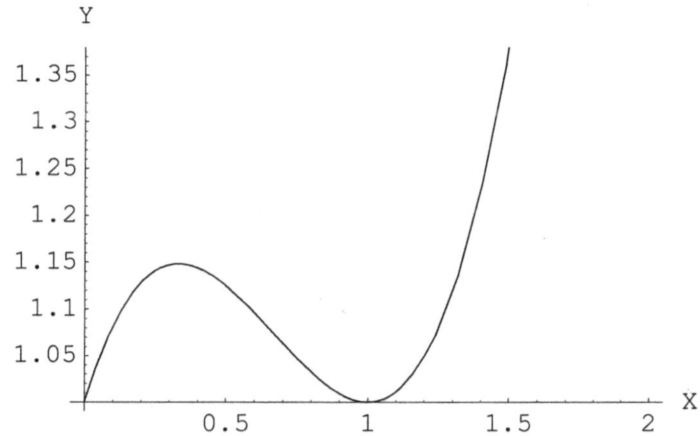

The condition f '(x) = 0 locates the extrema points of the function *f*, but it does not determine whether the points are maxima or minima. One convenient way to test whether an extremum is a maximum or a minimum is to use the second derivative test.

The second derivative of f at a point x_o shows the rate at which the slope of f(x) changes at x_o. In fact, it represents the slope of the slope function at x_o. If the function has a maximum at x_o, the slope of the curve at this point has to be horizontal or $f'(x_o) = 0$ and the slope of the slope at this point has to be negative, $f''(x_o) < 0$. That is any move from x_o at any direction should be a downward move. If the function has a minimum at x_o, the slope of the curve at this point has to be horizontal or $f'(x_o) = 0$ and the slope of the slope at this point has to be positive, $f''(x_o) > 0$. That is any move from x_o at any direction should be an upward move. If $f''(x_o) = 0$, then the function f may have an inflection point at x_o. The three cases are presented below:

Case 1: f '(x_o) = 0 and f "(x_o) < 0. At x_o, f(x) has a maximum.

Example: Test for the critical points of y = f(x) = 2x - x^2 in the interval 0 < x < 2.

Solution: The first derivative of the function is f '(x) = 2 - 2x and the second derivative is f "(x) = -2. The first derivative has a zero at x = 1. At this point slope of the function is zero, f '(1) = 0, and f "(1) = -2 < 0. Therefore, at x_o= 1, f has a maximum. The graph of the function below shows the critical point and the slope of the function in the neighborhood of the critical point. Using *Mathematica*,

In[4]:=
```
Clear[x, y, sol]
y = 2 x - x^2;
sol = Solve[D[y, x] == 0, x]
```

Out[4]= {{x → 1}}

In[5]:=
```
D[y, {x, 2}] /. sol
```

Out[5]= {-2}

In[6]:=
```
Plot[y, {x, 0, 2}, AxesLabel -> {"X", "Y"}]
```

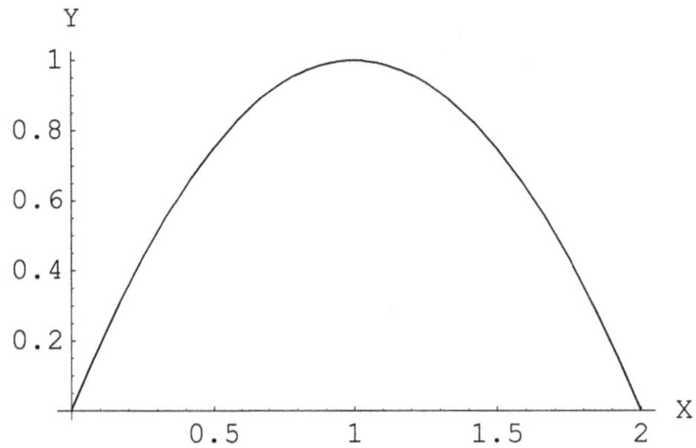

Case 2: f '(x_o) = 0 and f "(x_o) > 0. At x_o, f(x) has a minimum.

Example: Test for the critical points of y = f(x) = 2x + x^2 in the interval -2 < x < 0.

Solution: The first derivative of the function is f '(x) = 2 + 2x and the second derivative is f "(x) = 2. The first derivative has a zero at x_o= -1. At this point slope of the function is zero, f '(-1) = 0, and f "(-1) = 2 > 0. Therefore, at x_o= -1, *f* has a minimum. The graph of the function below shows the critical point and the slope of the function in the neighborhood of the critical point. Using *Mathematica,*

```
In[7]:= Clear[x, y, sol]
        y = 2 x + x^2 ;
        sol = Solve[D[y, x] == 0, x]

Out[7]= {{x → -1}}

In[8]:= D[y, {x, 2}] /. sol

Out[8] = {2}

In[9]:= Plot[y, {x, -3, 1}, AxesLabel -> {"X", "Y"}]
```

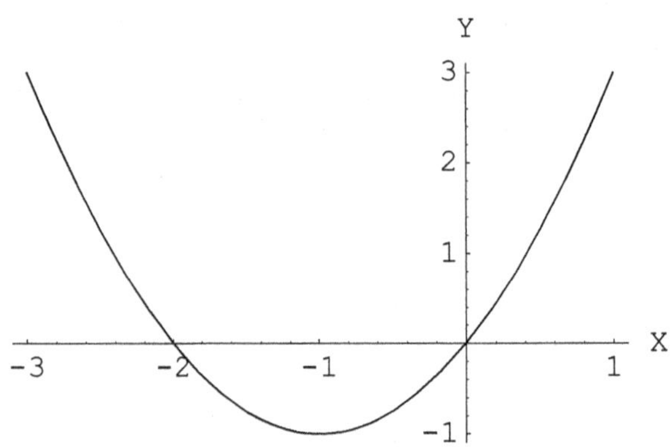

Case 3: f "(x_o) = 0. At x_o, f(x) may have an inflection point.

Example: Test for the critical points of y = f(x) = 3x - x^3 in the interval -3 < x < 3.

Solution: The first derivative of the function is f '(x) = 3 - 3x^2 and the second derivative is f "(x) = -6x. The first derivative has zeroes at x_1 = -1 and x_2 = 1. The signs of the second derivative of the function at these points rae f "(-1) = 6 > 0 and f "(1) = -6 < 0, respectively. Therefore, the function has a minima at x_1 = -1 and a maxima at

$x_2 = 1$. At $x_o = 0$, f "(0) = 0. The inspection of the graph below shows that f has an inflection point at (0, 0). Using *Mathematica*,

```
In[10]:= Clear[x, y, sol]
         y = 3 x - x^3;
         sol = Solve[D[y, x] == 0, x]

Out[10]= {{x → -1}, {x → 1}}

In[11]:= D[y, {x, 2}] /. sol

Out[11]= {6, -6}

In[12]:= D[y, {x, 2}] /. x -> 0

Out[12]= 0

In[13]:= Plot[y, {x, -3, 3}, AxesLabel -> {"X", "Y"}]
```

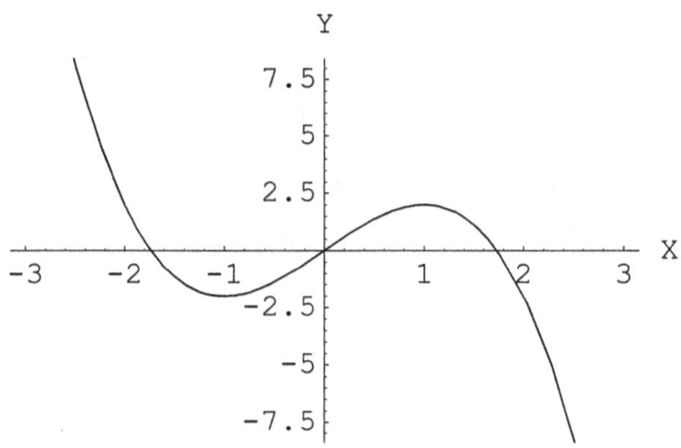

The preceding discussion shows that, f '(x) = 0 locates the relative extrema of a function and the sign of f "(x) decides whether the point is a maximum or a minimum at the extrema (x_o). If f '(x_o) = 0 and f "(x_o) < 0, the function f has a maximum at x_o. If f '(x_o) = 0 and f "(x_o) > 0, the function f has a minimum at x_o. If f "(x_o) = 0 the function f may have an inflection point at x_o. The first derivative test of extrema is called the **first order condition** and the second derivative test is called the **second order condition** for relative extrema.

Example: Find the critical points of $y = f(x) = x^3 - 3x^2 - 72x + 8$ and test for maxima and minima.

Solution: The first order condition is $y' = f'(x) = 3x^2 - 6x - 72 = 0$. The quadratic form has zeroes at x1 = 6 and x2 = -4. Therefore the extrema are (6, -316) and (-4, 184). To test whether these points are maxima or minima, find the second derivative and test for the second order condition. The second derivative is y" = f "(x) = 6x - 6. Evaluated at the critical points f "(6) = 30 > 0 and f "(-4) = -30 < 0. Thus the point (6, -316) is a relative minimum and the point (-4, -392) is a relative maximum. At x = 1, f"(1) = 0 and the point (1, -66) is an inflection point. The graph below shows the critical points. Using *Mathematica*, the solution is

```
In[14]:= Clear[x, y]
         y = x^3 - 3 x^2 - 72 x + 8;
         sol = Solve[D[y, x] == 0, x]

Out[14]= {{x → -4}, {x → 6}}

In[15]:= D[y, {x, 2}] /. sol

Out[15]= {-30, 30}
```

```
In[16]:= D[y, {x, 2}] /. sol
Out[16] = {-30, 30}
In[17]:= Plot[y, {x, -10, 12}, AxesLabel -> {"X", "Y"}]
```

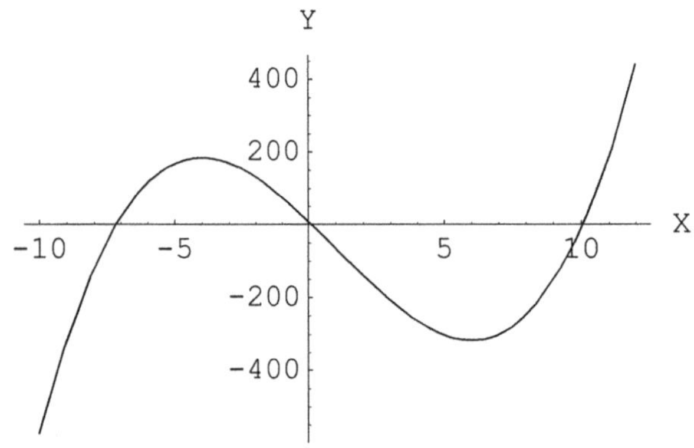

Geometrically, the first order condition, f ' (x) = 0, implies that a tangent to the curve at the relative maximum or relative minimum is parallel to the x-axis. For a function to have a maximum at x = x_o, the second order condition requires f "(x_o) < 0. This condition implies that, as x changes in the neighborhood of x_o, the function f(x) changes at a decreasing rate (increases at a decreasing rate and decreases at a decreasing rate). Such a function is said to be **concave upward.** Below, the graph of the y = f(x) = x^3 - $3x^2$ - 72x + 8 at the neighborhood of x_o = -4 illustrates this point.

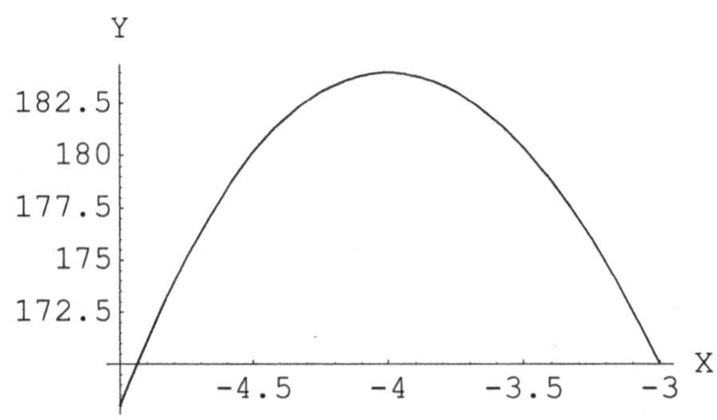

For a function f(x) to have a minimum at x = x_o, the second order condition requires f "(x_o) > 0. This condition implies that, as x changes in the neighborhood of x_o that contains x_o, the function f(x) changes at an increasing rate (decreases at an increasing rate and increases at an increasing rate). Such a function is said to be **convex upward.** Below, the graph of the y = f(x) = x^3 - $3x^2$ - 72x + 8 at the neighborhood of x_o = 6 illustrates this point.

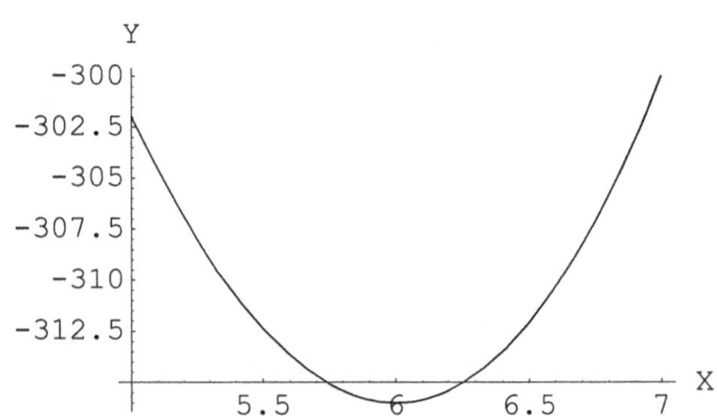

The second derivative test is a convenient way for finding the relative maxima and minima. However, the test fails if the second derivative vanishes at the critical point. If $f''(x_o) = 0$, one way to test for the relative extremum is to check the sign of $f''(x_o)$ in the neighborhood of x_o. If the sign of $f''(x)$ changes in the neighborhood of x_o, the point is a relative extremum. If the sign does not change, the point is an inflection point.

Example: Find the critical point of $y = f(x) = x^3$, and test for maxima or minima.

Solution: $y' = f'(x) = 3x^2$. Therefore, f(x) has a critical point at $x_o = 0$. The second derivative test shows that $f''(x) = 6x$ vanishes at $x = x_o$, $f''(0) = 0$. Therefore, the second-derivative test fails to determine maximum or minimum. To test for the sign of $f'(x)$ in the neighborhood of $x_o = 0$, it is clear that $f'(x) > 0$ for $x > 0$ and for $x < 0$. Since the sign of $f'(x)$, or the slope of the function does not change in the neighborhood of $x_o = 0$, the point (0, 0) is an inflection point and not an extrema. The graph of the function illustrates this fact.

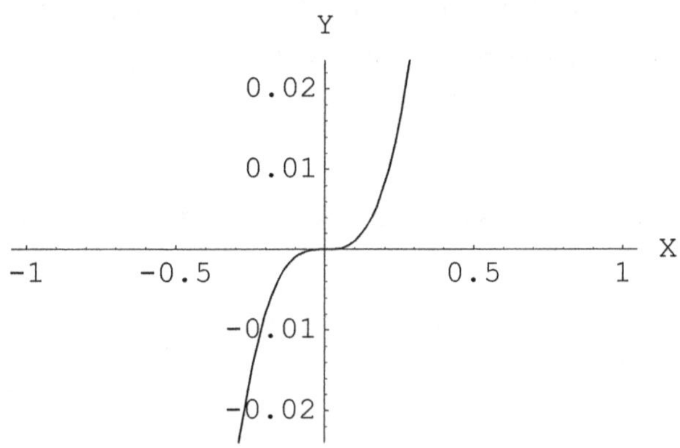

Example: Find the critical points of $y = f(x) = x^4$.

Solution: f(x) has a critical point at $x_o = 0$. The second order condition shows that the second derivative of f, $f''(x) = 12x^2$, vanishes at $x_o = 0$. Therefore, the second-derivative test fails to determine minimum or maximum points. However, the slope test in the neighborhood of $x_o = 0$ shows that $f'(x) < 0$ for $x < 0$ and $f'(x) > 0$ for $x > 0$. Since the slope of the function changes from negative to positive in the neighborhood of $x_o = 0$, the function has a relative minimum at $x_o = 0$. The graph of the function illustrates this point.

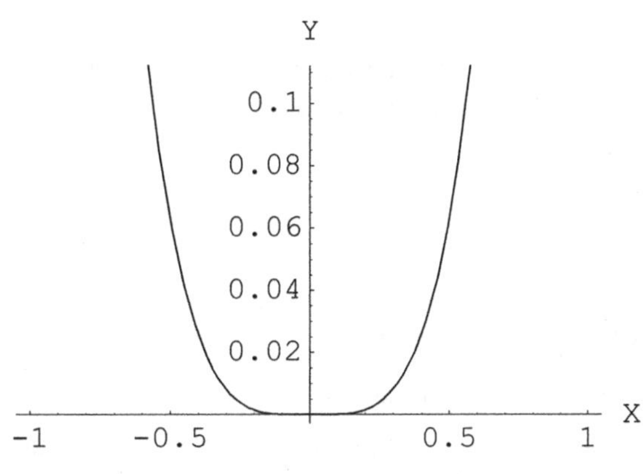

Example: Revenue Function

Demand for a product is given as Q = 100 - 2P. Find the price and the quantity at which profit is maximized. Prove that at this point price elasticity of demand equals -1.

Solution: Total revenue TR = P.Q. Using the given demand function, P = 50 - . 5Q and TR = 50Q - . $5Q^2$. To maximize TR, find Q by solving $\frac{dTR}{dQ} = 0$ and test for the second order condition.The solution of $\frac{dTR}{dQ} = 50 - Q = 0$

is Q = 50. Therefore, Q = 50 is the critical point. The second derivative of TR yields $\frac{d^2 \text{TR}}{\text{dQ}^2} = -1$. Since the second derivative is negative, at Q = 50 the revenue function has a maximum. At Q = 50, P = 25. To find the price elasticity of demand at this point, $\epsilon_p = \frac{\text{dQ}}{\text{dP}} \frac{P}{Q}$. Since $\frac{\text{dQ}}{\text{dP}} = -2$ (slope of the demand curve), $\epsilon_p = -2\frac{25}{50} = -1$.

Example: Theory of Firm

A firm in a perfectly competitive market has the cost function, TC = c(Q), where TC is the total cost and Q is output. What is the optimum level of output, if the objective of the firm is maximizing profit?

Solution: The total profit of the firm (TP) is total revenue (TR) minus total cost, where TR = P.Q. Therefore, TP = TR - TC = P.Q - C(Q). To maximize profit, the first order condition requires that derivative of TP with respect to Q to be equal to zero. That is, $\frac{\text{dTP}}{\text{dQ}} = \text{P} - \text{C}'(\text{Q}) = 0$ or C '(Q) = P. This means, to maximize the profit, a firm in a perfectly competitive market decides the output level by setting the marginal cost (MC = C '(Q)) equal to the market price (MC = P). The second order condition requires that $\frac{d^2 \text{TP}}{\text{dQ}^2} < 0$. Since $\frac{d^2 \text{TP}}{\text{dQ}^2} = -\text{C}''(\text{Q})$, the profit maximizing condition requires that C'(Q) > 0 or the marginal cost curve to be sloping upward at the profit maximizing point.

Example: Inventory Cost

A retailer has estimated that the cost for ordering and storing Q units of a products is TC(Q) = 3Q + $\frac{30000}{Q}$. The delivery truck can deliver at most 150 units per order, $1 \le \text{Q} \le 150$. Find the order size that will minimize the cost.

solution: Minimize TC with respect to Q. If the Q is in the range $1 \le \text{Q} \le 150$, the resulting number is the solution. Otherwise, look for an absolute minimum for TC at Q = 1 or Q 150. To minimize TC, set TC '(Q) = 0 and solve for Q. TC '(Q) = 3 + $\frac{-30000}{Q^2} = 0$. Solving for Q gives Q = 100. To test for second order condition, find TC "(Q) and evaluate it at Q = 100. TC "(Q) = $\frac{60000}{Q^3}$. Evaluated at Q = 100, TC "(100) = $\frac{3}{50} > 0$. That is at Q = 100, the function has a minimum.

Using *Mathematica,* the solution is found as,

```
In[18]:= Clear[TC, Q, sol]
         TC = 3 Q + 30000 / Q;
         sol = Solve[D[TC, Q] == 0, Q]

Out[18]= {{Q → -100}, {Q → 100}}
```

Q = 100 will be the solution, if it satisfies the second order condition. To test for the second order condition, find the second derivative of TC and evaluate it at Q = 100. If the resulting number is positive, Q = 100 is the solution to the problem.

```
In[19]:= D[TC, {Q, 2}] /. sol

Out[19]= {-3/50, 3/50}
```

Example: Total Cost and Perfectly Competitive Market

A firm in a perfectly competitive market has the cost function, TC(q) = $\frac{1}{3}q^3 - 3.5q^2 + 26q + 10$, where TC is the total cost and q is output. What is the optimum level of output, if the market price of the good is $20?

Solution: Total profit (TP), total cost (TC), and total revenue (TR) are all functions of q, TP(q) = TR(q) - TC(q). To maximize profit, set the first derivative of TP(q), or marginal profit, equal to zero. This will set the first derivative of TC(q) (marginal cost), equal to the first derivative of TR(q) (marginal revenue). In a perfectly competitive market, price and marginal revenue are the same. Thus, the profit maximizing quantity is decided by setting price equal to marginal cost. The marginal cost is TC '(q) = $q^2 - 7q + 26$. Solving $q^2 - 7q + 26 = 20$ gives q = 1 and q = 6. To test for the second order condition, the second derivative of TP(q) = -TC "(q) . To have maximum profit at the critical points, -TC "(q) has to be negative or TC "(q) > 0. TC"(1) = -5 < 0 and TC"(6) = 5 which is greater than zero. Therefore, the profit maximizing solution is q = 5. Using *Mathematica*,

```
In[20]:= Clear[TC, q, sol]
         TC = (1 / 3) q^3 - 3.5 q^2 + 26 q + 10;
         sol = Solve[D[TC, q] == 20, q]

Out[20]= {{q → 1.}, {q → 6.}}

In[21]:= D[TC, {q, 2}] /. sol

Out[21]= {-5., 5.}
```

The first-order condition results in two values for q, q = 1 and q = 6. It is obvious from the second-order condition that the only acceptable solution is q = 6.

Example: Monopoly

A firm operating in a monopoly market has the cost function, TC = c(q), where TC is the total cost and q is output. What is the optimum level of output, if the objective of the firm is to maximize profit?

Solution: If the firm is a monopoly, the price of the product is a function of output, and the total profit (TP) is TP(q) = P(q).q - C(q) = TR(q) - TC(q) where, TR(q) is total revenue as a function of q. The first order condition requires that TP'(q) = TR'(q) - TC'(q) = 0. TR'(q) is the marginal revenue (MR) and TC'(q) is the marginal cost (MC). Therefore, the first-order condition requires that MR = MC. Differentiating TR(q) with respect to q gives MR as, P(q) + q.P'(q). Factor out p(q) to get, p(q)(1 - $\frac{q}{p}$p'(q)). The term $\frac{q}{p}$p'(q) is the inverse of the price elasticity of demand (ϵ_p). Therefore, the profit-maximizing condition requires that (1 - $\frac{1}{\epsilon_p}$)P = MC. That is, to maximize profit, a monopolist sets its MC = MR = (1 - $\frac{1}{\epsilon_p}$)P. Since monopolies always sell at the elastic portion of their demand curve (1 - $\frac{1}{\epsilon_p}$)P < P. For monopolist this results in selling less goods than if the industry were a perfectly competitive industry and selling at a higher price than perfectly competitive market. The second order condition requires that (1 - $\frac{1}{\epsilon_p}$)$\frac{dp}{dq}$ - C"(q) < 0. $\frac{dp}{dq}$ is the slope of the demand curve and is negative and (1 - $\frac{1}{\epsilon_p}$) > 0. Therefore, the second order condition requires that C"(q) > 0, or the marginal cost curve slope upward at the profit maximizing level of output.

Example: Monopoly

A monopoly has the cost function, TC = $\frac{1}{3}$q^3 - 5q^2 + 23q + 25, where C is the total cost and q is output. What is the optimum level of output, if the market demand is given as q = -2p + 90.

Solution: With the given market demand, p = -. 5q + 45. Total revenue is TR = p.q = -. 5q^2 + 45q. Therefore, solve for q by setting MR = MC. That is, q^2 -10q + 23 = -q + 45, the solution of which is q = -2 and q = 11. The only acceptable solution is q = 11. To test for the second order condition find TC "(11). TC "(Q) = 2q - 10 and TC "(11) = 12. Since TC "(11) > 0, at q = 11 MC slopes upward and this quantity is the profit maximizing or loss minimizing quantity. Using *Mathematica*, the solution is

```
In[22]:= Clear[TR, TC, q, sol]
         TC = (1 / 3) q^3 - 5 q^2 + 23 q + 25;
         TR = -.5 q^2 + 45 q;

sol = Solve[D[TR, q] == D[TC, q], q]

Out[22]= {{q → -2.2915}, {q → 8.2915}}

In[23]:= D[TC, {q, 2}] /. sol

Out[23]= {-11.583, 9.58301}
```

PROBLEMS

1. Find the extreme points of the following functions and test for maximum and/or minimum.

   a) $y = x^2 - 6x + 5$

   b) $y = -x^2 + 6x - 5$

   c) $y = 2x^3 - 15x^2 + 36x + 10$

   d) $y = x^3 - 3x^2 + 3x$

   e) $y = (x - 1)^3$

   f) $y = (x^2 - 1)^2$

   g) $y = \ln\frac{x+1}{x^2+3}$

   h) $z = x^3 + 3x^2y^2 + y^3$

   i) $z = (1 - x)^2 + (1 - y)^2$

   j) $z = (x + y)^3 - 2xy$

2. A monopolistic firm faces the demand $P = 100 - 2Q$ for its product

a) Find the output level where the total revenue of the firm is maximum.

b) Prove that this point lies in the middle of the demand curve, where the price elasticity of demand is -1.

c) If marginal cost of the firm is $MC = 20$, what level of output will maximize the profit?

3. The output (Q) of a single-product firm is related to its labor input (L) through the production function $Q = 100L + 10L^2 - L^3$

a) Find the marginal product of labor.

b) Find the maximum level of output.

c) Find the inflection point of the function.

d) Show that for the output levels below the inflection point, an increase in labor input increases output at an increasing rate. For output levels higher than the inflection point, an increase in labor increases output at a deceasing rate.

4. The cost function of a single-product firm is given as $C = \frac{1}{3}Q^3 - 3.5Q^2 + 6Q$ where Q is output and C is total cost.

a) Find the inflection point of the cost function.

b) Show that for output levels below the inflection point an increase in output increases total cost at a decreasing rate. For output levels over the inflection point an increase in output increases cost at an increasing rate.

5. A profit-maximizing monopoly is facing market demand, $Q = 50 - \frac{1}{2}P$ and a total cost function, $TC = \frac{1}{12}Q^3 - 5Q^2 + 120Q + 30$. Find profit-maximizing price and quantity.

II. Critical Points of Multi-Variate Functions: The Solve[{D[z,x] == 0, D[z,y]==0, {x,y}] Command

In the previous section we studied the critical points of functions with only one independent variable. The results of that section can be extended to the case of functions of several variables. For the functions of two or more independent variables the first order and the second order conditions for the existence of extrema is analogous to the conditions of the previous section. However, here the functions represent surfaces in three or more dimensions. Therefore, the first order condition (parallel slope of tangent to f at the extrema) and the second order condition (upward convexity or concavity in the neighborhood of the extrema) should be satisfied not only in the direction of one axis but also in the direction of every relevant axes. For example, a function of two independent variables, z = f(x, y) represents a surface in three dimensions x, y, and z. For f to have a maximum at the critical point p(x_o, y_o, z_o), the tangent to the surface at this point should be parallel to x-y plane at every direction. The following figure illustrates a surface in three dimensions with a maximum.

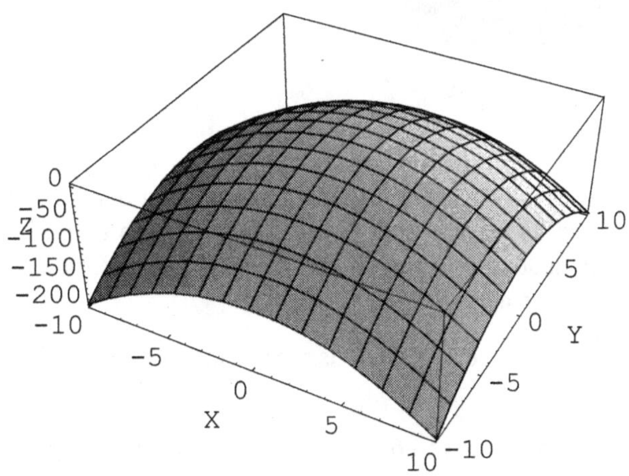

The fact that the tangent to the surface at the point p(x_o, y_o, z_o) should be parallel to x-axis implies that we are holding fixed the value of y and we are setting $\frac{\partial f}{\partial x} = f_x = 0$; and the fact that the tangent to the surface at p(x_o, y_o, z_o) should be parallel to y axis implies that we are holding fixed the value of x and we are setting $\frac{\partial f}{\partial y} = f_y = 0$. Therefore, for z = f(x, y) to have an extrema at p(x_o, y_o, z_o), the first order condition requires that $f_x = f_y = 0$ at this point. The second order condition for z = f(x, y) to have a maximum at the critical point p(x_o, y_o, z_o) is that d^2z to be negative definite at this point. For z = f(x, y) to have a minimum at p(x_o, y_o, z_o), the second order condition requires that d^2z to be positive definite at this point.

To find d^2z, find the total differential of $dz = f_x dx + f_y dy$. This gives $d^2z = f_{xx}dx^2 + f_{xy}dxdy + f_{yx}dydx + f_{yy}dy^2$. Using the matrix notations this equation can be expressed in a quadratic form as $\begin{pmatrix} dx & dy \end{pmatrix}\begin{pmatrix} f_{xx} & f_{xy} \\ f_{yx} & f_{yy} \end{pmatrix}\begin{pmatrix} dx \\ dy \end{pmatrix}$. A quadratic form is said to be **positive definite** if all the principal minors of the Hessian determinant, defined as H = $\begin{pmatrix} f_{xx} & f_{xy} \\ f_{yx} & f_{yy} \end{pmatrix}$ are positive. A quadratic form is said to be **negative definite** if the principal minors of the Hessian determinant alternate in sign, beginning with a negative first principal minor. Therefore, for $\begin{pmatrix} dx & dy \end{pmatrix}\begin{pmatrix} f_{xx} & f_{xy} \\ f_{yx} & f_{yy} \end{pmatrix}\begin{pmatrix} dx \\ dy \end{pmatrix}$ to be positive definite the following should be satisfied: $f_{xx} > 0$ and $f_{xx}f_{yy} - f_{yx}f_{xy} > 0$. The quadratic form will be negative definite if $f_{xx} < 0$ and $f_{xx}f_{yy} - f_{yx}f_{xy} > 0$. If the pattern of the principal minors of the quadratic form fits neither that of positive definite nor that of negative definite, then f is said to have a **saddle point** at the critical point. A saddle point is a point on the surface where it is a maximum when viewed against the background of xz-plane and is a minimum when viewed against the background of yz-plane, or vise versa. The following graphs shows a saddle point.

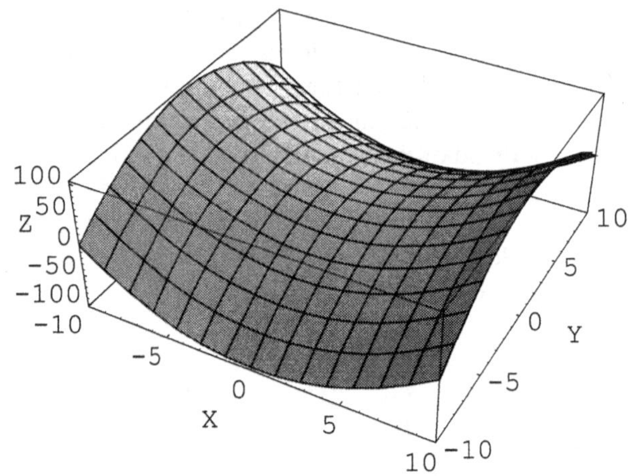

A second, may be an easier, way to test for the positive or negative definiteness of a quadratic form is to use the eigenvalues of the matrix. A quadratic form and its corresponding matrix **A** is said to be positive definite if and only if all the eigenvalues of **A** are positive. A quadratic form and its corresponding matrix **A** is said to be negative definite if and only if all the eigenvalues of **A** are negative.

The results of this section can be generalized to the case of functions with n independent variables z = f(x, y, w, ...).

Example: Find the critical points of $z = f(x, y) = xy - x^2 - y^2 - 6x$ and test for the second order condition.

Solution: The first order condition is $\frac{\partial f}{\partial x} = f_x = y - 2x - 6 = 0$ and $\frac{\partial f}{\partial y} = f_y = x - 2y = 0$. Solving the two equations together yields, x = -4, y = -2, and z = 12. Thus, the point (-4, -2, 12) is the extremum. To test for the second order condition, form the Hessian $\mathbf{H} = \begin{pmatrix} f_{xx} & f_{xy} \\ f_{yx} & f_{yy} \end{pmatrix} = \begin{pmatrix} -2 & 1 \\ 1 & -2 \end{pmatrix}$. The principal minors of the Hessian are -2 and 3, respectively. Since the principal minors alternate in sign, with the first principal minor negative, the matrix is negative definite and the function has a maximum at the extremum. The 3-D graph of the function shows the critical point.

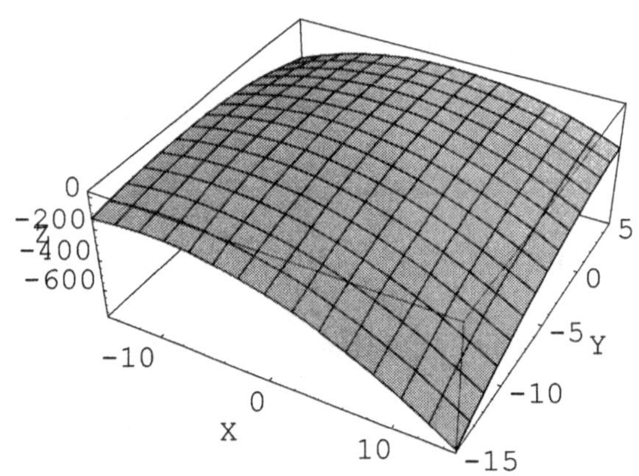

To solve the problem using *Mathematica*,

```
In[24]:= Clear[x, y, z, sol]
z = x y - x^2 - y^2 -6x;
sol = Solve[{D[z, x] == 0 , D[z, y]== 0}, {x, y}]
Out[24]= {{x → -4, y → -2}}
In[25]:= z/.sol
Out[25]= {12}
```

To test for the second order condition, evaluate the eigenvalues of the Hessian matrix at the critical point,

In[26]:= **Eigenvalues**$\left[\begin{pmatrix} \texttt{D[D[z, x], x]} & \texttt{D[D[z, x], y]} \\ \texttt{D[D[z, y], x]} & \texttt{D[D[z, y], y]} \end{pmatrix}\right]$ /. {sol}

```
Out[26]= {{{-3, -1}}}
```

Since both eigenvalues are negative, the quadratic form is negative definite and the critical point is a maximum.

Note that, in *Mathematica,* the second derivative of z with respect to x can be formulated as D[z, {x, 2}] or as D[D[z, x], x]. For the sake of harmony between mathematical and *Mathematica* notations I have adopted the latter form. To view the graph of the function at the critical point, use the "Plot3D" command as,

```
In[27]:= Plot3D[z, {x, -15, 15}, {y, -15, 5}, AxesLabel -> {"X", "Y", "Z"}]
```

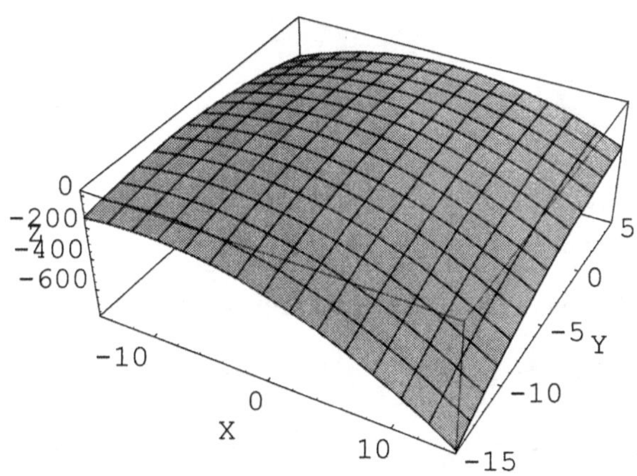

- SurfaceGraphics -

Example: Find the critical points of $z = f(x, y) = x^2 - y^2$ and test for maxima or minima.

Solution: The first order condition is $f_x = 2x = 0$ and $f_y = -2y = 0$. This gives x = y = z = 0 as the critical point. To test for the second order condition, form $\mathbf{H} = \begin{pmatrix} f_{xx} & f_{xy} \\ f_{yx} & fyy \end{pmatrix} = \begin{pmatrix} 2 & 0 \\ 0 & -2 \end{pmatrix}$ and do either an eigenvalue test or the principal minor test. The determinants of the principal minors are 2 and -4, respectively. Since the sign of the determinants does not match that of the positive or negative definiteness, the function has a saddle point at P(0, 0, 0). The graph below show this fact.

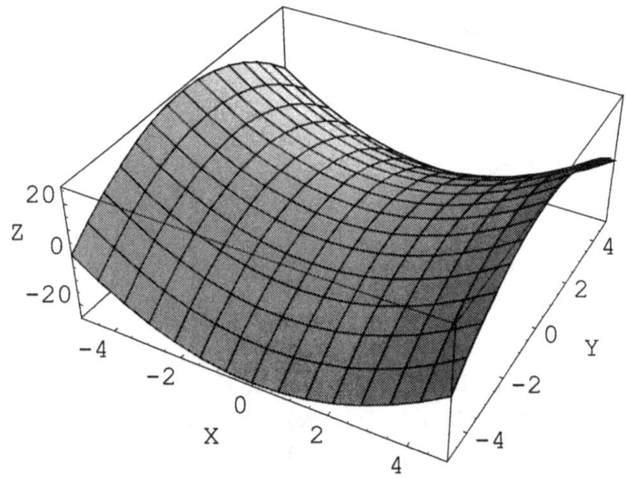

To solve the problem using *Mathematica*,

```
In[28]:= Clear[x, y, z, sol]
         z = x^2 - y^2;
         sol = Solve[{D[z, x] == 0, D[z, y] == 0}, {x, y}]
Out[28]= {{x → 0, y → 0}}
In[29]:= z /. sol
Out[29]= {0}
In[30]:= Eigenvalues[( D[D[z, x], x]  D[D[z, x], y]
                       D[D[x, y], x]  D[D[z, y], y] )] /. {sol}
Out[30]= {{{-2, 2}}}
```

The graph of the function use "Plot3D" command as,

```
In[31]:= Plot3D[x^2 - y^2, {x, -5, 5}, {y, -5, 5}, AxesLabel -> {"X", "Y", "Z"}]
```

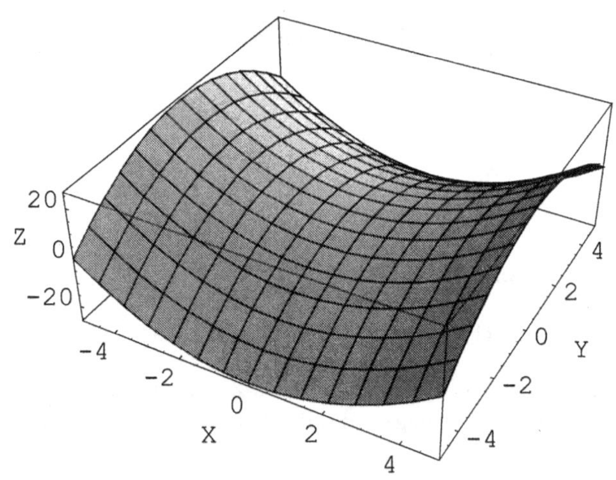


```
- SurfaceGraphics -
```

Do More Practice:

1- Change the function to z = $x^2 + y^2$, find the extremum, and graph the function.

2- Change the function to z = $-x^2 - y^2$, find the extremum, and graph the function.

Example: Find the critical points of z = f(x, y) = $-x^3 + 9xy - y^3$ and test for the second order condition.

Solution: The first order condition gives $f_x = -3x^2 + 9y = 0$ and $f_y = 9x - 3y^2 = 0$. Solving the two equations together results in two critical points (0, 0, 0) and (3, 3, 27). To test for the second order condition, form the **H** = $\begin{pmatrix} f_{xx} & f_{xy} \\ f_{yx} & f_{yy} \end{pmatrix} = \begin{pmatrix} -6x & 9 \\ 9 & -6y \end{pmatrix}$. Evaluate H at everyone of the critical points and do either an eigenvalues test or principal minor test to decide whether the critical points are maxima, minima, or saddle point. At (0, 0, 0), **H** = $\begin{pmatrix} 0 & 9 \\ 9 & 0 \end{pmatrix}$ and the determinant of the principal minors are 0 and -81, respectively. This doesn't fit any pattern of positive or negative definiteness. Therefore, at this point the function has a saddle point. At (3, 3, 27), **H** = $\begin{pmatrix} -18 & 9 \\ 9 & -18 \end{pmatrix}$. The determinant of the principal minors are -18 and 243. Since the principal minors alternate in sign, starting with a negative sign, at (3, 3, 27) **H** is negative definite and the function has a local maximum.

To solve the problem using *Mathematica*,

```
In[32]:= Clear[x, y, z, sol]
         z = -x^3 + 9 x y - y^3;
         sol = N[Solve[{D[z, x] == 0, D[z, y] == 0}, {x, y}]]

Out[32]= {{x → 0., y → 0.}}

In[33]:= z /. sol

Out[33]= {0.}

In[34]:= Eigenvalues[( D[D[z, x], x]  D[D[z, x], y]
                       D[D[z, y], x]  D[D[z, y], y] )] /. {sol}

Out[34]= {{{-2, 2}}}

In[35]:= Plot3D[z, {x, -5, 5}, {y, -5, 5}, AxesLabel -> {"X", "Y", "Z"}]
```

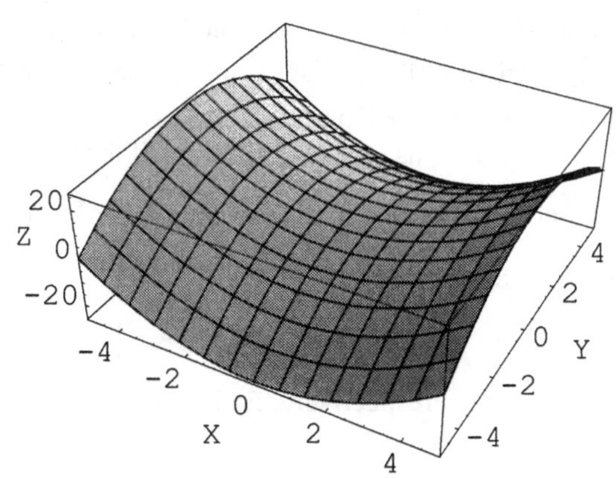

The eigenvalues test shows that at (0, 0, 0) the eigenvalues are -9 and +9. Therefore, at this point the function has a saddle point. At (3, 3, 27), the eigenvalues are -27 and -9. Since both eigenvalues are negative, H is negative definite and the function has a maximum at this point.

Example: Utility Function

A consumer has the utility function U = u(x, y), where x and y are the quantities of two consumption goods. Maximize the consumer utility, assuming that the consumer is not bound by the budget (has unlimited budget).

Solution: The first order condition for utility maximization yields Ux = Uy = 0, where Ux and Uy are the marginal utilities of x and y, respectively. Therefore, for an unconstrained maximization of utility, consumer consumes x and y up to the level that marginal utilities of the goods are zero, or the consumer is satiated. That is, with no budget constraint or with no prices paid for x and y, to maximize utility, the consumer consumes the goods as long as the marginal utilities received from the goods are positive. The second order condition requires that $\begin{pmatrix} Uxx & Uxy \\ Uyx & Uyy \end{pmatrix}$ to be negative definite. Since Uxy = Uyx, the principal minors rule for the second order condition requires Uxx < 0 and $(UxxUyy - Uxy^2) > 0$. These two conditions together imply that Uyy is also negative (Uyy < 0). That is, the marginal utility of x and y decrease as the consumer consumes more and more of x and y. In economics this is called the **law of diminishing marginal utility**.

PROBLEMS

1. A consumer has a hypothetical utility function U = lnxy, where x and y are quantities of two goods sold at market prices of $2 and $3 respectively. Suppose that the consumer's budget is non-binding (unlimited budget).

   a) Set up the consumer utility maximizing problem and find the optimal solution.

   b) Test for the second order condition.

2. The production function of a firm in a perfectly competitive market is given as Q = Q(K, L).

   a) Prove that, to maximize profit, the firm will hire inputs up to the level where the value of the marginal product of each input is equal to its price.

   b) Prove that the firm will have a profit if and only if the degree of homogeneity of production function is higher than one.

3. For the Cob-Douglas production function $Q = AL^{.75}K^{.25}$

   a) Find the marginal rate of technical substitution of labor for capital. (Hint: Marginal rate of technical substitution of labor for capital, $MRTSLK = \frac{MPK}{MPL}$ where, MP*K* and MP*L* are the marginal products of capital and labor, respectively. MRTS measures the reduction in one input per unit increase in the other input that is just sufficient to maintain a constant level of output). Show that, as capital is substituted for labor, MRTSLK declines.

4. For each of the following hypothetical utility functions U = lnxy and U = ln10xy - $(x^{.5} + y^{.5})$,

a) Find the values of x and y that will maximize utility.

b) Test for the second order conditions.

c) If the market price of x and y are given as px and py respectively, find the elasticity of substitution of y for x.

CHAPTER 8

CONSTRAINED OPTIMIZATION

The optimizing problems of business and economics are generally of a constrained nature. This is because every economic unit faces constraints in optimizing objectives. For example, every economy is constrained in production of goods and services (GDP) by the available economic resources or factors of production. Every consumer is constrained in buying goods and services by budget or income constraint, and every producer faces cost or input constraints.

Chapter 7 was concerned with the optimizing rules of functions with no constraints or restrictions on the variables. In such cases, the relative maximum or the relative minimum is the optimum solution. This chapter is concerned with the optimum solution of problems in which a function, called the objective function, is optimized subject to one or more constraints on variables. With a constraint imposed on the variables, the solution of the constrained optimizing problem may differ from the unconstrained problem, even if the objective functions are the same.

The problem of finding a constrained optimum in calculus assumes that the constraints hold with equality. The more general problem of maximizing or minimizing an objective function subject to inequality constraints is called mathematical programming. In both cases, however, the method of Lagrangian function can be used to solve constrained optimization problems.

TOOL BOX

Mathematica **Commands**	**Function**
z = f(x, y)	z is a function of x and y
sol = `Solve` [{D[z,x]==0,D[z,y]==0},{x,y}]	For z = f(x, y), solves first order condition & sets it to sol.
$\begin{pmatrix} D[D[z,x],x] & D[D[z,x,],y] \\ D[D[z,y],x] & D[D[z,y],y] \\ D[D[z,\lambda],x] & D[D[z,\lambda],y] \end{pmatrix}$/.{sol}	Finds the Hessian of Z = f(x, y) and evaluates it at sol.
`ViewPoint` ->{a, b, c}	Sets or changes the view point of a 3D graph.
`ContourPlot`[f(x, y), {x, a, b}, {y, c, d}]	Contourplots f(x,y) in the given x - y ranges.

I. Lagrangian Function: The sol = [{D[L,x]==0,D[L, λ]==0},{x, λ}] Command

Consider the following optimizing problem y = f(x) = $-x^2 + 6x + 1$, the graph of which is given below,

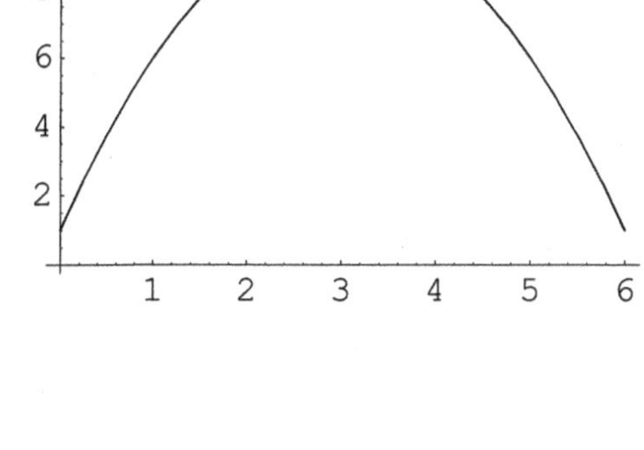

With no constraint on x, the problem is to find the value of x that will maximize f(x). This point is derived by finding the first derivative of the function, setting it equal to zero, and solving for x. That is, $y' = -2x + 6 = 0$, or $x = 3$. The maximum value of f(x) at x = 3 is, f(3) = 10. Testing for the second order condition yields $y'' = -2 < 0$, which confirms that the point (3, 10) is the maximum point and f(x) is maximized when x = 3.

Now suppose that the same function is to be maximized subject to the constraint $x \leq 2$. That is, the largest value that x could have is 2. This problem is a constraint optimizing problem, with the constraint on $x \leq 2$. The solution then, is x = 2 and f(2) = 9. That is the maximum that f(x) can attain is constrained by the value of x, which is x = 2.

The technique employed in solving constrained optimization is **Lagrangian function**. Consider the constrained maximization problem above:

Maximize: $y = f(x) = -x^2 + 6x + 1$

Subject to: $x \leq 2$

where = f(x) is the objective function, assumed to be differentiable, and $x \leq 2$ is the constraint. The problem is to maximize the objective function subject to the constraint. To solve the constrained optimizing problem, we form the Lagrangian function. The Lagrangian function is formed by augmenting the constraint together with the Lagrange multiplier (λ) to the objective function.

$$L = -x^2 + 6x + 1 + \lambda(2 - x)$$

Note that, in writing constraint you should always write the fixed value (c) first then x of constraint function. That is, the constraint should be written as c - x and not x - c. This is because, the Lagrangian multiplier λ has a special meaning in economics, which will be discussed later. In economics, the sign of λ is important. Changing the order, changes the sign of λ.

If the constraint is satisfied, no matter what value λ has, the second term in the right side of the Lagrangian equation will vanish and L = f(x). That is, maximizing L with respect to x is the same as maximizing f(x). Therefore, the trick in solving a constrained optimizing problem is to incorporate the constraint into the problem and insure that the constraint is satisfied with equality. This is done by treating λ as an additional variable and assuming L is a function of x and λ. That is,

$$L(x, \lambda) = -x^2 + 6x + 1 + \lambda(2 - x)$$

this function may be maximized with respect to the unknown variables x and λ. The first order conditions are:

$$L_x = -2x + 6 - \lambda = 0$$

$$L_y = 2 - x = 0$$

Solving the two equations together results in x = 2 and $\lambda = 2$. The graph below shows the objective function and the constraint.

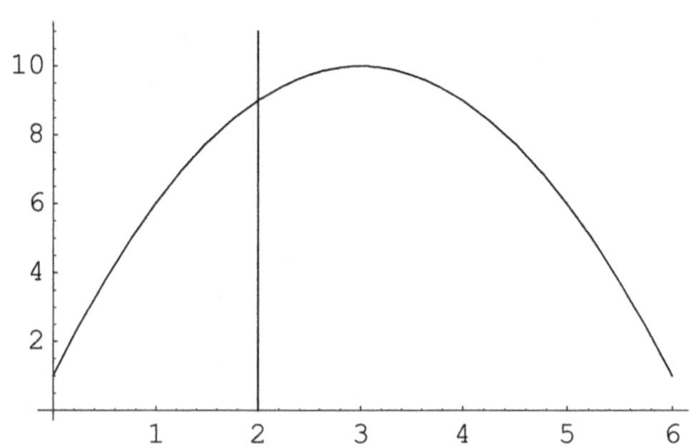

The constraint that doesn't allow the function f(x) achieve its full potential maximum (is the example above x = 3 and f(3) = 10) is called a **binding constraint.** If a constraint is binding, the corresponding Lagrangian multiplier (λ) will be positive, $\lambda > 0$. If a constraint is **non-binding,** then the value of its corresponding Lagrangian multiplier will be negative, $\lambda < 0$.

Let's solve the same problem above assuming that $x \leq 4$. That is,

Maximize: $y = f(x) = -x^2 + 6x + 1$

Subject to: $x \leq 4$

Form the Lagrangian function and find the first-order conditions,

$$L = -x^2 + 6x + 1 + \lambda(4 - x)$$

$$L_x = -2x + 6 - \lambda = 0$$

$$L_y = 4 - x = 0$$

Here, the solution to the first-order conditions is x = 4 and $\lambda = -2$. A negative λ, $\lambda < 0$, implies that the constraint is non-binding. A non-binding constraint implies that the function should be optimized as if there is no constant. In the case above, f(x) is optimized as an unconstrained optimizing problem when x = 3. Therefore, the solution to constrained optimizing problem above is x = 3 and f(3) = 10. The graph below illustrates why a non-binding optimizing problem should treated as unconstraint optimizing problem.

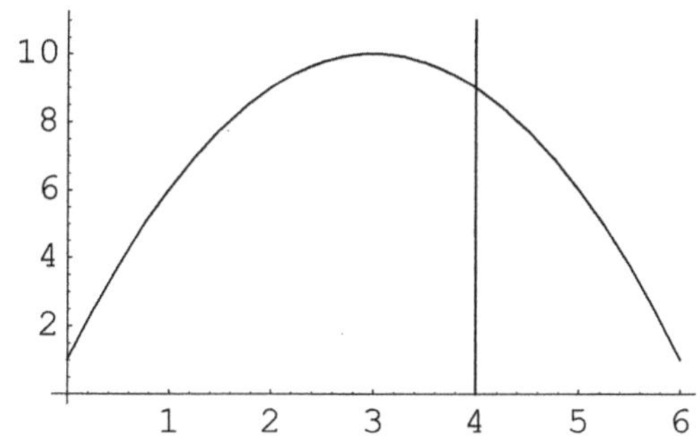

If the solution to the first-order condition of a constrained optimizing problem results in $\lambda = 0$, then the constrained optimizing solution coincide with unconstrained solution. The graph below shows the case where the objective is to maximize $y = f(x) = -x^2 + 6x + 1$ subject to the constraint $x \leq 3$. The solution to this problem is x = 0, f(3) = 10 and $\lambda = 0$.

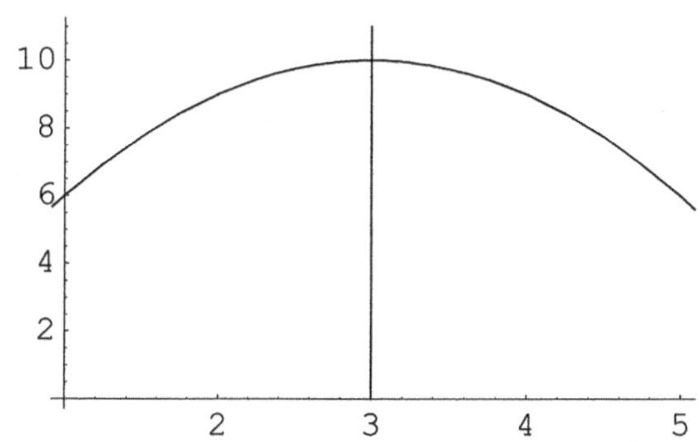

Example1: Production Function - The short-run production technology of a business is given as $Q = 14.4L + .6L^2 - .2L^3$. The firms number of labor input is limited to five units, $L \leq 5$. Maximize output of the business subject to the existing labor constraint.

Solution : The problem can be formulated as :

$$\text{Max}: \quad Q = f(L) = 14.4L + .6L^2 - .2L^3$$
$$\text{Subject To}: \quad L \leq 5$$

To solve the problem, form the Lagrangian function and find the firstorder conditions :

$$L = 14.4L + .6L^2 - .2L^3 + \lambda(5 - L)$$
$$\frac{\delta L}{\delta L} = 14.4 + 1.2L - .6L^2 - \lambda = 0$$
$$\frac{\delta L}{\delta \lambda} = 5 - L = 0$$

Solving for the first order condition gives L = 5 and $\lambda = 5.4$. Since $\lambda > 0$, the constraint is binding and L = 5 is solution, with maximum output at this level of input Q = f(5) =62 units. The graph below shows the relationship between the level of output in the production function and labor input. As the graph shows, the constraint is binding.

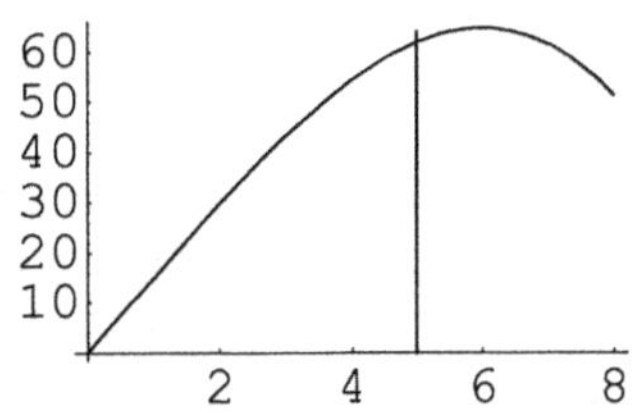

Solving the problem with *Mathematica,*

```
In[1] := Clear[L, l, λ]
         L = 14.4 l + .6 l^2 - .2 l^3 + λ(5 - λ);
         Solve[{D[L, l] == 0, D[L, λ] == 0}, {λ, λ}]
         Plot[14.4 l + .6 l^2 - .2 l^3, {l, 0, 8}];
Out[1] = {{λ → 5.4, l → 5.}}
```

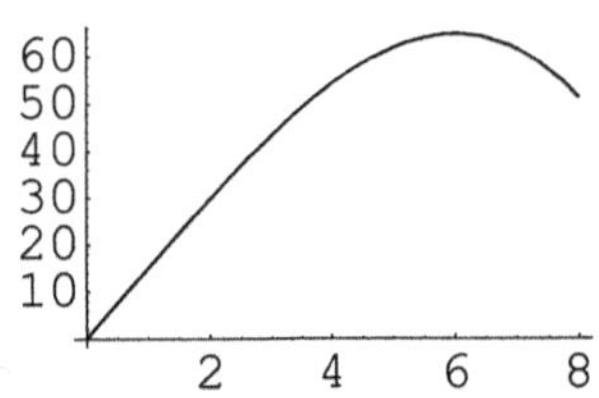


```
ListPlot[{{5, 0}, {5, 64}}, PlotJoined → True];
```

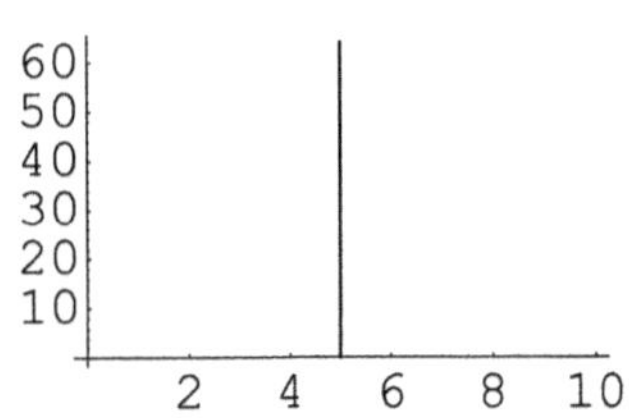


```
Show[%, %%]
```

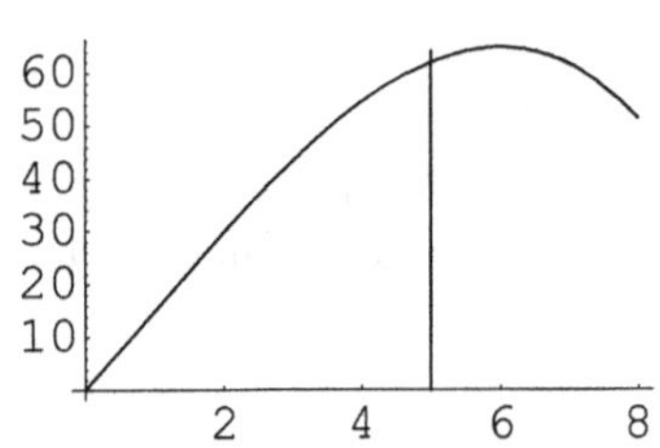

Example 2: Consumer Theory - The utility function of a consumer for good X is given as U = lnx where, U is utility and x is the number of units of X purchased by the consumer. The consumer has only \$60 to spend on X and the price of good X is \$3.

Solution: The problem can be formulated as:

Max: $U = f(x) = \ln x$

Subject To: $3x \leq 60$

To solve the problem, form the Lagrangian function and find the first-order conditions:

$$L = \ln x + \lambda(60 - 3x)$$

$$\frac{\delta L}{\delta L} = 1/x - 3\lambda = 0$$

$$\frac{\delta L}{\delta \lambda} = 60 - 3x = 0$$

Solving for the first-order condition gives $x = 20$ and $\lambda = 1/60$. Since $\lambda > 0$, the constraint is binding and $x = 20$ is the solution, with maximum utility $U = f(20) = 2.99$. Solving the problem with *Mathematica,*

```
In[2]:= Clear[L, x, λ]
        L = Log[E, x] + λ(60 - 3x);
        Solve[{D[L, x] == 0, D[L, λ] == 0}, {x, λ}]
        Plot[Log[E, x], {x, 0, 30}];
        ListPlot[{{20, 0}, {20, 4}}, PlotJoined → True];
        Show[%, %%]
        {{λ → 1/60, x → 20}}
```

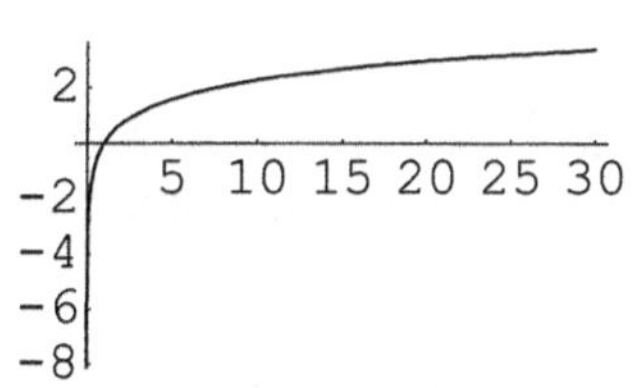

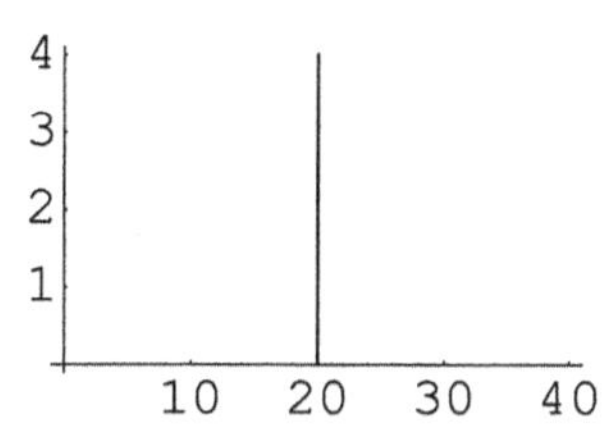

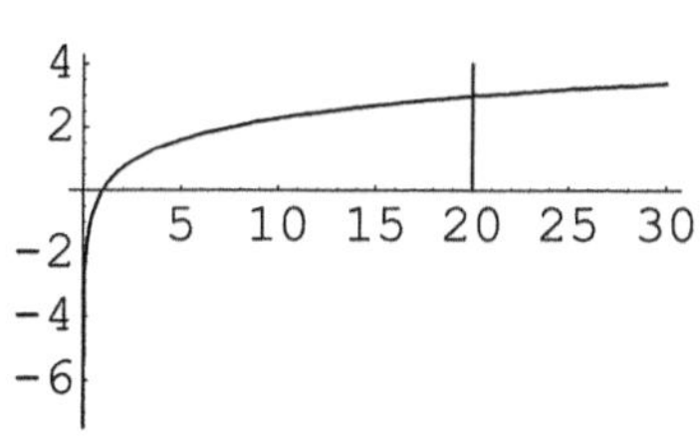

Problems:

1- Maximize y = f(x) = $-x^2 + 10x + 8$ subject to $x \leq 4$. Find the optimum values of y, x and λ.

2- Maximize y = f(x) = $-x^2 + 10x + 8$ subject to $x \leq 4$. Find the optimum values of y, x and λ.

3- Maximize y = f(x) = $-x^2 + 10x + 8$ subject to $x \leq 4$. Find the optimum values of y, x and λ.

4- Minimize y = f(x) = $.25x^2 - 4x + 20$ subject to $x \leq 6$. Find the optimum values of y, x and λ.

5- Minimize y = f(x) = $.25x^2 - 4x + 20$ subject to $x \leq 10$. Find the optimum values of y, x and λ.

6- Minimize y = f(x) = $.25x^2 - 4x + 20$ subject to $x \leq 8$. Find the optimum values of y, x and λ.

7- Maximize y = f(x) = $-\frac{1}{12}x^3 + 5x^2 + 5.0625x + 30$ subject to $x \leq 30$. Find the optimum values of y, x and λ.

8- Maximize y = f(x) = $-\frac{1}{12}x^3 + 5x^2 + 5.0625x + 30$ subject to $x \leq 45$. Find the optimum values of y, x and λ.

9- Maximize y = f(x) = $-\frac{1}{12}x^3 + 5x^2 + 5.0625x + 30$ subject to $x \leq 41$. Find the optimum values of y, x and λ.

10 - The short-run production technology of a firm is given as Q = $20L - .5L^2$. The firm has no more than 35 units of labor available. Maximize output of the business subject to the labor constraint. Find the optimum values of Q, L and λ.

11 - The short-run production technology of a firm is given as Q = $20L - .5L^2$. The firm has no more than 40 units of labor available. Maximize output of the business subject to the labor constraint. Find the optimum values of Q, L and λ.

12- The short-run production technology of a firm is given as Q = $20L - .5L^2$. The firm has no more than 50 units of labor available. Maximize output of the business subject to the labor constraint. Find the optimum values of Q, L and λ.

13- The short-run production technology of a business is given as Q = $6L + .25L^2 - .25L^3$. The firm has no more than 5 units of labor available. Maximize output of the business subject to the labor constraint. Find the optimum values of Q, L and λ.

14- A consumer's utility function for X is expressed as U = f(x) = $\sqrt{x}$. The consumer has no more than \$24 to spend on X. Maximize consumer's utility given that X is sold at a price of \$4. Find the optimum values of U, x and λ.

II. Lagrangian, The Case of Multi-Variate Functions: The sol = Solve [{D[L, x]==0, D[L, y]==0 D[L, λ]==0},{x,y,λ}] Command

Consider the following constrained maximization problem:

Maximize: U = U(X, Y)

Subject to: PxX + PyY = M

where the first equation is the objective function in which U(X, Y) is the utility function, assumed to be differentiable, and PxX + PyY = M is the budget constraint. The problem is to maximize the utility subject to the budget constraint. To solve the constrained optimizing problem, we form the Lagrangian function.

The Lagrangian function is formed by augmenting the constraint together with the Lagrange multiplier (λ) to the objective function.

$$L = U(X, Y) + \lambda(M - PxX - PyY)$$

If the constraint is satisfied, no matter what value λ has, the second term in the right side of the Lagrangian equation will vanish and L = U(X, Y). That is, maximizing L with respect to X and Y is the same as maximizing U(X, Y). Therefore, incorporate the constraint into the problem and insure that the constraint is satisfied with equality. This is done by treating λ as an additional variable and assuming L is a function of X, Y, and λ. That is,

$$L(X, Y, \lambda) = U(X, Y) + \lambda(M - PxX - PyY)$$

this function may be maximized with respect to the three unknown variables X, Y, and λ. The first order conditions are:

$$L_x = Ux - \lambda Px = 0$$

$$L_y = Uy - \lambda Py = 0$$

$$L_\lambda = M - PxX - PyY = 0$$

A point P(X*, Y*, λ*) that satisfies the three equations of the first-order condition is the extremum.

The second order condition for L(X, Y, λ) to have a maximum at P is that d^2L to be negative definite at the extremum. To find d^2L, find total differential of dL

$$dL = UxdX + UydY + d\lambda(M - PxX - PyY) - Px\lambda dX - Py\lambda dY$$

$$d^2L = UxxdX^2 + UxydXdY + UyxdXdY + UyydY^2 - 2Pxd\lambda dX - 2Pyd\lambda dY$$

Using the matrix notations, $d^2\,L$ can be expressed in a quadratic form as:

$$(\,dX \quad dY \quad d\lambda\,)\begin{pmatrix} Uxx & Uxy & -Px \\ Uyx & Uyy & -py \\ -Px & -Py & 0 \end{pmatrix}\begin{pmatrix} dX \\ dY \\ d\lambda \end{pmatrix}$$

The matrix of the quadratic form is called a **bordered Hessian matrix** which consists of a simple Hessian and two borders corresponding to the constraint.

The quadratic form of the second-order condition is said to be negative definite if the bordered principal minors of the bordered Hessian, $\mathbf{H} = \begin{pmatrix} Uxx & Uxy & -Px \\ Uyx & Uyy & -Py \\ -Px & -Py & 0 \end{pmatrix}$ alternate in sign, beginning with the first bordered principal minor negative. That is, the second-order condition for maximum utility at the point P* requires that $\text{Det}\begin{pmatrix} Uxx & -Px \\ -Px & 0 \end{pmatrix} < 0$ and $\text{Det}\begin{pmatrix} Uxx & Uxy & -Px \\ Uyx & Uyy & -Py \\ -Px & -Py & o \end{pmatrix} > 0$. In general, for a n-variable function y = f(x1, x2, ... xn) to be optimized subject to the constraint g(x1, x2, ..., xn) = c, the Lagrangian function is L(x1, ..., xn, λ) = f(x1, x2, ..., xn) + λ(c - g(x1, x2, ..., xn)) and the first-order conditions are:

$$\frac{\partial L}{\partial x1} = f_{x1} - \lambda g_{x1} = 0$$

$$\frac{\partial L}{\partial x2} = f_{x2} - \lambda g_{x2} = 0$$

....

$$\frac{\partial L}{\partial xn} = f_{xn} - \lambda g_{xn} = 0$$

$$\frac{\partial L}{\partial \lambda} = c - g(x1, ..., xn) = 0$$

The first order condition yields n + 1 equations in n + 1 variables which can be solved simultaneously to determine the extremum $P(x^*_1, x^*_2, ..., x^*_n, \lambda^*)$. To test for the maximum or minimum, the second order condition yields the following bordered Hessian matrix, (**Note** that, f_{11}, f_{12}, and f_{1n} below are representing f_{x1x1}, f_{x1x2}, and f_{x1xn} respectively. For the clarity of exposition the xs are deleted.)

$$\mathbf{H} = \begin{pmatrix} f_{11} & f_{12} & \cdot & \cdot & f_{1n} & -g_1 \\ f_{21} & f_{22} & \cdot & \cdot & f_{2n} & -g_2 \\ \cdot & \cdot & \cdot & \cdot & \cdot & \cdot \\ \cdot & \cdot & \cdot & \cdot & \cdot & \cdot \\ f_{n1} & f_{n2} & \cdot & \cdot & f_{nn} & -g_n \\ -g_1 & -g_2 & \cdot & \cdot & -g_n & 0 \end{pmatrix}$$

the extremum is a minimum if the bordered Hessian determinant and the corresponding matrix is positive definite. The bordered Hessian determinant is positive definite if the bordered principal minors, defined as follows and evaluated at the extremum, are all negative. That is:

$$\text{Det}\begin{pmatrix} f_{11} & -g_1 \\ -g_1 & 0 \end{pmatrix} < 0, \quad \text{Det}\begin{pmatrix} f_{11} & f_{12} & -g_1 \\ f_{21} & f_{22} & -g_2 \\ -g_1 & -g_2 & 0 \end{pmatrix} > 0, \quad \text{Det}\begin{pmatrix} f_{11} & f_{12} & f_{13} & -g_1 \\ f_{21} & f_{22} & f_{23} & -g_2 \\ f_{31} & f_{32} & f_{33} & -g_3 \\ -g_1 & -g_2 & -g_3 & 0 \end{pmatrix} < 0, \; ....$$

The extremum is a maximum if the bordered Hessian determinant and the corresponding matrix is negative definite. The bordered Hessian determinant is negative definite if the bordered principal minors, defined as above and evaluated at the extremum, alternate in sign, beginning with the first bordered principal minor negative. **Note** that, the principal minor rules for the positive definiteness and negative definiteness of the bordered-Hessian matrices are different from the rules for Hessians with no border (unconstrained optimization).

Example: Consumer Theory

A consumer has a utility function expressed as U(x, y) = lnxy, where x and y are quantities of two goods X and Y consumed by the consumer. The market price of X and Y are $2 and $3, respectively. The consumer has a budget of $120 to spend on the two goods. Solve the utility maximizing problem of the consumer and find the optimum quantities of x and y.

Solution: The utility maximizing problem can be expressed as

Maximize: U = lnxy

Subject to: 2x + 3y = 120

The Lagrangian function is L(x, y, λ) = lnxy + λ(120 - 2x - 3y) and the first order conditions are:

$\frac{\partial L}{\partial x} = \frac{1}{x} - 2\lambda = 0,$ or $\frac{1}{x} = 2\lambda$

$\frac{\partial L}{\partial y} = \frac{1}{y} - 3\lambda = 0,$ or $\frac{1}{y} = 3\lambda$

$\frac{\partial L}{\partial \lambda} = 48 - 2x - 3y = 0$

divide the first two equations of the first-order condition together to delete λ and to get y = $\frac{2}{3}$x. Substitute for x in the third equation to get x = 12 and y = 8. To test whether this point (x = 12, y = 8, and λ = $\frac{1}{24}$) is a minimum or maximum, the second-order conditions yields the Hessian, $\mathbf{H} = \begin{pmatrix} \frac{-1}{x^2} & 0 & -2 \\ 0 & \frac{-1}{y^2} & -3 \\ -2 & -3 & 0 \end{pmatrix}$, the determinants of the bordered principal minors of which are Det$\begin{pmatrix} \frac{-1}{x^2} & -2 \\ -2 & 0 \end{pmatrix}$ = -4 < 0 and Det $\begin{pmatrix} \frac{-1}{x^2} & 0 & -2 \\ 0 & \frac{-1}{y^2} & -3 \\ -2 & -3 & 0 \end{pmatrix} = \frac{9}{x^2} + \frac{4}{y^2} > 0$. Since the determinants of the bordered principal minors alternate in sign beginning with the first one negative, the Hessian is negative definite and the extremum is a maximum. (Note that the first bordered principal minor is always negative.) The solution implies that for the consumer to be at utility maximizing equilibrium, the consumer should buy 12 units of X and 8 units of Y. To solve the problem using *Mathematica*,

```
In[3]:= Clear[L, U, x, y, λ, sol]
  L = Log[x y] + λ (48 - 2 x - 3 y);
  sol = Solve[{D[L, x] == 0, D[L, y] == 0, D[L, λ] == 0}, {x, y, λ}]
  Out[3] =  (λ → 1/24), (x → 12), (y → 8)
```

The bordered Hessian matrix of the second order condition is:

```
In[4]:= MatrixForm[( D[D[L, x], x]  D[D[L, x], y]  D[D[L, x], λ]
                     D[D[L, y], x]  D[D[L, y], y]  D[D[L, y], λ]
                     D[D[L, λ], x]  D[D[L, λ], y]  D[D[L, λ], λ] )]
```

$$\text{Out[4]} = \begin{pmatrix} -\frac{1}{x^2} & 0 & -2 \\ 0 & -\frac{1}{y^2} & -3 \\ -2 & -3 & 0 \end{pmatrix}$$

The determinant of the bordered principal minors evaluated at the solution point are,

```
In[5]:= N[Det[( D[D[L, x], x]  D[D[L, x], λ]
                D[D[L, λ], x]  D[D[L, λ], λ] )]] /. {sol1}

  Out[5] =  {{-4}}
In[6]:= N[Det[( D[D[L, x], x]  D[D[L, x], y]  D[D[L, x], λ]
                D[D[l, y], x]  D[D[L, y], y]  D[D[L, y], λ]
                D[D[L, λ], x]  D[D[L, λ], y]  D[D[L, λ], λ] )]] /. {sol}
  Out[6] =  {{0.125}}
```

Since the determinants of the bordered principal minors alternate in sign beginning with the first determinant negative, the Hessian is negative definite and the extremum is a maximum.

Mathematica provides a very useful tool to observe the solution of the problem by plotting the functions in 3-D. As well, by using the command **ContourPlot[U(x, y), {x, a, b}, {y, c, d}]**, the indifference curves of the utility function and the budget constraint can be graphed in x-y axis. This allows us to find the solution to the consumer's utility maximizing problem at the tangency point of the highest feasible indifference curve and the budget constraint. To see the solution point graphically in three dimensions, plot the objective function and the budget constraint together as follows

In[7]:=
```
Clear[U, x, y]
U = Log[x y];
g1 = Plot3D[U, {x, .5, 40}, {y, .5, 40}]
```

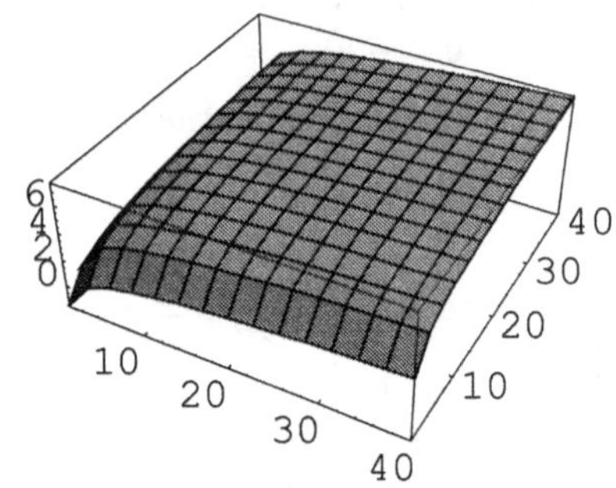

In[8]:=
```
g2 = Plot3D[2 x + 3 y - 48, {x, 0, 40}, {y, 0, 40}]
```

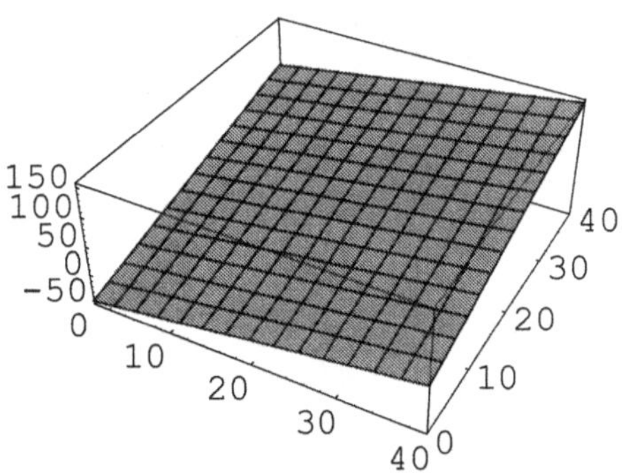

In[9]:=
```
Show[g1, g2, ViewPoint → {1.2, 1.2, 1.5}]
```

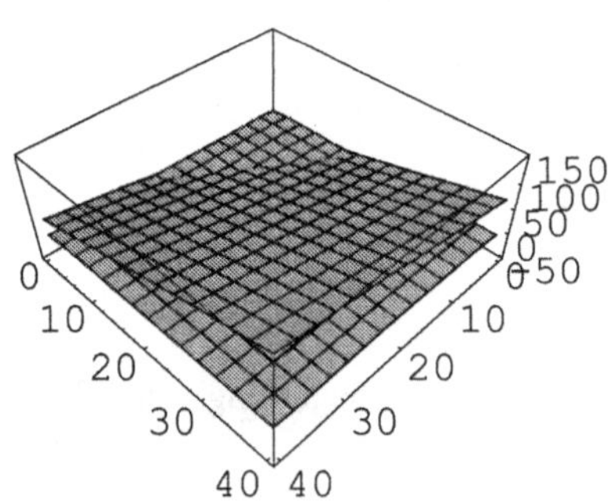

Note: you can always change the viewpoint in the "Show" command to have a better view of the solution point. To Change the viewpoint use the `ViewPoint` command.

To see the solution point graphically in x-y space, plot the indifference curves and the budget constraint using the `ContourPlot` command as follow,

```
In[10]:= con1 = ContourPlot[U, {x, .5, 40}, {y, .5, 20}, ContourShading -> False]
```

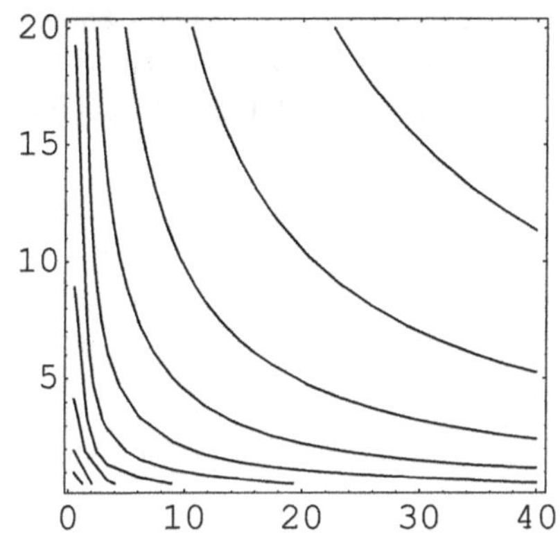

```
In[11]:= con2 = ContourPlot[2 x + 3 y - 48, {x, 0, 40}, {y, 0, 20}, Contours -> {0},
          ContourShading -> False]
```

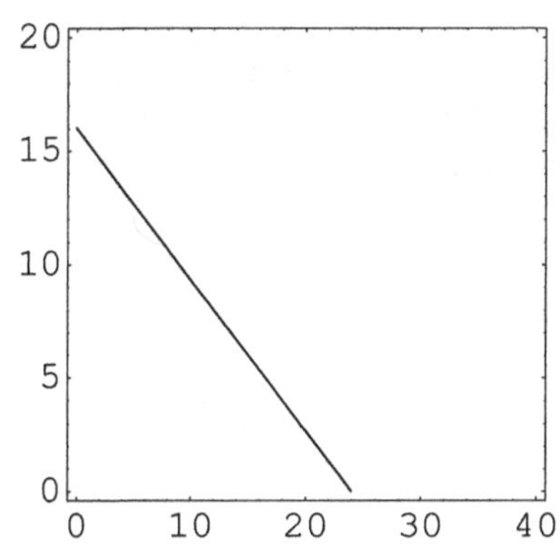

```
In[12]:= Show[con1, con2]
```

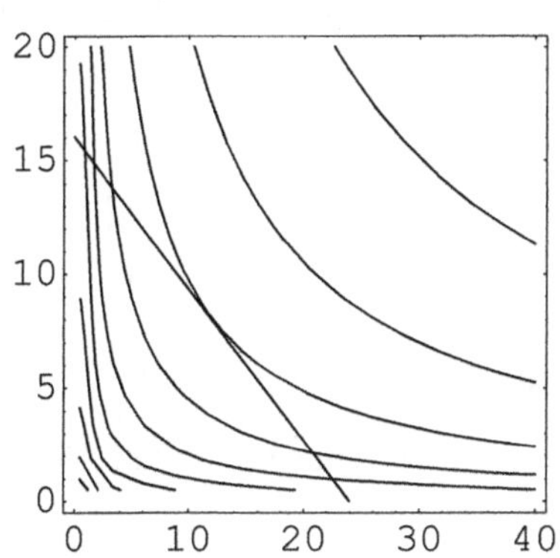

Do More Practice:

1- Change the budget constraint from \$120 to \$360 and re-run the program. Observe the changes in solution.

2- Change prices of x and y from \$2 and \$3 to \$2 and \$4, respectively. Hold the budget constraint at \$120. Re-run the program and compare the results.

Example: Theory of Production

A single-product firm has the production function $Q = (La^{.5} + K^{.5})^2$ where, La and K are labor and capital inputs, and Q is the output. The firm is hiring the services of labor and capital at prices \$12 and \$8 per unit, respectively. If the firm has a cost constraint of \$600, solve the output maximizing problem of the firm subject to its cost constraint.

Solution: The problem can be formulated as,

Maximize: $Q = (La^{.5} + K^{.5})^2$

Subject to: $12La + 8K = 600$

The Lagrangian function is $L(La, K, \lambda) = (La^{.5} + K^{.5})^2 + \lambda(600 - 12La - 8K)$ and the first order conditions are:

$$\frac{\partial L}{\partial La} = La^{-.5}(La^{.5} + K^{.5}) - 12\lambda = 0, \qquad \text{or } La^{-.5}(La^{.5} + K^{.5}) = 12\lambda$$

$$\frac{\partial L}{\partial K} = K^{-.5}(L^{.5} + K^{.5}) - 8\lambda = 0, \quad \text{or } K^{-.5}(L^{.5} + K^{.5}) = 8\lambda$$

$$\frac{\partial L}{\partial \lambda} = 600 - 12La - 8K = 0$$

divide the first two equations of the first-order condition together to delete λ and to get $\frac{La^{-.5}}{K^{-.5}} = \frac{3}{2}$ or $K = \frac{9}{4}La$. Substitute for K in the third equation to get La = 20 and K = 45. That is, to maximize its output, the firm should employ 20 units of labor and 45 units of capital. This will produce 125 units of output (this number is arrived at by substituting La and K in production function Q). To find λ, substitute for La and K in the first order condition to get $\lambda = .21$.

To test whether the extremum (La = 20, K= 45, and λ = .21) is a minimum or maximum, the second-order condition yields the Hessian, $\mathbf{H} = \begin{pmatrix} \frac{-.5(L^{.5}+K^{.5})}{La^{1.5}} + \frac{.5}{La} & \frac{.5}{K^{.5}La^{.5}} & -12 \\ \frac{.5}{K^{.5}La^{.5}} & \frac{-.5(L^{.5}+K^{.5})}{K^{1.5}} + \frac{.5}{K} & -8 \\ -12 & -8 & 0 \end{pmatrix}$. The determinants of the bordered principal minors evaluated at the critical point are $\text{Det}\begin{pmatrix} \frac{-.5(L^{.5}+K^{.5})}{La^{1.5}} + \frac{.5}{La} & -12 \\ -12 & 0 \end{pmatrix} = -144 < 0$ and Det[**H**] = 6.67 > 0, respectively. Since the determinants of the bordered principal minors alternate in sign beginning with the first one negative, the Hessian is negative definite and the extremum is a maximum. (**Note** that the first bordered principal minor is always negative.)

To solve the problem using *Mathematica*,

```
In[13]:= Clear[L, La, K, λ, sol]
         L = (La^.5 + K^.5)^2 + λ (600 − 12 La − 8 K);
         sol = Solve[{D[L, La] == 0, D[L, K] == 0, D[L, λ] == 0}, {La, K, λ}]
         Out[13] = {(La → 20.`), (K → 45.`), (λ → 0.20833333334`)}
```

The first order condition implies that the firm will hire 20 units of labor and 45 units of capital. To find Q, evaluate the production function at the optimal solution,

```
In[14]:= Q = (La^.5 + K^.5)^2;
         Q /. {sol}
         {{125.00000000000001` Null}}
```

To test for the second order condition, the bordered Hessian is:

```
In[15]:= MatrixForm[( D[D[L, La], La]  D[D[L, La], K]  D[D[L, La], λ]
                      D[D[L, K], La]   D[D[L, K], K]   D[D[L, K], λ]
                      D[D[L, λ], La]   D[D[L, λ], K]   D[D[L, λ], λ] )]
```

$$\text{Out[15]} = \begin{pmatrix} -\frac{0.5`(K^{0.5`}+La^{0.5`})}{La^{1.5`}} + \frac{0.5`}{La^{1.`}} & \frac{0.5`}{K^{0.5`}La^{0.5`}} & -12 \\ \frac{0.5`}{K^{0.5`}La^{0.5`}} & \frac{0.5`}{K^{1.`}} - \frac{0.5`(K^{0.5`}+La^{0.5`})}{K^{1.5`}} & -8 \\ -12 & -8 & 0 \end{pmatrix}$$

The bordered principal minors evaluated at the extremum are :

```
Det[( D[D[L, La], La]  D[D[L, La], λ]
      D[D[L, λ], La]   D[D[L, λ], λ] )] /. {sol}
Out[] = {{-144}}
Det[( D[D[L, La], La]  D[D[L, La], K]  D[D[L, La], λ]
      D[D[L, K], La]   D[D[L, K], K]   D[D[L, K], λ]
      D[D[L, λ], La]   D[D[L, λ], K]   D[D[L, λ], λ] )] /. {sol}
Out[] = {{6.666666666666666`}}
```

```
In[16]:= Q = (La^.5 + K^.5)^.5;
         graph1 = Plot3D[Q, {La, 1, 30}, {K, 1, 50}]
```

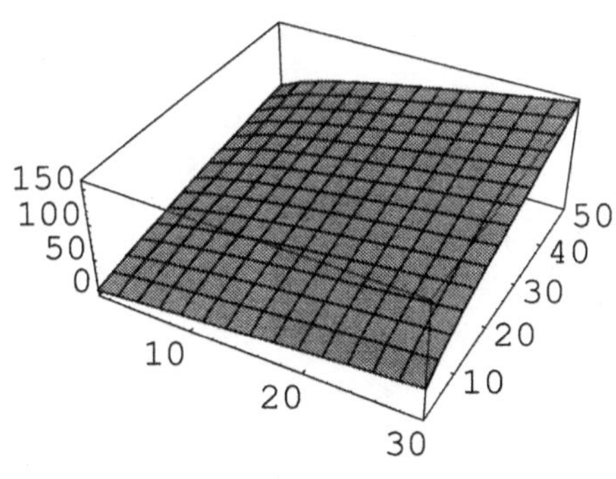


```
In[17]:= graph2 = Plot3D[12 La + 8 K - 600, {La, 1, 30}, {K, 1, 50}]
```

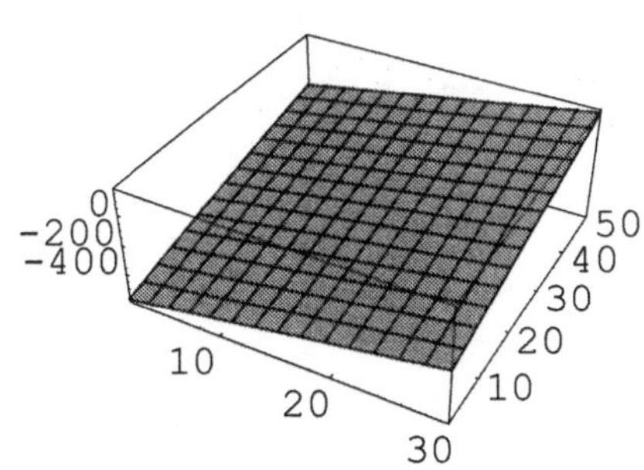

```
In[18]:= Show[graph1, graph2, ViewPoint → {1.2, 1.2, 1.6}]
```

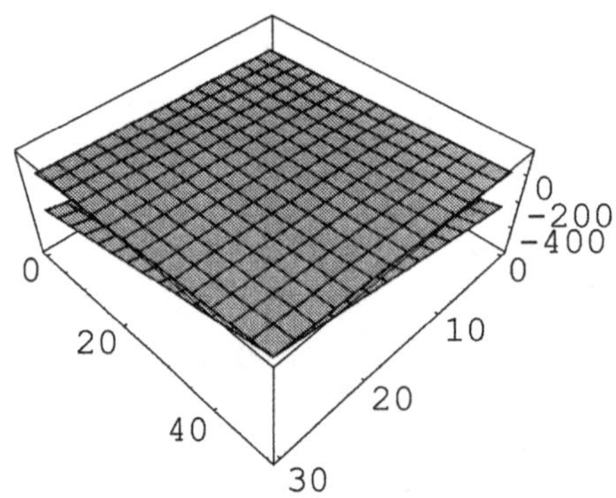

To see the solution point graphically in x-y space, plot the isoquants and the cost constraint using the "`ContourPlot`" command,

```
In[19]:= co1 = ContourPlot[Q, {La, 1, 50}, {K, 1, 100}, ContourShading -> False]
```

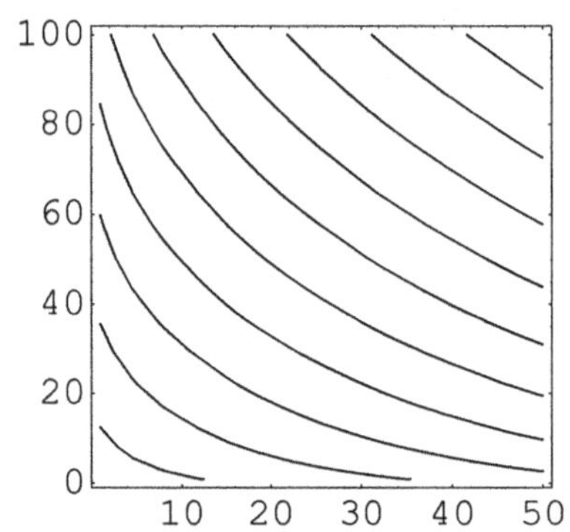


```
In[20]:= co2 = ContourPlot[12 La + 8 K - 600, {La, 1, 50}, {K, 1, 100},
             Contours -> {0}, ContourShading -> False]
```

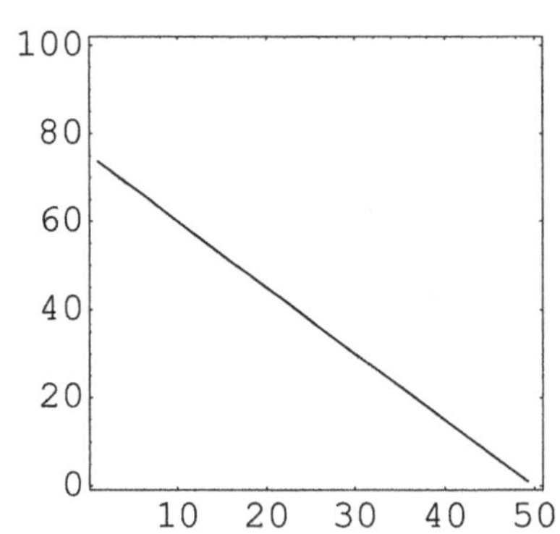

In[21]:=Show[col , co2]

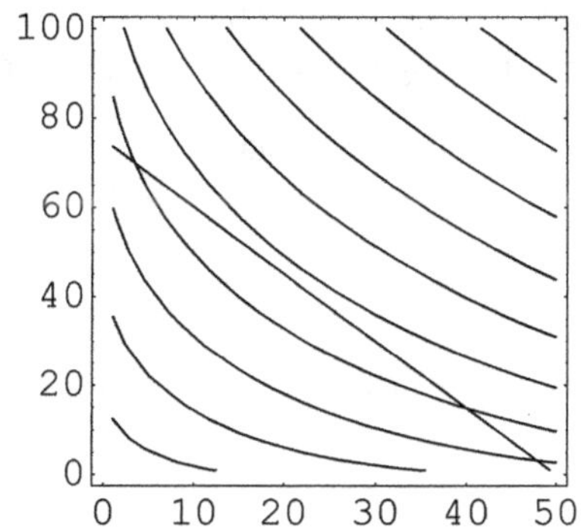

Do More Practice:

1- Change the cost constraint to $1200. What is the effect of doubling the cost budget on inputs and output?

2- Double the labor and capital costs and keep the total cost budget at $600. How does the solution change?

3- Change the capital price from $8 to $12, what is the effect on labor input, capital input and the output?

4- Change the production function to Q = $(La^2 + K^2)^{.5}$. Re-run the program. Is this a conventional utility function? Why?

Example: Cost and Production

Suppose that the objective of a firm employing two variable inputs labor and capital is to minimize cost (C) subject to a given level of output (Q_o). Solve the constrained optimizing problem.

Solution: The constrained optimizing problem of the firm can be formulated as,

Minimize: C = wLa + rK

Subject to: Q(La, K) = Q_o

where, La and K are labor and capital inputs used in production, and w and r are wages and rental cost of capital respectively. The Lagrangian function is L(La, K,) = wLa + rK + $\lambda(Q_o$ - Q(La, K)) and the first order condition is

$$\frac{\partial L}{\partial \text{La}} = \text{w} - \lambda\text{QLa} = 0$$

$$\frac{\partial L}{\partial K} = \text{r} - \lambda\text{QK} = 0$$

$$\frac{\partial L}{\partial \lambda} = Q_o - \text{Q(La, K)} = 0$$

The second order condition requires that the bordered-Hessian, derived from the total differential of the first-order condition, to be positive definite. The total differential of the first-order condition yields **H** = $\begin{pmatrix} -\lambda\text{QLaLa} & -\lambda\text{QLak} & -Q\text{La} \\ -\lambda\text{QkLa} & -\lambda\text{Qkk} & -Qk \\ -Q\text{La} & -Qk & 0 \end{pmatrix}$. The second order condition for a minimum is satisfied if Det$\begin{pmatrix} -\lambda\text{QLaLa} & -Q\text{La} \\ -Q\text{La} & 0 \end{pmatrix} <$ 0 and Det$\begin{pmatrix} -\lambda\text{QLaLa} & -\lambda\text{QLak} & -Q\text{La} \\ -\lambda\text{QkLa} & -\lambda\text{Qkk} & -Qk \\ -Q\text{La} & -Qk & 0 \end{pmatrix} < 0$.

Example: Cost and Production

A firm has a production function $Q = 10x^{.75}y^{.25}$, where Q is the output, x is the labor input, and y is the capital input. The firm has to produce 800 units to meet the market demand. The market prices of labor input and capital input are \$60 and \$172, respectively. Solve the cost minimizing problem of the firm and find the total cost.

Solution: The constrained optimizing problem of the firm can be formulated as

Minimize: $C = 60x + 172y$

Subject to: $10x^{.75}y^{.25} = 800$

the Lagrangian function can be formulated as $L(x, y, \lambda) = 60x + 172y + \lambda(800 - 10x^{.75}y^{.25})$. The first order condition results in the following three equations:

$$\frac{\partial L}{\partial x} = L_x = 60 - 7.5\lambda\frac{y^{.25}}{x^{.25}} = 0, \quad \text{or } 60 = 7.5\lambda\frac{y^{.25}}{x^{.25}}$$

$$\frac{\partial L}{\partial y} = L_y = 172 - 2.5\lambda\frac{x^{.75}}{y^{.75}} = 0, \text{or } 172 = 2.5\lambda\frac{x^{.75}}{y^{.75}}$$

$$\frac{\partial L}{\partial \lambda} = L_\lambda = 800 - 10x^{.75}y^{.25} = 0, \quad \text{or } 800 = 10x^{.75}y^{.25}$$

Divide the first two equations of the first-order condition together to delete λ. This gives $y = \frac{5x}{43}$. By substituting for y in the third equation of the first-order condition we get x = 137, y = 16, and λ = 13.70.

The first-order conditions of the constrained optimizing problem yields a solution with 1367 units of labor input and 16 units of capital input. To test for the second-order condition, form the bordered Hessian matrix and evaluate it at the critical point (x = 137, y = 16, λ = 13.70),

$$\mathbf{H} = \begin{pmatrix} L_{xx} & L_{xy} & L_{x\lambda} \\ L_{yx} & L_{yy} & L_{y\lambda} \\ L_{\lambda x} & L_{\lambda y} & L_{\lambda\lambda} \end{pmatrix} = \begin{pmatrix} \frac{1.875\, y^{0.25}\,\lambda}{x^{1.25}} & -\frac{1.875\,\lambda}{x^{0.25}\,y^{0.75}} & -\frac{7.5\,y^{0.25}}{x^{0.25}} \\ 0 & \frac{1.875\,x^{0.75}\,\lambda}{y^{1.75}} & -\frac{2.5\,x^{0.75}}{y^{0.75}} \\ -\frac{7.5\,y^{0.25}}{x^{0.25}} & -\frac{2.5\,x^{0.75}}{y^{0.75}} & 0 \end{pmatrix} = \begin{pmatrix} 0.11 & -0.94 & -4.38 \\ 0 & 8.10 & -12.55 \\ -4.38 & -12.55 & 0 \end{pmatrix}$$

The determinant of the bordered principal minors are -19.18 and -276.14, respectively. Since the determinants of the bordered principal minors are both negative, the Hessian is positive definite and the extremum is a minimum point. To find the total cost, substitute for x and y in the objective function to get C = 10972.

To solve the problem using *Mathematica*, formulate the Lagrangian function and use the "Solve" command as,

```
In[22]:= Clear[x, y, λ, sol, cont1, cont2]
         L = 60 x + 172 y + λ (800 – 10 x^.75 y^.25);

         sol = Solve[{D[L, x] == 0, D[L, y] == 0, D[L, λ] == 0}, {x, y, λ}]
Out[22] = {{x → 136.9981175864378`, y → 15.930013672841602`, λ → 13.699811758643778`}}
```

To test for the second-order condition, form the bordered Hessian matrix and evaluate it at the solution "sol",

```
                     ( D[D[L, x], x]  D[D[L, x], y]  D[D[L, x], λ] )
In23]:= MatrixForm[  ( D[D[L, K], x]  D[D[L, y], y]  D[D[L, y], λ] ) ] /. sol
                     ( D[D[L, λ], x]  D[D[L, λ], y]  D[D[L, λ], λ] )

             ( 0.109490555521`   -0.94161877748`   -4.3796222208`  )
Out[23] = {  ( 0                 8.0979214864`     -12.5549170331` ) }
             ( -4.3796222208`    -12.5549170331`   0               )
```

The determinants of the bordered principal minors of the Hessian evaluated at the solution point "sol" are,

```
In[24]:= N[Det[( D[D[l, x], x]  D[D[L, x], λ]
                 D[D[L, λ], x]  D[D[L, λ], λ] )]] /. sol

Out[24] = {-19.181090797607066`}

In[25]:= Det[( D[D[L, x], x]  D[D[L, x], y]  D[D[L, x], λ]
               D[D[L, y], x]  D[D[L, y], y]  D[D[L, y], λ]
               D[D[L, λ], x]  D[D[L, λ], y]  D[D[L, λ], λ] )] /. sol

Out[25] = {-276.13683076016366`}
```

Since the determinants of the bordered principal minors are both negative, the Hessian is positive definite and the extremum is a minimum point. To find the cost at the optimum solution,

```
In[26]:=cost = 60 x + 172 y;
        cost /. sol
        {cost}
```

Therefore, the firm produces 800 units of output at a cost of $10959.80, using 136.6 units of labor input and 16 units of capital input.

To see the solution point graphically, plot the iso-quants and the iso-cost together,

```
In[27]:=cont1 = ContourPlot[60 x + 172 y - 10943.9, {x, 0, 200},
          {y, 0, 30}, Contours -> {0}, ContourShading -> False]
```

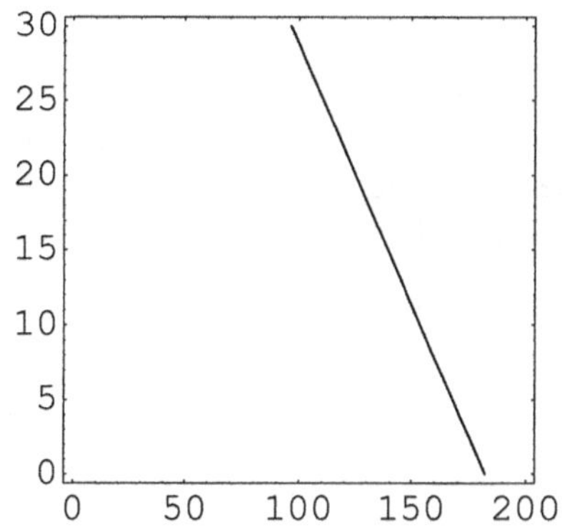


```
In[28]:= cont2 = ContourPlot[x^.75 y^.25, {x, 0, 200}, {y, 0, 30},
           ContourShading -> False]
```

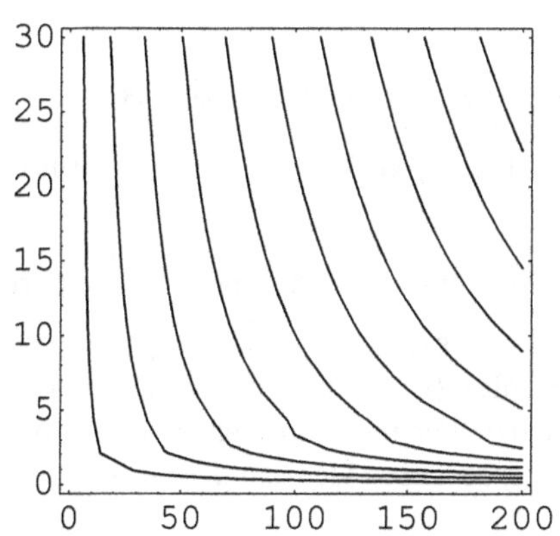

In[29]:= `Show[cont1, cont2]`

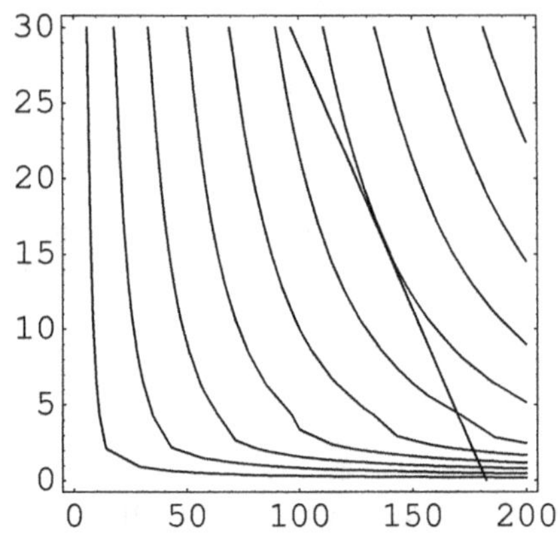

the optimum solution is the point where the isoquant is tangent to the lowest feasible isocost.

Do More Practice: Change the production function to Q = $10x^{.5}y^{.5}$, keeping other assumptions the same. Rerun the program and observe changes in the solution.

2- Change the production function to Q = $10x^{.5}y^{.5}$ and the cost constraint to \$2000. Re-run the program and compare the results to those of practice 1 above.

m-Constraint Case: The Lagrangian multiplier technique may be generalized to cases in which a function is optimized subject to more than one constraint, provided that the constraints are consistent and the number of constraints is less than the number of variables. (If the number of the constraints is equal to the number of the variables, locating the extremum will become a trivial matter of finding the intersection point of the constraints.)

For the following n-variable m-constraint problem (m < n), a maximizing problem can be formulated as:

Maximize: $y = f(x1, \ldots, xn)$

Subject to: $g^1(x1, \ldots, xn) = c1$

....

$g^i(x1, \ldots, xn) = ci$

....

$g^m(x1, \ldots, xn) = cm$

where the superscript " i " identifies the constraint i, i =1, ..., m. The Lagrangian function is,

$L = f(x1, \ldots, xn) + \lambda 1(c1 - g^1(x1, \ldots, xn)) + \lambda 2(c2 - g^2(x1, \ldots, xn)) + \ldots + \lambda m(cm - g^m(x1, \ldots, xn))$

and the first order condition is, $\frac{\partial L}{\partial xj} = f_j + \sum_{j=1}^{m} \lambda i g_j^i = 0$ and $\frac{\partial L}{\partial \lambda i} = ci - g^i = 0$, where f_j is the partial derivative of $f(x1, \ldots, xn)$ with respect to xj and gj^i is the partial derivative of $g^i(x1, \ldots, xn)$ with respect to xj. The first order condition yields n + m equations with n + m unknown variables (n independent variables and m Lagrangian multipliers) that can be solved to find the extremum. To test whether the extremum is a relative maximum or a relative minimum, the second order condition yields the following bordered Hessian determinant, with m borders

$$\mathbf{H} = \begin{pmatrix} f11 & f12 & . & . & f_{1n} & -g_1^1 & -g_1^2 & . & . & -g_1^m \\ f21 & f22 & . & . & f_{2n} & -g_2^1 & -g_2^2 & . & . & -g_2^m \\ . & . & . & . & . & . & . & . & . & . \\ fn1 & fn2 & . & . & fnn & -g_n^1 & -g_n^2 & . & . & -g_n^m \\ -g_1^1 & -g_2^1 & . & . & -g_n^1 & 0 & 0 & . & . & 0 \\ -g_1^2 & -g_2^2 & . & . & -g_n^2 & 0 & 0 & . & . & 0 \\ . & . & . & . & . & . & . & . & . & . \\ . & . & . & . & . & . & . & . & . & . \\ -g_1^m & -g_2^m & . & . & -g_n^m & 0 & 0 & . & . & 0 \end{pmatrix}$$

As can be seen, the Hessian can be partitioned into four parts; a plain Hessian corresponding to objective function *f*, two border matrices corresponding to the constraints, and a null matrix.

For the extremum to be a maximum, the bordered Hessian determinant should be negative definite. For Det[**H**] to be

negative definite, the determinants of the bordered principal minors of **H** should alternate in sign, beginning with the

first principal minor having the sign $(-1)^{m+1}$. For the extremum to be a minimum, the bordered Hessian should be positive definite. For **H** to be positive definite, the determinants of the bordered principal minors should all have the sign $(-1)^m$.

Economic Interpretation of the Lagrangian Multiplier : The Lagrange multiplier can be thought of as the rate of change of the objective variable with respect to the constraint. That is, it shows the marginal valuation, or the implicit price of the constraint. For example, in a utility maximizing problem subject to the budget constraint the objective variable is U and the constraint is M (Money or budget.) Therefore, $\lambda = \frac{dU}{dM}$, which is the marginal utility of money (marginal valuation of money) or the implicit price of money for the consumer.

For a utility maximizing problem, the first order conditions $Ux = \lambda\ Px$, or $\frac{Ux}{Px} = \lambda$, and $Uy = \lambda\ Py$, or $\frac{Uy}{Py} = \lambda$, imply that consumer is in equilibrium if the marginal utility received from spending one dollar on X or Y is equal to the marginal utility of that one dollar for the consumer. If the budget constraint is binding, i.e., if the consumer spends all of her money on X and Y, then λ is positive ($\lambda > 0$), which implies that money has a positive marginal utility or implicit price for the consumer. In this case, the optimum solution will be somewhere on the budget constraint.

If the budget constraint is non-binding, i.e. the consumer's budget is not a factor in her buying decision, then $\lambda = 0$, which means that the consumer has so much money that the marginal valuation of money or the implicit price of

money is zero for the consumer. In this case, the solution will be the same as the unconstrained maximization. That is, the consumer buys those quantities of X and Y which satiates the consumer.

In general, if a resource is limited and its constraint is binding, that resource will have a positive implicit price, ($\lambda > 0$). On the contrary, if a resource is abundant and its constraint is non-binding, then the implicit price of the resource will be zero, ($\lambda = 0$).

To prove that $\frac{dU}{dM} = \lambda$, totally differentiate the objective function $U = U(x, y)$ and the constraint $M = PxX + Py\ Y$ to get $dU = Uxdx + Uydy$ and $dM = Pxdx + Pydy$. Divide both sides of the two equations together to get $\frac{dU}{dM} = \frac{Uxdx + Uydy}{PxX + PyY}$. Substitute for Ux and Uy from the first order condition, where $Ux = \lambda Px$ and $Uy = \lambda Py$, to get $\frac{dU}{dM} = \frac{\lambda Pxdx + \lambda Pydy}{Pxdx + Pydy} = \lambda$.

Example: For a constrained cost minimizing problem prove that the Lagrangian multiplier λ is the marginal cost of production.

Solution: A cost minimizing problem can be formulated as Minimize C = wL + rK subject to Q(L, K) = Q_o, where L and K are labor and capital inputs, w and r are wages and rental cost of capital and Q is the output. Marginal cost is defined as cost of additional unit of output or $\frac{dC}{dQ}$. Totally differentiate the objective function and the constraint to get dC = wdL + rdK and dQ = QLdL + QKdK . Divide the two equations together and substitute for w and r from the first order condition (w = λQL and r = λQK), $\frac{dC}{dQ} = \frac{\lambda QLdL + \lambda QKdK}{QLdL + QKdK} = \lambda$. Therefore, λ is the marginal cost of production $\frac{dC}{dQ} = \lambda$.

Problems

1. A consumer has a hypothetical utility function U = lnxy, where x and y are quantities of two goods sold at market prices of $2 and $3, respectively. Given that consumer has an income of $60:

a) Set up the mathematical programming problem of maximizing consumer's utility subject to his budget constraint.

b) Find the optimum levels of x and y.

c) Test for the second order condition.

2. Suppose that the objective of the consumer in problem 1 is to minimize expenditures subject to a fixed level of utility, Uo= 150. Given that market prices are the same as in problem 1,

a) set up the proper optimizing problem.

b) find the optimum levels of x and y.

c) test the second order conditions.

3. A consumer has a hypothetical utility function U = ln xy + xy, where x and y are quantities of goods sold at market price of $2 and $1, respectively. Given that consumers has an income of $60:

a) Set up the mathematical programming problem of maximizing utility subject to the budget constraint.

b) Find the optimum levels of x and y.

c) Test for the second order condition.

4. Solve the following problem of minimizing expenditure(E), subject to the given utility levels of Uo = 25

Minimize: E = 2x + 3y

Subject to: U = $(x^{.5} + y^{.5})^2$

5. A firm operating in an environment of pure competition has the output function Q = f(L, K), where L and K stand for labor and capital, respectively. L and K are purchasable in competitive markets at prices w, r, respectively.

a)Set up the problem of minimizing cost (C) subject to the given level of output Qo.

b) Find the necessary and sufficient conditions for the constrained minimizing problem.

c) Write the expression for the second order condition.

6. Suppose that the objective of the firm in problem 5 is to maximize output subject to the budget constraint C = Co.

a) Set up the problem of maximizing output subject to given level of cost.

b) Find the necessary and sufficient conditions for the constrained maximizing problem.

c) Write the expression for the second order condition.

7. For the utility maximizing problem: Maximize U = lnxy, subject to 2x + 3y = 60,

a) Write an expression presenting the Lagrange multiplier.

b) What variable does the Lagrangian multiplier represent?

c) Find the derivative of the Lagrange multiplier and show that it is negative.

8. For the expenditure-minimizing problem: Minimize C = 2x + 3y subject to Uo = 150,

a) Write an expression presenting the Lagrange multiplier.

b) What variable does Lagrange multiplier represent? What is the sign of the derivative of the Lagrangian Multiplier?

9. For the cost-minimizing problem: Minimize C = wL + rK subject to Qo = f(L, K),

a) What variable in the problem represents the marginal cost?

b) Identify the expression for the rate of change of marginal cost with respect to output.

c) What can you infer about the sign of the rate of change of marginal cost?

10. For the output-maximizing problem: Maximize Q = f(L, K) subject to wL + rK = Co,

a) What does Lagrangian multiplier represent?

b) Find the derivative of the Lagrange multiplier. What can you infer about the sign of this derivative?

III. Parametric Programming

In the previous section, computational techniques for a constrained optimization problem were introduced. It was assumed that certain parameters, such as market prices and the consumer's budget were fixed. But economic parameters do change, and much of economic theory deals with the reaction of economic units or economic variables to changes in economic conditions. Using parametric programming techniques, variations of the optimum solutions in response to changing economic conditions can be studied.

The technique is to find the total differential of the first order conditions not only with respect to the independent variables, but also with respect to those parameters which their change are to be studied. For example, in the simple two-goods constrained utility maximizing problems of the previous section, the total differential of the first order condition with respect to the independent variables, prices, and the consumer's budget yields:

$$Uxxdx + Uxydy - d\lambda Px - \lambda dPx = 0$$

$$Uyxdx + Uyydy - d\lambda Py - \lambda dPy = 0$$

$$dM - xdPx - dxPx - ydPy - dyPy = 0$$

Suppose that the price of X is changing, and we would like to find the effect of this price change on demand for X and Y. Assuming that price of Y and the consumer's budget are still fixed, (dPy = dM = 0), we can write

$$\begin{pmatrix} Uxx & Uxy & -Px \\ Uyx & Uyy & -Py \\ -Px & -Py & 0 \end{pmatrix}\begin{pmatrix} \frac{dx}{dPx} \\ \frac{dy}{dPx} \\ \frac{d\lambda}{dPx} \end{pmatrix} = \begin{pmatrix} \lambda \\ 0 \\ x \end{pmatrix}$$. Using the Cramer's rule, $\frac{dx}{dPx}$ can be derived as,

$$\frac{dx}{dPx} = \frac{Det\begin{pmatrix} \lambda & Uxy & -Px \\ 0 & Uyy & -Py \\ x & -Py & 0 \end{pmatrix}}{Det\begin{pmatrix} Uxx & Uxy & -Px \\ Uyx & Uyy & -Py \\ -Px & -Py & 0 \end{pmatrix}} = \frac{\lambda Det\begin{pmatrix} Uyy & -Py \\ -Py & 0 \end{pmatrix}}{Det\begin{pmatrix} Uxx & Uxy & -Px \\ Uyx & Uyy & -Py \\ -Px & -Py & 0 \end{pmatrix}} + \frac{x\begin{pmatrix} Uxy & -Px \\ Uyy & -Py \end{pmatrix}}{Det\begin{pmatrix} Uxx & Uxy & -Px \\ Uyx & Uyy & -Py \\ -Px & -Py & 0 \end{pmatrix}}$$

This equation is known as the **Slutsky equation**, and can be written as, $(\frac{dx}{dPx})_{dM=0} = (\frac{dxc}{dPx})_{dU=0} - x\,(\frac{dx}{dM})_{dPx=0}$.

This equation expresses the full effect of a price change of good X on the slope of the demand for X, ($\frac{dx}{dPx}$), in terms of the slope of the compensated demand ($\frac{dxc}{dPx}$), and the income effect of a price change ($\frac{dx}{dM}$).

Economic theory predicts that, in general, the full effect of a price change on demand is negative. But the full effect is the sum of the two terms. The first term on the right side of the above equation, $(\frac{dxc}{dPx})_{dU=0}$, captures the **pure substitution effect**, which is always negative, $\frac{dxc}{dPx} < 0$. It is the slope of the **compensated demand** curve, when utility and Py are held constant. The negative definiteness of the bordered Hessian matrix guarantees a positive value for the determinants of the bordered Hessians in the above equations, while the numerator of the $\frac{dxc}{dPx}$ is equal to $-Py^2$ and is negative. The second term is $-\ x\,(\frac{dx}{dM})_{dPx=0}$ which is the **income effect of a price change**. The sign of the income effect of price change is the opposite sign of the pure income effect, which is described below.

Suppose that prices are held constant and only income has changed. From the total differentiation of the first-order condition we can write $\begin{pmatrix} Uxx & Uxy & -Px \\ Uyx & Uyy & -Py \\ -Px & -Py & 0 \end{pmatrix}\begin{pmatrix} \frac{dx}{dM} \\ \frac{dy}{dM} \\ \frac{d\lambda}{dM} \end{pmatrix} = \begin{pmatrix} 0 \\ 0 \\ -1 \end{pmatrix}$. Using the Cramer's rule $\frac{dx}{dM} = \frac{Det\begin{pmatrix} 0 & Uxy & -Px \\ 0 & Uyy & -Py \\ -1 & -Py & 0 \end{pmatrix}}{Det\begin{pmatrix} Uxx & Uxy & -Px \\ Uyx & Uyy & -Py \\ -Px & -Py & 0 \end{pmatrix}}$.

$\frac{dx}{dM}$ is called the **pure income effect**. The sign of the pure income effect depends on whether a good is a normal good or an inferior good. If X is a normal good, an increase in M increases consumption of X, and $\frac{dx}{dM}$ will be positive. If X is an inferior good, then pure income effect will be negative. Note that, the denominator of $\frac{dx}{dM}$ is negative, based on the assumption of the negative definiteness of the bordered Hessian matrix. The numerator, however, is an off-diagonal determinant and there is no way to determine the sign of an off-diagonal matrix from the original matrix.

You may have noticed that there are two differences between the pure income effect and the income effect of price change. First, they have different signs; second, income effect of price change is multiplied by x. Hence, from the pure income effect, we can get information about the sign of the income effect of price change.

If a good is a normal good, income effect of price change is negative, and the sign of the second term in $(\frac{dx}{dPx})_{dM=0} = (\frac{dxc}{dPx})_{dU=0} - x\,(\frac{dx}{dM})_{dPx=0}$ will be negative, resulting in negative slope for demand for X. If a good is an inferior good, the income effect of price change is positive, and the sign of $\frac{dx}{dPx}$ will depend on the magnitude of the substitution effect and income effect of price change. If the substitution effect outweighs the income effect of price change, then $\frac{dx}{dPx} < 0$ and demand for X will be a standard downward-sloping demand curve. If income effect of price change outweighs the substitution effect, then demand for X will have a positive slope. A good with a positively-sloped demand curve is called **Giffen good**. Note that, for a good to be a Giffen good it should be an inferior good and the income effect of the price change should be so strong that it more than offsets the substitution effect.

Example: Consumer Theory

A consumer's utility function for X and Y is given as U(x, y) = $(x^{.5} + y^{.5})^2$. The consumer is spending \$60.00 on the two goods which are sold at market prices of Px = \$3.00 and Py = \$1.00.

a- Find the utility maximizing quantities of X and Y and test for the second order condition.

b- Find the total effect, substitution effect and the income effect of price change if Px changes from \$3.00 to \$3.50.

Solution: The utility maximizing problem can be formulated as:

Maximize: $U = (x^{.5} + y^{.5})^2$

subject to: $3x + y = 60$

The lagrangian function is $L(x, y, \lambda) = (x^{.5} + y^{.5})^2 + \lambda(60 - 3x - y)$ and the first order condition can be derived as:

$$\frac{\partial L}{\partial x} = L_x = x^{-.5}(x^{.5} + y^{.5}) - 3\lambda = 0, \quad \text{or} \quad x^{-.5}(x^{.5} + y^{.5}) = 3\lambda$$

$$\frac{\partial L}{\partial y} = L_y = y^{-.5}(x^{.5} + y^{.5}) - \lambda = 0, \quad \text{or} \quad y^{-.5}(x^{.5} + y^{.5}) = \lambda$$

$$\frac{\partial L}{\partial \lambda} = L_\lambda = 60 - 3x - y = 0$$

Divide the first two equations of the first-order conditions to delete λ. This gives $\frac{y^{0.5}}{x^{0.5}} = 3$, or y = 9x. Substitute for y in the third equation and use other equations to find x = 5, y = 45, and λ = 1.33. This point is the extremum that has to be tested for maximum or minimum.

To test for second-order condition, form the bordered Hessian matrix, $\mathbf{H} = \begin{pmatrix} L_{xx} & L_{xy} & L_{x\lambda} \\ L_{yx} & L_{yy} & L_{y\lambda} \\ L_{\lambda x} & L_{\lambda y} & L_{\lambda\lambda} \end{pmatrix} =$

$\begin{pmatrix} \frac{.5\,y^{.5}}{x^{1.5}} & \frac{.5}{x^{.5}\,y^{.5}} & -3 \\ \frac{.5}{x^{.5}\,y^{.5}} & \frac{-.5\,x^{.5}}{y^{1.5}} & -1 \\ -3 & -1 & 0 \end{pmatrix}$. Evaluated at the extremum, the bordered Hessian is $\mathbf{H} = \begin{pmatrix} -.3 & .033 & -3 \\ .033 & -.0037 & -1 \\ -3 & -1 & 0 \end{pmatrix}$.

The determinants of the bordered principal minors of **H** are $\text{Det}\begin{pmatrix} -.3 & -3 \\ -3 & 0 \end{pmatrix} = -9$ and $\text{Det}\begin{pmatrix} -.3 & .033 & -3 \\ .033 & -.0037 & -1 \\ -3 & -1 & 0 \end{pmatrix} = .53$, respectively. Since the determinants of the bordered principal minors alternate in sign beginning with a negative, the Hessian is negative definite and the solution is a maximum.

To find the total effect of the change in Px find $\frac{dx}{dPx}$ through the Slutsky equation above. Note that, the denominators of the terms in the Slutsky equation are the determinant of the Hessian which is 0.533. Therefore, the total effect can be calculated as $(\frac{dx}{dPx})_{dM=0} = \frac{1}{\text{Det}[\boldsymbol{H}]}\text{Det}\begin{pmatrix} \lambda & L_{xy} & -3 \\ 0 & L_{yy} & -1 \\ x & -1 & 0 \end{pmatrix} = \frac{1}{.533}\,\text{Det}\begin{pmatrix} 1.33 & .033 & -3 \\ 0 & -.0037 & -1 \\ 5 & -1 & 0 \end{pmatrix} =$ -2.92. Using the Slutsky equation, the substitution effect or the slope of the compensated demand can be calculated as $(\frac{dx}{dPx})_{dU=0} = \frac{1}{.533}\lambda\text{Det}\begin{pmatrix} L_{yy} & -1 \\ -1 & 0 \end{pmatrix} = \frac{1.33}{.533}\text{Det}\begin{pmatrix} -.007 & -1 \\ -1 & 0 \end{pmatrix} = -2.50$. Finally, using the Slutsky equation, the income effect of price change can be calculated as $-x(\frac{dx}{dM})_{dPx=0} = \frac{1}{.533}x\text{Det}\begin{pmatrix} L_{xx} & -3 \\ L_{yy} & -1 \end{pmatrix} = \frac{5}{.533}\text{Det}\begin{pmatrix} .033 & -3 \\ -.0037 & -1 \end{pmatrix} = -.42$

The final results can be summarized as, $(\frac{dx}{dPx})_{dM=0} = -2.91$, $(\frac{dxc}{dPx})_{dU=0} = -2.49$, and $-x\,(\frac{dx}{dM})_{dPx=0} = -.42$. Since dPx = 3.50 - 3.00 = .50, the total effect of an increase in Px on quantity demanded of X is a decrease in quantity by dx = .50(-2.91) = -1.45 units. From this quantity, -1.24 units is a reduction due to substitution effect, and -.21 units is due to income effect of price increase.

To solve the problem using *Mathematica*, formulate the Lagrangian as follows and find the solution to first-order conditions,

```
In[30]:= Clear[L, U, x, y, λ, sol]
         L = (x^.5 + y^.5)^2 + λ (60 - 3 x - y);
         sol = Solve[{D[L, x] == 0, D[L, y] == 0, D[L, λ] == 0}]

Out[30]= {{x → 5., y → 45., λ → 1.33333}}
```

The first-order condition yields the solution x = 5, y = 45, and λ = 1.33. To test for second-order condition, form the Hessian and find the determinants of the bordered Hessian

```
In[31]:= MatrixForm[( D[D[L, x], x]  D[D[L, x], y]  D[D[L, x], λ]
                      D[D[L, y], x]  D[D[L, y], y]  D[D[L, y], λ]
                      D[D[L, λ], x]  D[D[L, λ], y]  D[D[L, λ], λ] )] /. sol

Out[31]= {( -0.3       0.0333333   -3
            0.0333333  -0.0037037  -1
            -3         -1           0 )}
```

The determinant of the principal minors are,

```
In[32]:=Det[( D[D[L, x], x]  D[D[L, x], λ]
              D[D[L, λ], x]  D[D[L, λ], λ] )] /. sol

Out[32]= {-9}

In[33]:=Det[( D[D[L, x], x]  D[D[L, x], y]  D[D[L, x], λ]
              D[D[L, y], x]  D[D[L, y], y]  D[D[L, y], λ]
              D[D[L, λ], x]  D[D[L, λ], y]  D[D[L, λ], λ] )] /. sol

Out[33]= {0.533333}
```

Since the determinants of the bordered principal minors alternate in sign beginning with a negative, the Hessian is negative definite and the solution is a maximum.

To find the total effect of the change in Px, find the $\frac{dx}{dPx}$ through the Slutsky equation formulate the terms of the equations to *Mathematica* commands. Note that, the denominators of the terms in the Slutsky equation are the determinant of the Hessian which is 0.533. Therefore, the total effect can be calculated as,

```
In[34]:= (1 / .533) Det[( λ  D[D[L, x], y]  -3
                          0  D[D[L, y], y]  -1
                          x  -1              0 )] /. sol

Out[34]= {-2.91849 Null}
```

using the Slutsky equation, the substitution effect or the slope of the compensated demand can be calculated as,

```
In[35]:= (1 / .533) λ Det[( D[D[L, y], y]  -1
                            -1              0 )] /. sol

Out[35]= {-2.50156}
```

finally, using the Slutsky equation, the income effect of price change can be calculated as,

```
In[36]:= (1/.533) x Det[( D[D[L, x], y]  -3
                          D[D[L, y], y]  -1 )] /. sol

Out[36]= {-0.416927}
```

The final results can be summarized as, $(\frac{dx}{dPx})_{dM=0}$= -2.91, $(\frac{dxc}{dPx})_{dU=0}$= -2.49, and - x $(\frac{dx}{dM})_{dPx=0}$ = -.42. Since dPx = 3.50 - 3.00 = .50, the total effect of an increase in Px on quantity demanded of X is a decrease in quantity by dx = .50(-2.91) = -1.45 units. From this quantity, -1.24 units is a reduction due to substitution effect, and -.21 units is due to income effect of price increase.

Finally, the graphic capabilities of *Mathematica* allow us to see the solution visually. To graph the utility function and the budget constraint in 3D, use the following commands:

```
In[37]:= Clear[x, y, U, p1, p2, con1, con2, con3]
         U = (x^.5 + y^.5)^2;
         p1 = Plot3D[U, {x, 0, 10}, {y, 0, 50}, AxesLabel → {"X", "Y", "U"}]
         p2 = Plot3D[y + 3 x − 60, {x, 0, 10}, {y, 0, 50}, AxesLabel → {"X", "Y", "U"}]
         Show[p1, p2]
```

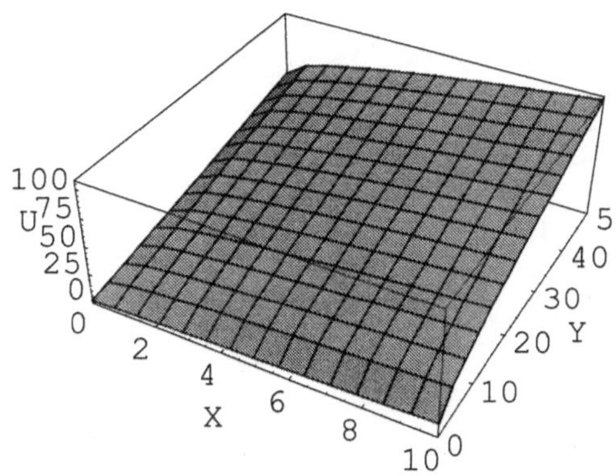

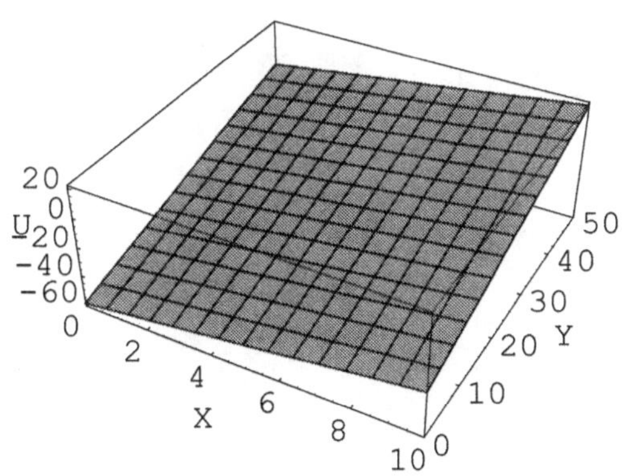

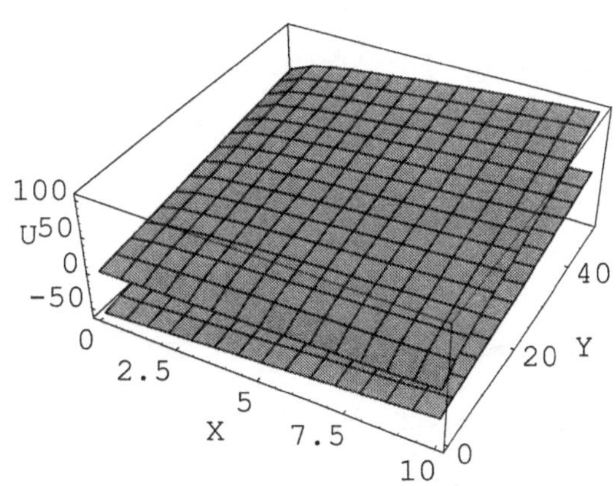

To graph the indifference curves and the budget constraints use the following commands:

```
In[38]:=con1 = ContourPlot[U, {x, 0, 10}, {y, 0, 80}, ContourShading → False]
        con2 = ContourPlot[3 x + y – 60, {x, 0, 10}, {y, 0, 80}, Contours → {0}, ContourShading → False]
        con3 = ContourPlot[5 x + y – 60, {x, 0, 10}, {y, 0, 80}, Contours → {0}, ContourShading → False]
        Show[con1, con2, con3]
```

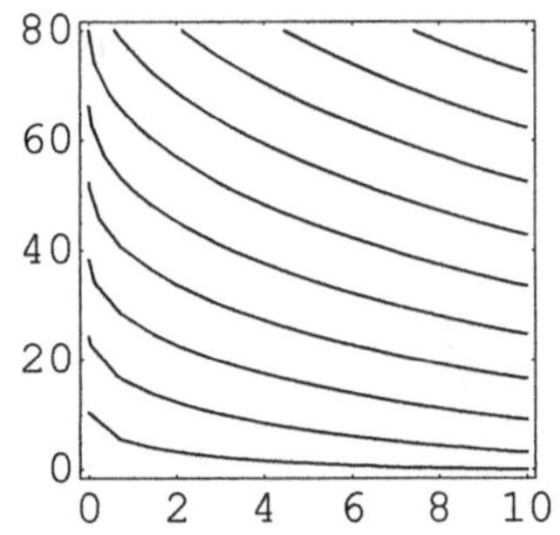

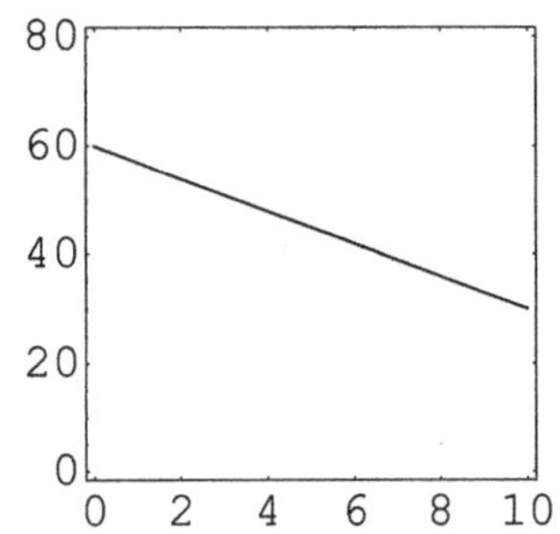

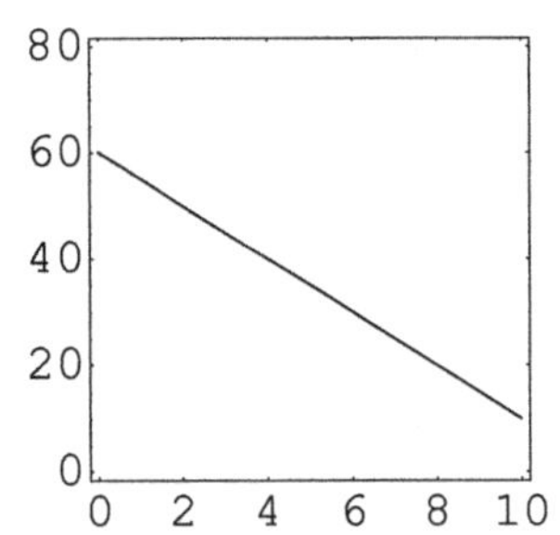

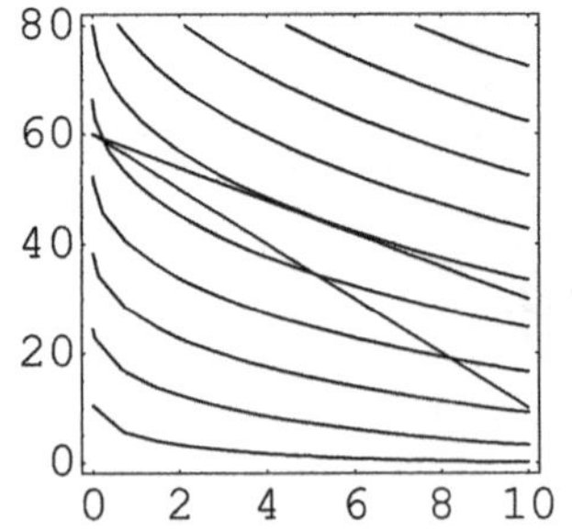

Do More Practice:

1- Double the prices and the budget constraint. Re-run the program and observe the changes in solution.

2- Find the total effect, substitution effect and the income effect of price change if Px changes from \$3.00 to \$2.50 (keep all the other assumptions the same as in the text.). Re-run the program and observe the changes in solution. Is the effect of a 50 cents decrease in price of x symmetric to the effect of a 50cents increase in price of x?

Problems

1. For the utility maximizing problem 1 of section I, derive the demand function for X. Analyze the sign of the slope of the demand curve. If the market price of X and Y are given as Px and Py respectively, find the elasticity of substitution of X for Y. (Hint: Elasticity of substitution of X for Y is defined as the ratio of percent change in $\frac{x}{y}$ to percent change in $\frac{Px}{Py}$).

2. For the expenditure minimizing problem 2 of section I, derive the demand function for X. Analyze the sign of the slope of the demand curve. What differences are there between the slopes of the demand curves in problems 1 and 2?

IV. Constrained Optimization When Some Constraints Are Inequalities

The more general problem of finding a maximum or minimum subject to equality and inequality constraints is called mathematical programming. A general mathematical programming problem is one of finding $x_1,.. x_n$ so as to

$$\text{Maximize: } f(x_1, \ldots\ldots, x_n)$$

$$\text{Subject to : } g^i(x_1, \ldots\ldots, x_n) < b_i \quad \text{for i} = 1, \ldots\ldots, \text{r}$$

$$g^i(x_1, \ldots\ldots, x_n) > b_i \quad \text{for i} = \text{r}+1, \ldots, \text{s}$$

$$g^i(x_1, \ldots\ldots, x_n) = b_i \quad \text{for i} = \text{s}+1, \ldots, \text{m}$$

$$x_1, \ldots\ldots, x_n > 0$$

In 1951, Kuhn and Tucker developed a set of necessary and sufficient conditions for an optimal solution to the general programming problem. The technique is to transform inequality constraints to equality, by adding **slack or surplus variables**, and using the Lagrangian multiplier technique of the previous section.

A slack variable is a positive variable that can be added to an inequality constraint of "less than" form (<) to set it to equality. A surplus variable is a positive variable that can be subtracted from an inequality constraint of "greater than" form (>) to set it to equality.

In general, the solution to the mathematical programming problem can be summarized as: If $f(x_1, \ldots, x_n)$ is a concave function and $g^i(x_1, \ldots\ldots, x_n)\{< = >\}b_i$, i = 1, ... , m, forms a convex set, the necessary and sufficient conditions for an extremum to the mathematical programming problem are as follows: (for a review of convex and concave functions, see appendix at the end of this chapter.)

I- $f_j(.) - \sum_{i=1}^{m}\lambda_i g^i x_j(.) \geq 0 \quad$ i = 1, ..., m j = 1, ..., n

II- $\sum_{j=1}^{n} x_j\big(f_j(.) - \sum_{i=1}^{m}\lambda_i g^i x_j(.)\big) = 0$

$b_i - g^i(.) > 0 \quad$ i = 1, ..., r

III- $b_i - g^i(.) \leq 0 \quad$ i = r + 1 ,...., s

$b_i - g^i(.) = 0 \quad$ i = s + 1, ..., m

IV- $\sum_{i=1}^{m}\lambda_i(\ b_i - g^i(.)\ ) = 0$

Where $f(.)$ and $g(.)$ are f and g functions, evaluated at the extremum. The point $P(x_1^*, ..., x_n^*, \lambda_1^*, ..., \lambda_m^*)$ that satisfies the conditions I-IV guarantees a relative maximum for the mathematical programming problem as stated above.

The necessary and sufficient conditions for a point to be an extremum also are necessary and sufficient conditions for the Lagrangian function $L = f(x_1, ..., x_n) + \sum_{i=1}^{m} \lambda_i(b_i - g^i(x_1, ..., x_n))$ to have a saddle point at that point, provided that the concavity condition of the objective function and the convexity of the constraint set are satisfied. A saddle point is a point is a point that $L(x_1, ..., x_n, \lambda_1, ..., \lambda_m)$ is a maximum with respect to $(x_1^*, ..., x_n^*)$ and is a minimum with respect to $(\lambda_1^*, ..., \lambda_m^*)$.

No attempt is made to prove the Kuhn-Tucker theorem here. Instead, using the interpretation of the Lagrangian Multiplier described in the previous section of this chapter and an economic example an attempt is made to facilitate the understanding of the theorem and economic interpretation of the Kuhn-Tucker conditions.

Example: Consumer Theory

A consumer has a conventional twice differentiable utility function, U = U(x, y, z), where X and Y are purchased in the market at fixed prices and subject to the budget constraint. Z, on the other hand, is the services of a durable consumer good, which is assumed to be fixed over the decision period. Set up the proper mathematical programming problem and derive the necessary and sufficient conditions.

Maximize: $U = U(x, y, z)$

Subject to: $xPx + yPy \leq M$

$$z \leq z_o$$

The first step is to form the Lagrangian function and to find the first order conditions. The Lagrangian function is

$$L(x, y, z, \lambda 1, \lambda 2) = U(x, y, z) + \lambda_1(M - xPx - yPy) + \lambda_2(z_o - z)$$

The first order condition is

$$\frac{\partial L}{\partial x} = Ux - \lambda_1 Px \geq 0, \quad \text{or} \quad \frac{Ux}{Px} > \lambda_1$$

$$\frac{\partial L}{\partial y} = Uy - \lambda_1 Py \geq 0, \quad \text{or} \quad \frac{Ux}{Py} > \lambda_1$$

$$\frac{\partial L}{\partial z} = Uz - \lambda_2 \geq 0, \quad \text{or} \quad Uz > \lambda_2$$

$$\frac{\partial L}{\partial \lambda 1} = M - xPx - yPy \geq 0$$

$$\frac{\partial L}{\partial \lambda 2} = z_o - z \geq 0$$

The first three equations above are maximizing L with respect to x, y, and z. The last two equations above are minimizing L with respect to λ_1 and λ_2. Using the economic interpretation of the Lagrangian multiplier, described in the previous section, the Kuhn-Tucker condition can be interpreted as follows:

1- The first equation, which corresponds to condition I, states that, to maximize utility subject to the constraints, the consumer buys x as long as the marginal utility received from spending one dollar on x is greater or equal to the marginal valuation of that one dollar. However, if $\frac{Ux}{Px} < \lambda_1$, then x = 0. The same is true for y in the second equation.

2- Condition II follows from condition I. For this example it can be written as $x(Ux - \lambda_1 Px) + y(Uy - \lambda_1 Py) + z(Uz - \lambda_2) = 0$. The three terms in this equation follow expressions in the first-order condition. Either $Ux - \lambda_1 Px = 0$, that is, the consumer is buying positive amounts of x (x > 0), or $Ux - \lambda 1 \ Px < 0$, where x = 0. The same is true for y and z in the second and third terms.

3- Condition III follows the inequality constraints in the optimizing problem, which should be satisfied.

4- Condition IV, which in this example can be written as $\lambda_1(M - xPx - yPy) + \lambda_2(z_o - z) = 0$ states that, if the budget constraint is binding, i.e., if xPx + yPy = M, then marginal valuation of money is positive, $\lambda_1 > 0$. If the budget constraint is not binding, xPx + yPy < M, then $\lambda_1 = 0$. The same is true for the constraint on the services of durable good z. Either the consumer uses all the services of durable good, ($z = z_o$), In which case, the implicit price of z will be positive, $\lambda_2 > 0$; or the consumer is not using all the services of z, ($z < z_o$), and the implicit price of z will be zero, $\lambda_2 = 0$.

If all these conditions are satisfied, then the first order condition will guarantee a global maximum. If the sufficiency conditions of the Kuhn-Tucker theorem are not satisfied, it may be necessary to evaluate all the feasible solutions. In this case, the computational effort may be enormous.

Example: Consumer Theory

A consumer has the utility function, U = U(x, y, z) = lnxyz, where X and Y are purchased at the market prices of \$2 and \$3 respectively, and are subject to a \$120 budget constraint. Z, on the other hand, is the services of a durable good, which consumer needs at least 8 units of it over the decision period. Set up the proper mathematical programming problem and derive the necessary and sufficient conditions.

Solution: The problem can be formulated as,

Maximize: U = lnxyz

Subject to: $2x + 3y \leq 120$

$z \leq 8$

The Lagrangian function is $L(x, y, z, \lambda_1, \lambda_2) = \ln xyz + \lambda_1(120 - 2x - 3y) + \lambda_2(8 - z)$. The first-order conditions are:

$\frac{\partial L}{\partial x} = \frac{1}{x} - 2\lambda_1 = 0,$ or $\frac{1}{x} = 2\lambda_1$

$\frac{\partial L}{\partial y} = \frac{1}{y} - 3\lambda_1 = 0,$ or $\frac{1}{y} = 3\lambda_1$

$\frac{\partial L}{\partial z} = \frac{1}{z} - \lambda_2 = 0,$ or $\frac{1}{z} = \lambda_2$

$\frac{\partial L}{\partial \lambda 1} = 120 - 2x - 3y = 0$

$\frac{\partial L}{\partial \lambda 2} = 8 - z = 0$

Use the first two equations of the first-order conditions to delete λ_1. This yields $y = \frac{2x}{3}$. Substitute y in the fourth equation to find x = 30 and y = 20. Using the first equation, λ_1 is found to be $\lambda_1 = \frac{1}{60}$. The third and fifth equations of the first-order conditions give z = 8 and $\lambda_2 = \frac{1}{8}$.

To test for the Kuhn-Tucker conditions, follow the requirements as follows:

Condition I is satisfied through the first order condition, which results in solution set ($x = 30$, $y = 20$, $z = 8$, $\lambda_1 = \frac{1}{60}$, $\lambda_2 = \frac{1}{8}$).

Condition II can be tested for by evaluating $x(Ux - 2\lambda_1) + y(Uy - 3\lambda_1) + z(Uz - \lambda_2)$ at the solution set ($x = 30$, $y = 20$, $z = 8$, $\lambda_1 = \frac{1}{60}$, $\lambda_2 = \frac{1}{8}$). Here, $Ux = \frac{1}{x}$, $Uy = \frac{1}{y}$, and $Uz = \frac{1}{z}$ are marginal utilities of x, y, and z, respectively. Evaluated at the solution set, $x(Ux - 2\lambda_1) + y(Uy - 3\lambda_1) + z(Uz - \lambda_2) = 0$ and satisfies the condition II.

Condition III follows the inequality constraints, which is satisfied in this problem.

Condition IV can be tested for by evaluating the numerical value of $\lambda_1(120 - 2x - 3y) + \lambda_2(8 - z)$ at the solution set. Evaluated at the solution set $\lambda_1(120 - 2x - 3y) + \lambda_2(8 - z) = 0$ and the condition IV.

To solve the problem using *Mathematica*, formulate Lagrangian function and proceed with the steps as follows,

```
In[39]:= Clear[L, U, x, y, z, λ1, λ2, sol]
   U = Log[x y z]; L = Log[x y z] + λ1 (120 - 2 x - 3 y) + λ2 (8 - z);
   sol = Solve[{D[L, x] == 0, D[L, y] == 0,
      D[L, z] == 0, D[L, λ1] == 0, D[L, λ2] == 0}, {x, y, z, λ1, λ2}]

Out[39] = {{λ2 → 1/8, λ1 → 1/60, x → 30, y → 20, z → 8}}
```

Condition I is satisfied through the first order condition, which is the solution above. Condition II can be tested for through the following command.

```
In[40]:= N[x (D[U, x] - 2 λ1 ) + y (D[U, y] - 3 λ1 ) + z (D[U, z] - λ2)] /. sol

Out[40]= {0.}
```

Condition III follows the inequality constraints, which is satisfied in this problem. Condition IV can be test for as follows,

```
In[41]:= N[λ1 (120 - 2 x - 3 y) + λ2 (8 - z)] /. sol

Out[41]= {0.}
```

Do More Practice: Change the utility function to U = $x^{\frac{1}{3}}y^{\frac{1}{3}}z^{\frac{1}{3}}$. Keep the other assumptions the same. Re-run the program and compare the results.

2- Change the utility function to U = lnx + lny + lnz. Keep the other assumptions the same. Re-run the program and compare the results.

Problems

1. Solve the following constrained maximizing problem and test for the Kuhn-Tucker conditions.

   Maximize: f(x, y) = x + y

   Subject to: $x^2 + y^2 < 4$

   x > 0, y > 0

2. A single product firm with two processes, v and w, employs two fixed inputs, labor and capital. In the short run, the firm is constrained to employing services of labor and capital of 150 and 850 units each, respectively. The technological relationship between output and the processes given as Q = $(v^{.5} + w^{.5})^2$. Process v employs two units of labor and one unit of capital. Process w employs one unit of labor and four units of capital.

   a) Set up the mathematical programming problem of maximizing output subject to the constraints.

   b) Find the optimum levels of v and w.

   c) Test for the Kuhn-Tucker conditions.

   d) What economic variables do Lagrangian multipliers represent?

First Order Conditions and Derivation of Special Functions: Once the optimum solution to a mathematical programming problem is derived, the first order condition will yield itself to parametric programming, as described in the previous section. At the optimum point, all the inequality relationships are dropped, and all the equations are held to derive the unknown variables. This follows the conditions II and IV of the Kuhn-Tucker theorem that any constraint satisfied with inequality will have a zero Lagrange multiplier or implicit price. Therefore, the system of equations in the first order conditions can be used to study behavior of variables, when other variables or parameters are changing. The procedure, however, is the same as described in the previous section.

First order conditions can also be used to derive certain functional relationships between variables. For example, first order conditions of a utility maximization problem can be used to prove the existence of demand and to derive demand functions.

To prove the existence of demand functions, and to derive them, we need to employ the Implicit Function Theorem. The Implicit Function theorem states that a subset of variables can be expressed as explicit functions of the remaining variables, provided that the Jacobian determinant with respect to that subset of variables does not vanish.

The first step is to write the first order conditions as implicit functions, including all the variables and parameters. For the utility maximizing example above, the first order condition gives the following implicit functions:

$$F^1(\text{x, y, z, Px, Py}, \lambda_1, \lambda_2, \text{M}, z_o) = 0$$

....

....

$$F^5(\text{x, y, z, Px, Py}, \lambda_1, \lambda_2, \text{M}, z_o) = 0$$

The requirement for application of the Implicit Function Theorem is that all the functions have continuous first partial derivatives in the neighborhood of the optimal point. The total differential of the first order condition with respect to all variables can be summarized in Jacobian

$$\text{J} = \frac{d(\text{F1} \ldots \text{F5})}{d(x, y, z, \text{px}, \text{py}, \lambda 1, \lambda 2, M, \text{zo})} = \begin{pmatrix} \text{Uxx} & \text{Uxy} & \text{Uxz} & -\text{Px} & 0 & -\lambda_1 & 0 & 0 & 0 \\ \text{Uyx} & \text{Uyy} & \text{Uyz} & -\text{Py} & 0 & 0 & -\lambda_1 & 0 & 0 \\ \text{Uzx} & \text{Uzy} & \text{Uzz} & 0 & -1 & 0 & 0 & 0 & 0 \\ -\text{Px} & -\text{Py} & 0 & 0 & 0 & -x & -y & 1 & 0 \\ 0 & 0 & -1 & 0 & 0 & 0 & 0 & 0 & 1 \end{pmatrix}$$

If all the entries in the matrix are continuous, demand for x, y, and z can be expressed as explicit functions of prices, income, and z_o. That is, demand for x is x = X(Px, Py, M, z_o), demand for y is y = Y(Px, Py, M, z_o), and demand for z is z = Z(Px, Py, M, z_o).

Problems

1. For the cost-minimizing problem 3 of this chapter, give a precise, detailed derivation of total cost as a function of output.

2. For the output-maximizing problem 4 of this chapter,

   a) Give a precise, detailed derivation of the output function.

   b) What does the Lagrangian multiplier represent?

   c) Write the first order condition as an implicit function. What special economic functions can be derived from the first order condition?

   d) Give a precise, detailed derivation of the total cost and output functions.

3. In general, the consumer's behavior in the theory of consumption can be formulated either as:

   Maximize: U = U(x1, ..., xn)

   Subject to : p1x1 + ... + pnxn ≤ M

Or as:

Minimize: E = p1x1 + ... + pnxn

Subject to: U(x1, ..., Un) ≥ Uo

Where x1, ..., xn are n commodities purchasable at market prices p1, ..., pn, U is utility, and E stands for expenditure. For a hypothetical two-commodity model:

a)Find the necessary and sufficient conditions for the existence of an extremum in each case.

b) What do the Lagrangian multipliers represent in each case?

c) Write the first order conditions in implicit function form. Derive the demand for x in each case. Analyze the slope of the demand curve.

d) What differences are there between the two demand curves?

4. In general, the behavior of producers in the theory of the firm can be formulated as:

Maximize: Q = f(x1, ..., xn)

Subject to: c1x1 + ...+ cnxn ≤ Co

Or as:

Minimize: C = c1x1 + ... + cnxn

Subject to: f(x1, ..., xn) ≥ Qo

Where x1, ..., xn are n inputs bought in the market at prices c1, ..., cn, C stands for total cost, and Q is output. For a hypothetical case of a two-input firm:

a) Find the necessary and sufficient conditions for the existence of an extremum in each case.

b) What do Lagrangian multipliers represent in each case? What can you say about the sign of the first derivative of Lagrangian multipliers?

c) Write the first order conditions in implicit function form. What special economic functions can be derived from the first order conditions in each case? Give a precise, detailed derivation of total cost and output functions.

Appendix: Concavity and Convexity of Functions

A function f(x) is said to be convex over the interval (x_1, x_2) if for the two points x_1 and x_2 and for all, $0 < \alpha < 1$,

$$f[\alpha x_2 + (1- \alpha)x_1] \leq [\alpha f(x_2) + (1- \alpha)f(x_1)]$$

A function f(x) is said to be concave if -f(x) is convex. A function f(x) is said to be strictly convex if the equation above is a strict inequality.

Graphically, a function f(x) is convex if a line segment joining any two points $[x_1, f(x_1)]$, $[x_2, f(x_2)]$ of f(x) lies on or above the surface of f(x). An easier way to show that a function is convex is to use the second order derivative of the function, if the function is twice differentiable. A univariate continuous function f(x) is convex in the neighborhood of x_o if and only if f "(x_o) > 0. A multivariable continuous function $f(x_1,......x_n)$ is convex if and only if the Hessian determinant of f is positive definite.

Example: Prove that y = f(x) = $2x^2$ + x + 3 is a convex function.

Solution: f '(x) = 4x + 1 and f "(x) = 4. Since f "(x) > 0, the function is convex over the entire real number. The graph of the function is shown below.

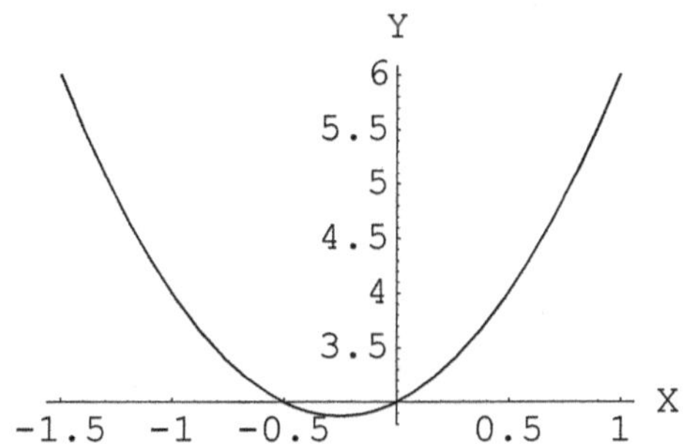

Example: Discuss the concavity or convexity of y = f(x) = x^3 - $3x^2$ + x + 5.

Solution: f '(x) = $3x^2$-6x + 1 and f "(x) = 6x - 6. f "(x) = -6 + 6x is positive for all the values of x > 1 and is negative for all the values of x < 1. Therefore, f(x) will be convex for x > 1 and concave for x < 1. The graph of the function is shown below.

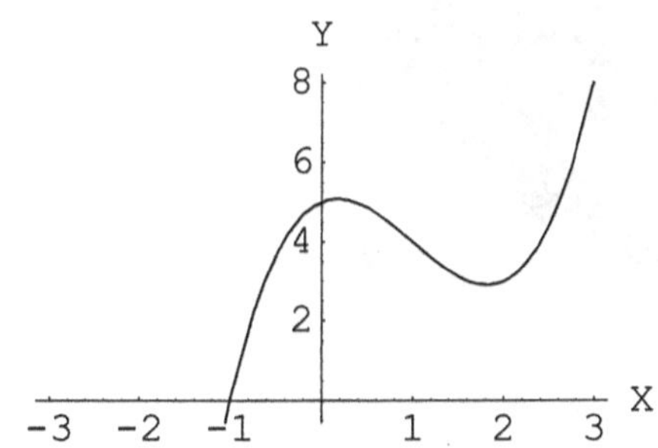

Example: Discuss the convexity or concavity of the z = f(x, y) = 2 x^2 + y^2.

Solution: The partial derivative of z with respect to x and y are $\frac{\partial z}{\partial x}$= 4x and $\frac{\partial z}{\partial y}$= 2y, respectively. The second derivatives results in the quadratic form the Hessian matrix of which is **H** = $\begin{pmatrix} 4 & 0 \\ 0 & 2 \end{pmatrix}$. Since the determinants of the principal minors of **H** are all positive, **H** is positive definite and f(x, y) is convex over the entire Euclidean E2 space. The graph of the function is presented below.

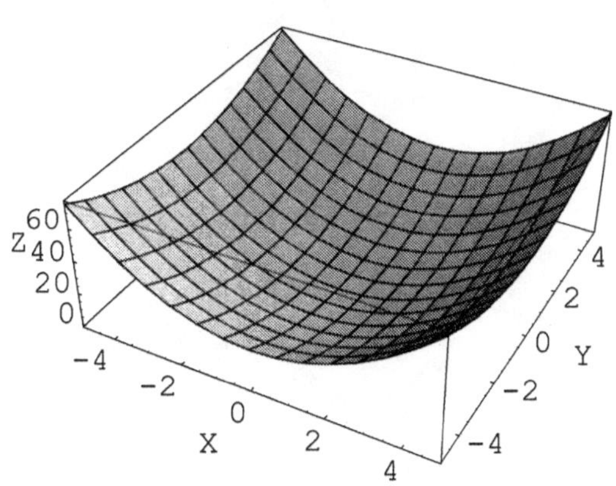

To solve this problem using *Mathematica,* use the following commands.

```
In[42]:= Clear[x, y, z]
         z = 2x^2 + y^2;
         H = MatrixForm[( D[D[z, x], x]  D[D[z, x], y] )]
                        ( D[D[z, y], x]  D[D[z, y], y] )
         Det[%]
Out[42]= ( 4  0 )
         ( 0  2 )

         8
```

To graph the function use the following command.

```
In[43]:= Plot3D[2 x^2 + y^2, {x, -5, 5}, {y, -5, 5},
           AxesLabel -> {"X", "Y", "Z"}]
```

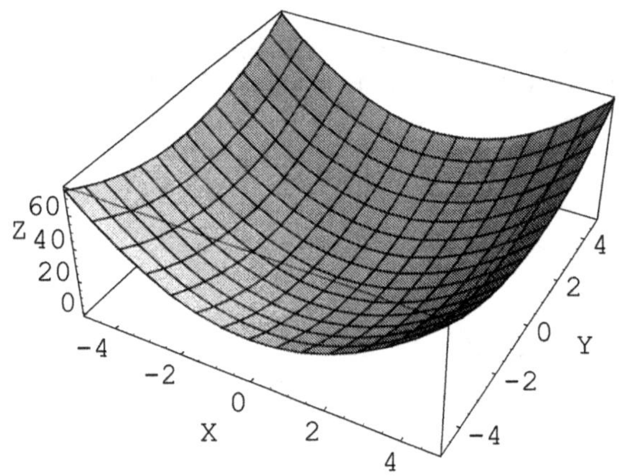

Do More practice:

1- Change the z function to z = f(x, y) = $-2x^2 - y^2$ by changing the sign of the coefficients. Re-run the program and analyze the results.

2- Change the z function to z = f(x, y) = $2x^2 - y^2$ by changing the sign of the coefficients. Re-run the program and analyze the results.

CHAPTER 9

LINEAR PROGRAMMING

One of the best known mathematical programming models is the linear programming model, where all the functions in the model are linear in all variables. Linear programming models are used for optimum allocation of fixed inputs among competing needs. Linear programming has been successfully used in solving many business and economic problems. The model may be written as:

Optimize: $Z = c_1x_1 + \ldots + c_nx_n$

Subject to: $a_{11}x_1 + \ldots + a_{1n}x_n \ \{<=>\} \ b_1$

$\ldots\ldots$

$a_{m1}x_1 + \ldots + a_{mn}x_n \ \{<=>\} \ b_m$

$x_1, \ldots, x_n \geq 0$

Where, x_1, ..., x_n are the decision variables and a's, b's, and c's are the known constants. The objective is to optimize (maximize or minimize) the objective function Z subject to m + n constraints (m constraints and n non-negativity restrictions on decision variables).

TOOL BOX

Mathematica Commands	Function
Polygon[{{2,0}, {2,2}, {2,4}, {0,2}}]	Draws a polygon with vertices at (2,0), (4,2), (2,4), and (0, 2).
Table[Plot[2x+i], {x, 0, 3}], {i, 0, 5}]	Plots 2x+i for i =0 to 5. Double click on the graph to see its movements.
ConstrainedMax[2x1+3x2,	Solves the linear programming problem Max: Z =2x1+3x2,
{x1+2x2 <= 7,3x1+x2 <= 6},{x1,x2}]	Subject To: x1+ 2x2 ≤ 7 and 3x1 + x2 ≤ 6.
ConstrainedMin[2x1+ 3x2,	Solves the linear programming problem Max: Z = 2x1+3x2,
{x1+2x2 >= 7,3x1+x2 >= 6},{x1,x2}]	Subject To: x1+ 2x2 ≥ 7 and 3x1 + x2 ≥ 6.

In matrix form the model may be written as:

Optimize: $Z = \mathbf{c'x}$

Subject to: $\mathbf{A.x} \ \{<=>\} \ \mathbf{b}$

$\mathbf{x} \geq \mathbf{0}$

Where **c** is a n.1 vector of constant coefficients of objective function, **A** is a m.n matrix of constant coefficients of constraint equations, **b** is a n.1 vector of constraints and Z is the objective variable.

There are different ways of solving a linear programming model. The best known, however, is the **Simplex method**, developed by George B. Dantzig during the World War II.

In the next section, the graphic solution of the linear programming problems is presented. The following sections present the Simplex solution of the linear programming models, the solution to minimizing problems, and the duality.

I. Graphic Solution of the Linear Programming Models: The `ConstrainedMax` Command

If the linear programming problem involves only two variables, the simplest way to solve the problem is using graphs of the functions. The graphic solution is illustrated by the following examples.

Example: A firm produces two goods, x1 and x2, which it sells at prices $2 and $3 each, respectively. The firm employs two fixed inputs labor and capital, available to the firm at fixed quantities of 7 and 8 units respectively. Each x1 requires one unit of labor and three units of capital. Each x2 requires two units of labor and one unit of capital. Set up a linear programming problem of maximizing total revenue subject to the constraints and find the optimum solution.

Solution: Set up the objective function and the constraints based on the given information

Maximize: $Z = 1.5x1 + x2$

Subject to: $x1 + 2x2 \leq 7$

$3x1 + x2 \leq 6$

$x1, x2 \geq 0$

the first step in solving the problem is to find the **constraint region**. The constraint region is the set of points that simultaneously satisfies the constraints. To find the constraint region, graph both constraints

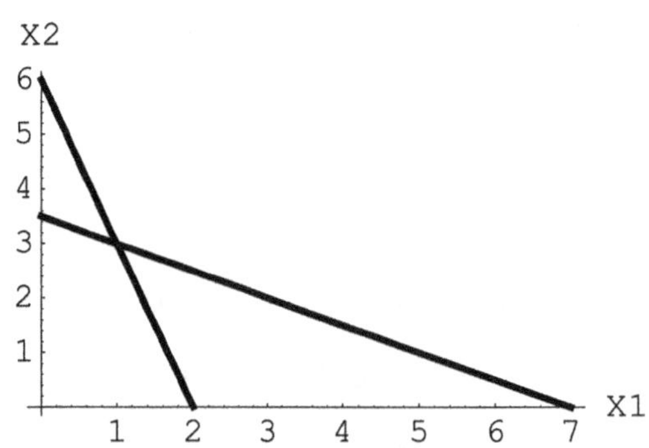

and find the region that satisfies both constraints. In a maximizing problem, this generally results in a polygon with the vertices at the intersections of the constraints, the constraints and the axes, and the origin. Here, the two constraints meet each other at (1, 3) and the constraining region is a polygon with the vertices at (0, 0), (0, 3.5), (2, 0), and (1, 3).

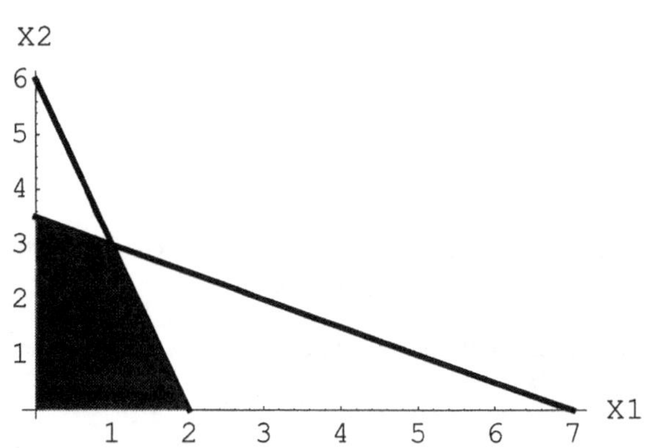

The Simplex method provides us with a prove that the optimum solution of the linear programming problems always lies on one of the corner points of the polygon or on one of the sides of the polygon, if there are multiple solutions. The optimum solution, however, depends on the slope of the objective function and the point at which the objective function is tangent to the constraint region. The constraint region and a group of objective lines are graphed below

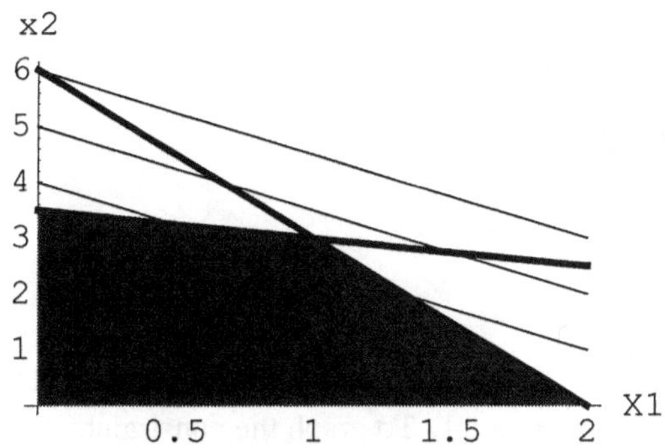

as the objective line is moved from the north east corner of the quadrant toward the constraint region, the first point where the objective line touches the constraint region (the point of tangency) provides us with the optimum solution to the problem. In this case, the solution is (1, 3), which is the intersection of the two constraints, and the maximum value of Z is Z = 11.

The optimum solution can be tested for using the *Mathematica* command **ConstrainedMax** as follows,

```
In[1]:= Clear[x1, x2]
        ConstrainedMax[2 x1 + 3 x2, {x1 + 2 x2 <= 7, 3 x1 + x2 <= 6}, {x1, x2}]

Out[1]= {11, {x1 → 1, x2 → 3}}
```

It may happen that the slope of the objective function is the same as the slope of one of the constraints. In that case, the linear programming problem will have **multiple solutions**. Any point on the constraint function which also lies on the constraint region will be a solution to the problem. For example, if the objective function of the preceding example were $Z = x_1 + 2x_2$, which has the same slope as the first constraint, the point (0, 3.5), or (1,3), or any point on the line joining these two points is a solution to the problem. All of these points will maximize Z with a value of Z = 7. The following figure shows the case of multiple solutions.

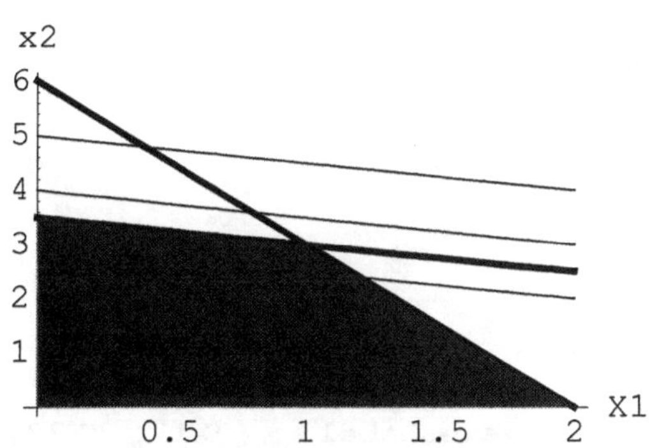

If the objective line intersects the constraint region in one of the vertices lying on the axes (point (2, 0) or (0, 3.5)), the firm will produce positive amounts of one good and none of the other. Such a solution is called a **degenerate solution**. For example, if the objective function was Z = 4x1 + x2, the solution would be x1 = 2 and x2 = 0. If the objective function was Z = x1 + 3x2, the solution would be x1 = 0 and x2 = 3.5. The following figure shows the degenerate solution (2, 0).

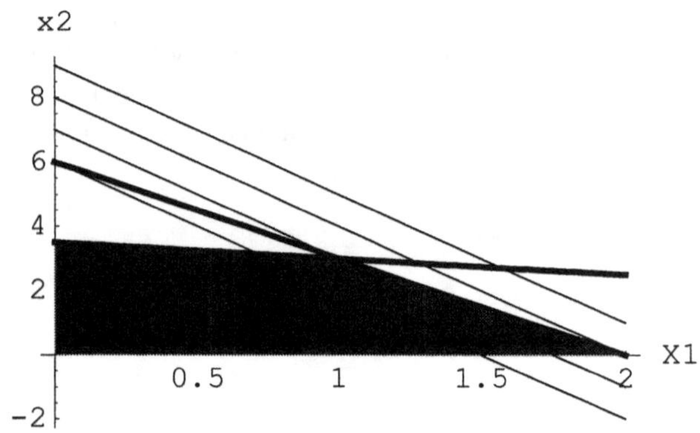

<u>Graphic Solution of the Example Using *Mathematica*:</u> To solve the problem using *Mathematica*, graph the constraints and the objective function to find the optimum point. To graph the constraints, first write x_2 of each equation in terms of x_1 and graph them. For the first constraint, this is

```
In[2]:= Clear[x1, x2, equ1, cons1]
        equ1 = Solve[x1 + 2 x2 == 7, x2]
        eq1[x1_] := x2 /. equ1[[1]]
        cons1 = Plot[eq1[x1], {x1, 0, 7},
          PlotStyle -> {AbsoluteThickness[2]}, AxesLabel -> {"X1", "X2"} ]
```

Out[2]= $\{\{x2 \to \frac{7-x1}{2}\}\}$

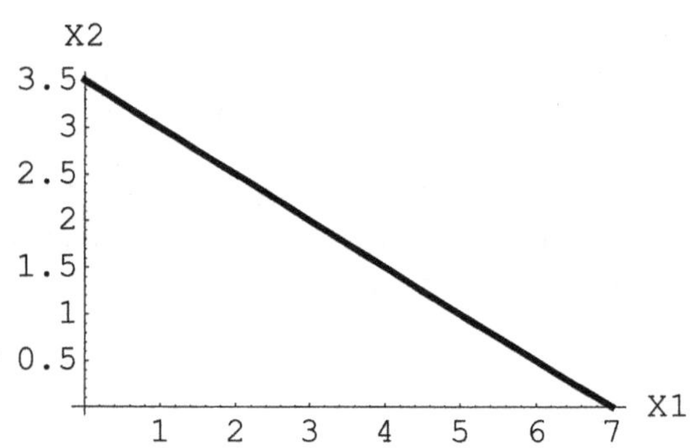

for the second constraint this is

```
In[3]:= Clear[x1, x2, equ2, cons2]
        equ2 = Solve[3 x1 + x2 == 6, x2]
        eq2[x1_] := x2 /. equ2[[1]]
        cons2 = Plot[eq2[x1], {x1, 0, 2},
          PlotStyle -> {AbsoluteThickness[2]}, AxesLabel -> {"X1", "X2"}]
```

Out[3]= {{x2 → -3 (-2 + x1)}}

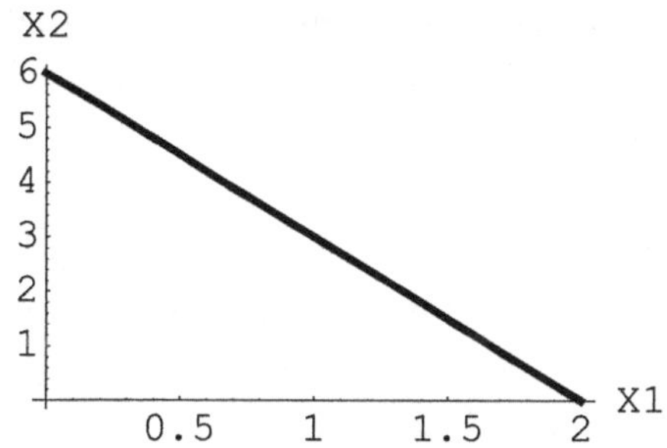

To graph the two constraints together, use the show command as below,

```
In[4]:= cons = Show[cons1, cons2]
```

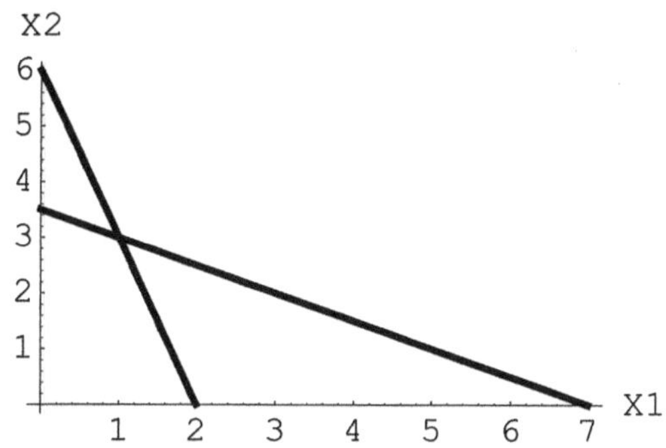

The constraint region is the shaded area shown below, with the vertices (0, 0), (0, 3.5), (1, 3), and (2, 0)

```
In[5]:= consregion = Show[Graphics[ {Hue[.7], Polygon[{{0, 0}, {0, 3.5},
          {1, 3}, {2, 0}}]}]]
```

To show the constraints and the constraint region together, use the show command

```
In[6]:= Show[cons , consregion]
```

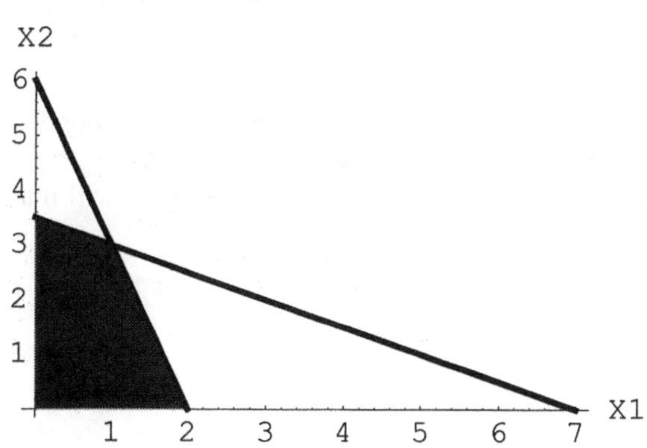

To graph the constraints and the objective together use the show command, listing all three equations. However, listing the objective as a "Table" with an index "i" allows for graphing several parallel lines of objective and for movement of them to visually pinpoint the optimum solution.

```
In[7]:= Show[Plot[{eq1[x1], eq2[x1]}, {x1, 0, 2},
          PlotStyle -> {AbsoluteThickness[2]}, AxesLabel -> {"X1", "x2"}],
        Table[Plot[-1.5 x1 + i, {x1, 0, 2}, DisplayFunction -> Identity],
          {i, 3, 6}],
        Graphics[{Hue[.7], Polygon[{{0, 0}, {0, 3.5}, {1, 3}, {2, 0}}]}]]
```

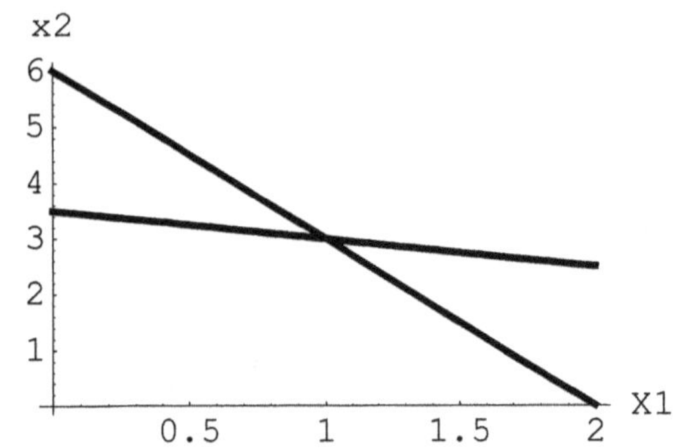

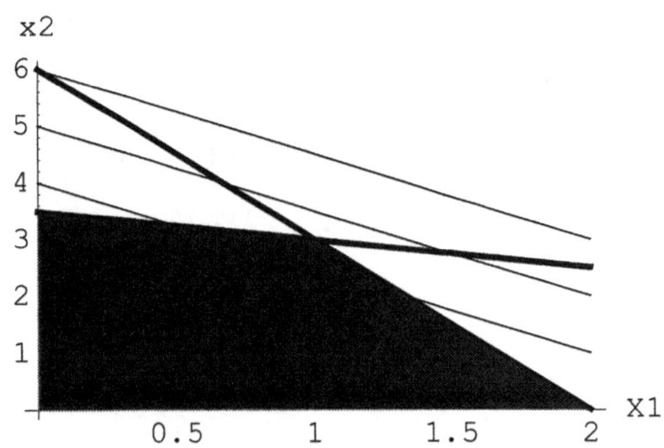

Note: To see the motion of the objective lines double click on the graph.

When the objective line is moved from the north east portion of the first quadrant toward the constraint region, the first point where the objective line touches the constraint region will be the optimizing point. In this case, the solution is (1, 3), which is the intersection of the two constraints. The solution can be tested for using the Mathematica command "ConstrainedMax" as below,

```
In[8]:= ConstrainedMax[2 x1 + 3 x2, {x1 + 2 x2 <= 7, 3 x1 + x2 <= 6}, {x1, x2}]

Out[8]= {11, {x1 → 1, x2 → 3}}
```

For the multiple solutions the objective function is changed to It may happen that the slope of the objective function is the Z = x1 + 2x2, which has the same slope as the first constraint, the point (0, 3.5), or (1,3), or any point on the line joining these two points is a solution to the problem. All of these points will maximize Z with a value of Z = 7. The following figure shows the multiple solution.

```
In[9]:= Show[Plot[{eq1[x1], eq2[x1]}, {x1, 0, 2},
         PlotStyle -> {AbsoluteThickness[2]},  AxesLabel -> {"X1", "x2"}],
        Table[Plot[-.5 x1 + i, {x1, 0, 2}, DisplayFunction -> Identity],
         {i, 3, 5 }], Graphics[
         {Hue[.7], Polygon[{{0, 0}, {0, 3.5}, {1, 3}, {2, 0}}]}]]
```

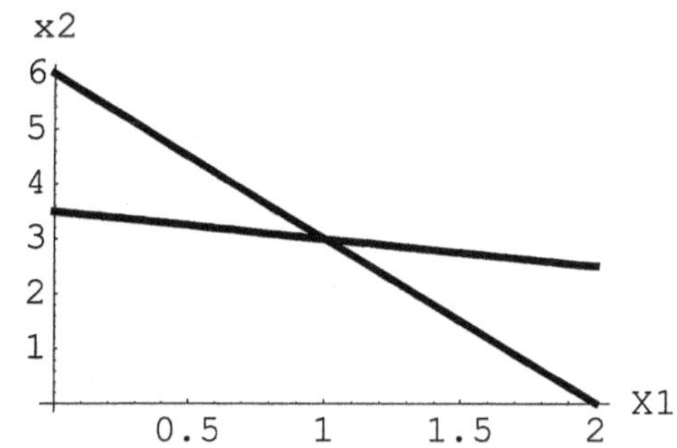

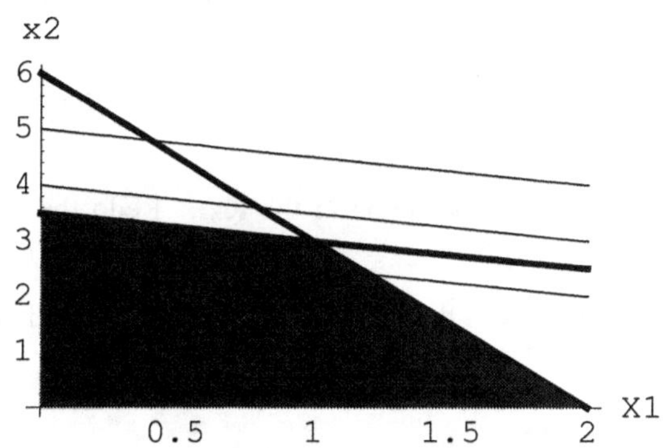

Note: To see the motion of the objective function double click on the graph.

If the objective function intersects the constraint region in the corner points (2, 0) or (0, 3.5), the solution is called degenerate solution. The following graph shows degenerate solution with an objective function of Z = 4x1 + x2, the solution is x1 = 2 and x2 = 0. If the objective function is Z = x1 + 3x2, the solution is x1 = 0 and x2 = 3.5. The following figure shows the degenerate solution (2, 0).

```
In[10]:= Show[Plot[{eq1[x1], eq2[x1]}, {x1, 0, 2},
          PlotStyle -> {AbsoluteThickness[2]}, AxesLabel -> {"X1", "x2"}],
         Table[Plot[-4 x1 + i, {x1, 0, 2}, DisplayFunction -> Identity],
          {i, 6, 9 }], Graphics[
          {Hue[.7], Polygon[{{0, 0}, {0, 3.5}, {1, 3}, {2, 0}}]}]]
```

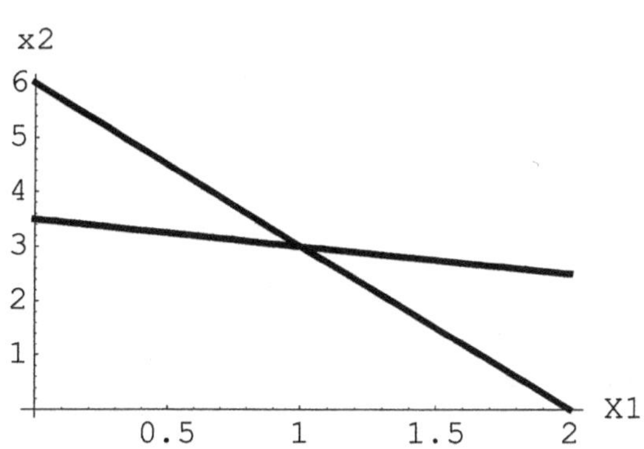

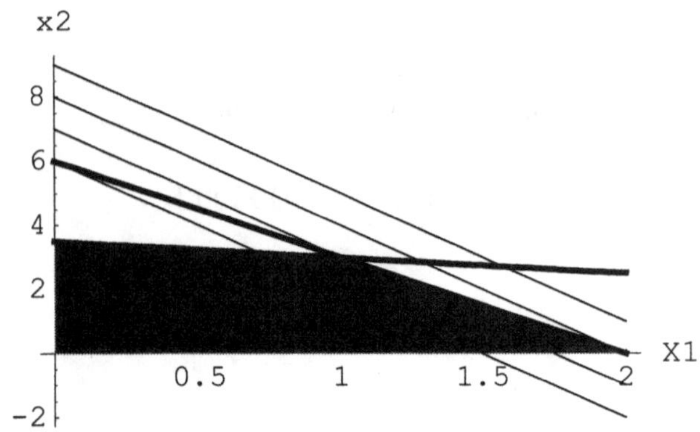

Note: To see the motion of the objective function double click on the graph.

Do More Practice:

1- Double the labor and capital inputs to 14 and 12, respectively. Redo the problem and compare the solution to the one in the text.

2- Double the objective function coefficients to 3 and 2, respectively. Keep other parameters the same as in the text. Redo the problem and compare the solution to the one in the text.

3- Let the objective be Z = .5x1 + x2. Keep other parameters the same as in the text. Redo the problem and compare the solution to the one in the text.

4- Let the objective be Z = x2. Keep other parameters the same as in the text. Redo the problem and compare the solution to the one in the text.

Example: Minimizing Problem

Suppose that the objective of a firm is to minimize cost subject to the constraints given as follows:

Minimize: C = 7x1 + 6x2

Subject to: x1 + 3x2 $\geq$ 2

2x1 + x2 $\geq$ 3

x1, x2 $\geq$ 0

where, x1 and x2 are the prices of two inputs, labor and capital, and C is the cost. Find the optimum solution to the problem.

Solution: The first step is to find the constraint region. Graph the constraints and find the region that satisfies both constraints (the blue region in the graph below).

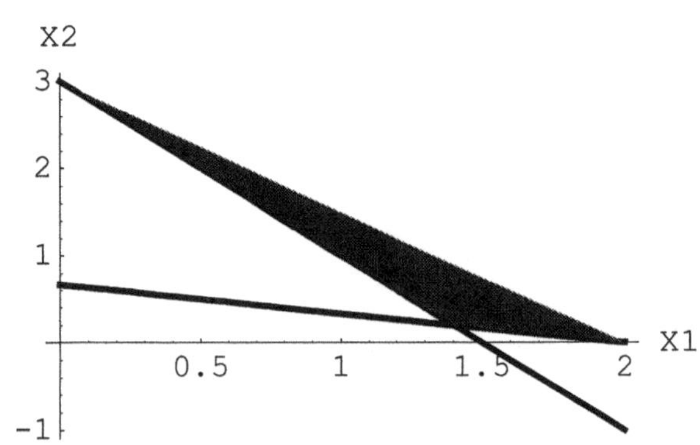

Since the objective is to minimize C, the optimum solution will be the point where the objective line touches (is tangent to) the lowest point of the constraint region. The optimum solution, however, depends on the slope of the objective function, as shown in the following graph.

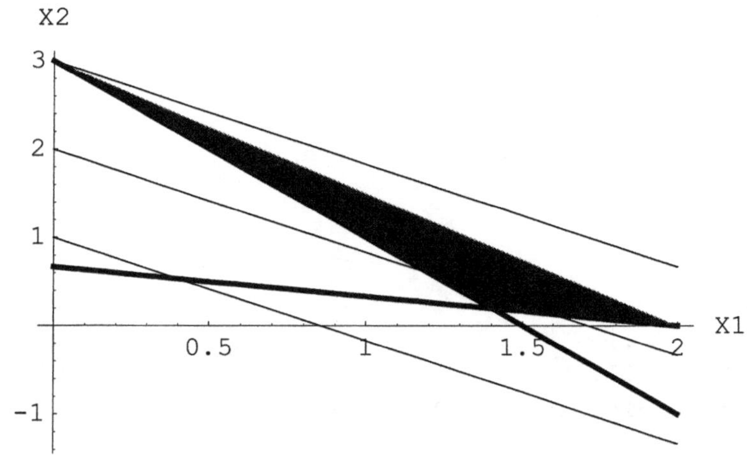

The two constraints intersect at the point $(\frac{7}{5}, \frac{1}{5})$ and the graph above shows that the objective line is tangent to the constraint region at this point. Hence, the optimum solution is the intersection of the two constraints at $(\frac{7}{5}, \frac{1}{5})$, with a minimum cost of C = 11.

If the slope of the objective function is the same as the slope of one of the constraints, then the problem will have a multiple solutions and any point on that constraint that also satisfies the constraint region will be a solution. If the slope of the objective function is such that it is tangent to the constraint region on one of the corner points on the x-axis, then the problem will have a degenerate solution (see, the Do More Practice at the end of this section).

To check the solution using the *Mathematica*, use the command **ConstrainedMin** as follows

```
In[11]:= Clear[x1, x2]
         ConstrainedMin[7 x1 + 6 x2, {x1 + 3 x2 >= 2, 2 x1 + x2 >= 3}, {x1, x2}]

Out[11]= {11, {x1 → 7/5, x2 → 1/5}}
```

<u>Graphic Solution of the Example Using *Mathematica*:</u> To generate the lines and solve the problem using *Mathematica*, first write each equation in terms of one variable, say x_1 and plot them. For the first constraint, this is done as

```
In[12]:= Clear[x1, x2, equ1, equ2]
         equ1 = Solve[x1 + 3 x2 == 2, x2]
         eq1[x1_] = x2 /. equ1[[1]];
         Plot[eq1[x1], {x1, 0, 2},
          PlotStyle -> {AbsoluteThickness[2]}, AxesLabel -> {"X1", "X2"} ]

Out[12] = {{x2 → (2-x1)/3}}
```

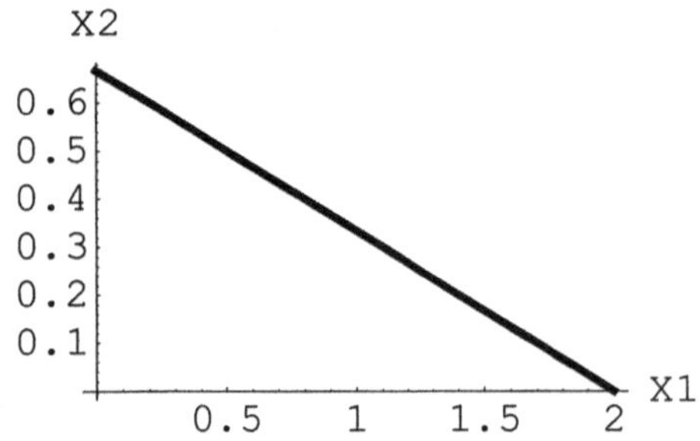

for the second constraint, this is done as

```
In[13]:= Clear[x1, x2]
         equ2 = Solve[2 x1 + x2 == 3, x2]
         eq2[x1_] = x2 /. equ2[[1]];
         Plot[eq2[x1], {x1, 0, 2},
          PlotStyle -> {AbsoluteThickness[2]},  AxesLabel -> {"X1", "x2"}]
```

```
Out[13]= {{x2 → 3 - 2 x1}}
```

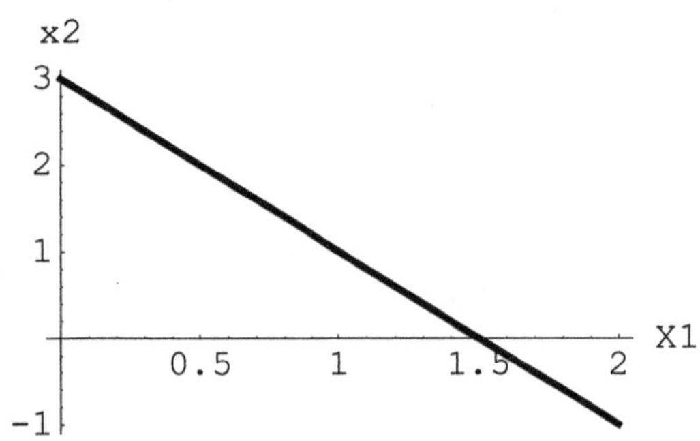

The two constraints intersect at the solution of the following two equations,

```
In[14]:= Solve[{x1 + 3 x2 == 2, 2 x1 + x2 == 3}, {x1, x2}]
```

Out[14]= $\{\{x1 \to \frac{7}{5},\ x2 \to \frac{1}{5}\}\}$

The constraint region is the polygon with the vertices at (0, 3), (2, 0), and $(\frac{7}{5}, \frac{1}{5})$. The polygon is shown below together with the constraint lines.

```
In[15]:= Show[Plot[{eq1[x1], eq2[x1]}, {x1, 0, 2},
           PlotStyle -> {AbsoluteThickness[2]}, AxesLabel -> {"X1", "X2"}],
          Graphics[{Hue[.7], Polygon[{ {0, 3}, {2, 0}, {7 / 5, 1 / 5}}]}]]
```

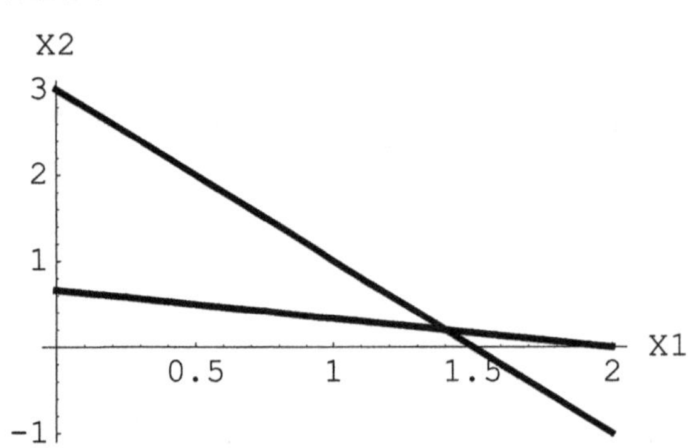

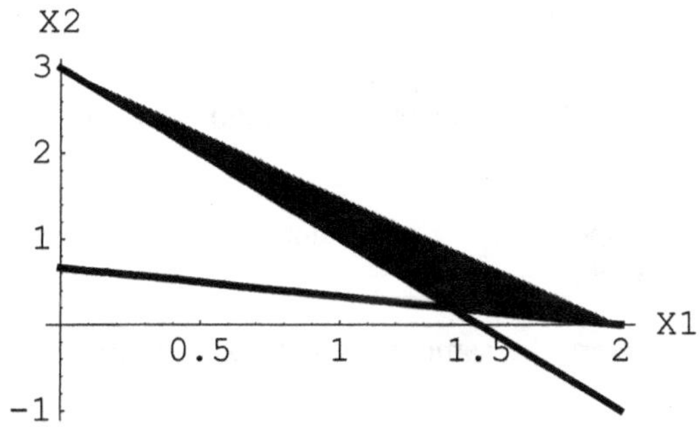

The optimum solution, however, depends on the slope of the objective function, as shown in the following graph. The following commands plot the constraint region and a group of objective lines.

Input[16]:=
```
Show[Plot[{eq1[x1], eq2[x1]}, {x1, 0, 2},
  PlotStyle -> {AbsoluteThickness[2]}, AxesLabel -> {"X1", "X2"}],
 Table[Plot[(-7 / 6) x1 + i, {x1, 0, 2}, DisplayFunction -> Identity],
   {i, 1, 3}], Graphics[{Hue[.7], Polygon[{ {0, 3},
     {2, 0}, {7 / 5, 1 / 5}}]}]]
```

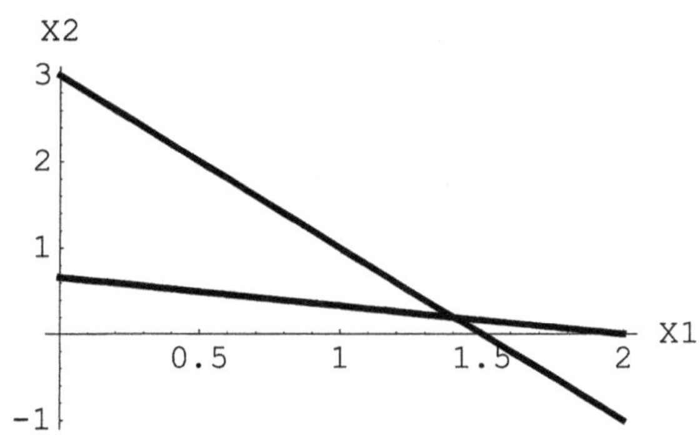

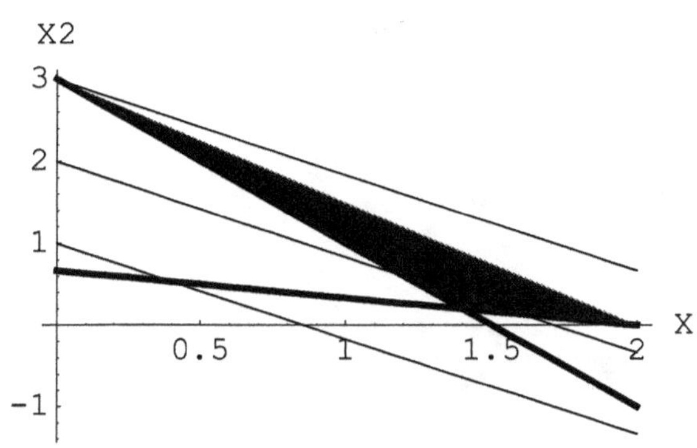

Note: To see the motion of the objective function double click on the graph.

Hence, the optimum solution is the intersection of the two constraints at $(\frac{7}{5}, \frac{1}{5})$, with a minimum cost of C = 11. The optimum solution can also be derived using *Mathematica* command **ConstrainedMin** as follows,

In[17]:=
```
ConstrainedMin[7 x1 + 6 x2, {x1 + 3 x2 >= 2, 2 x1 + x2 >= 3}, {x1, x2}]
```

Out[17]= $\{11, \{x1 \to \frac{7}{5}, x2 \to \frac{1}{5}\}\}$

The graphic solution is limited to the cases where there are no more than two decision variables. However, graphic solutions illustrate two important features of linear programming problems. First, the solution will occur at a corner point of the constraint region (or along the constraint line, if the solution is not unique); Second, the constraint region forms a convex set. (for definition of the convex sets see appendix at the end of Chapter 8.)

Do More Practice: Double the constraining values to 4 and 6, respectively. Redo the problem and compare the solution to the one in the text.

2- Double the objective function coefficients to 14 and 12, respectively. Keep other parameters the same as in the text. Redo the problem and compare the solution to the one in the text.

3- Let the objective be C = 2x1 + 6x2. Keep other parameters the same as in the text. Redo the problem and compare the solution to the one in the text.

4- Let the objective be Z = 4x1 + 2x2. Keep other parameters the same as in the text. Redo the problem and compare the solution to the one in the text.

PROBLEMS

Using graphic solution, find the optimum values of the decision and objective variables of the following problems.

1. Maximize: $z = x1 + 2x2$

   Subject to: $x1 + 4x2 \leq 12$

   $3x1 + 2x2 \leq 18$

   $x1, x2 \geq 0$

2. Solve problem 1 when the objective function is $z = 10x1 + 3x2$

3. Solve problem 1 when the objective is $z = 6x1 + 4x2$.

4. Maximize: $z = 2x1 + 3x2$

   Subject to: $x1 + 3x2 \leq 10$

   $4x1 + x2 \leq 12$

   $x1 + x2 \leq 6$

   $x1, x2 \geq 0$

5. Solve problem 4 when the objective function is $z = 3x1 + 2x2$.

6. Maximize: $z = x1 + 2x2$

   Subject to: $x1 + x2 \leq 10$

   $2x1 + 3x2 \leq 20$

   $x1, x2 \geq 0$

7. Solve problem.6, when the objective function is $z = x1$.

8. Minimize: $C = 13y1 + 18y2$

   Subject to: $y1 + 3y2 \geq 3$

   $4y1 + 2y2 \geq 2$

   $y1, y2 \geq 0$

9. Minimize: $C = y1 + 5y2$

Subject to: $4y1 + y2 \geq 6$

$y1 + 2y2 \geq 5$

$y1 + 10y2 \geq 13$

$y1, y2 \geq 0$

10. Solve problem 9, when the objective function is $C = 6y1 + y2$.

II. Simplex Solution of the Linear Programming Models

The Simplex algorithm is an efficient procedure to find the optimum solution of linear programming problems. The preceding section showed that the solution of a linear programming problem occurs at a corner point of the constraint region, which is the intersection or the solution set of a system of equations. However, there is more to finding a solution to a linear programming problem than finding a solution to a system of equations. For a student of economics, Simplex method is a step by step practice in economics of resource allocation. The solution starts from the corner point origin, which could be the worst or one of the non-optimal solutions, and proceeds to the optimum solution. In each step, the question is what to produce next and how much to produce, given the constraints on available resources. These questions are all economic questions that, in the Simplex method, are answered through the marginal cost marginal benefit analysis and other economic criterion. The procedure sorts out from the totality of solutions to the system of linear equations a specific solution that is optimal.

A **standard linear programming model** is a linear programming model in which the objective function is maximized subject to the constraints of the " $\leq$ " form. That is,

Maximize: $Z = c1x1 + \dots\dots\dots + cnxn$

Subject to: $a11x1 + \dots\dots\dots + an1xn \leq b1$

....

$am1x1 + \dots\dots\dots + amnxn \leq bm$

$x1, \dots\dots, xn \geq 0$

The Simplex method starts with a solution which is feasible, but not optimum. It then continues with iterations in such a way that each new solution is at least as good as the previous one.

A **Feasible Solution** to a linear programming problem is a set of x_s that satisfy the constraints and the non-negativity restrictions. An **Optimal Solution** is a feasible solution that optimizes the objective function.

To use the Simplex method, the first step is to set the inequality constraint to equality by adding **slack variables.** A slack variable is a positive variable to be added to the inequality constraints of the "<" form to set them to equality. The following example illustrates the application of Simplex algorithm.

Example: Solve the following linear programming problem, using the Simplex algorithm.

Maximize: $Z = 2x1 + 3x2$

Subject to: $x1 + 2x2 \leq 7$

$3x1 + x2 \leq 6$

$x1, x2 \geq 0$

Solution: The first step is to add slack variables to the inequality constraints to set them to equalities.

Maximize: Z = 2x1 + 3x2

Subject to: x1 + 2x2 + s1 = 7

3x1 + x2 + s2 = 6

x1, x2, s1, s2 ≥ 0

Here, s1 and s2 are the slack variables. For the clarity of exposition, the coefficients are written in the table of coefficients. Then the elimination procedure is used to set the matrix of coefficients into reduced form.

Table 1. Table of coefficients

Variable:	x1	x2	s1	s2	Constant
	1	2	1	0	7
	3	1	0	1	6
Objective	2	3	0	0	Z

The initial basic feasible solution to the problem is s1= 7 and s2 = 6. The basic feasible solution is determined by the standard basis (unit matrix of the constraint column in table of coefficients, highlighted yellow). This solution implies that both inputs go to slack (they are idle or are not used), with no production taking place. Hence, s1 = 7, s2 = 6, x1 = x2 = 0, and Z = 0.

Certainly, this solution is not a proper solution and the value of Z can be improved by entering one of the decision variables into the basic solution. The Simplex criteria for selecting such a variable is to choose the variable which gives the largest per unit increase to the objective function(highest marginal revenue). Here, that variable is x2, with a coefficient of 3 in the objective row. The column presenting x2 (the new entry to the basic solution) in the table of coefficients is called the **pivot column**.

Since the inputs set a restriction on the values of the decision variables, the next step is to determine which constraint is more restrictive. This is decided by dividing available inputs, (7 units of labor and 6 units of capital), by the corresponding input requirements for x2, (2 units of labor and 1 unit of capital). The ratios for labor and capital are $\frac{7}{2}$ and $\frac{6}{1}$, respectively. Hence, the more restrictive constraint is labor. The row representing the more restrictive constraint is called the **pivot row**. Here, the pivot row is the first row (labor).

The intersection of the pivot column and the pivot row represents the **pivot element**. Here, the pivot element is 2. Once the pivot element is decided, the elimination technique can be applied to bring x2 into the basic solution by transforming the pivot element to one, and all the other entries in the pivot column to zero.

Table 2. Table of coefficients

Variable:	x1	x2	s1	s2	Constant
	.5	1	.5	0	3.5
	2.5	0	-.5	1	2.5
Objective	.5	0	-1.5	0	Z - 10.5

The basic variables, decided by the unit matrix in the constraint rows (highlighted by yellow), in table 2 are x2 = 3.5 and s2 = 2.5, which results in a solution, x1 = 0, x2 = 3.5, s1 = 0, s2 = 2.5, and an objective value of Z = 10.5. Note that, non-basic variables (here, x1 and s1) have a value of zero.

The elimination technique to get from table 1 to table 2 is as follows,

1- Divide all the elements of pivot row by pivot element, (2).

2- Multiply the first row by -1 and add it to the second row.

3- Multiply the first row by -3 and add it to the third row.

As long as there is a positive value in the objective row, there is a chance to increase the objective variable by changing the basic variables. Here, the variable that can increase the objective is x1, with a positive coefficient of .5 in the objective row. Hence, x1 will represent the pivot column.

The pivot row is selected by finding the more restrictive constraint. Divide elements of the constant column in table 2 by the corresponding input coefficients in the pivot column (x1 column), ($\frac{3.5}{.5}$ = 7, $\frac{2.5}{2.5}$ = 1). The more restrictive constraint is capital. The pivot element is 2.5, in the intersection of the first column and the second row.

Therefore, x1 enters the basic solution. Table 3 represents the new basic feasible solution.

Table 3. Table of coefficients

Variable:	x1	x2	s1	s2	Constant
	0	1	.6	-.2	3
	1	0	-.2	.4	1
Objective	0	0	-1.4	-.2	Z - 11

The basic feasible solution is x1 = 1, x2 = 3, which corresponds to the unit matrix (highlighted yellow) with an objective value of Z = 11. This solution is also the optimum solution because there is no positive entry in the objective row. Therefore, the optimum solution is x1 = 1, x2 = 3, s1 = s2 = 0 and Z = 11. Z can be calculated either by using the objective function and substituting for decision variables, or by setting the last entry in the objective row equal to zero, (Z - 11 = 0).

To solve the problem using *Mathematica*, use the command "ConstrainedMax".

```
In[18]:= Clear[x1, x2]
         ConstrainedMax[2 x1 + 3 x2, {x1 + 2 x2 <= 7, 3 x1 + x2 <= 6}, {x1, x2}]

Out[18]= {11, {x1 → 1, x2 → 3}}
```

Example: Solve the following linear programming problem, using the Simplex algorithm.

Maximize: Z = 4x1 + x2

Subject to: x1 + 2x2 ≤ 7

3x1 + x2 ≤ 6

x1, x2 ≥ 0

Solution: Set the inequality constraint to equality by adding the slack variables (s1, s2).

Maximize: Z = 4x1 + x2

Subject to: x1 + 2x2 + s1 = 7

3x1 + x2 + s2 = 6

x1, x2, s1, s2 > 0

The table of coefficients is

Table 1. Table of coefficients

Variable:	x1	x2	s1	s2	Constant
	1	2	1	0	7
	3	1	0	1	6
Objective	4	1	0	0	Z

The initial solution is x1 = x2 = 0, s1 = 7, s2 = 6, and Z = 0. Both inputs are going slack (s1 and s2 are in the basic solution) and no production is taking place. The pivot column is column x1with the largest revenue margin of 4. The pivot row is selected by dividing elements of the constant column (constraints on inputs) to the corresponding elements of the pivot column ($\frac{7}{1}$, $\frac{6}{3}$). Since the second constraint row is setting the constraining limit, it is selected as the pivot row. The intersection of the pivot column and the pivot row decides the pivot element, which is "3" in the x1 column. Therefore x1 enters the basic solution.

Table 2. Table of coefficients

Variable:	x1	x2	s1	s2	Constant
	0	$\frac{5}{3}$	1	$-\frac{1}{3}$	5
	1	$\frac{1}{3}$	0	$\frac{1}{3}$	2
Objective	0	$-\frac{1}{3}$	0	$-\frac{4}{3}$	Z - 8

Since all the entries in the objective row are negative, there is no way to increase the objective variable with further iteration. Therefore, the optimum solution x1 = 2, x2 = 0, s1 = 5, s2 = 0 and Z = 8. With x2 = 0, the solution is also a degenerate solution.

It is also clear from table 2 that s1 is in the basic solution, with a value of s1 = 5. This implies that from the total 7 units of labor available for the firm, only 2 units are used and 5 units go slack.

To solve the problem using *Mathematica*, use the following command

```
In[19]:= Clear[x1, x2]
         ConstrainedMax[4 x1 + x2, {x1 + 2 x2 <= 7, 3 x1 + x2 <= 6}, {x1, x2}]

Out[19]= {8, {x1 → 2, x2 → 0}}
```

PROBLEMS

Solve the following problems through the Simplex method

.1,. Maximize: $z = x1 + 2x2$

Subject to: $x1 + 4x2 \leq 12$

$3x1 + 2x2 \leq 18$

$x1, x2 \geq 0$

2. Maximize $z = 10x1 + 3x2$

Subject to: $x1 + 4x2 \leq 12$

$3x1 + 2x2 \leq 18$

$x1, x2 \geq 0$

3. Maximize: $z = 2x1 + 3x2$

Subject to: $x1 + 3x2 \leq 10$

$4x1 + x2 \leq 12$

$x1 + x2 \leq 6$

$x1, x2 \geq 0$

4. Maximize: $Z = 3x1 + 5x2$

Subject to: $x1 + 2x2 \leq 8$

$3x1 + x2 \leq 9$

$x1, x2 \geq 0$

5. Maximize: $Z = 10x1 + 5x2 + 6x3$

Subject to: $3x1 + x2 + 3x3 \leq 60$

$2x1 + 2x2 + x3 \leq 50$

$x1, x2, x3 \geq 0$

6. A firm manufactures three products, x1, x2, and x3, which are sold at market prices of \$3, \$5,and \$4,respectively. The firm has limitations on its machine time and available labor. Total machine time available is 180 hours, and total labor available is 40 man-hours. Each unit of product x1 requires 2 hours of machine time and one man-hour. Each unit of product x2 requires 3 machine hours and .5 man-hour. Each unit of product x3 requires 1.5 hours of machine time and 3 man-hours. Set up the linear programming problem of maximizing total revenue subject to the constraints.

III. The Minimizing Problem: The `ConstrainedMin` Command

The Simplex algorithm is devised for standard linear programming problems, where the objective function is to be maximized subject to inequalities of the " $\leq$" form. When the problem is a minimizing one with the inequalities of the "$\geq$" form, there is no initial standard basis or basic feasible solution and the Simplex algorithm cannot get started. To solve this problem, a method has been devised which is called Phase I of the Simplex method. The procedure is illustrated in the following example.

Example: Solve the following linear programming problem, using the Simplex algorithm.

Minimize: $C = 7y1 + 6y2$
Subject to:
$y1 + 3y2 \geq 2$
$2y1 + y2 \geq 3$
$y1,\ y2 \geq 0$

The first step is to set the inequality constraints to equality by adding **surplus variables.** A surplus variable (su) is a positive variable that is subtracted from the inequality constraints of the ">" form to set them to equality.

Minimize: $C = 7y1 + 6y2$
Subject to: $y1 + 3y2 - su1 = 2$
$2y1 + y2 - su2 = 3$
$y1,\ y2 \geq 0$

Where, su1 and su2 are surplus variables. The Simplex process, however, cannot get started, because there is no standard basis or basic feasible solution. To get the process started, an initial basic feasible solution is created by adding **artificial variables** to the constraints.

Unlike the slack and surplus variables, artificial variables do not have any meaning. Once artificial variables are added, the Simplex method will have an initial basic feasible solution and can get started. However, so long as any artificial variable remains at the positive level, we do not have a set of admissible variables that forms a basic feasible solution.

The objective, thus, is to minimize the value of the artificial variables, and if possible, reduce their values to zero. Therefore, the Phase I problem can be written as Table 1 below, where A1 and A2 are artificial variables.

Minimize: $W = A1 + A2$
Subject to: $y1 + 3y2 - su1 + A1 = 2$
$2y1 + y2 - su2 + A2 = 3$
$y1, y2, su1, su2 \geq 0$

Where, . Table.1, below, represents the table of coefficients.

Table 1 The Minimizing Problem

Variable:	y1	y2	su1	su2	A1	A2	Constant
	1	3	-1	0	1	0	2
	2	1	0	-1	0	1	3
Objective	7	6	0	0	1	1	W

The objective in Phase I is to force W to zero. If the original problem has a basic feasible solution, Phase I will terminate with W = 0.

The criteria for choosing the pivot column is the same as in maximizing problems. After all, any minimizing problem can be transformed to a maximizing problem by multiplying it by (-1). Therefore, in a minimizing problem the negative coefficient of the largest numerical value determines the entering variable. Hence, the pivot column is the column corresponding to y1, with an objective coefficient of 7. The pivot row is the second constraint.

Table 1 of Phase I

Variable:	y1	y2	su1	su2	A1	A2	Constant
	0	$\frac{5}{2}$	-1	$\frac{1}{2}$	1	$-\frac{1}{2}$	$\frac{1}{2}$
	1	$\frac{1}{2}$	0	$-\frac{1}{2}$	0	$\frac{1}{2}$	$\frac{3}{2}$
Objective	0	$\frac{5}{2}$	0	$\frac{7}{2}$	1	$-\frac{5}{2}$	$W - \frac{21}{2}$

The next variable to enter the basic solution is su2, which has the largest coefficient in the objective, ($\frac{7}{2}$). The pivot row is the first constraint row.

Table 2 of Phase I

Variable:	y1	y2	su1	su2	A1	A2	Constant
	0	5	-2	1	2	-1	1
	1	3	-1	0	1	0	2
Objective	0	-15	7	0	-6	1	W - 14

Since A1 and A2 are not in the basic feasible solution, the Phase I terminates here, with A1 = A2 = 0. As well, W = A1 + A2 = 0.

If Phase I ends up with a W > 0, it implies that either the system of equations in the constraints are inconsistent and have no solution, or the equations have a solution with at least one of the decision variables being negative.

When Phase I terminates in a basis containing only decision and surplus variables, the artificial variables can be dropped from the true objective, and Phase II starts.

Table 1 of Phase II

Variable:	y1	y2	su1	su2	Constant
	0	5	-2	1	1
	1	3	-1	0	2
Objective	0	-15	7	0	C - 14

Based on the Simplex criteria, the variable that contributes to minimizing cost (C) is y2 with a negative marginal contribution of (-15). Therefore, y2 will enter the basic solution. The pivot row is the row corresponding to the first constraint.

Table 2 of Phase II

Variable:	y1	y2	su1	su2	Constant
	0	1	-.4	.2	.2
	1	0	.2	-.6	1.4
Objective	0	0	1	3	C - 11

The basic feasible solution is y1 = 1.4 and y2 = .2, with the value of C = 11. Since all the coefficients in the objective row are positive or zero, any further iteration will either increase C or will have no effect on C. Therefore, the above solution is a basic feasible optimal solution. It is also clear from the solution that the value of the surplus variables su1 = su2 = 0.

To solve the same problem using *Mathematica*, use the command **ConstrainedMin** as follows

```
In[20]:= Clear[y1, y2]
         ConstrainedMin[7 y1 + 6 y2, {y1 + 3 y2 >= 2, 2 y1 + y2 >= 3}, {y1, y2}]
Out[20]= {11, {y1 → 7/5, y2 → 1/5}}
```

Problems

Solve the following problems through the Simplex method.

1. Minimize: C = 13y1 + 18y2

   Subject to: y1 + 3y2 ≥ 3

   4y1 + 2y2 ≥ 2

   y1, y2 ≥ 0

2. Minimize: $C = y1 + 5y2$

Subject to: $4y1 + y2 \geq 6$

$y1 + 2y2 \geq 5$

$y1 + 10y2 \geq 13$

$y1, y2 \geq 0$

3. Solve problem 1, when the objective function is $C = 6y1 + y2$.

4. Minimize: $C = y1 + y2$

Subject to: $3y1 + y2 \geq 9$

$y1 + 2y2 \geq 8$

$y1, y2 \geq 0$

5. Minimize: $C = 2y1 + 3y2$

Subject to: $3y1 + y2 \geq 12$

$y1 + y2 \geq 7$

$y1 + 3y2 \geq 9$

$y1, y2 \geq 0$

6. Solve problem 5 for the case where the objective function is $C = 3y1 + 2y2$.

7. Minimize: $C = 18y1 + 15y2 + 4y3$

Subject to: $4y1 + 3y2 + 1.5y3 \geq 6$

$12y1 + 4y2 + y3 \geq 10$

$12y1 + 10y2 + y3 \geq 9$

$y1, y2, y3 \geq 0$

IV. Duality

The general form of the linear programming problem in matrix form can be written as:

Optimize: $z = \mathbf{c}'.\mathbf{x}$

Subject to: $\mathbf{A.x}\ \{<=>\}\ \mathbf{b}$

$\mathbf{x} \geq \mathbf{0}$

Where, **c** is a nx1 vector of objective coefficients, **c'** is **c** transposed, **x** is a nx1 vector of decision variables, **A** is a mxn matrix of constraint coefficients, and **b** is a mx1 vector of constraints. The standard form of the linear programming problem is:

Maximize: $Z = \mathbf{c}'.\mathbf{x}$

Subject to: $\mathbf{A.x} \leq \mathbf{b}$

$\mathbf{x} \geq \mathbf{0}$

For every linear programming problem there is a corresponding problem, which is called the **dual**, the solution of which is closely related to the original problem. The original problem is called **primal.** To construct the dual from a primal, take the following steps:

1- Write the primal in standard linear programming form.

2- Change the decision variables x to new variables, say y. The new variables y are the dual variables or the **implicit price** of the constraints (b). Therefore, the number of y variables will be as many as the number of the constraints; i.e., **y** is a mx1 vector.

3- Change the objective from a maximizing problem to a minimizing problem.

4- Change the constraint functions to **A'.y**, where **A'** is **A** transposed.

5- Change the inequalities from "≤" to "≥" form.

6- Change the place of **b** with **c**.

Hence, the dual can be written as:

Minimize: $C = \mathbf{b'.y}$

Subject to: $\mathbf{A'.y} \geq \mathbf{c}$

$\mathbf{y} \geq \mathbf{0}$

The following examples demonstrate the procedure.

Example: Construct the dual problem of the following primal.

Maximize: $Z = x1 + 3x2$

Subject to: $6x1 + 2x2 \leq 15$

$3x1 + 5x2 \leq 40$

$2x1 + 3x2 \leq 24$

$x1, x2 \geq 0$

The coefficients in matrix form are: $\mathbf{c'} = \begin{pmatrix} 1 & 3 \end{pmatrix}$, $\mathbf{x} = \begin{pmatrix} x1 \\ x2 \end{pmatrix}$, $\mathbf{A} = \begin{pmatrix} 6 & 2 \\ 3 & 5 \\ 2 & 3 \end{pmatrix}$, and $\mathbf{b} = \begin{pmatrix} 15 \\ 40 \\ 24 \end{pmatrix}$. Since the problem is in standard linear programming form, no change in the sign of the constraints is required. The dual is:

Minimize: $C = \mathbf{b'.y} = 15y1 + 40y2 + 24y3$

Subject to: $\mathbf{A'.y} = \begin{pmatrix} 6 & 3 & 2 \\ 2 & 5 & 3 \end{pmatrix} \begin{pmatrix} y1 \\ y2 \\ y3 \end{pmatrix} \geq \begin{pmatrix} 1 \\ 3 \end{pmatrix}$

$\mathbf{y} \geq \mathbf{0}$

Or:

Minimize: $C = 15y1 + 40y2 + 24y3$

Subject to: $6y1 + 3y2 + 2y3 \geq 1$

$2y1 + 5y2 + 3y3 \geq 3$

$y1, y2, y3 \geq 0$

Example: Write the dual of the following primal:

Minimize: $C = 3x1 + 2x2$

Subject to: $3x1 + 4x2 \leq 30$

$2x1 + x2 \geq 10$

$2x1 + 3x2 \geq 15$

$x1, x2 \geq 0$

Since the primal is not in standard form, the first step is to standardize the primal.

Maximize: $-Z = -3x1 - 2x2$

Subject to: $3x1 + 4x2 \leq 30$

$-2x1 - x2 \leq -10$

$-2x1 - 3x2 \leq -15$

$x1, x2 \geq 0$

The dual problem corresponding to this standard form is

Minimize: $C = 30y1 - 10y2 - 15y3$

Subject to: $3y1 - 2y2 - 2y3 \geq -3$

$4y1 - y2 - 3y3 \geq -2$

$y1, y2, y3 \geq 0$

Dual problems and their solution are of great importance in linear programming models. In a revenue-maximizing problem, the dual variables represent the implicit prices or the **shadow prices** of the fixed inputs (**b**). Contrary to variable inputs which are bought by the firms at the market prices, there is no market price for fixed inputs. Dual variables are interpreted as the prices at which the management leases the services of fixed inputs to production managers.

Mathematically, dual variables are the Lagrangian multipliers of the constraints in the Lagrangian function of a linear programming problem. Therefore, they represent the rate of change of the objective variable with respect to the corresponding constraints. That is, in a linear programming problem of maximizing revenue with respect to the given input constraints, the dual variable $yi = \frac{dZ}{dbi}$. This implies that yi is the marginal revenue product of input bi. According to the condition IV of the Kuhn-Tucker condition in Chapter 8, if bi is fully utilized (the constraint is binding), its marginal revenue product or its implicit price will be positive, $yi > 0$. On the other hand, if bi is not fully utilized (the constraint is non-binding), its marginal revenue product will be zero, $yi = 0$.

One excellent feature of the Simplex method is that the optimal solution of the primal provides information about the values of the dual variables. The positive values of the entries for the slack variables in the objective row of the final table, or the basic feasible optimal solution, represents the implicit price of corresponding inputs (constraints). For example, in the final table (Table 3) of the first example of section II, The Simplex Solution of the Linear Programming Models, which is reproduced in the next page

Table 3. Table of coefficients

Variable:	x1	x2	s1	s2	Constant
	0	1	.6	-.2	3
	1	0	-.2	.4	1
Objective	0	0	-1.4	-.2	Z - 11

the absolute values under the slack variables s1 and s2 in the objective row are 1.4 and .2, respectively. These coefficients represent the implicit prices of labor (7 units) and capital (6 units), respectively. The total cost of producing x1 and x2, then is C = 1.4(7) + .2(6) = 11, which is the same as the optimum value of Z.

Furthermore, the optimal solution of the primal is related to optimal solution of the dual in such a way that if both the primal and the dual problems have feasible solutions, both have optimal solutions, and the optimal value of the two objective variables will be equal. In any linear programming problem, x_s are the dual variables for the dual problem, and y_s are the dual variables for the primal problem.

Based on the Kuhn-Tucker conditions, if an input, say bi, goes to slack (si > 0), its implicit price or Lagrange multiplier must be zero (yi = 0); and if an input is fully utilized (si = 0), then its implicit price is positive (yi > 0). Also, if a surplus variable is positive ($su_i > 0$), the corresponding product incurs a loss or needs a price subsidy and should not be produced (xi = 0); and if a product does not need a surplus or price subsidy ($su_i = 0$), it should be produced (xi > 0). In a linear programming problem, this guarantees that in the optimal solution C = Z.

The following example is designed to show how the solution sets of a primal and a dual in a linear programming are related and how by solving one the solution of the other can be derived.

Example: Revenue-Maximizing Problem

Suppose the following problem represents a revenue-maximizing problem of a firm that hires the services of three fixed inputs labor, land, and capital at 15, 40, and 24 units each, respectively. The firm produces two goods, x1 and x2, which are sold at market prices of \$1 and \$3 each, respectively. With the given input requirements, the revenue maximizing problem of the firm is modeled as:

Primal: Maximize: $Z = x1 + 3x2$

Subject to: $6x1 + 2x2 \leq 15$

$3x1 + 5x2 \leq 40$

$2x1 + 3x2 \leq 24$

$x1, x2 \geq 0$

The optimal solution of this problem yields the following table:

Optimal Solution Table of Primal

Variable:	x1	x2	s1	s2	s3	Constant
	3	1	.5	0	0	$\frac{15}{2}$
	-12	0	-2.5	1	0	$\frac{5}{2}$
	-7	0	-1.5	0	1	$\frac{3}{2}$
Objective	-9	0	$-\frac{3}{2}$	0	0	$Z - \frac{45}{2}$

The primal has a degenerate solution with x1 = 0, x2 = $\frac{15}{2}$, and Z = $\frac{45}{2}$. Since s2 and s3 are also in the basic solution of the final table (unit matrix highlighted yellow), s2 = $\frac{5}{2}$ and s3 = $\frac{3}{2}$. That is, $\frac{5}{2}$ units of the second input and $\frac{3}{2}$ units of the third input are idle (go to slack).

The positive values of the slack variables in the objective row of the final table are y1 = $\frac{3}{2}$, y2 = 0, and y3 =0 (y_s are denoting the internal prices of the inputs). This implies that, the first constraint (labor) is binding and all of the labor input is used in the production process. Therefore, labor has a positive internal price or shadow price of y1 $=\frac{3}{2}$. However, the second and third constraint are non-binding and not all of the land and capital inputs are used in the production. Therefore, their internal price or shadow prices are zero, y2 = y3 = 0. Also, the total cost of production is C = ($\frac{3}{2}$) 15 + (0) 40 + (0) 24 = $\frac{45}{2}$, which equals to Z.

The dual problem to this primal changes the problem to a cost-minimizing problem, where the objective is to minimize the cost of inputs given the constraints on the market prices of x1 and x2 which are $1 and $3, respectively. That is,

Dual: Minimize: C = 15y1 + 40y2 + 24y3

Subject to: $6y1 + 3y2 + 2y3 \geq 1$

$2y1 + 5y2 + 3y3 \geq 3$

$y1, y2, y3 \geq 0$

The constraints of the dual problem compare marginal costs and marginal revenues of the products. The left side of each constraint represents the marginal cost of one product and the right side represents the corresponding marginal revenue or the market price. If the marginal cost of x_i is greater than its marginal revenue, x_i's production requires a subsidy and a profit-maximizing or cost-minimizing firm will produce zero amounts of x_i. The optimal solution of this problem is presented in the following table (for the complete solution of the dual, see the Appendix at the end of the chapter).

Final Table of the Optimum Solution to Dual

Variable:	y1	y2	y3	su1	su2	Constant
	1	$\frac{5}{12}$	1.5	0	-.5	$\frac{3}{2}$
	0	12	7	1	-3	8
Objective	0	$\frac{5}{2}$	$\frac{3}{2}$	0	$\frac{15}{2}$	C - $\frac{45}{2}$

The feasible optimal solution to the dual is, y1 = $\frac{3}{2}$, y2 = y3 = 0, and C = $\frac{45}{2}$. That is, to minimize cost, 15 units of labor input will be used in the production at internal price of y1 = $\frac{3}{2}$. Of course, land and capital are also used in production of x1 and x2. But, since these inputs are available at excessive amounts, their internal prices are zero, y2 = y3 = 0. Also, since su1 is in the basic solution (the unit matrix, highlighted yellow), its value is su1 = 8. That is, the marginal cost of producing the second output (x2) is greater than its marginal revenue by 8 points, or its production requires subsidy. Therefore, the second product will be produced at zero quantity. The positive values under the surplus variables (su1 and su2) in the objective row of the final table show the quantities of x1 and x2, respectively. That is, x1 = 0 and x2 = $\frac{15}{2}$.

The fact that the solution of a primal provides all the information on the solution to its dual or vise versa, can help in solving some dual or primal in an easier way by transforming a primal to dual or a dual to primal, as the case may require. For example, the dual of the previous example with two constraints and three variables can't be solved using graphic method. However, by transforming it to its primal (three constraint and two variable) the problem can easily be solved through the graphic method.

The solutions to the previous example can be checked using *Mathematica* as follows,

```
In[21]:= Clear[x1, x2]
         ConstrainedMax[x1 + 3 x2,
          {6 x1 + 2 x2 <= 15, 3 x1 + 5 x2 <= 40, 2 x1 + 3 x2 <= 24}, {x1, x2}]
```

Out[21]= $\{\frac{45}{2}, \{x1 \to 0, x2 \to \frac{15}{2}\}\}$

The solution to dual is,

```
In[22]:= Clear[y1, y2, y3]
         ConstrainedMin[15 y1 + 40 y2 + 24 y3,
          {6 y1 + 3 y2 + 2 y3 >= 1, 2 y1 + 5 y2 + 3 y3 >= 3}, {y1, y2, y3}]
```

Out[22]= $\{\frac{45}{2}, \{y1 \to \frac{3}{2}, y2 \to 0, y3 \to 0\}\}$

Problems

1. The following table summarizes the technological relationships between fixed inputs and outputs of a firm.

	x1	x2	x3	x4	Fixed Inputs
One	5	1	3	1	90
Two	2	6	2	4	120

The products are sold at market prices of p1 = 10, p2 = 6, p3 = 8, and p4 = 3. Set up linear programming problem and find the optimal solution.

2. Construct the dual of the problem 1.

3. Solve the primal and the dual in problems 1 an 2 and check for the correspondence of the solution between the dual and primal.

4. The optimal feasible solution to a linear programming problem of a firm with three outputs x1, x2, x3, and two fixed inputs labor (L = 45 units) and capital (K = 50 units) is given as follows: (s1 and s2 are slack variables for labor and capital respectively).

	x1	x2	x3	s1	s2	constant
	$-\frac{1}{4}$	0	1	$\frac{3}{4}$	2	150
	1	1	0	$-\frac{5}{2}$	$\frac{1}{4}$	100
objective	-2	0	0	-8	-11.8	Z - 950

The market prices of the three products are 3, 2, and 5 respectively.

a) How many units of x1, x2, and x3 will be produced?

b) What is the total revenue of the firm at the optimal solution?

c) What are the internal prices of labor and capital?

d) What is the total cost?

5. The optimal feasible solution to a linear programming problem of a firm with three outputs x1, x2, x3, and three fixed inputs, labor (L = 30 units), capital (K = 45 units), and raw material (M = 80) is given as follows: (S1, S2, and S3 are slack variables for labor, capital, and raw material respectively).

Variable:	x1	x2	x3	s1	s2	s3	Constant
	0	3	1	$\frac{3}{4}$	0	$\frac{5}{8}$	20
	0	5	0	$-\frac{1}{2}$	1	3	30
	1	$\frac{1}{6}$	0	2	0	$-\frac{1}{4}$	25
Objective	0	$-\frac{1}{2}$	0	$-2\frac{1}{3}$	0	-.5	Z - 110

the market prices of the three products are 2, 1, and 3 respectively.

a) How many units of x1, x2, and x3 will be produced?

b) What is the total revenue of the firm?

c) What are the internal prices of labor, capital, and raw material?

d) What is the total cost?

V. Sensitivity Analysis

Linear programming deals with constrained optimization. That is, at least in the short-run, the constraints are fixed and they can't be exchanged in the market. For example, in the revenue-maximizing problem of the section I,

Maximize: $Z = 2x1 + 3x2$

Subject to: $x1 + 2x2 \leq 7$

$3x1 + x2 \leq 6$

$x1, x2 \geq 0$

we are assuming that labor and capital inputs are fixed at quantities of 7 and 6 units, respectively and the firm can't add to inputs or dispose of inputs. Input sensitivity analysis deals with relaxing the assumption of fixed constraints and analyzes the effect of a change in one constraint on the optimal solution of the problem. Let's, for example, relax the assumption of fixed capital input and assume that capital input can be increased. Given that the first input (labor) is fixed, the question is what will be the effect of increase in capital on total revenue, on outputs (X1 & X2), and what is the limit of the increase in capital. Formulating the increments in capital to the optimizing problem gives,

```
In[23]:= Clear[z, u, x1, x2]
         fu = Table[ConstrainedMax[ 2 x1 + 3 x2,
              {x1 + 2 x2 <= 7, 3 x1 + x2 <= 6 + i}, {x1, x2}], {i, 20}];
         u = N[Table[List[i, fu[[i, 1]]], {i, 20}]];
         ListPlot[u,  PlotStyle -> {RGBColor[1, 0, 0], PointSize[.03]},
          AxesLabel -> {"Capital", "Revenue"}]
```

```
Out[23] = {{56/5, {x1 → 7/5, x2 → 14/5}}, {57/5, {x1 → 9/5, x2 → 13/5}},
  {58/5, {x1 → 11/5, x2 → 12/5}}, {59/5, {x1 → 13/5, x2 → 11/5}}, {12, {x1 → 3, x2 → 2}},
  {61/5, {x1 → 17/5, x2 → 9/5}}, {62/5, {x1 → 19/5, x2 → 8/5}}, {63/5, {x1 → 21/5, x2 → 7/5}},
  {64/5, {x1 → 23/5, x2 → 6/5}}, {13, {x1 → 5, x2 → 1}}, {66/5, {x1 → 27/5, x2 → 4/5}},
  {67/5, {x1 → 29/5, x2 → 3/5}}, {68/5, {x1 → 31/5, x2 → 2/5}}, {69/5, {x1 → 33/5, x2 → 1/5}},
  {14, {x1 → 7, x2 → 0}}, {14, {x1 → 7, x2 → 0}}, {14, {x1 → 7, x2 → 0}},
  {14, {x1 → 7, x2 → 0}}, {14, {x1 → 7, x2 → 0}}, {14, {x1 → 7, x2 → 0}}}
```

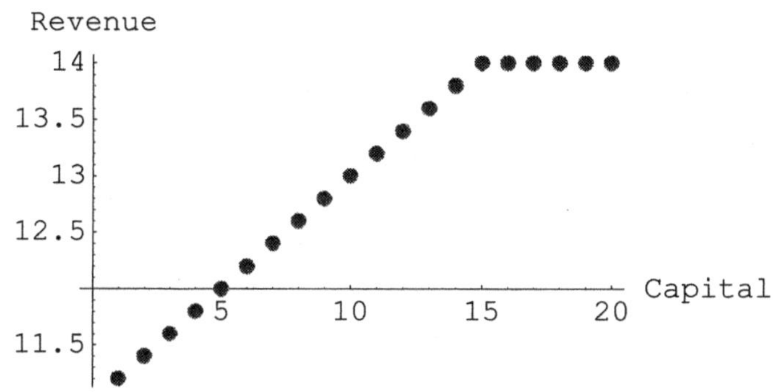

the solution and the graph above show that as capital input increases total revenue will increase. With labor input fixed, maximum total revenue is achieved when capital input is 22 units, where X1=7, X2 = 0, and the total revenue is $14.

Now let the first input (labor) be variable and assume that the second input (capital) is fixed. Formulating the problem into the constrained optimizing problem gives,

```
In[24]:= Clear[z, u, x1, x2]
         fu = Table[ConstrainedMax[ 2 x1 + 3 x2,
              {x1 + 2 x2 <= 7 + i, 3 x1 + x2 <= 6}, {x1, x2}], {i, 10}];
         u = N[Table[List[i, fu[[i, 1]]], {i, 10}]];
         ListPlot[u,  PlotStyle -> {RGBColor[1, 0, 0], PointSize[.03]},
          AxesLabel -> {"Labor", "Revenue"}]
```

```
Out[24]= {{1.,12.4},{2.,13.8},{3.,15.2},{4.,16.6},{5.,18.},{6.,18.},{7.,18.}
         {8., 18.},{9.,18.},{10.,18.}}
```

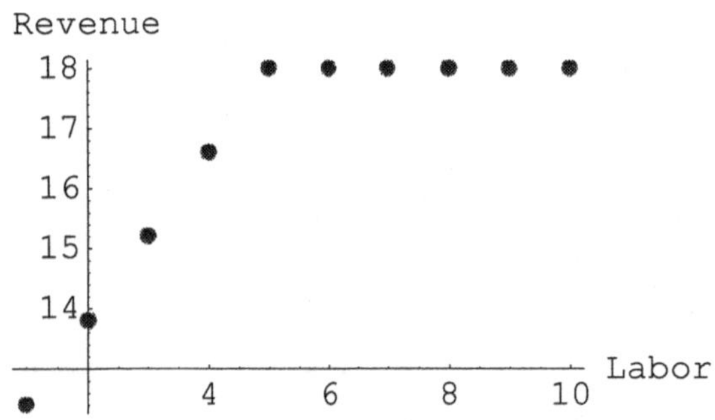

the solution and the graph above show that as labor input increases total revenue will increase. With capital input fixed, maximum total revenue is achieved when labor input is 12 units, where X1= 0, X2 = 6, and the total revenue is $18.

Problems

1. For the revenue-maximizing problem below,

Maximize: $R = 5x1 + 3x2$

Subject to: $x1 + 3x2 \leq 40$

$2x1 + x2 \leq 30$

$x1, x2 \geq 0$

where the two constraints represent labor and capital, respectively.

a- Solve the linear programming problem of maximizing revenue subject to the given constraints and find the optimum values of x1, x2 and R.

b- Find the shadow prices of labor and capital.

c- Hold the capital input constant. Find the optimum values of x1, x2 and R by adding 2 units, 4 units, and 6 units to labor input. At what level of labor input the revenue is maximized?

d- Graph the relationship between labor input and the revenue.

e- Hold the labor input constant. Find the optimum values of x1, x2 and R by adding 2 units, 4 units, and 6 units to capital input. At what level of capital input the revenue is maximized?

f- Graph the relationship between capital input and the revenue.

2. For the revenue-maximizing problem below,

Maximize: $R = 4x1 + x2$

Subject to: $x1 + 3x2 \leq 40$

$2x1 + x2 \leq 30$

$x1, x2 \geq 0$

where the two constraints represent labor and capital, respectiveyl.

a- Solve the linear programming problem of maximizing revenue subject to the given constraints and find the optimum values of x1, x2 and R.

b- Find the shadow prices of labor and capital.

c- Hold the capital input constant. Find the optimum values of x1, x2 and R by adding 2 units, 4 units, and 6 units to labor input. At what level of labor input the revenue is maximized?

d- Graph the relationship between labor input and the revenue.

e- Hold the labor input constant. Find the optimum values of x1, x2 and R by adding 2 units, 4 units, and 6 units to capital input. At what level of capital input the revenue is maximized?

f- Graph the relationship between capital input and the revenue.

Appendix: Solution Tables of the Dual Problem

Minimize: $C = 15y1 + 40y2 + 24y3$

Subject to:

$$6y1 + 3y2 + 2y3 \geq 1$$

$$2y1 + 5y2 + 3y3 \geq 3$$

$$y1, y2, y3 \geq 0$$

Note: Pivot columns are highlighted yellow, pivot rows are highlight green.

Su1 and su2 are surplus variables, A1 and A2 are artificial variables.

Table 1. The Minimizing Problem

Variable:	y1	y2	y3	su1	su2	A1	A2	Constant
	6	3	2	-1	0	1	0	36
	2	5	3	0	-1	0	1	20
Objective	15	40	24	0	0	1	1	W

Table 1 of Phase I

Variable:	y1	y2	y3	su1	su2	A1	A2	Constant
	2	1	$\frac{2}{3}$	$-\frac{1}{3}$	0	$\frac{1}{3}$	0	$\frac{1}{3}$
	-8	0	$\frac{-1}{3}$	$\frac{5}{3}$	-1	$\frac{-5}{3}$	1	$\frac{4}{3}$
Objective	-65	0	$\frac{-8}{3}$	$\frac{40}{3}$	0	$\frac{-40}{3}$	0	$W-\frac{40}{3}$

Table 2 of Phase I

Variable:	y1	y2	y3	su1	su2	A1	A2	Constant
	$\frac{2}{5}$	1	$\frac{3}{5}$	0	$-\frac{1}{5}$	0	$\frac{1}{5}$	$\frac{3}{5}$
	$-\frac{24}{5}$	0	$\frac{-1}{5}$	1	$-\frac{3}{5}$	-1	$\frac{3}{5}$	$\frac{4}{5}$
Objective	-1	0	0	0	8	0	-8	W - 24

The Phase one terminates here with A1 = A2 = W = 0

Table 1 of Phase II

Variable:	y1	y2	y3	su1	su2	Constant
	$\frac{2}{5}$	1	$\frac{3}{5}$	0	$-\frac{1}{5}$	$\frac{3}{5}$
	$-\frac{24}{5}$	0	$\frac{-1}{5}$	1	$-\frac{3}{5}$	$\frac{4}{5}$
Objective	-1	0	0	0	8	C - 24

Table 2 of Phase II, The Final Table

Variable:	y1	y2	y3	su1	su2	Constant
	1	$\frac{5}{2}$	$\frac{3}{2}$	0	$-\frac{1}{2}$	$\frac{3}{2}$
	0	12	7	1	-3	8
Objective	0	$\frac{5}{2}$	$\frac{3}{2}$	0	$\frac{15}{2}$	C - $\frac{45}{2}$

The optimum solution is y1 = $\frac{3}{2}$, y2 = y3 = 0, and C = $\frac{45}{2}$. The solution to the dual problem reveals the solution of the primal also. The values under su1 and su2 in the objective row are the values of x1 and x2, respectively. That is, x1 = 0 and x2 = $\frac{15}{2}$. Since x1and x2 are valued at \$1 and \$3 respectively, Z = (0) 1 + ($\frac{15}{2}$) 3 = $\frac{45}{2}$. That is, in the final solution C = Z = $\frac{45}{2}$.

CHAPTER 10

APPLICATION OF LINEAR PROGRAMMIG

Linear programming has found many applications in business and economics. Even a summary of important applications would require several pages. Economic models of the firm, theory of production, and others have been reformulated using linearity assumptions and linear programming techniques. Two major areas in which linear programming has made significant impacts are input-output analysis and the theory of games. In this chapter input-output systems and the theory of games are studied.

TOOL BOX

Mathematica Commands	Function
ConstrainedMax[2x1+3x2, {x1+2x2 <= 7,3x1+x2 <= 6},{x1,x2}]	Solves the linear programming problem Max: Z =2x1+3x2, Subject To: x1+ 2x2 $\leq$ 7 and 3x1 + x2 $\leq$ 6.
Solve[{D[z, x]==0, D[z, y]==0},{x, y}]	Solves $\frac{\partial Z}{\partial x}$=0 and $\frac{\partial Z}{\partial y}$ = 0 for x and y simultaneously.
N[MatrixForm[$\begin{pmatrix} a11 & a12 \\ a21 & a22 \end{pmatrix} * \begin{pmatrix} 1/a1 & 1/a2 \\ 1/a1 & 1/a2 \end{pmatrix}$]]	Finds $\begin{pmatrix} \frac{a11}{a1} & \frac{a12}{a2} \\ \frac{a21}{a1} & \frac{a22}{a2} \end{pmatrix}$

Note that, star product of matrices using "*" is different from dot product using "." (See, Chapter 4)

I. Input-Output Systems: The **ConstrainedMax** Command

The Leontief's input-output system was designed to be a simplified model of the economy, describing the interrelationships within the producing sectors and the relationships between producing and consuming sectors of the economy. (See, Wassily W. Leontief, The Structure of American Economy 1919-1939, 2nd ed., Oxford University Press, Fair Lawn, N.J., 1951.)

Every industry in the economy buys variable inputs from other industries and sells variable inputs to other industries and final goods to the consuming sector. The producing sector also sells to government and to the foreign countries. The main objective of the input-output analysis is to determine the level of output that each industry should produce in order to satisfy the total demand for the product of the industry in the economy.

The following table illustrates the input-output model of an economy producing n goods with an available labor force of L_o units. For the simplicity of exposition, however, sales to the government and foreign countries will be regarded as consumer goods. (Input-output models represent considerable simplifying assumptions, such as treating consumer demand as exogenous; the industry, rather than the firm being the unit of production; and assuming simple production technology so that each industry uses a single linear production process to produce a single output.)

Table 1. Input-Output Model of the Economy

To Industry :

From Industry :	1	2	.	.	j	.	.	n	Unsold Inventory	Consumption	Total
1	x11	x12	.	.	x1j	.	.	x2n	u1	c1	x1
2	x21	x22	.	.	x2j	.	.	x2n	u2	c2	x2
.	.	.	.	.	.	.	.	.	.	.	.
.	.	.	.	.	.	.	.	.	.	.	.
i	xi1	xi2	.	.	xij	.	.	xin	ui	ci	xi
.	.	.	.	.	.	.	.	.	.	.	.
.	.	.	.	.	.	.	.	.	.	.	.
n	xn1	xn2	.	.	xnj	.	.	xnn	un	cn	xn
Labor	L_1	L_2	.	.	L_j	.	.	L_n	L_{un}	C	L_o

In this table, x_{ij} represents the quantity of goods sold by industry i to industry j; u_i is the unsold inventory of industry i; C_i is the quantity of goods sold by industry i to consumers; and x_i is the total output of industry i. Labor input of the industry i is denoted by L_i, and L_{un} represents unemployment.

Let $a_{ij} = \frac{xij}{xj}$, for j = 1, ..., n. a_{ij} is the amount of good i needed to produce a unit of good j. The a_{ij} coefficients, which are called **input-output coefficients**, are assumed to be constant and do not change over time, unless, there is a significant technological change in the industry. Therefore, table1 can be converted to table 2, below, where the a_{ij}'s are input-output coefficients, and l_i is the amount of labor needed to produce a unit of good i, $l_i = \frac{Li}{xi}$.

Table 2. Input-Output Model of the Economy with Input-Output Coefficients

To Industry :

From Industry :	1	2	.	.	j	.	.	n	Unsold Inventory	Consumption	Total
1	a11	a12	.	.	a1j	.	.	a2n	u1	c1	x1
2	a21	a22	.	.	a2j	.	.	a2n	u2	c2	x2
.	.	.	.	.	.	.	.	.	.	.	.
.	.	.	.	.	.	.	.	.	.	.	.
i	ai1	ai2	.	.	aij	.	.	ain	ui	ci	xi
.	.	.	.	.	.	.	.	.	.	.	.
.	.	.	.	.	.	.	.	.	.	.	.
n	an1	an2	.	.	anj	.	.	ann	un	cn	xn
Labor	l_1	l_2	.	.	l_j	.	.	l_n	L_{un}	C	L_o

In this form, the Leontief's input-output system is a useful tool for planning and resource allocation. The model is mostly a descriptive and predictive model of the economy and does not have a criteria for optimization. However, if combined with linear programming techniques, it can be used to optimize an objective function subject to certain constraints.

Using the matrix notations of the chapter 4, the production and consumption activity of table 2 can be summarized as, $\mathbf{Ax} + \mathbf{U} + \mathbf{C} = \mathbf{x}$ or $(\mathbf{I} - \mathbf{A})\mathbf{x} = \mathbf{U} + \mathbf{C}$, and the labor activity (the last row of the table) can be summarized as $\mathbf{l'x} + l_{un} = L_o$, where $\mathbf{A} = \begin{pmatrix} a11 & a12 & . & . & a1n \\ a21 & a22 & . & . & a2n \\ . & . & . & . & . \\ . & . & . & . & . \\ an1 & an2 & . & . & ann \end{pmatrix}$, $\mathbf{x} = \begin{pmatrix} x1 \\ x2 \\ . \\ . \\ xn \end{pmatrix}$, $\mathbf{U} = \begin{pmatrix} u1 \\ u2 \\ . \\ . \\ un \end{pmatrix}$, $\mathbf{C} = \begin{pmatrix} c1 \\ c2 \\ . \\ . \\ cn \end{pmatrix}$, $\mathbf{l} = \begin{pmatrix} l1 \\ l2 \\ . \\ . \\ ln \end{pmatrix}$, and L_{un} and L_o are scalars.

The following example illustrates the use of input-output analysis to predict the total output and resource requirements for a given change in consumer demand.

Example: The input-output table of a two-sector economy is given as,

To Industry:

From Industry	1	2	Consuption	Total
1	15	10	25	50
2	20	8	12	40
Labor	30	10	□	40

it is expected that the demand for the products of industries 1 and 2 to increase by 5% and 15% respectively. Find the level of output that will just be sufficient to satisfy the consumer demand (no inventory). Find also the required labor input.

Solution: The input-output coefficients can be derived by finding $a_{ij} = \frac{xij}{xj}$. This will give the input-output coefficients as $\begin{pmatrix} \frac{15}{50} & \frac{10}{40} \\ \frac{20}{50} & \frac{8}{40} \end{pmatrix} = \begin{pmatrix} .30 & .25 \\ .40 & .20 \end{pmatrix}$. The labor input requirements per unit of output is $(\frac{30}{50} \quad \frac{10}{40})$. Therefore, the input-output model can be written as,

To Industry:

From Industry	1	2	Consuption	Total
1	.30	.25	25	50
2	.40	.20	12	40
Labor	.60	.25	□	40

with the expected increase in demand for the products of the two industries, the consumption will change to $\mathbf{C^*} = \begin{pmatrix} 25 * 1.05 \\ 12 * 1.15 \end{pmatrix}$ or $\begin{pmatrix} 26.25 \\ 13.80 \end{pmatrix}$. Therefore, the total production to satisfy the increased demand and to feed the inter-industry needs can be derived by solving $(\mathbf{I} - \mathbf{A}).\mathbf{x} = \mathbf{C^*}$ or $[\begin{pmatrix} 1 & 0 \\ 0 & 1 \end{pmatrix} - \begin{pmatrix} .30 & .25 \\ .40 & .20 \end{pmatrix}.\begin{pmatrix} x1 \\ x2 \end{pmatrix}] = \begin{pmatrix} 26.25 \\ 13.80 \end{pmatrix}$. The solution of this equation results in $\begin{pmatrix} x1 \\ x2 \end{pmatrix} = \begin{pmatrix} 116.30 \\ 115.65 \end{pmatrix}$.

To find the labor requirement for the new production level, multiply the per unit labor requirement by total output, $L = \mathbf{l'x} = (.60 \quad .25).\begin{pmatrix} 116.30 \\ 115.65 \end{pmatrix} = 98.70$. That is, to satisfy the new consumption level, 116.30 units of x1 and 115.65 units of x2 is needed with a labor input of 99 units.

To solve the problem using *Mathematica*, define the variables and solve the $(\mathbf{I} - \mathbf{A})\mathbf{x} = \mathbf{C^*}$ equation for x1 and x2 as follows,

```
In[1]:= Clear[x1, x2, A]
A = (15 10
     20  8) * (1/50 1/40
               1/50 1/40);

sol = Solve[(IdentityMatrix[2] - A) . {x1, x2} == {50 * 1.05,
   40 * 1.15}, {x1, x2}]
```

to find the labor requirement, find **l'.x** as follows,

```
In[2]:= ( .60  .25) . (x1
                       x2) /. {sol}

Out[2]= {{{{0.6 x1 + 0.25 x2}}}}
```

The necessary and sufficient conditions for the existence of a general solution to an input-output system is provided by the Hawkins-Simon conditions. (See, David Hawkins and Herbert A. Simon, "Note: Some Conditions of Macroeconomic Stability," Econometrica, Vol. 17, July-October, 1949, PP. 245-48.) The Hawkins-Simon conditions require that all of the principal minors of the coefficient matrix **(I - A)** to be non negative. That is, $(1 - a11) \geq 0$, $\text{Det}\begin{pmatrix} 1 - a11 & -a12 \\ -a21 & 1 - a22 \end{pmatrix} \geq 0$, $\text{Det}\begin{pmatrix} 1 - a11 & -a12 & -a13 \\ -a21 & 1 - a22 & -a23 \\ -a31 & -a32 & 1 - a33 \end{pmatrix} \geq 0$, .., and $\text{Det}[(\mathbf{I} - \mathbf{A})] \geq 0$.

The first and subsequent inequalities above require that $a_{ii} < 1$, (i = 1, ..., n). If $a_{ii} \geq 1$ for some i, one or more units of i would be required to produce one unit of i. No output can take place under such conditions.

The preceding example illustrates the descriptive and predictive features of the input-output analysis. Combined with the linear programming technique, the input-output analysis can be used to include the optimizing criteria.

Allowing for the possibilities of inventories and unemployment, the two equations of the input-output system can be written as,

$$(\mathbf{I} - \mathbf{A}).\mathbf{x} \geq \mathbf{C}$$

$$\mathbf{l}'.\mathbf{x} \leq L_o$$

These two equations are the constraints based on the inter-industry input requirements, final consumer demand, and the resource requirements. The objective can be set as maximizing gross national product, GNP = **P'.x**, where **P** is a vector of n average prices for outputs of the industries, **P'** = (p1 . . . pn). Therefore, the optimization problem can be written as:

Maximize: $\text{GNP} = \mathbf{P}'.\mathbf{x}$

Subject to: $(\mathbf{I} - \mathbf{A}).\mathbf{x} \geq \mathbf{C}$

$$\mathbf{l}'.\mathbf{x} \leq L_o$$

$$\mathbf{x} > \mathbf{0}$$

Example: Suppose that a simple economy is described by the following input-output model.

To Industry:

From Industry :	1	2	3	Consumption	Unsold Inventory	Total
1	24	10	14	30	2	80
2	32	20	7	40	1	100
3	40	5	14	10	1	70
Labor	60	80	42	□	18	200

It is expected that demand for the three products will increase by 10% for the next year, while the labor input will increase by 1%. If the unit prices for the products are $15, $25, and $10 respectively, set up the problem of maximizing GNP subject to the constraints, and test for the existence of a solution.

Solution: Find the input-output coefficients by finding $a_{ij} = \frac{xij}{xj}$ to convert the input-output table to the table of input-output coefficients. The conversion yields a matrix of input-output coefficients $\mathbf{A} = \begin{pmatrix} \frac{24}{80} & \frac{10}{100} & \frac{14}{70} \\ \frac{32}{80} & \frac{20}{100} & \frac{7}{70} \\ \frac{40}{80} & \frac{5}{100} & \frac{14}{70} \end{pmatrix} = \begin{pmatrix} .30 & .10 & .20 \\ .40 & .20 & .10 \\ .50 & .05 & .20 \end{pmatrix}$ and a vector of per unit labor requirements $\mathbf{l'} = (\frac{60}{800} \;\; \frac{80}{100} \;\; \frac{42}{200})$. The input-output model can be written as,

To Industry:

From Industry :	1	2	3	Consumption	Unsold Inventory	Total
1	.30	.10	.20	30	2	80
2	.40	.20	.10	40	1	100
3	.50	.05	.20	10	1	70
Labor	.75	.80	.60	□	18	200

For the coming period, demand is expected to be $\mathbf{C^*} = 1.10^*\mathbf{C} = (33 \;\; 44 \;\; 11)$ and the labor force is expected to be $L_o = 202$. The problem can be formulated as: (**Note** that Inventories are not explicitly included in the formulation. However, they are included implicitly by setting inequalities **(I - A)x ≥ C**, rather than **(I - A)x = C.)**

Maximize: $(15 \;\; 25 \;\; 10) . \begin{pmatrix} x1 \\ x2 \\ x3 \end{pmatrix}$

Subject To: $\left(\begin{pmatrix} 1 & 0 & 0 \\ 0 & 1 & 0 \\ 0 & 0 & 1 \end{pmatrix} - \begin{pmatrix} .3 & .1 & .2 \\ .4 & .2 & .1 \\ .5 & .05 & .2 \end{pmatrix} \right) . \begin{pmatrix} x1 \\ x2 \\ x3 \end{pmatrix} \geq \begin{pmatrix} 33 \\ 44 \\ 11 \end{pmatrix}$

$$(.75 \;\; .8 \;\; .6) . \begin{pmatrix} x1 \\ x2 \\ x3 \end{pmatrix} \leq 202$$

$$x1, x2, x3 \geq 0$$

This is a linear programming problem that can be solved through Simplex method discussed in the previous chapter. The solution yields x1 = 85.07, x2 = 117.06, and x3 = 74.24, with a GNP of 4949.07.

Testing for the existence of a solution, the three principal minors of the **(I - A)** matrix are, (1 - .3) = .7 > 0, $\text{Det}\begin{pmatrix} .7 & -.1 \\ -.4 & .8 \end{pmatrix} = .52 > 0$, and Det[(**I** - **A**)] = .323 > 0. Since all the three principal minors are positive, the system has a solution with all three outputs positive.

To solve the problem using *Mathematica*, set up the linear programming problem, define the variables, and use the **ConstrainedMax** command.

```
In[3]:= Clear[x1, x2, x3, A]
        A = (24 10 14)   (1/80 1/100 1/70)
            (32 20  7) * (1/80 1/100 1/70) ;
            (40  5 14)   (1/80 1/100 1/70)
        MatrixForm[IdentityMatrix[3] - A]
```

$$\text{Out[3]=} \begin{pmatrix} \frac{7}{10} & -\frac{1}{10} & -\frac{1}{5} \\ -\frac{2}{5} & \frac{4}{5} & -\frac{1}{10} \\ -\frac{1}{2} & -\frac{1}{20} & \frac{4}{5} \end{pmatrix}$$

```
In[4]:=
        ConstrainedMax[15 x1 + 25 x2 + 10 x3, {.7 x1 - .1 x2 - .2 x3 >= 33,
          -.4 x1 + .8 x2 - .1 x3 >= 44, -.5 x1 - .05 x2 + .8 x3 >= 11,
          .75 x1 + .8 x2 + .6 x3 <= 202}, {x1, x2, x3}]

Out[4]= {4945.07, {x1 → 85.0771, x2 → 117.061, x3 → 74.2395}}
```

Problems

1- For the following input-output system, find the input-output coefficients and determine whether the system has a general solution. Suppose that consumers will be increasing their consumption of the three goods by 8% . Determine the output levels and the labor needed to produce them.

To Industry:

From Industry :	1	2	3	Consumption	Unsold Inventory	Total
1	160	100	420	100	20	800
2	400	400	70	120	10	1000
3	80	400	140	50	30	700
Labor	500	600	400	□	100	1600

2- The following input-output model represents an economy that the consumption for goods is growing at an annual rate of 5% and the labor force is growing at an annual rate of 1.5% . Given that prices are, P1 = 3, P2 = 8, and P3 = 4, find the GDP, production levels and the consumption levels of the three goods for the next year.

To Industry:

From Industry :	1	2	3	Consumption	Unsold Inventory	Total
1	360	200	360	250	30	1200
2	600	100	180	100	20	1000
3	360	300	90	140	10	900
Labor	1000	850	700	□	50	2600

3- The following input-output model represents an economy that the consumption of good 1 is growing at an annual rate of 5% and the consumption of good 2 is growing at a rate of 15%. The labor force is growing at an annual rate of 1.5% . Given that prices are, P1 = 24, and P2 = 18, find the GDP, production levels and the consumption levels of the three goods for the next year.

To Industry:

From Industry :	1	2	Consumption	Unsold Inventory	Total
1	640	225	700	35	1600
2	480	150	100	20	750
Labor	800	325	□	25	1150

4- The following input-output model represents an economy that the consumption for good 1, 2 , and 3 are growing at an annual rate of 5%, 8%, and 12%, respectively. The labor force is growing at an annual rate of 1.8% . Given that prices are, P1 = 10, P2 = 8, and P3 = 14, find the GDP, production levels and the consumption levels of the three goods for the next year.

To Industry:

From Industry :	1	2	3	Consumption	Unsold Inventory	Total
1	340	180	320	220	18	1078
2	540	85	160	80	20	885
3	340	320	80	130	12	882
Labor	960	850	720	□	54	2584

II. Theory of Games: The **Solve**[{D[z, x]==0, D[z, y]==0},{x, y}] Command

In 1944, Von Neumann and Morgenstern published their Theory of Games and Economic Behavior. (See, Von Neumann, J. and O. Morgenstern, 1964, Theory of Games and Economic Behavior, 3rd ed., Science Edition, Wiley, New York.) The main topic of the book was to determine standards of "rational behavior" in situations in which the outcomes depend on the actions of interdependent players. The players may behave cooperatively or competitively. In each case, the concept of solution is a set of rules for each player which tells the player how to behave in every situation which may arise.

Experience with game theory models shows that the theory of games is more relevant to the study of specific business and economic problems than the classical approach and solutions. One of the topics in economics in which the interaction of decisions has an important role is imperfect competition, situations such as bilateral monopoly, duopoly, and oligopoly.

Two-Person, Zero-Sum Games: Every game consists of a number of strategies on the part of each of its participants. A **strategy** is the specification of a particular move for one of the participants. Each player has a finite number of strategies under her control. The outcome of a game, i.e., the profit or the utility earned by each participant, is determined by the interaction of the selected strategies by participants. A game may be a one-person, two-person, or n-person game. It may also be a zero-sum game or non-zero-sum game.

A **zero-sum game** is one in which the algebraic sum of the outcomes for all the participants equals zero for every possible combinations of strategies. That is, the gain of one player equals the loss of other players. If the net outcome of a game is other than zero for at least one strategy combination, it is called a **non-zero-sum game**. The analysis below is restricted to two-person zero-sum games.

Suppose that the payoff matrix for player I is given by the following table,

Payoff Matrix of Player I

Player II

		B1	B2	.	.	Bj	.	.	Bm	Min. Row
	A1	a11	a12	.	.	a1j	.	.	a1m	a1
	A2	a21	a22	.	.	a2j	.	.	a2m	a2
Player I	.	.	.	.	.	.	.	.	.	.
	Ai	ai1	ai2	.	.	aij	.	.	aim	ai
	.	.	.	.	.	.	.	.	.	.
	An	an1	an2	.	.	anj	.	.	anm	an
	Max. Column	b1	b2	.	.	bj	.	.	bm	

player I chooses from strategies A1, ..., An, and player II chooses from strategies B1, ..., Bm. a_{ij} is the payoff to player I if he chooses Ai strategy when player II selects Bj strategy. The last column in the table shows the minimum value across row i and the last row shows the maximum value down column j.

The game is said to have a **saddle point**, if there is a strategy for each player at which the game settles. If the game has a saddle point, the value of the game is determined by the **minimax criterion**. The minimax criterion is based on the following relationship, $a^* = \text{Max}_i\ (\text{Min}_j a_{ij}) \leq Z \leq \text{Min}_j\ (\text{Max}_i a_{ij}) = b^*$, where, a* is the maximum value among a_i, b* is the minimum value among b_j, and Z is the value of the game.

This relationship holds because $\text{Min}_j a_{ij}$ is a vector of the smallest values, of which the largest is $\text{Max}_j\ (\text{Min}_j a_{ij})$. On the other hand, $\text{Max}_i a_{ij}$ is a vector of the large values, of which the smallest is $\text{Min}_j\ (\text{Max}_i a_{ij})$. If a* = b*, $\text{Max}_i\ (\text{Min}_j a_{ij}) = Z = \text{Min}_j(\text{Max}_i a_{ij})$ and the game has a saddle point, with the value of the game equal to Z.

Example: The payoff matrix in dollars for player I is given by the following table. Find the player I's payoff.

Payoff Matrix of Player I

Player II

		B1	B2	B3	Min. Row
	A1	8	6	5	5
Player I	A2	9	7	8	[7]
	A3	6	5	4	4
	Max. Column	9	[7]	8	

Solution: By the minimax criterion, the solution exists and the saddle point is at the intersection of strategies A2 and B2, with a payoff of $7 for player I.

This solution is arrived at by the minimax criterion. However, if different options of player I are tested for, the solution will be no different from that of the minimax criterion. For example, player I may select strategy A1 to receive a payoff of 8, the largest payoff in that strategy. However, player II's reaction will be to select B3, to give the player I a payoff of 5, the smallest payoff in row 1. Player II's strategy of B3 will force player I to select strategy A2, to get a payoff of 8. This will make player II move to strategy B2 to give player I a payoff of 7. With player II taking strategy B2, player I will not have any incentive to move away from strategy A2. Any further move by player I will hurt that player and will benefit player II. Any further move by player II will also not benefit that player. It will hurt player II and will benefit player I. Therefore, A2B2 is the saddle point.

Example: The payoff matrix in market shares for player I is given by the following table. Find the solution of the game.

Payoff Matrix of Player I

Player II

		B1	B2	B3	Min. Row
Player I	A1	.7	.3	.5	.3
	A2	.6	.4	.8	.4
	A3	.5	.6	.9	.5
	Max. Column	.7	.6	.9	

Solution: This game does not have a saddle point. However, a solution may be found using the mixed strategies of the next section.

Mixed Strategies: Players may select pure strategies and play. The game, however, may or may not have a solution. If the game does not have a saddle point, players will select their strategies on a probabilistic basis. Each player will randomize the strategy selection by assigning probabilities to strategies. The players are now concerned with the expected, rather than the actual payoff. The following example illustrates the case.

Example: The payoff matrix for a game is given below. Find the solution of the game.

Payoff Matrix of Player I

Player II

		B1	B2	Min. Row
Player I	A1	4	2	2
	A2	1	3	1
	Max. Column	4	3	

Solution: This game does not have a saddle point. Now, suppose that player I selects strategy A1, x% of times, and strategy A2, (1 - x)% of times. Player II selects strategy B1, y% of times, and strategy B2, (1 - y)% of times. Since strategies are selected randomly, player I can not anticipate player II's choice of strategy, even if she knows y, and vice versa.

Based on the given probabilities, each payoff depends on the joint probability of selected strategies by the two players. The following table shows the payoffs and their probabilities.

Payoff Matrix of Player I

Player II

			B1	B2
		Probabilities	y	(1 - y)
Player I	A1	x	4 xy	2 x (1 - y)
	A2	(1 - x)	1 (1 - x) y	3 (1 - x) (1 - y)

The expected payoff to player I is the sum of the payoffs times their corresponding probabilities,

$$Z(x,y) = 4xy + 2x(1 - y) + (1 - x)y + 3(1 - x)(1 - y)$$

Player I's objective is to maximize Z(x,y) with respect to x, and for a given y. This is achieved by finding derivative of Z(x, y) with respect to x, setting it equal to zero, and solving it for y.

$$\frac{\partial Z}{\partial x} = 4y + 2(1 - y) - y - 3(1 - y) = 0, \quad \text{or } y = \frac{1}{4}$$

Player II's objective is to minimize Z(x,y) with respect to y, and for a given x. This is achieved by finding derivative of Z(x, y) with respect to y, setting it equal to zero, and solving for x.

$$\frac{\partial Z}{\partial y} = 4x - 2x + (1 - x) - 3(1 - x) = 0, \quad \text{or } x = \frac{1}{2}$$

Substitute for x and y in Z(x, y) to get the payoff $Z = \frac{5}{2}$. The solution implies that player I will adopt strategy A1 50% of times and strategy A2 50% of times in a given period to maximize her payoff, regardless of what strategies player II selects. To minimize player I's payoff, player II will adopt strategy B1 and B2 at the rates of 25% and 75%, respectively, regardless of what strategy player I selects.

To solve the problem using *Mathematica*, define the payoff equation, set its derivatives with respect to x and y equal to zero, and solve the resulting equations simultaneously for x and y. To find the payoff for player I, evaluate Z for the solution.

```
In[5]:= Clear[x, y, z, game]
        z = 4 x y + 2 x (1 - y) + (1 - x) y + 3 (1 - x) (1 - y);
        game = Solve[{D[z, x] == 0, D[z, y] == 0}, {x, y}]

Out[5]= {{Null (x → 1/2), Null (y → 1/4)}}

In[6]:= z /. {game}

Out[6]= {{5/2}}
```

The mixed strategies solution is the maximum that player I will receive if player II plays optimally and both players have complete information. The solution assures player I of a minimum level of return no matter how player II plays. However, if player II does not play optimally, then player I will receive more.

Example: The game in the following table does not have a saddle point. Solve the mixed-strategy game and find the payoff of player I.

Payoff Matrix of Player I

Player II

			B1	B2	B3	Min. Row
		Probabilities	y1	y2	(1 - y1 - y2)	
Player I	A1	x1	5	2	3	2
	A2	x2	1	3	4	1
	A3	(1 - x1 - x2)	2	6	3	2
	Max. Column		5	6	4	

Solution: To find the mixed strategies solution, assume that player I selects A1, A2, and A3 with probabilities x1, x2, and (1 - x1 - x2), respectively, and player II selects B1, B2, and B3 with probabilities y1, y2, and (1 - y1 - y2), respectively. The payoff for the player I is Z(x1, x2, y1, y2) = 5x1y1 +2x1y2 + 3x1(1 - y1 - y2) + x2y1 + 3x2y2 + 4x2(1 - y1 - y2) + 2(1 - x1 - x2)y1 + 6(1 - x1 - x2)y2 + 3(1 - x1 - x2)(1-y1-y2). Player I maximizes Z assuming given levels of y1 and y2. That is, she sets $\frac{\partial Z}{\partial x1} = \frac{\partial Z}{\partial x2} = 0$ and solves them for y1 and y2. That is,

$$\frac{\partial Z}{\partial x1} = 3y1 - 4y2 = 0$$

$$\frac{\partial Z}{\partial x2} = 1 - 2y1 - 4y2 = 0$$

Solving these two equations together results in y1 = $\frac{1}{5}$ and y2 = $\frac{3}{20}$. Player II minimizes Z assuming given levels of x1 and x2. That is, she sets $\frac{\partial Z}{\partial y1} = \frac{\partial Z}{\partial y2}$ = 0 solves them for x1 and x2. That is,

$$\frac{\partial Z}{\partial y1} = -1 + 3x2 + 2x2 = 0$$

$$\frac{\partial Z}{\partial y2} = 3 - 4x2 - 4x2 = 0$$

Solving the above two equations together results in x1 = $\frac{1}{2}$ and x2 = $\frac{1}{4}$. The mixed strategies solution to the game is found by substituting for x1, x2, y1, and y2 in the Z equation,. The payoff of player I is found to be Z = $\frac{13}{4}$.

To solve the problem using *Mathematica*, formulate Z, find its derivatives with respect to x1, x2, y1, y2, and set them equal to zero. Solve the equations simultaneously for x1, x2, y1, and y2.

```
In[7]:= Clear[x1, x2, y1, y2, z, game2]
        z = 5 x1 y1 + 2 x1 y2 + 3 x1 (1 - y1 - y2) + x2 y1 + 3 x2 y2 + 4 x2 (1 - y1 - y2)
          + 2 (1 - x1 - x2) y1 + 6 (1 - x1 - x2) y2 + 3 (1 - x1 - x2) (1 - y1 - y2);
        game2 = Solve[{D[z, x1] == 0, D[z, x2] == 0,
          D[z, y1] == 0, D[z, y2] == 0}, {x1, x2, y1, y2}]
```

Out[7]= {{Null (x1 → $\frac{1}{2}$), Null (x2 → $\frac{1}{4}$), Null (y1 → $\frac{1}{5}$), Null (y2 → $\frac{3}{20}$)}}

To find Z,evaluate Z at the solution set of the equations.

```
In[8]:= z /. {game2}
```

Out[8] = {{$\frac{13}{4}$}}

The mixed strategies solution to the game is where the player I adopts strategies A1, 50% of times, A2, 25% of times, and A3, 25% of times. The payoff for player I is Z = $\frac{13}{4}$. This is the maximum that player I will receive if player II plays optimally, taking B1, B2, and B3 with probabilities of 20%, 15%, and 65%, respectively. However, if player II does not play optimally, say for example adopts B1and stays with that strategy only, then player I can play the pure strategy A1 and receive a payoff of 5, which is more than the mixed strategies payoff of Z = $\frac{13}{4}$.

It is important to note that the payoff of the mixed strategies solution ($\frac{13}{4}$) lies between the minimax solution (4), which is the minimum payoff in the "Max. Column" of the payoff matrix above and the maximin solution (2), which is the maximum payoff in the "Min. Row" column of the payoff matrix above.

Dominated Strategies: Occasionally, the solution of a mixed strategies game can be simplified by deleting dominated or inferior strategies. A dominated strategy for player I is the one for which all the elements of the strategy are no more than the corresponding elements of some other strategy. A dominated strategy for player II is the one that all elements of the strategy are no less than the corresponding elements of some other strategy.

It is obvious that player I's objective is to select strategies that will maximize her payoff. Therefore, if all elements of a strategy are less than or equal to the corresponding elements of another strategy, player I will never play or select the strategy with lower payoffs. On the contrary, player II's objective is to adopt strategies that will minimize player I's payoff. Therefore, if all elements of a strategy are more than or equal to the corresponding elements of other strategy, player II will never select the strategy with higher payoffs for player I.

Deleting dominated strategies will not effect the saddle point or the solution of mixed strategies games. The following example illustrates the point.

Payoff Matrix of Player I

Player II

Player I

	B1	B2	B3	B4	Min. Row
A1	10	16	20	13	10
A2	16	18	19	18	16
A3	13	1	0	8	0
A4	17	6	9	8	6
Max. Column	17	18	20	18	

Solution: This game does not have a saddle point. A mixed strategies solution without deleting the dominated strategies will be cumbersome. However, a close look at the table reveals that all the elements of strategy A3 are less than or equal to the corresponding elements of A4. For player I this strategy is dominated by strategy A4 and can be deleted. Once A3 is deleted, all the elements of B3 turn out to be greater than B4. Therefore, player II will never play strategy B4. For player II, strategy B3 is a dominated strategy and can be deleted. The reduced payoff matrix will be as follows,

Reduced Payoff Matrix of Player I

Player II

Player I

	B1	B2	B4	Min. Row
A1	10	16	13	10
A2	16	18	18	16
A4	17	6	8	6
Max. Column	17	18	18	

The reduced payoff matrix can be reduced further more. All elements of A1 are less than A2. For player I strategy A1 is dominated by A2 and can be deleted. Once A1 is deleted, elements of B4 become greater than or equal to elements of B2. Therefore, for player II strategy B4 is a dominated strategy and can be deleted. The reduced game matrix and the probabilities of each payoff are as follows,

Payoff Matrix of Player I

Player II

Player I

		B1	B2
	Probablities	y	(1 – y)
A2	x	16	18
A4	(1 – x)	17	6

assuming that player I selects strategies A1 and A2 with probabilities x and (1 - x) respectively, and player II selects strategies B1 and B2 with probabilities y and (1 - y) respectively, the payoff matrix can be found as:

$Z = 16zy + 16x(1 - y) + 17y(1 - x) + 6(1 - x)(1 - y)$

To find the optimal solution, maximize z with respect to x by setting $\frac{\partial Z}{\partial x} = 12 - 13y = 0$, or $y = \frac{12}{13}$ and minimize Z with respect to y by setting $\frac{\partial Z}{\partial y} = 11 - 13x = 0$, or $x = \frac{11}{13}$. Substitute for x and y in the payoff equation Z to find $Z = \frac{210}{13}$.

To solve the problem using *Mathematica*, formulate Z and find the optimal solution as follows,

```
In[9]:= Clear[x, y, z, game3]
        z = 16 x y + 18 x (1 - y) + 17 (1 - x) y + 6 (1 - x) (1 - y);
        game3 = Solve[{D[z, x] == 0, D[z, y] == 0}, {x, y}]
```

```
Out[9]= {{Null (x → 11/13), Null (y → 12/13)}}
In[10]:= z /. {game3}
Out[10]= {{210/13}}
```

Problems

1- For the following game matrices, find the saddle point and strategies that provide the saddle point. Find the payoff of each game.

a- Payoff Matrix of Player I

Player I / Player II

	B1	B2
A1	4	8
A2	3	5

b- Payoff Matrix of Player I

Player I / Player II

	B1	B2	B3	B4
A1	12	7	9	12
A2	8	6	11	22

c- Payoff Matrix of Player I

Player I / Player II

	B1	B2	B3
A1	8	4	2
A2	-2	1	-1
A3	10	5	3

d- Payoff Matrix of Player I

Player I / Player II

	B1	B2	B3	B4
A1	1	2	3	12
A2	7	8	6	10
A3	5	11	4	3

2- For the following games, the saddle point can not be determined if the players select pure strategies. Find the probabilities of mixed strategies which will provide a saddle point. Find the payoff of the games.

a- Payoff Matrix of Player I

Player I / Player II

	B1	B2
A1	6	3
A2	2	8

b- Payoff Matrix of Player I

Player II (columns), Player I (rows)

	B1	B2	B3
A1	19	23	11
A2	15	19	20
A3	27	18	19

c- Payoff Matrix of Player I

Player II (columns), Player I (rows)

	B1	B2	B3	B4
A1	12	8	9	12
A2	8	9	11	7

CHAPTER 11

INTEGRAL CALCULUS

Integration is the inverse of differentiation. The problem here is to find $f(x)$, when $f'(x)$ is given. We call $f(x)$ an antiderivative of $f'(x)$, and the method of finding it is called integral calculus.

Integral calculus will be used for finding solutions of differential equations in chapter 12. As well, integral calculus has many applications in business and economic problems. One such problem is to find the original function when the marginal function, or the rate of change of one variable with respect to other variable is known. For example, If marginal cost function is known, the total variable cost function can be found by finding the anti-derivative of the marginal cost function.

While the antiderivative of a function is a function, the integral of a function at a closed interval is, generally, a number. For example, given the demand or marginal revenue functions, the increase or decrease in total revenue resulting from a price change can be found by integrating the marginal revenue function over the given price range.

In the next section, indefinite integrals are discussed and some simple rules of integration are presented. It is followed by the definite integrals and application of integral calculus to economic problems.

Tool Box

Mathematica **Commands**	**Function**
Integrate[f(x), x]	Finds the indefinite integral of f(x) with respect to x, $\int f(x)dx$.
Integrate[f(x), {x, a, b}]	Finds the definite integral of f(x) in the interval [a, b], $\int_a^b f(x)\,dx$.

I. Indefinite Integrals: The Integrate[f(x), x] Command

Indefinite integration (antiderivative) is the process of reverse differentiation. That is, given the derivative of a function with respect to some variable, often called **derived function**, the indefinite integral (denoted by $\int$) is used to find the **primitive function,** or the function that gave rise to that derivative. For example, $\int x^2\,dx = \frac{x^3}{3} + c$ because $\frac{d}{dx}(\frac{x^3}{3} + c) = x^2$.

If the derived function F'(x) = f(x), F(x) is an integral of f(x). However, the general form of the integral of f(x) is F(x) + c, where c is an arbitrary constant. Such a general form of an integral is called indefinite integral and is denoted by $\int f(x)dx = F(x)+c$ where, f(x)dx is the differential of F(x), f(x) is the integrand function, and c is the constant of integration.

Example: Cost Function

A firms's marginal cost (MC) as a function of output (Q) is given as $MC(Q) = 30 - 2Q + .1Q^2$. Find the total cost function (TC).

Solution: What we know is the rate of change of total cost with respect to output, or marginal cost (the derived function), and we wish to uncover the primitive function TC. This is the inverse of differentiation, $TC(Q) = \int MC\,dQ = 30Q - Q^2 + \frac{1}{30}Q^3 + C$, where C is the constant of integration, here interpreted as the fixed cost of the firm.

TC(Q) is the primitive function, which shows more than one function because C can be any number. Therefore, additional information is needed to identify the function. This additional information is called **initial condition** or **boundary condition**. If we had the information that the fixed cost is 100, then TC(0) = C = 100, and TC(Q) = $30Q - Q^2 + .033Q^3 + 100$.

To find the indefinite integral of a function using *Mathematica*, use the command **Integrate** as follows.

```
In[1]:= Clear[Q]
        Integrate[30 - 2Q + .1Q^2, Q]
```

Out[1]= $30.\,Q - 1.\,Q^2 + 0.0333333\,Q^3$

Rules of Integration: The following are some useful rules of integration. It should be noted that there are integrands for which exact integrals cannot be found. In those cases, only approximations will be available.

Rule 1. Integration of a constant k is kx + c: $\int k\,dx = kx + c$.

Example: Find $\int 5dx$

Solution: Here, k = 5. Therefore, $\int 5dx = 5x + c$.

In *Mathematica*, use the command "Integrate" as follows,

```
In[2]:= Clear[x]
        Integrate[5, x]
```

Out[2]= 5 x

Rule 2. The Power Rule: Integral of x^n is $\frac{1}{n+1}x^{n+1} + c$, $\int x^n dx = \frac{1}{n+1}x^{n+1} + c$, $n \neq -1$

Example: Find $\int x^3 dx$

Solution: Here, n = 3. Therefore, $\int x^3 dx = \frac{1}{4}4^x + c$.

In *Mathematica*,

```
In[3]: = Clear[x]
         Integrate[x^3, x]
```

Out[3]= $\frac{x^4}{4}$

Example: Find $\int \sqrt[3]{x^2}\,dx$.

Solution: Since $\sqrt[3]{x^2} = x^{\frac{2}{3}}$, $\int \sqrt[3]{x^2}\,dx = \frac{3}{5}x^{\frac{5}{3}} + c$.

In *Mathematica*,

```
In[4]:= Clear[x]
        Integrate[x^(2/3), x]
```

Out[4]= $\frac{3x^{5/3}}{5}$

Example: Find $\int \frac{1}{x^3}dx$.

Solution: Since $\frac{1}{x^3} = x^{-3}$, we have n = -3. Therefore, $\int \frac{1}{x^3}dx = \frac{-1}{2}x^{-2} + c = \frac{-1}{2x^2} + c$.

Using *Mathematica*,

```
In[5]:= Clear[x]
        Integrate[1/x^3, x]
```

Out[5]= $-\frac{1}{2x^2}$

Example: Variable Cost Function

The marginal cost function (MC) of a single product firm t is given as $MC(Q) = Q^2 - 7Q + 26$, where Q is the quantity of output. Find the total variable cost function (TVC).

Solution: Total variable cost function represents the total variable cost of producing Q units of output. Since the integral of a rate of change (marginal) is the total change, given the marginal cost MC(Q) the total variable cost $TVC(Q) = \int MC\,dQ = \int (Q^2 - 7Q + 26)dx. = 26Q - \frac{7Q^2}{2} + \frac{Q^3}{3}$.

Using *Mathematica*,

```
In[6]:= Clear[Q]
        Integrate[Q^2 - 7Q + 26, Q]
```

Out[6]= $26Q - \frac{7Q^2}{2} + \frac{Q^3}{3}$

Example: Capital Formation

The net investment flow of a firm (I) at time t is given as $I(t) = 3t^{.8}$, and the initial capital stock (K) at t = 0 is K(0) = 1500. Find the time path of the capital formation.

Solution: If the capital formation is continuous over time, the rate of capital formation at time t is $\frac{dk}{dt} = I(t)$ or $dk = I(t)dt$. Therefore, $K(t) = \int I(t)dt$. Substituting for I(t) yields, $K(t) = \int 3t^{.8}dt = 1.66667t^{1.8} + c$. Since K(0) = 1500, solve for c by setting t = 0 and K(t) = 1500. That is c = 1500 and $K(t) = 1.66667t^{1.8} + 1500$.

Using *Mathematica*, integrate the function and set the integral equal to sol. Solve sol+c = 1500 for t = 0.

```
In[7]:= Clear[t, sol, c]
        sol = Integrate[3 t^.8, t]
```

Out[7]= $1.66667t^{1.8}$

```
In[8]:= Solve[sol + c == 1500, c] /. t -> 0
```

Out[8]= {{c → 1500.}}

Therefore, $K(t) = 1.66667t^{1.8} + 1500$. You can plot the time path of capital growth by plotting the K(t) function as,

```
In[9]:= Plot[1.66667t^1.8 + 1500, {t, 0, 20}]
```

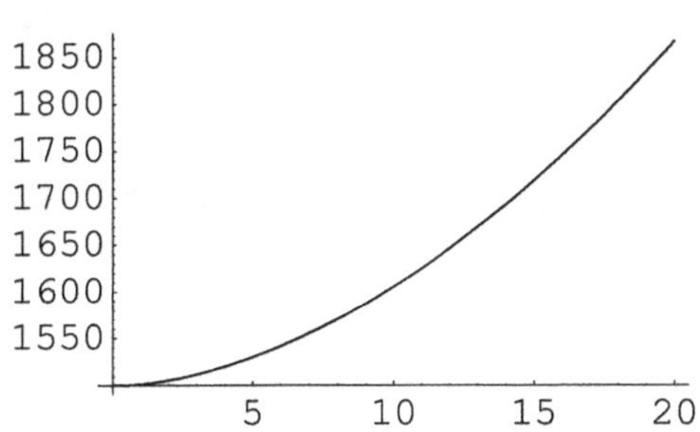

Rule 2a. Special Case of Power Rule: If n = -1, then $\int x^n dx = \int \frac{1}{x} dx = \ln x + c$

Example: Find $\int \frac{5}{x} dx$

Solution: $\int \frac{5}{x} dx = 5\ln x + c.$

Using *Mathematica*,

```
In[10]:= Clear[x]
         Integrate[5 / x, x]
Out[10]= 5 Log[x]
```

Rule 3. Constant Multiple of a Function: $\int kf(x)dx = k\int f(x)dx.$

Example: Find $\int 5x^4 dx.$

Solution: $\int 5x^4 dx = 5 \int x^4 dx = 5(\frac{1}{5}x^5 + c) = x^5 + C$ where, C = 5c.

Using *Mathematica*,

```
In[11]:= Clear[x]
         Integrate[5 x^4, x]
Out[11]= x^5
```

Rule 4. Integral of a sum of Functions: $\int (u + v + w)dx = \int u dx + \int v dx + \int w dx$, where u, v, and w are a function of x.

Example: Find $\int (5x^3 + 3x^2 + x)dx.$

Solution: $\int (5x^3 + 3x^2 + x)dx = \int 5x^3 dx + \int 3x^2 dx + \int x dx = \frac{5}{4}x^4 + c1 + x^3 + c2 + \frac{1}{2}x^2 + c3 = \frac{5}{4}x^4 + x^3 + \frac{1}{2}x^2 + C,$ where C = c1 + c2 +c3.

Using *Mathematica*,

```
In[12]:= Clear[x]
         Integrate[5 x^3 + 3 x^2 + x, x]
Out[12]= x^2/2 + x^3 + 5 x^4/4
```

Example: Find $\int \frac{x^3 + 2x^2 - 1}{10x} dx.$

Solution: $\int \frac{x^3 + 2x^2 - 1}{10x} dx. = \int \frac{x^3}{10x} dx + \int \frac{2x^2}{10x} dx. + \int \frac{-1}{10x} dx = \frac{x^3}{30} + \frac{x^2}{10} - \frac{1}{10}\ln x + C.$

Using *Mathematica*,

```
In[13]:= Clear[x]
         Integrate[(x^3 + 2 x^2 - 1) / (10 x), x]
Out[13]= 1/10 (x^2 + x^3/3 - Log[x])
```

Rule 5. Integration by Substitution or Integration by Change of Variable: $\int u^n du = \frac{1}{n+1}u^{n+1}du + c$, where u = f(x) and $n \neq 1$.

Example: Find $\int 2x(x^2 + 3)^5 dx.$

Solution: Set $x^2 + 3 = u$ so that, $(x^2 + 3)^5 = u^5$, and $\frac{du}{dx} = 2x$ or $du = 2xdx$. Therefore, $\int 2x(x^2 + 3)^5 dx = \int (x^2 + 3)^5 (2x)dx = \int u^5 du = \frac{1}{6}u^6 + c = \frac{1}{6}(x^2 + 3)^6 + c.$ Integration by substitution is applicable whenever the original integrand function, f(x), can be transformed as, $\int f(x)dx = \int f(g(x))g'(x)dx$ so that we can set $f(g(x)) = f(u)$ where, $u = g(x)$ and $g'(x) = \frac{du}{dx}$. The integral is then transformed to $\int f(x)dx = \int u du$.

Using *Mathematica*, the solution to the example is,

```
In[14]:= Clear[x]
         Integrate[2 x (x^2 + 3)^5, x]
```

Out[14]= $2\left(\frac{243 x^2}{2} + \frac{405 x^4}{4} + 45 x^6 + \frac{45 x^8}{4} + \frac{3 x^{10}}{2} + \frac{x^{12}}{12}\right)$

```
In[15]:= Factor[%]
```

Out[15]= $\frac{1}{6} x^2 (6 + x^2)(9 + 3 x^2 + x^4)(27 + 9 x^2 + x^4)$

Example: Find $\int x^2 (2 x^3 + 1)^4 dx$.

Solution: Set $2x^3 + 1 = u$ so that, $(2 x^3 + 1)^4 = u^4$, and $\frac{du}{dx} = 6x^2$ or $du = 6x^2dx$. Therefore, $\int x^2 (2 x^3 + 1)^4 dx = \int \frac{1}{6} u^4 du = \frac{1}{30} u^5 + c = \frac{1}{30}(2 x^3 + 1)^4 + c$.

Using *Mathematica*,

```
In[16]:= Clear[x]
         Integrate[x^2 (2 x^2 + 1)^4, x]
```

Out[16]= $\frac{x^3}{3} + \frac{8 x^5}{5} + \frac{24 x^7}{7} + \frac{32 x^9}{9} + \frac{16 x^{11}}{11}$

```
In[17]:= Factor[%]
```

Out[17]= $\frac{x^3 (1155+5544 x^2+11880 x^4+12320 x^6+5040 x^8)}{3465}$

Rule 5a. Special Case of Integration by Substitution: If $n = -1$, then $\int u^n du = \int \frac{1}{u} du = \ln u + c$.

Example: Find $\int \frac{1}{z+5} dx$.

Solution: Set $x + 5 = u$, and note that $\frac{du}{dx} = 1$ or $du = dx$. Hence, $\int \frac{1}{x+5} dx = \int \frac{1}{u} du = \ln u + c = \ln(x + 5) + c$.

Using *Mathematica*,

```
In[18]:= Clear[x]
         Integrate[1 / (x + 5), x]
Out[18]= Log[5 + x]
```

Example: Find $\int \frac{3x}{x^2+1} dx$.

Solution: Set $x^2 + 1 = u$. Since $\frac{du}{dx} = 2x$, $du = 2xdx$. Therefore, $\int \frac{3x}{x^2+1} dx = \int \frac{3}{2u} du = \frac{3}{2} \ln u + c = \frac{3}{2}\ln(x^2 + 1)$.

Using *Mathematica*,

```
In[19]:= Clear[x]
         Integrate[3 x / (x^2 + 1), x]
```

Out[19]= $\frac{3}{2}$ Log[1 + x^2]

Example: Find $\int \sqrt{1+x}\, x^3\, dx$.

Solution: Set $u = 1 + x$. This gives $du = dx$ and $x^3 = (u-1)^3 = u^3 - 3u^2 + 3u - 1$. Therefore, $\int \sqrt{1+x}\, x^3\, dx = \int \sqrt{u}\,(u^3 - 3u^2 + 3u - 1)\, du = \int (u^{\frac{7}{2}} - 3u^{\frac{5}{2}} + 3u^{\frac{3}{2}} - u^{\frac{1}{2}})\, du = \frac{2u^{\frac{9}{2}}}{9} - \frac{6u^{\frac{7}{2}}}{7} + \frac{6u^{\frac{5}{2}}}{5} - \frac{2}{3}u^{\frac{3}{2}}$. Substituting for $u = 1 + x$ gives $\int \sqrt{1+x}\, x^3\, dx = \sqrt{1+x}\left(-\frac{32}{315} + \frac{16x}{315} - \frac{4x^2}{105} + \frac{2x^3}{63} + \frac{2x^4}{9}\right)$.

Using *Mathematica*,

```
In[20]:= Clear[x]
         Integrate[x^3 Sqrt[x + 1], x]
```

Out[20]= $\sqrt{1+x}\left(-\frac{32}{315}+\frac{16x}{315}-\frac{4x^2}{105}+\frac{2x^3}{63}+\frac{2x^4}{9}\right)$

Rule 6. Integration of Exponential Functions of Base e: $\int e^{nx}dx = \frac{1}{n}e^{nx} + c$.

Example: Find $\int 2e^x\, dx$.

Solution: $\int 2e^x\, dx = 2e^x + c$.

```
In[21]:= Clear[x]
         Integrate[2 E^x, x]
```

Out[21]= $2e^x$

Rule 6a. $\int e^u du = e^u + c$ where, $u = f(x)$.

Example: Find $\int e^{5x}dx$.

Solution: Let $5x = u$ so that, $du = 5dx$, or $dx = \frac{1}{5}du$. Therefore, $\int e^{5x}dx = \int \frac{1}{5}e^u du = \frac{1}{5}e^u + c = \frac{1}{5}e^{5x} + c$.

Using *Mathematica*,

```
In[22]:= Clear[x]
         Integrate[E^(5 x), x]
```

Out[22]= $\frac{e^{5x}}{5}$

Example: Find $\int 3xe^u dx$, where $u = x^2 + 5$.

Solution: Since $du = 2xdx$, we can write $\int 3xe^u dx = \int \frac{3}{2}e^u du = \frac{3}{2}e^u + c$. Substituting for u results in $\int 3xe^u dx = \frac{3}{2}e^{(x^2+5)} + c$.

```
In[23]:= Clear[x]
         Integrate[3 x E^(x^2 + 5), x]
```

Out[23]= $\frac{3e^{5+x^2}}{2}$

Example: Cost Function

The marginal cost function of a single-product firm (MC) is given as $MC(Q) = 2e^{(2+.1Q)}$. Find the total variable cost function (TVC) of the firm.

Solution: The integral of a rate of change (marginal) is the total change. Therefore, $TVC(Q) = \int MC\,dQ = \int 2e^{(2+.1Q)}\,dx$.

Using *Mathematica*,

```
In[24]:= Clear[x]
         Integrate[2 E^(2 + .1 Q), Q]
```

Out[24]= $147.781\, e^{0.1Q}$

Rule 7. Integration of Exponential Functions of Base b: $\int b^{nx}dx = \frac{1}{lnb}b^{nx} + c$.

Example: Find $\int 5^x dx$.

Solution: $\int 5^x dx = \frac{5^x}{lnx} + c$.

Using *Mathematica*,

```
In[25]:= Clear[x]
         Integrate[5^x, x]
Out[25]= 5^x/Log[5]
```

Rule 8. Integration by Parts: $\int u dv = uv - \int v du$, where u = f(x) and v = g(x). The formula follows from the derivative of the product uv with respect to x. We know that, d(uv) = udv + vdu or udv = d(uv) - vdu. Integrating both sides yields, $\int u dv = uv - \int v du$. Basically, integration by parts is used when the integral of an expression involving products or logarithms cannot be evaluated directly. The integrand should then be transformed to standard integrable parts.

The application of integration by parts depends on the appropriate choice of u and dv, so that $\int v du$ and $\int dv$ can be evaluated, even though $\int u dv$ cannot. Therefore, it is advisable to choose u and v in such a way that: 1) dv is integrable; 2) when the integrand is a product of two functions, choose the more complicated one as part of dv, in order to make $\int v du$ as easily integrable as possible.

Example: Find $\int xe^x dx$.

Solution: Since this integral cannot be evaluated directly, set u = x and dv = e^xdx, so that, du = dx and v = $\int e^x dx = e^x + c$. Hence, $\int xe^x dx = \int u dv = uv - \int v du = xe^x - e^x dx = xe^x - e^x + c = e^x(x - 1)$.

Using *Mathematica*,

```
In[26]:= Clear[x]
         Integrate[x E^x, x]
Out[26]= e^x (-1 + x)
```

Example: Find $\int x lnx dx$.

Solution: Set x = dv and lnx = u so that, $v = \frac{1}{2}x^2$ and $du = \frac{1}{x}$. Using Rule 8, $\int x lnx dx = \int u dv = uv - \int v du = \frac{1}{2}x^2 lnx - \int \frac{x}{2} dx = \frac{1}{2}x^2 lnx - \frac{x^2}{4} + c$.

```
In[27]:= Clear[x]
         Integrate[x Log[x], x]
Out[27]= -x^2/4 + 1/2 x^2 Log[x]
```

Problems

1- Find integrals of the following functions:

a) $\int \frac{1}{2}x^3 dx$ b) $\int 3x^{\frac{1}{2}} dx$

c) $\int 3x^2(x^3 - 4)^2 dx$ d) $\int \frac{1}{x-2} dx$

e) $\int (2x^3 + \frac{1}{x^4} + 2x) dx$ f) $\int 2e^{4x} dx$

g) $\int \frac{1}{2}(e^2 + 1)^2 e^{2x} dx$ h) $\int \ln 5^{(2-2x)} dx$

i) $\int x e^x dx$ j) $\int x^3 e^{3x} dx$

k) $\int x e^{-x} dx$ l) $\int \ln x dx$

m) $\int 5 \ln x^4 dx$ n) $\int x^2 \ln x dx$

0) $\int \sqrt{1+x^2}\, x^5\, dx$ p) $\int \sqrt{1+x^3}\, x^2\, dx$

2. The marginal cost function (MC) of a single product firm is given as MC = 5 + 16Q, where Q is the quantity of output. Find the total variable cost function (TVC).

3. Suppose that the net investment flow of a firm (I) at time t is given as $I(t) = 3t^{.5}$, and the initial capital stock (K) at t = 0 is K(0) = 500. Find the time path of the capital (Note that, $I = \frac{dK}{dt}$).

4. The marginal propensity to save (MPS) of a country is given as MPS = f(Y) = $.3 - .1y^{.5}$. Find the saving function, given that S = 0 when Y = 81.

5. The marginal cost function of a single-product firm (MC) is given as MC = $2e^{(2+.1Q)}$, and the fixed cost is FC = 90. Find the total cost function.

II. Definite Integrals: The **Integrate**[f(x), {x, a, b}] Command

Contrary to the indefinite integrals, which do not possess definite numerical values, definite integrals have specific numerical values free of the variable x and the arbitrary constant c.

Definite integrals are the result of what is famous to the **Fundamental Theorem of Integral Calculus**. This theorem connects integral calculus to differential calculus, and in turn, it simplifies the solution of many integral problems. The Fundamental Theorem enables us to compute areas under the functions and find integrals of functions without having to compute them as limits of sums. The Fundamental Theorem consists of the following two parts:

Part 1. If f(x) is continuous on [a, b], then function g(x) defined by

$$g(x) = \int_a^x f(x)\, dx, \quad a \le x \le b$$

is continuous on [a, b] and differentiable on (a, b), and g'(x) = f(x).

Part2. If f(x) is continuous on [a, b], then

$$\int_a^b f(x)\, dx = F(a) - F(b)$$

where F(x) is any antiderivative of f(x). That is, a function that F'(x) = f(x).

Proof of the fundamental theorem of integral calculus can be found in any standard calculus text book. From expressions above it is clear that a definite integral can be evaluated by first finding the indefinite integral of the integrand function, and then substituting the upper and lower limits of integration for x to get F(*a*) and F(*b*).

Example: Evaluate the definite integral $\int_1^6 x^2\, dx$.

Solution: The indefinite integral of the integrand function is $F(x) = \frac{1}{3}x^3 + c$. Substitute the upper and lower limits of integration for x to find the definite integral, $F(x)_1^6 = F(6) - F(1) = 72 + c - (\frac{1}{3} + c) = \frac{215}{3}$. In *Mathematica,* use the command **Integrate** as follows,

```
In[28]:= Clear[x]
         Integrate[x^2, {x, 1, 6}]
```

Out[28]= $\frac{215}{3}$

Example: Evaluate $\int_0^5 \frac{x}{x^2+1}\, dx$.

Solution: The indefinite integral of the integrand function is $F(x) = \frac{1}{2}\ln(x^2 + 1) + c$. Therefore, $\int_0^5 \frac{x}{x^2+1}\, dx = F(5) - F(0) = \frac{1}{2}(\ln 26 - \ln 1) = \frac{1}{2}\ln 26$.

In Mathematica,

```
In[29]:= Clear[x]
         Integrate[x/(x^2 + 1), {x, 0, 5}]
```

Out[29]= $\frac{\text{Log}[26]}{2}$

Example: Application to Business

Marginal cost of a product is estimated to be $MC(Q) = \frac{25}{\sqrt[3]{10Q+1}} + \frac{1}{3}Q^2$. Find the total cost of increasing output from 20 to 25 units. Find average variable cost (AVC) function and graph AVC and MC functions on the same set of axes.

Solution: Since $MC(Q) = \frac{dTC(Q)}{dQ}$, to find total cost find the antiderivative of MC(Q) as, $TC(Q) = \int MC(Q)\, d dQ = \int\left(\frac{25}{\sqrt[3]{10Q+1}} + .3\,Q^2\right) dQ = \frac{Q^3}{9} + \frac{15}{4}(1 + 10Q)^{\frac{2}{3}} + C$, where $\frac{Q^3}{9} + \frac{15}{4}(1 + 10Q)^{\frac{2}{3}}$ is the total variable cost (TVC) and C can be interpreted as the fixed cost. Total cost of increasing output from 20 units to 25 units is evaluated as $\left(\frac{Q^3}{9} + \frac{15}{4}(1 + 10Q)^{\frac{2}{3}} + C\right)_{20}^{25} = 867.76$. The average variable cost (AVC) is , $AVC(Q) = \frac{TVC}{Q} = \frac{Q^2}{9} + \frac{15}{4Q}(1 + 10Q)^{\frac{2}{3}}$. The graph of MC(Q) and AVC(Q) are shown below.

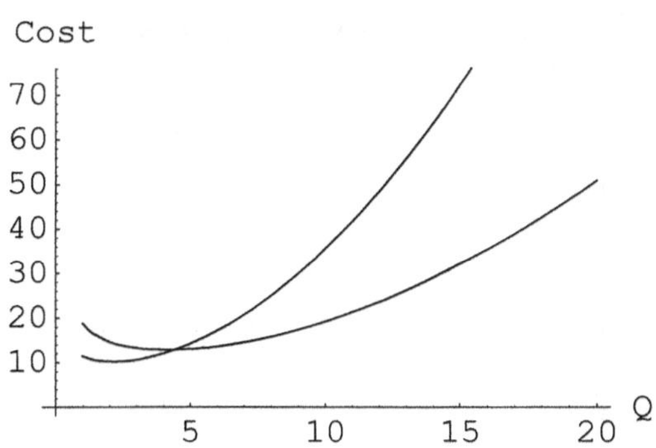

As any standard MC and AVC graphs, MC passes through the lowest point of AVC and continues to rise.

Using *Mathematica*, set TVC equal to indefinite integral of MC

```
In[30]:= Clear[Q, MC, AVC, TVC]
         MC = 25/((10 Q + 1)^(1/3)) + (1/3) Q^2;
         TVC = Integrate[ 25/((10 Q + 1)^(1/3)) + (1/3) Q^2, Q]
```

Out[30]= $\frac{Q^2}{3} + \frac{25}{(1+10\,Q)^{1/3}}$

$\frac{Q^3}{9} + \frac{15}{4}(1 + 10\,Q)^{2/3}$

Increase in total cost for increase in Q from 20 to 30 units can be evaluated as the definite integral of MC in the range Q = 20 to Q = 30.

```
In[31]:=
        N[Integrate[25/((10 Q + 1)^(1/3)) + (1/3) Q^2, {Q, 20, 25}]]

Out[31]= 867.762
```

The average variable cost is AVC = $\frac{TVC}{Q}$. Use the "Plot" command to graph the functions.

```
In[32]:= AVC = TVC/Q;
         Plot[{MC, AVC}, {Q, 1, 20}, AxesLabel -> {"Q", "Cost"}]
```

Out[32]= $\frac{\frac{Q^3}{9} + \frac{15}{4}(1+10\,Q)^{2/3}}{Q}$

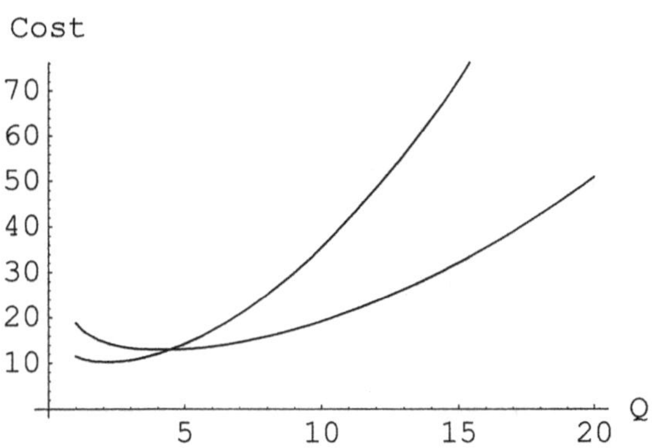

Definite Integral As the Area Under the Curve: The numerical value of a definite integral may be interpreted as the area under, or enclosed by, a curve. Such an area is considered to be the limit of the sum of the subareas which result from a certain partitioning of the area in question. For example, the graph of the continuous function $f(x) = x^2$, in the closed interval [0, 4] is represented below,

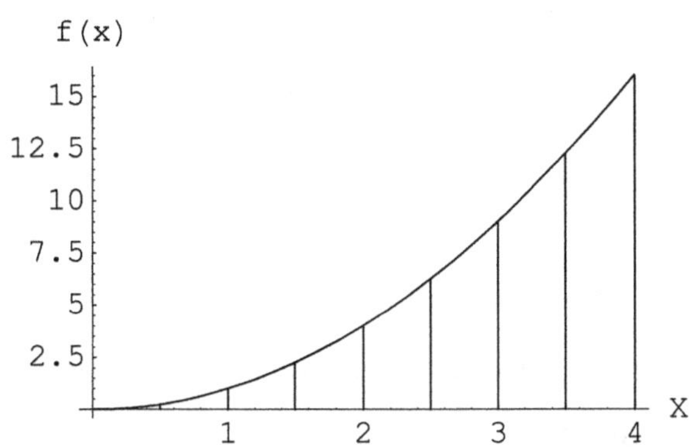

the area in the interval [0, 4] can be divided into n equal parts, with each subinterval $\Delta x = \frac{4-0}{n}$. Suppose now that we designate the middle points of the subintervals by $m_1, \ldots\ldots, m_n$; then the following sum, S_n, is an approximation to the area under the curve, over the interval [0, 4], $S_n = f(m_1)\Delta x_1 + \ldots\ldots\ldots + f(m_n)\Delta x_n = \sum_{i=1}^{n} f(m_1)\,\Delta x_i$. The size of n determines the goodness of the measurement of S_n. The larger n, the better is the approximation. Denoting the exact area by S, we can write, $\lim_{n\to\infty} S_n = S$. If the $\lim_{n\to\infty} S_n$ exits, then

It is interesting to note the similarities between the discrete summation, $\sum_{i=a}^{b} f(mi)\,\Delta xi$, and continuous summation, $\int_a^b f(x)\, dx$. f(x) and dx are continuous counterparts of f(mi) and Δxi, with continuous summation ($\int$) replacing the discrete summation (Σ).

Example: Find the area under the curve $f(x) = x^2 - 8x + 7$ and the x-axis in the interval x = 0 to x = 1.

Solution: The numerical value of the definite integral $\int_0^1 f(x)\, dx$ is the area under the curve and the x axis. Therefore, $\int_0^1 f(x)\, dx = \int_0^1 (x^2 - 8x + 7)\, dx = (\frac{1}{3}x^3 - 4x^2 + 7x + c)_0^1 = \frac{10}{3}$.

In *Mathematica*,

```
In[33]:= Clear[x]
         Integrate[x^2 - 8 x + 7, {x, 0, 1}]
Out[33]= 10/3
```

One important property of definite integrals is that they can be expressed as a sum of a finite number of integrals. That is, if $a < b < c < d$, then $\int_a^d f(x)\, dx = \int_a^b f(x)\, dx + \int_b^c f(x)\, dx + \int_c^d f(x)\, dx$. This property of definite integrals is used in evaluating the area under a curve, when a portion of the curve under consideration lies below the x axis.

If a portion of the curve under consideration lies below the x axis, the total area should be broken into the sum of two or more integrals by appropriate changes of the limits of integration. The total area for the interval will be the sum of the absolute values of the sub-areas calculated by definite integral for the sub-intervals.

Example: Application to Production

A manufacturer of sods has estimated that the sales of sods follows a seasonal pattern given by $Q = 12000(1 + \text{Sin}\frac{2\pi(t-60)}{365})$ where Q is measured in square feet per day, t is time in days, with t = 1 representing january 1. The manufacturer wants to set up a schedule to produce a uniform amount of sods each day. Graph the Q for a year and find the average daily production of sods.

Solution: By substituting values for t, this sine function can be graphed as,

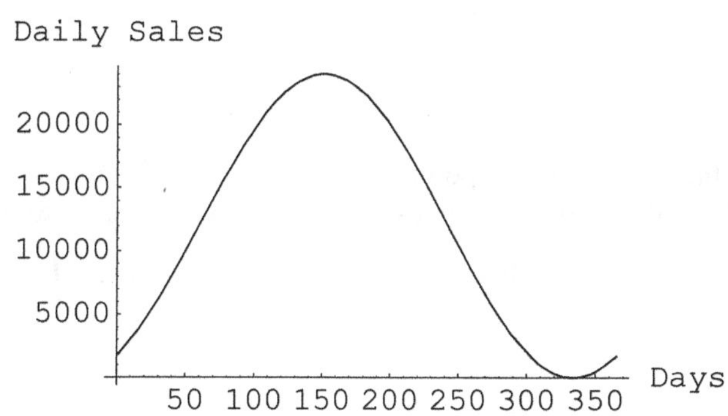

the total sale is the area under the curve in the range t = 1 to t = 365. Therefore, find the definite integral of the Q in this range then, divide it by 365 to find the average daily production. Annual Sales = $\int_1^{365} 12000(1 + \text{Sin}\frac{2\pi(t-60)}{365})\, dt = 12000(t - \frac{\text{Cos}[2\pi(-60+t)]}{2\pi})_1^{365} = 4.37809\times 10^6$. Average Daily Production = $\frac{4.37809\times 10^6}{365} \approx 12000$ feet per day.

Note: $\int \text{Sin}\, x\, dx = -\text{Cos x} + \text{c}$ and $\int \text{Sin}\, u\, du = \frac{-1}{u'}\text{Cos u} + \text{c}$ where, u is a function of x and $u' = \frac{du}{dx}$.

$\int \text{Cos}\, x\, dx = \text{Sin x} + \text{c}$ and $\int \text{Cos}\, u\, du = \frac{1}{u'}\text{Sin u} + \text{c}$ where, u is a function of x and $u' = \frac{du}{dx}$.

In[34]:=
```
Clear[q]
q = 12000 (1 + Sin[2 Pi (t - 60) / 365]);
Plot[q, {t, 1, 365}, AxesLabel -> {"Days", "Daily Sales"}]
```

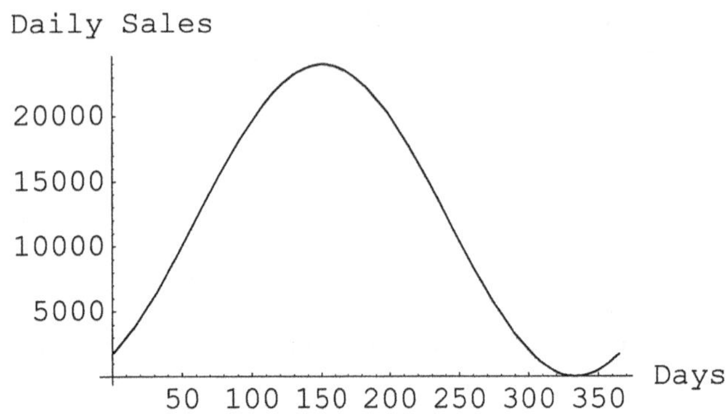

In[35]:= `N[(1 / 365) Integrate[q, {t, 0, 365}]]`

Out[35]= 12000.

Example: Find the area under the curve for $f(x) = x^2 - 8x + 7$, for the interval x = 0 to x = 10.

Solution: A look at the graph of the function, below, shows that a part of the graph in the interval of the interest falls below the x-axis.

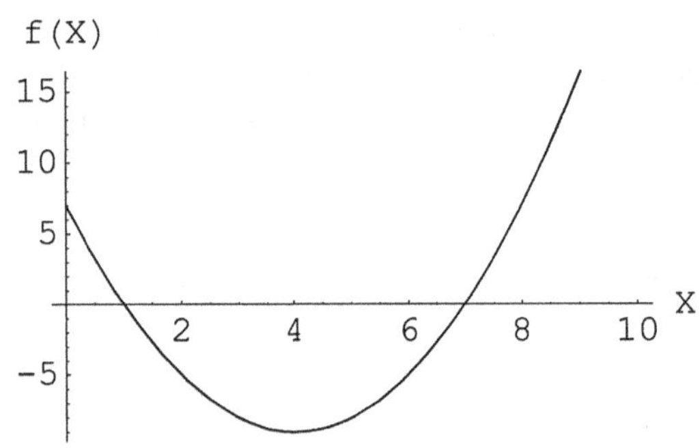

Set f(x) = 0 to find the points where f(x) crosses the x-axis. These points are x =1 and x =7. The interval [0, 10] can be broken into three sub-intervals, [0, 1], [1,7], and [7, 10]. The areas in the sub-intervals [0, 1] and [7, 10] lie above the x-axis and the area in the interval [1, 7] lies below the x-axis. Therefore, $\int_0^{10} f(x)\, dx = \int_0^1 f(x)\, dx + \int_1^7 f(x)\, dx + \int_7^{10} f(x)\, dx$.

The three areas are calculated as, $(\frac{1}{3}x^3 - 4x^2 + 7x + c)_0^1$, $(\frac{1}{3}x^3 - 4x^2 + 7x + c)_1^7$, and $(\frac{1}{3}x^3 - 4x^2 + 7x + c)_7^{10}$, the numerical values of which are $\frac{10}{3}$, -36, and 36, respectively. However, the area of interest is the sum of the absolute value of these three number, $\frac{10}{3} + 36 + 36 = \frac{226}{3} = 75.33$.

To solve the problem using *Mathematica*, first plot the function and find zeroes of the function. Then find the total area as the sum of the absolute values of the integrals.

```
In[36]:= Clear[x, area1, area2, area3]
         Plot[x^2 - 8 x + 7, {x, 0, 10}, AxesLabel → {"X", "f(X)"}]
```

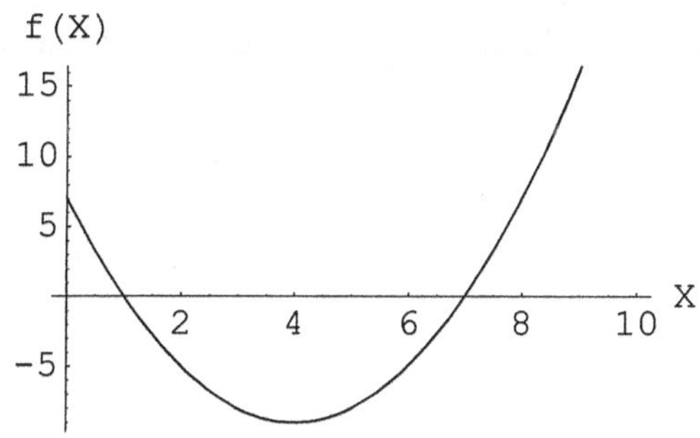


```
In[37]:= Solve[x^2 - 8 x + 7 == 0, x]

Out[37]= {{x → 1}, {x → 7}}

In[38]:= area1 = Integrate[x^2 - 8 x + 7, {x, 0, 1}];
         area2 = Integrate[x^2 - 8 x + 7, {x, 1, 7}];
         area3 = Integrate[x^2 - 8 x + 7, {x, 7, 10}];
         area = N[Abs[area1] + Abs[area2] + Abs[area3]]

Out[38]= 75.3333
```

Example: Revenue of the Firm

Marginal revenue of a firm is estimated to be MR = 100 - .2Q. A price change in the market results in a quantity change of 8 units from the current Level of 34 units. Find the change in total revenue of the firm.

Solution: Total revenue generated from the price change is the area under the marginal revenue in the interval [34, 42], (the sum of all marginal revenues between 34 and 48 units, the graph below).

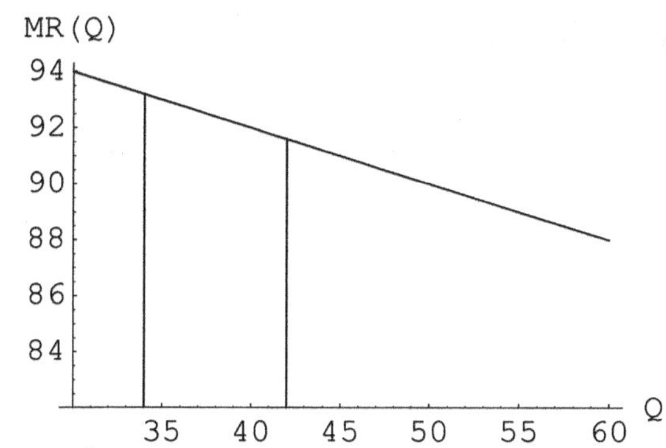

Therefore, find $\int_5^6 (1000 - .2\,Q)\,dQ \;=\; (1000\,Q - .1\,Q^2 \;+\; C)_5^6 = 739.20.$

Using *Mathematica*,

```
In[39]:= Clear[Q]
         Integrate[100 - .2 Q, {Q, 34, 42}]

Out[39]= 739.2
```

Example: Total Cost

Marginal cost of manufacturing Q units of goods is MC(Q) = 300 - Q + $.06Q^2$. Find the increase in total cost if the production level is increased from 20 units to 30 units.

Solution: Total cost of increasing Q from Q_1 to Q_2 is the area under the MC(Q) function in the interval $[Q_1, Q_2]$ (the graph below).

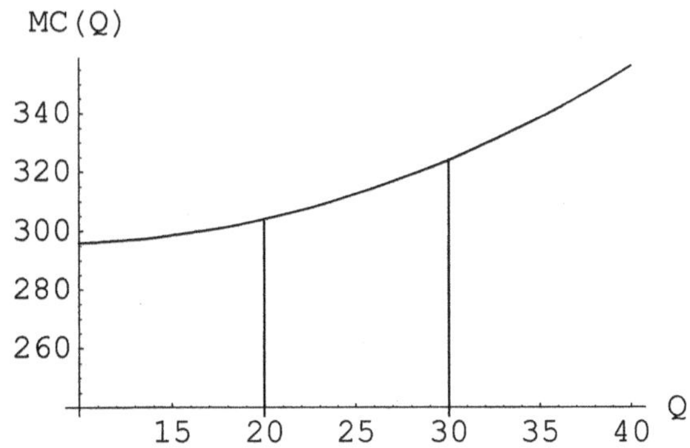

Therefore, total cost of producing ten more units is $\int_{20}^{30}(300 - Q + .06\,Q^2)\,dQ = (300\,Q - \frac{1}{2}\,Q^2 + .02\,Q^3 + C)_{20}^{30} =$ 3130.

```
In[40]:= Clear[Q]
         Integrate[300 − Q + .06 Q^2, {Q, 20, 30}]

Out[40]= 3130.
```

Area of the Region Between a Curve and Y-axis: To find the area of a region between a curve and the y-axis write x as a function of y, integrate the new function with respect to y, and evaluate it for the defined boundaries.

Example: Find the area of the region between y = $\sqrt[4]{x}$, y =1, and the y-axis.

Solution: The graph below shows the region between y = $\sqrt[4]{x}$, y = 1, and the y-axis.

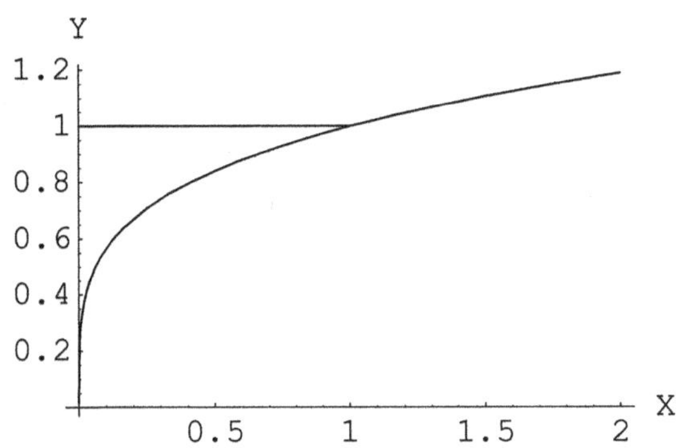

To find the region of the interest, write x in terms of y as x = y^4 and find $\int_0^1 y^4\,dy = (\frac{y^5}{5} + c)_0^1 = \frac{1}{5}$.

Using *Mathematica*,

```
In[41]:= Clear[y]
         Integrate[y^4, {y, 0, 1}]

Out[41]= 1/5
```

Example: Application to Market

A consumer's demand for a product, which is selling at the market price of $P_e = 10$, is given as $P = \frac{100}{Q+4}$. Find the consumer's surplus.

Solution: Consumer surplus for every unit of a good is the difference between the price that consumer is willing to pay for that unit and the price that consumer actually pays. Therefore, for a demand function $P = f(Q)$, consumer surplus is the region between the demand curve and P-axis that lies above the market price P_e. The graph below shows that region.

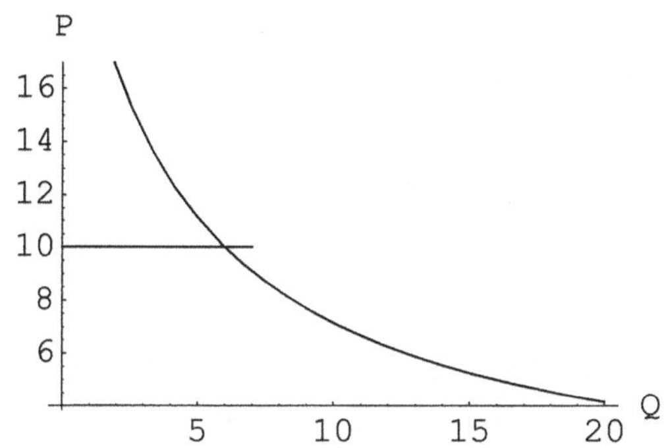

At Q = 0, P = 25. Since we are interested in the region between the vertical axis (P-axis) and the function, write Q in terms of P, as $Q = \frac{100-4P}{P}$, integrate it with respect to P, and evaluate it in the boundaries P = 10 and P = 25.

Consumer surplus = $\int_{10}^{25} \frac{100-4P}{P}\, dP = (100 \ln P - 4P)_{10}^{25} = 31.63$

```
In[42]:= Clear[P]
         N[Integrate[(100 - 4 P)/P, {P, 10, 25}]]

Out[42]= 31.6291
```

Example: Learning Curve

Learning curve shows the increase in productivity of labor as labor learns by doing the same job over time. The number of units produced between t_1 and t_2 periods is the area under the learning curve and bounded by t_1 and t_2. Suppose that the learning curve of new employees for producing a certain product is estimated to be $\frac{dQ}{dt} = 50(1 - \frac{100}{(t+10)^2})$ where, Q is the number of units and t is the time. Find the number of units produced by a new employee between day 1 and day 10. Find the number of units produced by the same employee between day 11 and day 20. What is the increase in average daily productivity between the two periods?

Solution: The following graph shows the learning curve $\frac{dQ}{dt} = 50(1 - \frac{100}{(t+10)^2})$.

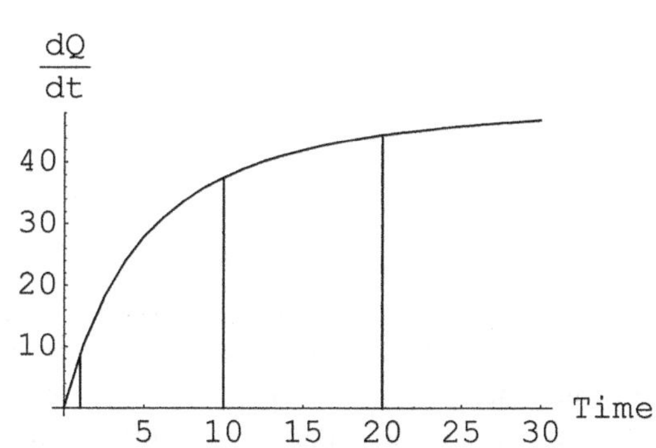

The number of units produced between day 1 and day 10 is $Q1 = \int_1^{10} 50\left(1 - \frac{100}{(t+10)^2}\right) dt = 245.5$ and the number of units produced by the same employee between the days 11 and 20 is $Q2 = \int_{11}^{20} 50\left(1 - \frac{100}{(t+10)^2}\right) dt = 378.6$. The increase in production in the second period is 133.1 units, with an average daily productivity increase of 13.31 units. Using *Mathematica*,

```
In[43]:= Clear[t, Q1, Q2, AP]
         Q1 = N[Integrate[50 (1 - 100/(t + 10)^2), {t, 1, 10}]]

Out[43]= 245.455 Null

In[44]:= Q2 = N[Integrate[50 (1 - 100/(t + 10)^2), {t, 11, 20}]]

Out[44]= 378.571

In[45]:= AP = (Q2 - Q1)/10

Out[45]= 13.3117
```

Area of a Region Between Two Functions: The area (A) of the region bounded by the curves y = f(x), y = g(x), and the lines x = a, x = b, where f(x) and g(x) are continuous and $f(x) \geq g(x)$ for all x in [a, b], is $A = \int_a^b [f(x) - g(x)]\, dx$.

Example: Find the area of the region bounded between $y = e^{1.2x}$, $y = 10 + 18x - 2x^2$, and bounded on the sides by x = 1 and x = 2.

Solution: The following graph shows the graph of the functions and the lines of boundary.

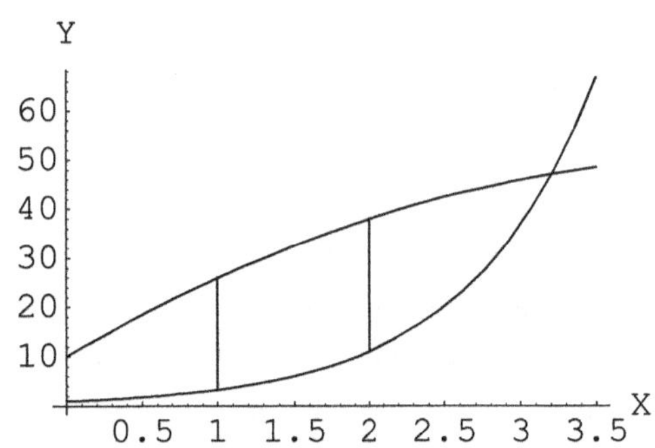

Here, $y = 10 + 18x - 2x^2$ lies above $y = e^{1.2x}$ in [1, 2]. Therefore, we write the area as:

$$A = \int_1^2 [(10 + 18x - 2x^2) - e^{1.2x}]\, dx = \left(10x + 9x^2 - \frac{2x^3}{3} - \frac{e^{1.2x}}{1.2}\right)_1^2 = 25.91$$

Using *Mathematica*,

```
In[46]:= Clear[x]
           Integrate[10 + 18 x - 2 x^2 - E^(1.2 x), {x, 1, 2}]

Out[46]= 25.9141
```

Example: Income Distribution

Lorenz curve is a continuous, increasing, and concavely upward function that passes through points (0, 0) and (1, 1) and describes income distribution of households in a country. The x-axis represents the percentage of households and the y-axis represents the percentage of income. A point such as (.50, .25) on the curve means that the bottom 50% of the households earn 25% of total national income. Absolute equality of income is represented by a straight line y = x. The area between the Lorenz curve and the y = x line measures how much the income distribution differs from absolute equality. The so called coefficient of inequality is the ratio of the area between the lorenz curve and the line y = x to the area under y = x line.

Suppose that the Lorenz curve for a country is given as $L(x) = .4x^2 + .6x$. What is the percentage of total income received by the bottom 50% of the households? What is the coefficient of inequality?

Solution: To find the percentage of income received by the bottom 50% of the population find L(.50) = .40. That is, the bottom 50% of the population earn 40% of the income. To find the coefficient of inequality, graph L(x) and y = x.

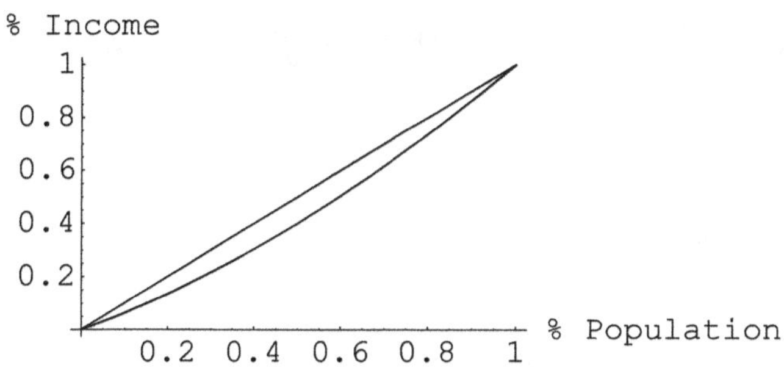

The coefficient of inequality is the ratio of the area between the curves to the area under y = x and bounded by [0, 0] and [1, 1]. Since y = x lies above the L(x) curve, to find the area between the curves, write the area as the integral of x - L(x) between 0 and 1.

$$A1 = \int_0^1 [x - (.4x^2 + .6x)]\,dx = \left(.2x^2 - \tfrac{.4x^3}{3} + c\right)\Big|_0^1 = .067$$

The total area under y = x and bounded by [0, 0] and [1, 1] is

$$A2 = \int_0^1 x\,dx = \left(\tfrac{x^2}{2} + c\right)\Big|_0^1 = \tfrac{1}{2}.$$

Therefore, the coefficient of inequality is $\frac{.067}{.5} = .133$. Using *Mathematica*,

```
In[47]:= Clear[L, x]
         L = .4 x^2 + .6 x;
         y = x;
         A1 = Integrate[y - L, {x, 0, 1}]

Out[47]= 0.0666667

In[48]:= A2 = Integrate[y, {x, 0, 1}]

Out[48]= 1/2

In[49]:= coef = A1 / A2

Out[49]= 0.133333
```

Symmetry Rule: The following symmetry rules simplify the calculation of integrals of functions that posses symmetry properties:

1) If f(x) is even function [f(-x) = f(x)], then $\int_{-a}^{a} f(x)\,dx = 2\int_0^a f(x)\,dx$.

2) If f(x) is an odd function [f(-x) = -f(x)], then $\int_{-a}^{a} f(x)\,dx = 0$.

Example: Find $\int_{-3}^{3} x^4 \, dx$

Solution: Here, f(x) = x^4 is an even function, f(-x) = f(x). Therefore, $\int_{-3}^{3} x^4 \, dx = 2\int_0^3 x^4 \, dx = \frac{486}{5}$.

Using *Mathematica*,

```
In[50]:= Clear[x]
         Integrate[x^4, {x, -3, 3}]

Out[50]= 486/5
```

Example: Find $\int_{-3}^{3} \frac{x}{x^4 + x^2 + 1} \, dx$.

Solution: Here, f(x) = $\frac{x}{x^4 + x^2 + 1}$ is an odd function with f(x) = f(-x). Therefore, $\int_{-3}^{3} \frac{x}{x^4 + x^2 + 1} \, dx = 0$.

Using *Mathematica*,

```
In[51]:= Clear[x]
         Integrate[x/(x^4 + x^2 + 1), {x, -3, 3}]

Out[51]= 0
```

Problems

1. Find the following definite integrals:

a) $\int_{-1}^{2} (2x^2 - 5x + 3) \, dx$

b) $\int_{0}^{2} (2x^2 - 1)^2 \, dx$

c) $\int_{2}^{\infty} 3x^{-2} \, dx$

d) $\int_{1}^{3} \frac{1}{1+x} \, dx$

e) $\int_{-1}^{1} x(x^2 - 10)^4 \, dx$

f) $\int_{1}^{3} e^{x^2} dx$

g) $\int_{0}^{2} xe^{x} dx$

h) $\int_1^2 e^{(x-1)}dx$

i) $\int_2^4 \frac{10x}{x^2-1}dx$

2. For the following demand functions find consumer surplus, given that market price of the product is \$5.

   a) $p = -3q + 10$

   b) $p = e^{(2-q)}$

   c) $p = \frac{25}{1+q}$

   d) What will the consumer surplus be if the product price changes to \$4?

3. The marginal cost function of a single-product firm is given as, $MC = Q^2 - 8Q + 18$,

   a) Find the producer surplus, if the market price of the product is \$6.

   b) What will the increase in producer surplus be if the market price changes to \$7?

4. A monopolist has the following demand and total cost functions:

   $P = 2Q + 23$

   $TC = \frac{1}{3}Q^3 - 4Q^2 + 18Q + 1000$

   a) Find the profit-maximizing level of output and the price.

   b) Find the social loss of monopoly. (Hint: the social loss of monopoly is the difference between the consumer surplus and the producer surplus at the price that monopolist is selling the product.)

5. A monopolist has the following demand and total cost functions:

   $P = 95 - 2Q$

   $TC = \frac{1}{3}Q^3 - 4Q^2 + 18Q + 1000$

a) Find profit-maximizing levels of price and quantity.

b) Find the social loss of the monopoly.

c) Suppose the government's regulatory agency regulates the monopolist based on the marginal cost pricing (hint: here, price and quantity are derived by setting MC = P, as in case of perfect competition). Find the price and quantity of the regulated monopoly. Find the social loss of the monopoly.

d) Suppose the government's regulatory agency regulates the monopolist based on the average cost pricing (hint: here, price and quantity are derived by setting P = ATC, normal profit pricing). Find the price and quantity of the regulated monopoly. Find the social loss of the monopoly.

6. Suppose that the capital formation of a business is continuous over time, with $\frac{dk}{dt} = I(t) = 3t^{.5}$. If the initial stock of the capital at t = 0 is k(0) = 100,

a) What is the time path of the capital k?

b) What is the capital formation during the time interval t1 and t2?

7. The profit of a business in thousands of dollars is estimated to be $p = 8(\sqrt{t} + 45)$. Find the total profit for the next 10 years and the average profit of the business over the next ten years.

8. A manufacturer depreciates the value of machines used in production by the following rule $\frac{dv}{dt} = 200000(t - 6)$ where v is the value of the machines after t years. What is the total loss of the value of the machines over the first 3 years? What is the total loss of the value of the machines over the first five years?

CHAPTER 12
ORDINARY DIFFERENTIAL EQUATIONS

Differential equations provide a bridge between abstract mathematics, particularly linear algebra and calculus, and their application in dynamic economic models. The basic problem studied in this chapter is one of finding functions satisfying a given relationship between a variable and its derivative. More specifically, the problem is to find function y(t) such that y' (t) = ay(t), for every t in some interval.

Ordinary differential equations contain only ordinary derivatives of order one or higher. The order of a differential equation is the order of the highest-order derivative that it contains. The degree of a differential equation is the highest power to which any of the derivatives in the equation is raised. For example, the differential equation $y''(t) + 4y'(t) + y(t) = 0$ is of second order, third degree.

A differential equation where the variable and its derivatives are both of the first degree is called a linear differential equation. For example, the differential equation y' (t) + 2y(t) = 0 is a first order, linear differential equation.

TOOL BOX

Mathematica **Commands**	**Function**
DSolve[y' [t] == ay[t] + b, y[t], t]	Solves the differential equation for y(t).
DSolve[{y' [t] == ay[t] + b, y[0] ==yo}, y[t], t]	Solves the initial-value differential equation for y(t).
sol = DSolve[y' [t] == ay[t] + b, y[t], t]	Solves the problem and sets the solution equal to sol.
y[t_] = y[t]/.sol[[1]]	uses the replacement rule " /. "to define y[t] as the solution.
NDSolve[{y' [t] == ay[t] + b, y[0]==yo}, y, {t, 0, 10}]	Finds numerical approximations to the solution.
Plot[Evaluate[y[t]/.%], {t, 0, 10}**]**	Plots the values of the InterpolatingFunction.

I. First order, Homogenous Linear Differential Equations

In many cases, it is observed that the rate of change of some economic variable is proportional to the value of the variable. For example, the annual investment (the capital growth) of a business may be proportional to its stock of capital, or the aggregate investment in an economy may be proportional to the stock of capital in the economy. That is, I(t) = K'(t) = aK(t), where I is the investment and K is the stock of capital. This, is the simplest form of the differential equations, which is the first-order homogenous linear differential equation. The problem here is to find the function that gives rise to such a relationship between the variable and its differential.

In general, the problem of first order homogenous linear differential equations is that of finding the function y(t) such that it satisfies, y'(t) = ay(t). The theory or experiment may supply us with the form of the equation or even the value of the "a". Any differentiable function that satisfies the above equation is said to be the solution of the differential equation. To find such a function, divide both sides of the equation y'(t) = ay(t) by y(t) and integrate, $\int \frac{y'(t)}{y(t)} dt = \int a\, dt$. This will yield, lny(t) + c1 = at + c2 or lny(t) = at + c, where c = c1 + c2. By transforming this solution into its exponential form we can write, $y(t) = e^c e^{at}$. Set the $e^c = A$, so that the solution will be

$y(t) = Ae^{at}$. This solution provides a general solution to the first order, homogenous linear differential equation $y'(t) = ay(t)$.

In practice, as well as the differential equation, the value of y(t) at some specified time, t = to may be available. In this case, the set of information, y'(t) = ay(t) and y(to) = yo, is known as the initial-value problem. Here, the problem is to find a differentiable function y(t), such that y'(t) = ay(t) for all t and y(to) = yo. Given the general solution $y(t) = Ae^{at}$, $y(to) = Ae^{ato}$ =yo or $A = yoe^{-ato}$. Therefore, the initial-value problem y'(t) = ay(t), y(to) = yo has the unique solution, $y(t) = yoe^{a(t-to)}$.

In *Mathematica,* the two commands used to solve the differential equations are "DSolve" and "NDSolve". The command DSolve solves a differential equation for the function y(t). The command NDSolve finds numerical approximations to the solution of the differential equation as InterpolatingFunction. The InterpolatingFunction, then stores a table of values for y(ti) and interpolates this table to find an approximation to y(t) when a particular value of t is given. The command NDSolve is more appropriate for solving nonlinear differential equations.

To solve the initial-value, first order linear differential equation y'(t) = ay(t), y(to) = yo using *Mathematica,*

```
In[1]:= Clear[y]
        DSolve[{y'[t] == a y[t], y[to] == yo}, y[t], t]

Out[1]= {{Null (y[t] → e^(a t-a to) yo)}}
```

Example: Solve y'(t) + 4y(t) = 0, for y(o) = 1.

Solution: The general solution of the problems is, $y(t) = Ae^{-4t}$. To find the solution to the initial-value problem, set t = 0 to find A = 1. Therefore, the solution of the initial-value problem is $y(t) = e^{-4t}$.

To solve the same problem using *Mathematica,*

```
In[2]:= Clear[y]
        DSolve[{y'[t] + 4 y[t] == 0, y[0] == 1}, y[t], t]

Out[2]= {{Null (y[t] → e^(-4 t))}}
```

Example: Using *Mathematica,* solve y'(t) - .4y(t) = 0, for y(o) = 2. Find the value of y at t = 5 and plot the time path of y.

```
In[3]:= Clear[y, sol]
        sol = DSolve[{y'[t] - .4 y[t] == 0, y[0] == 2}, y[t], t]

Out[3]= {{y[t] → 2. e^(0.4 t)}}

In[4]:= sol /. t -> 5

Out[4]= {{y[5] → 14.7781}}
```

Once the problem is solved, use the replacement rule " /. "to define the function y[t] and graph the time path of y.

```
In[5]:= y[t_] = y[t] /. sol[[1]];
        Plot[y[t], {t, 0, 10}]
```

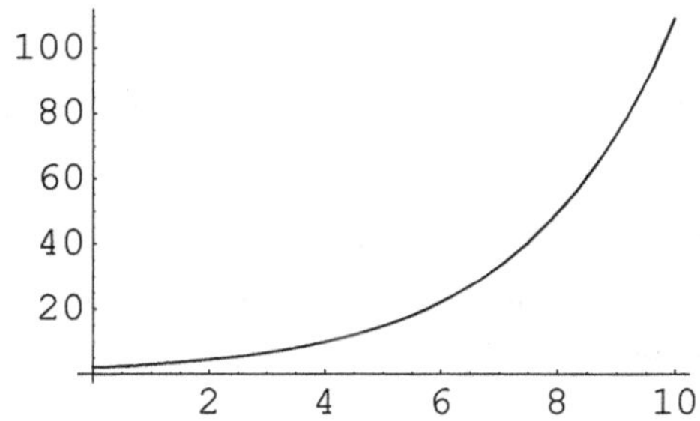

Do More Practice:

1- Change the coefficient of y(t) from -.4 to .4. Keep the other assumptions the same. Re-run the program and compare the results.

2- Change the coefficient of y(t) from -.4 to -4. Keep the other assumptions the same. Re-run the program and compare the results.

3- Change the coefficient of y(t) from -.4 to 4. Keep the other assumptions the same. Re-run the program and compare the results.

Note that, DSolve solves the differential equation for the function y[t]. To find the value of y at t = 5, evaluate the solution as t ->5. The NDSolve command finds numerical approximations to the solution of the differential equation as InterpolatingFunction solution.

```
In[6]:= Clear[y, sol2]
        sol2 = NDSolve[{y'[t] - .4 y[t] == 0, y[0] == 2}, y, {t, 0, 10}]

Out[6]= {{y → InterpolatingFunction[{{0., 10.}}, <>]}}
```

NDSolve represents numerical approximations to the solution of the differential equation. To find a particular value of y at time t, use the InterpolatingFunction to estimate y[t]. For example to find the value of y at t = 5,

```
In[7]:= y[5] /. sol2

Out[7]= {14.7782}
```

and to plot the values y[t] for the range t = 0 to t = 10, type

```
In[8]:= Plot[Evaluate[y[t] /. sol2], {t, 0, 10}]
```

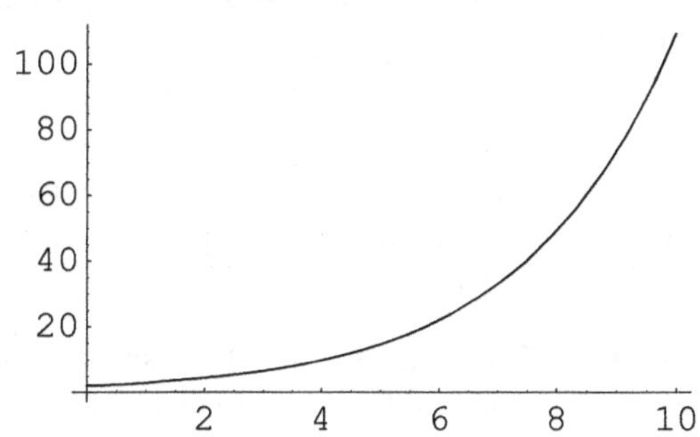

Do More Practice:

1- Change the coefficient of y(t) form -.4 to .4. Solve the problem and observe changes in the solution and the plot.

2- Change the coefficient of y(t) form -.4 to 4. Solve the problem and observe changes in the solution and the plot.

Example: Application to Business

The net capital investment of a firm is .2 proportion of the capital stock. Find the time path of investment. When will the capital stock double?

Solution: The problem can be written as $I(t) = \frac{dK(t)}{dt} = K'(t) = .2k(t)$. The solution of this linear homogenous differential equation is $K(t) = Koe^{.2t}$. To find the time period after which the capital stock will double set, $2Ko = Koe^{.2t}$ or $2 = e^{.2t}$. This will yield $t = 5\ln 2 = 3.47$. In *Mathematica,*

```
In[9]:= Clear[k]
        DSolve[{k'[t] == .2 k[t], k[0] == ko}, k[t], t]

Out[9]= {{k[t] → e^(0.2t) ko}}

In[10]:= Solve[2 ko == E^(.20 t) ko, t]

Solve::ifun : Inverse functions are being used by Solve, so some solutions may not be found.

Out[10]= {{t → 3.46574}}
```

Do More Practice:

1- Change the growth rate to .4. Re-run the program. How long does it take for capital stock to double at this rate?

2- How long does it take for capital stock to triple or quadruple at the proportionality rate of .2?

Example: Application in Macroeconomics

Domar's growth model, in a simple form, is a set of equations as, $S = \alpha Y$, $I = \beta \frac{dy}{dt}$, and $S = I$. The first equation expresses actual saving (S) as a constant proportion α of the economy's total output (Y); the second equation assumes investment (I) as constant proportion β of the change in total output over time; and the third equation expresses actual saving equal to the economy's desired investment. Given this information, find the time path of output.

Solution: Solving the three equations for Y yields, $Y = \frac{\beta}{\alpha}\frac{dy}{dt}$, or $\frac{dy}{dt} = \frac{\alpha}{\beta}y$. The general solution of the problem is $Y = Ae^{\frac{\alpha t}{\beta}}$. If the level of the output is known for an initial period, say t = to, we can estimate the time path of output for the subsequent periods indefinitely (See, Evsey D. Domar, "Capital Expansion, Rate of Growth, and Employment, " Econometrica, April, 1946, pp. 137-147; reprinted in Domar, Essays in the Theory of Economic Growth, Oxford University Press, Fair Lawn, N.J., 1957, pp. 70-82.)

Example: Application to Macroeconomics

Suppose that the actual saving of an economy, S(t), is proportional to its total income, Y(t); and the desired investment, I(t), is a constant proportion 4 of the change in income. Given that marginal propensity to consume (MPC) is .75, and Y(0) = 3000 billion, find the time path of income. Find y(6).

Solution: Since MPC = .75, MPS (marginal propensity to save) is .25 and S(t) = .25Y(t). Also I(t) = 4Y'(t) and Y(0) = 3000. Using Domar's model above, $y'(t) = \frac{.25}{4}y(t)$. Using *Mathematica,* the problem can be solved as

```
In[11]:= Clear[y, rat]
         rat = DSolve[{y'[t] == (.25/4) y[t], y[0] == 3000}, y[t], t]
         y[t_] = y[t] /. rat[[1]];
         Plot[y[t], {t, 0, 10}]

Out[11]= {{y[t] → 3000. e^(0.0625t)}}
```

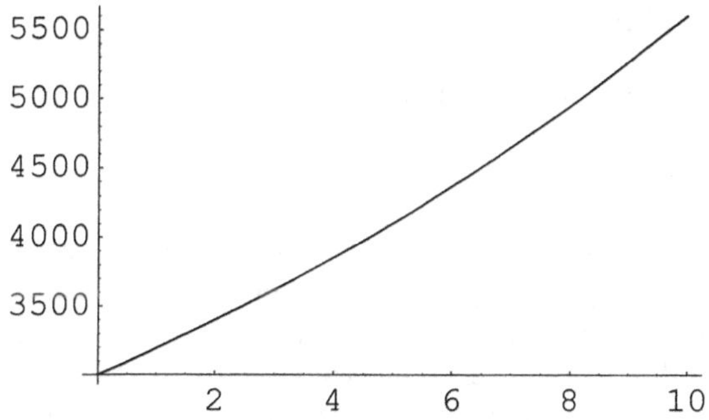


```
In[12]:= y[6]

Out[12]= 4364.97
```

Do More Practice: Change the MPC to .90 and re-run the program. Compare the results to those in the text.

2- Suppose that the desired investment, I(t), is a constant proportion .5 of the change in income. Keep other assumptions the same as in the example. Re-run the program and compare the results to those in the text.

Problem

1. Classify each of the following differential equations as to order and degree.

a) $y'(t) - y(t) = 4$

b) $ty'(t) + x^2(t) - y(t) = 2$

c) $y''(t) - 4y'(t) + y(t) = 0$

d) $x'^3(t) - y'(t) = 1$

e) $t^3 y'''(t) + x'^2 = 27$

2. Solve the following differential equations.

a) $y'(t) + 4y(t) = 0$

b) $y'(t) - 2y(t) = 0$

c) $3y'(t) - 2e^2 y(t) = 0$

d) $y'(t) - 5y(t) = 0, \quad y(0) = 10$

e) $3y'(t) - 4y(t) = 0, \quad y(2) = 4$

f) $y'(t) - 5y(t) = 0, \quad y(10) = 0$

3. Suppose that the GDP of a country, y, is growing at a rate proportional to the size of capital stock, k, $\frac{dy}{dK} = .3k$. Find y as a function of k.

4. The population of a country, p(t), is growing at a rate of 1.5 percent a year. Find the time path of the population growth, given that, population at time t = 0, is 200 million. How long will it take for the population to double in size?

5. Suppose that the actual saving of an economy, S(t), is proportional to its total income, Y(t); and the desired investment, I(t), is a constant proportion 3 of the change in income. Given that marginal propensity to consume is .80, and Y(0) = 200 billion, find the time path of income and y(5).

II. First Order, Nonhomogeneous Linear Differential Equations

A problem of the form y' (t) = ay(t) + f(t), where f(t) is any continuous function of t, is called first order, nonhomogeneous linear differential equation. Associated with any nonhomogeneous equation is the complementary homogenous equation, y'c (t) = ay(t), obtained by setting f(t) = 0.

A solution of the complementary equation is yc(t) = e^{at}, the generalization of which provides a solution to the nonhomogeneous equation. To find the particular solution, yp, search for a function u(t) such that, y(t) = u(t)e^{at}. Differentiating this equation yields, y'(t) = au(t)e^{at} + u' (t)e^{at}. Since the first term in the right hand side of this equation, au(t)e^{at}, is equal to ay(t), (by the assumption that y(t) = u(t)e^{at}), the second term, u' (t)e^{at} should be equal to f(t). Therefore, we can write f(t) = u' (t)e^{at}, and we select u(t) such that u'(t) = e^{-at}f(t). Having found u(t) by integrating this equation, we obtain the particular solution yp(t) = u(t)e^{at}. However, this solution is only one of many particular solutions to nonhomogeneous equation. To find the general solution, write y(t) = kyc(t) + yp(t), where k is any constant. This equation provides a general solution to the nonhomogeneous equation y' (t) = ay(t) + f(t), where yc(t) = e^{at} , yp(t) = u(t)e^{at}, and u(t) is obtained by integrating u'(t) = $\int u'(t)\,dt = \int f(t)\, e^{-at} dt$.

In the special case where f(t) is a constant and the nonhomogeneous equation is in the form, y' (t) = ay(t) + c, the general solution is, y(t) = $ke^{at} - \frac{c}{a} e^{at} + \frac{c}{a}$.

Example: Solve the nonhomogeneous linear differential equation y' (t) = 2y(t) + e^{2t}.

Solution: The complementary homogenous equation y'(t) = 2y(t) has a solution yc(t) = e^{2t}. Select u(t) such that u' (t) = e^{-2t}f(t) = $e^{-2t} e^{2t}$ = 1. To find u(t), integrate u'(t) in the interval 0 to t to get, u(t) = $\int_0^t 1\, dt$ = t. Therefore, yp(t) = u(t)e^{at} = te^{2t}, and the general solution to the problem is, y(t) = kyc(t) + yp(t) = ke^{2t} + te^{2t}. Using *Mathematica*,

```
In[13]:= Clear[y]; DSolve[y'[t] == 2 y[t] + E^(2 t), y[t], t]

Out[13]= {{y[t] → e^(2 t) t + e^(2 t) C[1]}}
```

Do More Practice: Change the nonhomogeneous part of the problem from e^{2t} to e^{-2t} and re-run the program. Compare the results to those in the text.

2- Change the nonhomogeneous part of the problem from e^{2t} to 7 and re-run the program. Compare the results to those in the text.

Example: Solve y'(t) = 2y(t) + e^{2t} + 6.

Solution: The solution follows the solution of the preceding example. Here, u' (t) = e^{-2t}f(t) = e^{-2t}(e^{2t} + 6) = 1 + 6e^{-2t}, and u(t) = $\int_0^t (1 + 6\, e^{-2t})\, dt$ = t - 3e^{-2t} + 3. The particular solutions is, yp = u(t)e^{at} = (t - 3e^{-2t} + 3)e^{2t} = te^{2t} + 3e^{2t} - 3, and the general solution of the problem is, y(t) = kyc + yp = ke^{2t} + te^{2t} + 3e^{2t} - 3 = (k + 3)e^{2t} + te^{2t} -3.

```
In[14]:= Clear[y]
         DSolve[y'[t] == 2 y[t] + E^(2 t) + 6, y[t], t]
Out[14]= {{y[t] → -3 + e^(2 t) t + e^(2 t) C[1]}}
```

Example: Application to Markets

Given the following demand and supply functions, Qd = a - bP, Qs = c + dP, b & d > 0, find the equilibrium price and the equilibrium quantity. Suppose that the market price P differs from the equilibrium price Pe and price changes in the market are proportional to excess demand or excess supply in the market. Find the time path of P.

Solution: The equilibrium price and quantity are $Pe = \frac{a-c}{b+d}$, and $Qe = \frac{ad+bc}{b+d}$, respectively. Assume that the rate of change of price with respect to time is v proportional of excess demand in market, $\frac{dP}{dt} = v(Qd - Qs)$, where v is the coefficient of adjustment and $\frac{dP}{dt} = 0$ if Qd =Qs. Substituting for Qd and Qs from the demand and supply equations we get, $\frac{dP}{dt} = v(a - bp(t) - c - dp(t)) = -v(b + d)p(t) + v(a - c)$. This equation is a first order, nonhomogeneous linear differential equation the solution of which is $P(t) = ke^{-v(b+d)t} + \frac{a-c}{b+d}$. Since $\frac{a-c}{b+d} = Pe$, $P(t) = ke^{-v(b+d)t} + Pe$. If the initial value of the P is given, P(o) = Po, then k = Po - Pe and $P(t) = (Po - Pe)e^{-v(b+d)t} + pe$.

To solve the same problem using *Mathematica*,

```
In[15]:= ClearAll[Qd, Qs, p]
         Qd = a - b p[t]; Qs = c + d p[t];
         DSolve[{p'[t] == v (Qd - Qs), p[0] == po}, p[t], t]
```

$$\text{Out[15]= } \left\{\left\{\text{Null}^2 \left(p[t] \to \frac{a-c+b\,e^{t(-b v-d v)}\left(\frac{-a+c}{b+d}+po\right)+d\,e^{t(-b v-d v)}\left(\frac{-a+c}{b+d}+po\right)}{b+d}\right)\right\}\right\}$$

Note that since b, d, and v are assumed to be positive, as $t \to \infty$, $e^{-v(b+d)t} \to 0$. This implies that, as long as demand and supply have the conventional slopes (negative for demand and positive for supply), the price will always converge to its steady-state equilibrium point. The steady-state or stationary value of P is the value that $\frac{dP}{dt} = 0$. However, if demand or supply have unconventional slopes, then the behavior of the price in market over time will depend on whether b+d is positive or negative. If $b + d > 0$, price will be converge to its steady-state value. If $b + d < 0$, then the price will diverge from the steady-state value.

Example: Application to Markets

The market demand and supply for a product are given as Qd = 10 - .2p and Qs = 4 + .1p, respectively. Assume that the price adjustment in the market follows the rule $\frac{dp}{dt} = .3(Qd - Qs)$. Given that the market price at time t = 0 is \$4, find the solution of the differential equation and plot the time path of the price..

```
In[16]:= ClearAll[p, bat]
         Qd = 10 - 2 p[t];  Qs = 4 + .1 p[t];
         bat = DSolve[{p'[t] == .3 (Qd - Qs), p[0] == 4}, p[t], t]
```

$$\text{Out[16]= } \{\{p[t] \to 0.142857\, e^{-0.63 t}\,(8. + 20.\, e^{0.63 t})\}\}$$

```
In[17]:= p[t_] = p[t] /. bat[[1]];
         Plot[p[t], {t, 0, 10}]
```

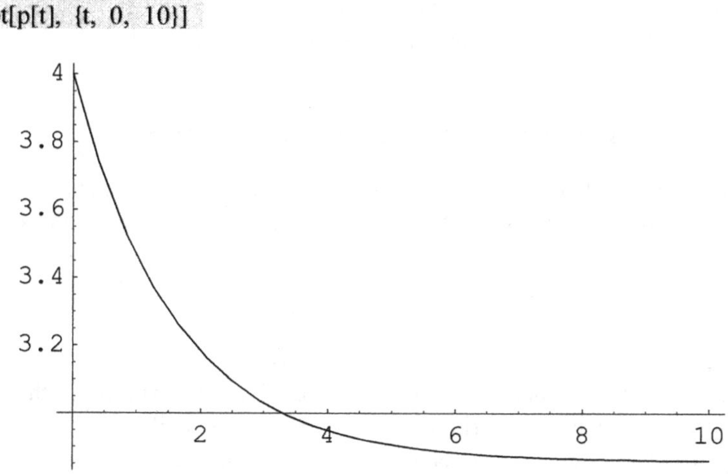

Do More Practice:

1- Change the speed of adjustment from .3 to 3. Re-run the program and compare the results.

2- Change the supply to Qs = -4 + 2p and keep the other assumptions the same as in the text. Re-run the program and compare the results.

3- Assume that demand has an unconventional slope and is expressed as Qd = 10 + .2p and supply is Qs =14 + .1p. Re-run the program and compare the results.

4- Assume that demand has an unconventional slope and is expressed as Qd = 14+ .1p and supply is expressed as Qs = 10 + .1p. Re-run the program and compare the results.

Problems

1. Solve the following differential equations.

a)	y' (t) + 2 y(t) = 6,	y(0) = 10
b)	y' (t) + 10 y(t) = 15,	y(0) = 0
c)	2y' (t) + 3y(t) = 12,	y(0) = 1
d)	y" (t) - 2y(t) = e^t,	y(1) = 2
e)	y' (t) - y(t) = t,	y(1) = 10
f)	y' (t) - y(t) = t + 2,	y(1) = 10

2. Suppose that the GDP growth of country follows, $\frac{dy}{dk}$= .3k + 100. Find the growth path of y as a function of k.

3. The market demand and supply for a product are given as, Qd = 20 - 2p and Qs = 8 + p, respectively.

a) Find the equilibrium output and price.

b) Suppose that the market price at time t = 0 is $6. Find the excess supply in the market.

c) Assume that the price adjustment in the market follows the rule $\frac{dp}{dt}$ = .003(Qd - Qs). Find the solution of the differential equation and plot the time path of the price..

4. The demand and supply functions for a product, with a market price of p = 12, are given as, Qd = 700 - 45p and Qs = -100 + 35p, respectively. Assuming that price adjustment in the market follows the rule $\frac{dp}{dt}$ = .01(Qd - Qs). Find the solution and plot the time path of the price adjustment.

III. Systems of First Order, Homogenous Linear Differential Equations

Many economic variables are related to each other through a system of several interrelated variables and equations. For example, a dynamic IS-LM model of the economy can be expressed by the following system of equations

$$\frac{dY}{dt} = \alpha[D(Y,r) - Y]$$

$$\frac{dr}{dt} = \beta[L(Y,r) - M]$$

Where Y is the output or income, r is the interest rate, M is the money supply, and α and β are the coefficients of adjustment. The first equation, $\frac{dY}{dt}$, represents equilibrium in the output market, IS. It expresses changes in output as α proportion of the gap between aggregate demand, D(Y, r), and aggregate output, Y. The second equation, $\frac{dr}{dt}$, represents the equilibrium in the money market, LM. It expresses changes in the interest rate as a β proportion of the gap between demand for money, L(Y,r), and the money supply, M.

The laws that govern the behavior of variables in this model and the system can be estimated by a linear differential equation model. That is, a model which assumes that the rate of growth of each variable is a linear combination of other variables. The system in the general form can be written as, **Y'**(t) = **AY**(t) where, **Y'**(t) is a vector with elements $\frac{dYi}{dt}$, **A** is a matrix of constant coefficients, and **Y**(t) is the vector of yi(t)s.

The resemblance of **Y'**(t) = **AY**(t) to y'(t) = ay(t) suggests that a trial solution of the form y(t) = e^{at} may be applicable to the system of differential equations. Let us study the possibility by searching for a constant vector **u** and a number k such that, **y**(t) = e^{kt}**u** satisfies **Y'**(t) = **AY**(t). Since **y'**(t) = ke^{kt}**u**, for **y**(t) to be a solution it is necessary and sufficient that ke^{kt}**u** = e^{kt}**Au.** This equation implies that, **Au** - k**u** = (**A** - k**I**)**u** = **0.** Therefore, k is the eigenvalue and **u** is the eigenvector of the matrix **A** (see, Chapter 5, Special Topics in Matrix Algebra), and the solution to **y'**(t) = **A**y(t) is, **y** = e^{kt}**u.**

Example: Solve the following system of linear homogeneous differential equations, y1' (t) = 2y1(t) + y2(t) and y2' (t) = 3y1(t).

Solution: The system in the matrix form is, $\begin{pmatrix} y1'(t) \\ y2'(t) \end{pmatrix} = \begin{pmatrix} 2 & 1 \\ 3 & 0 \end{pmatrix} \begin{pmatrix} y1(t) \\ y2(t) \end{pmatrix}$. Find the eigenvalues and eigenvectors of $\mathbf{A} = \begin{pmatrix} 2 & 1 \\ 3 & 0 \end{pmatrix}$,

```
In[18]:= Eigensystem[(2 1
                      3 0)]

Out[18]= {{-1, 3}, {{-1, 3}, {1, 1}}}
```

the two eigenvalues are -1 and 3, with the corresponding eigenvectors $\begin{pmatrix} -1 \\ 3 \end{pmatrix}$ and $\begin{pmatrix} 1 \\ 1 \end{pmatrix}$, respectively. Therefore, the system of equations have two solutions, **y**1 = $\begin{pmatrix} y1(t) \\ y2(t) \end{pmatrix} = e^{-t}\begin{pmatrix} -1 \\ 3 \end{pmatrix}$ and **y2** = $\begin{pmatrix} y1(t) \\ y2(t) \end{pmatrix} = e^{3t}\begin{pmatrix} 1 \\ 1 \end{pmatrix}$.

Using *Mathematica*, the problem can be solved as,

```
In[19]:= DSolve[{y1'[t] == 2 y1[t] + y2[t], y2'[t] == 3 y1[t]}, {y1[t], y2[t]}, t]

Out[19]= {{y1[t] → 1/4 e^-t (C[1] + 3 e^4t C[1] - C[2] + e^4t C[2]),
          y2[t] → 1/4 e^-t (-3 C[1] + 3 e^4t C[1] + 3 C[2] + e^4t C[2])}}
```

For the existence of the solutions of the systems of first order linear differential equations, see appendix 1.

Do More Practice:

1- Change the coefficient of the y1(t) in the first equation from 2 to -2. Re-run the program and find the eigenvalues. What is the effect of the change in the coefficient on the eigenvalues and on the solution.

2- Change the coefficient of the y1(t) in the first equation from 2 to -2 and the coefficient of the y1(t) in the second equation to -3. Re-run the program and find the eigenvalues. What is the effect of the changes in the coefficients on the eigenvalues and on the solution.

Problems

1. For the following matrices, find a solution to **y'**(t) = **A**y(t).

a) $\mathbf{A} = \begin{pmatrix} 1 & -3 \\ 0 & 1 \end{pmatrix}$ b) $\mathbf{A} = \begin{pmatrix} 2 & 0 \\ 1 & 0 \end{pmatrix}$

c) $\mathbf{A} = \begin{pmatrix} 1 & 2 \\ 2 & 1 \end{pmatrix}$ d) $\mathbf{A} = \begin{pmatrix} 2 & 3 \\ 1 & 0 \end{pmatrix}$

e) $\mathbf{A} = \begin{pmatrix} 0 & 1 \\ \frac{5}{3} & \frac{2}{3} \end{pmatrix}$ f) $\mathbf{A} = \begin{pmatrix} 1 & 4 & 1 \\ 0 & 2 & 3 \\ 0 & 0 & 3 \end{pmatrix}$

g) $\mathbf{A} = \begin{pmatrix} 1 & 0 & 1 \\ 0 & 2 & 1 \\ 4 & 0 & 1 \end{pmatrix}$ h) $\mathbf{A} = \begin{pmatrix} 1 & 1 & 0 \\ 0 & 0 & 1 \\ 0 & 1 & 1 \end{pmatrix}$

2. Solve problems a, b, c, and d of the problem 1 for the initial value, $\mathbf{y}(0) = \begin{pmatrix} 1 \\ 2 \end{pmatrix}$. Plot the time path of $\mathbf{y}(t)$.

3. Solve problems a and b of the problem 1 for the initial value, $\mathbf{y}(0) = \begin{pmatrix} 0 \\ 3 \end{pmatrix}$. Plot the time path of $\mathbf{y}(t)$.

4. Solve problems f and g of the problem 1 for the initial value, $\mathbf{y}(0) = \begin{pmatrix} 1 \\ 2 \\ 1 \end{pmatrix}$.

5. Solve the differential equation $\mathbf{y}'(t) = \begin{pmatrix} 1 & 1 & 0 \\ 0 & 1 & 1 \\ 0 & 0 & 1 \end{pmatrix} \mathbf{y}(t)$ for the initial value $\mathbf{y}(0) = \begin{pmatrix} 1 \\ 1 \\ 1 \end{pmatrix}$.

IV. Systems of First Order, Nonhomogeneous Linear Differential Equations

A system of first order, nonhomogeneous linear differential equations can be expressed in general form as, $\mathbf{y}'(t) = \mathbf{A}\mathbf{y}(t) + \mathbf{f}(t)$ where $\mathbf{f}(t)$ is any continuous vector function of t. The resemblance of this function to the first order, nonhomogeneous linear differential equation suggests that a solution similar to the solution of y'(t) = ay(t) + f(t) may be applicable here. An investigation of the solution shows that the equation $\mathbf{y}(t) = \mathbf{M}(t)\mathbf{u}(t) + \mathbf{M}(t)C$ is the solution of $\mathbf{y}'(t) = \mathbf{A}\mathbf{y}(t) + \mathbf{f}(t)$ for any choice of C. In this solution, $\mathbf{IM}(t)$ is the fundamental matrix of the complementary homogeneous system and $\mathbf{u}(t)$ is a vector such that, $\mathbf{u}'(t) = \mathbf{M}^{-1}(t)\mathbf{f}(t)$. For definition of the fundamental matrix and the solution of the system of first order nonhomogeneous differential equations, the interested reader is referred to appendix 2. Here, using *Mathematica*, the application of the system of first order nonhomogeneous linear differential equations to economic problems is presented by using several microeconomic and macroeconomic problems.

Example: Application to Markets

One of the major characteristics of a perfectly competitive industry is that the firms have easy entry or exit to the industry. Suppose that the firms' entry or exit in a perfectly competitive industry (the change in the number of firms, $\frac{dn}{dt}$) is proportional to the gap between the market price (p) and the minimum average total cost (atc), $\frac{dn}{dt}$ = n' = λ(p - atc), $\lambda > 0$. That is, if p - atc > 0, the positive economic profit attracts more entry to the industry and results in lowers. The opposite occurs when p - atc < 0. In addition, assume that changes in the market price is proportional to excess demand in market, $\frac{dP}{dt}$ = p' = μ(Qd - Qs), $\mu > 0$, where Qd and Qs represent market demand and market supply, respectively. λ and μ represent the speed of adjustment for the number of firms and price, respectively. Given that Qd = 100 - 2p, Qs = 5n, atc = 20, λ = .1 and μ = .2 formulate the price behavior in the market, solve the resulting system of differential equations, and plot the time path of the price. The initial values of p and n are given as po = 25 and no = 15. Note that, in this problem, it is assumed that the firms in the industry are of the uniform size and that Qs depends on the number of firms in the industry.

Solution: The system of the differential equations can be expressed as p' = .2(100 -2p -5n) and n' = .1(p - 20). In matrix form the system can be written as $\begin{pmatrix} p' \\ n' \end{pmatrix} = \begin{pmatrix} -.4 & -1 \\ .1 & 0 \end{pmatrix} \begin{pmatrix} p \\ n \end{pmatrix} + \begin{pmatrix} 20 \\ -2 \end{pmatrix}$. The steady-state value of p and n are obtained by setting p' = n' = 0. Using *Mathematica*,

In[20]:= Clear[Qd, Qs, p, n, chimp]

Solve[$\begin{pmatrix} -.4 & -1 \\ .1 & 0 \end{pmatrix}.\begin{pmatrix} p \\ n \end{pmatrix}+\begin{pmatrix} 20 \\ -2 \end{pmatrix}==\begin{pmatrix} 0 \\ 0 \end{pmatrix}$, {p, n}]

Out[20]= {{Null (p → 20.), Null (n → 12.)}}

the steady-state values of p and n are 20 and 12, respectively. To solve the system of differential equations using *Mathematica*,

```
In[21]:= Qd = 100 - 2 p[t]; Qs = 5 n[t];
         chimp = DSolve[{p'[t] == .2 (Qd - Qs), n'[t] == 10 (p[t] - 20),
            p[0] == 25, n[0] == 15}, {p[t], n[t]}, t]
```

Out[21]= {{p[t] → (0. − 0.0100402 i) ((63.1189 + 249. i) $e^{(-0.2-3.15595 i) t}$ − (63.1189 − 249. i) $e^{(-0.2+3.15595 i) t}$ + (0. + 1992. i) $e^{(0.+0. i) t}$),

n[t] → (0. + 0.00200803 i) ((3992.27 − 747. i) $e^{(-0.2-3.15595 i) t}$ − (3992.27 + 747. i) $e^{(-0.2+3.15595 i) t}$ − (0. + 5976. i) $e^{(0.+0. i) t}$)}}

```
In[22]:= p[t_] = p[t] /. chimp[[1]];
         Plot[p[t], {t, 0, 10}]
```

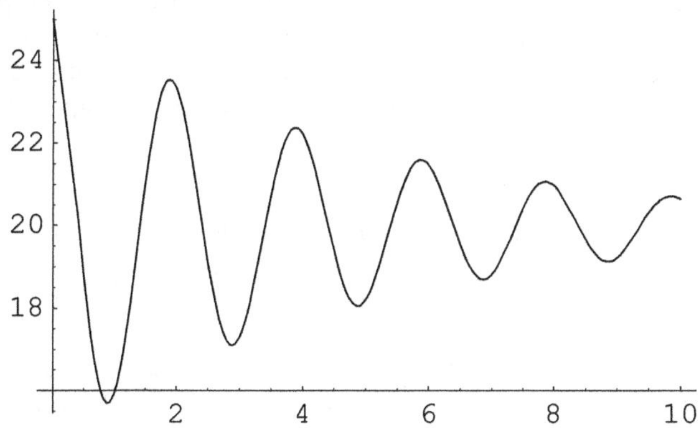

Do More Practice:

1- Change the speed of adjustment for the price equation from .2 to .02. Keep other assumptions the same. Re-run the program and compare the results.

2- Change the speed of adjustment for the number of firms equation from .1 to 10. Keep other assumptions the same. Re-run the program and compare the results (Note that, when the speed of adjustment for n is way higher than speed of adjustment for n, at some point, there may be an overshooting of the price. This happens because the price adjustment depends on the entry and exit of the firms as well as the excess demand).

Problems

1. Solve the following systems of differential equations:

a) $\mathbf{y}'(t) = \begin{pmatrix} 1 & 0 \\ -1 & 2 \end{pmatrix} \mathbf{y}(t) + \begin{pmatrix} 1 \\ 2 \end{pmatrix}$

b) $\mathbf{y}'(t) = \begin{pmatrix} 1 & 0 \\ -1 & 2 \end{pmatrix} \mathbf{y}(t) + \begin{pmatrix} e^t \\ 2 \end{pmatrix}$

c) $\mathbf{y}'(t= \begin{pmatrix} 1 & 0 \\ -1 & 1 \end{pmatrix} \mathbf{y}(t) + \begin{pmatrix} t \\ 1 \end{pmatrix}$

d) $\mathbf{y}'(t) = \begin{pmatrix} 1 & 0 & 0 \\ 0 & 2 & 0 \\ 0 & 0 & 1 \end{pmatrix} \mathbf{y}(t) + \begin{pmatrix} e^t \\ t \\ 1 \end{pmatrix}$

e) $\mathbf{y}'(t) = \begin{pmatrix} 1 & 0 & 1 \\ 0 & 1 & 0 \\ 0 & 0 & 0 \end{pmatrix} \mathbf{y}(t) + \begin{pmatrix} e^t \\ t \\ 1 \end{pmatrix}$

2. Solve problems a, b, and c of the problem for the initial value $\mathbf{y}(0) = \begin{pmatrix} 1 \\ 1 \end{pmatrix}$.

3. Solve problems d and e of the problem 1 for the initial value $\mathbf{y}(0) = \begin{pmatrix} 1 \\ 1 \\ 1 \end{pmatrix}$.

4. Suppose that the goods market and the money market of an economy are described by the following system of differential equations: C = .8Y + 100, I = .1Y - 200R + 50, Y = C + I, Md = .8Y - 100R, Ms = 200, Md = Ms, where C is consumption, Y is income, I is investment, R is interest rate, Md is money demand, and Ms is money supply. Output (income) and interest rate adjust to disequilibrium in the market with a speed proportional to excess demand or excess supply in the corresponding markets. The coefficients of adjustment for the goods market and money market are .01 and .001 respectively. Build a system of differential equations describing the adjustment of output and interest rate in the markets, and analyze the time path of the output and interest rate.

V. Linear Differential Equations of Higher order

The general form of the linear differential equations of higher order is, $y^{(n)} + an\text{-}1y^{(n-1)} + \ldots a1y^{(1)} + aoy = f(t)$, where $y^{(n)}$ is the nth order derivative of y with respect to t; ais, i = 0,......, n-1 are constant coefficients; and f(t) is a differentiable function of t.

Solutions of the nth order differential equations are closely related to the solutions of the systems of differential equations, which were studied in the previous section. In fact, every nth order equation of the form above can be expressed as a system of differential equations by setting the first order variable $y^{(1)}$ equal to a variable, such as z, which is not differentiated. The following examples show how this transformation can be done.

Example: Solve the second order differential equation $y'' - 2y' - 3y = e^t$.

Solution: Set y' = z. This implies that, $y'' = z' = 2z + 3y + e^t$. These two equations give rise to a system of differential equations as, y' = z and $z' = 2z + 3y + e^t$. This system, in a matrix form, can be written as $\mathbf{y}'(t) = \mathbf{A}\mathbf{y}(t) + \mathbf{f}(t)$, where $\mathbf{y}'(t) = \begin{pmatrix} y' \\ z' \end{pmatrix}$, $\mathbf{A} = \begin{pmatrix} 0 & 1 \\ 3 & 2 \end{pmatrix}$, $\mathbf{y} = \begin{pmatrix} y \\ z \end{pmatrix}$, and $\mathbf{f}(t) = \begin{pmatrix} 0 \\ e^t \end{pmatrix}$. That is, $\begin{pmatrix} y' \\ z' \end{pmatrix} = \begin{pmatrix} 0 & 1 \\ 3 & 2 \end{pmatrix} \begin{pmatrix} y \\ z \end{pmatrix} + \begin{pmatrix} 0 \\ e^t \end{pmatrix}$. Note that, the second element of the vector **y** is z, which is the first derivative of y.

The solution of this system of differential equations follows the solution of the systems of first order nonhomogeneous linear differential equations of the previous section. That is, using the two equations $\mathbf{u}'(t) = \mathbf{M}^{-1}(t)\mathbf{f}(t)$ and y(t) = yp(t) + yc(t)C = **M**(t)**u**(t) + **M**(t)C of the previous section, the solution is $\mathbf{y}(t) = \begin{pmatrix} y \\ z \end{pmatrix} = \frac{1}{8} \begin{pmatrix} e^{3t} - 2e^t + e^{-t} \\ 3e^{3t} - 2e^t - e^{-t} \end{pmatrix}$. Using this solution, it can be shown that the second term of the vector **y**(t) , (z), is the derivative of the first term, (y), and that the solution satisfies the equation $y'' - 2y' - 3y = e^t$. Since the initial values are not given, the solution is also one of the general solutions to the problem.

To solve the same problem using *Mathematica*,

```
In[23]:= ClearAll[y]
         DSolve[y''[t] - 2 y'[t] - 3 y[t] == E^t, y[t], t]
```

Out[23]= $\{\{y[t] \to -\frac{e^t}{4} + e^{-t} C[1] + e^{3t} C[2]\}\}$

Example: Write the third-order differential equation $2y''' + y'' - y = 2t$ as a system of differential equations.

Solution: Set y' = z and y" = z' = x. This yields the following equations: y' = z, z' = x, and $x' = -\frac{1}{2}x + \frac{1}{2}y + t$. In matrix form, these three equations can be written as $\begin{pmatrix} y' \\ z' \\ x' \end{pmatrix} = \begin{pmatrix} 0 & 1 & 0 \\ 0 & 0 & 1 \\ \frac{1}{2} & 0 & -\frac{1}{2} \end{pmatrix} \begin{pmatrix} y \\ z \\ x \end{pmatrix} + \begin{pmatrix} 0 \\ 0 \\ t \end{pmatrix}$. The problem can be solved using the method described in the previous section. To solve the problem using *Mathematica*,

```
In[24]:= ClearAll[y]
         DSolve[2 y'''[t] + y''[t] - y[t] == 2 t, y[t], t]
```

Out[24]= $\{\{y[t] \to -2t + e^{t\,\mathrm{Root}[-1+\#1^2+2\#1^3\&,1]} C[1] + e^{t\,\mathrm{Root}[-1+\#1^2+2\#1^3\&,2]} C[2] + e^{t\,\mathrm{Root}[-1+\#1^2+2\#1^3\&,3]} C[3]\}\}$

Companion Matrix: In general, for every linear differential equation of higher order of the from $y^{(n)}$ + an-1$y^{(n-1)}$ + ... a1$y^{(1)}$ + aoy = f(t) there is a corresponding or companion system of first-order linear differential equations, the matrix **A** of which has the following standard form, $\mathbf{A} = \begin{pmatrix} 0 & 1 & 0 & 0 & 0 & . & . \\ 0 & 0 & 1 & 0 & 0 & . & . \\ 0 & 0 & 0 & 1 & 0 & . & . \\ . & . & . & . & . & . & . \\ . & . & . & . & . & . & . \\ -ao & -a1 & -a2 & -a3 & . & . & . \end{pmatrix}$. This matrix is called the companion matrix of higher-order equations. Once the companion matrix of the higher-order differential equation is constructed, the solution of the corresponding system follows the instructions of the previous section.

Example: Write the companion matrix of the following 4th order differential equation $2\,y^{(4)} + 3y^{(2)} - 4\,y^{(1)} - y = 4e^t$.

Solution: Write the equation in the standard form , by dividing both sides of equation by 2, $y^{(4)} + \frac{3}{2}y^{(2)} - 2\,y^{(1)} - \frac{1}{2}y = 2e^t$. The companion matrix of this 4th order differential equation is a 4x4 matrix $\mathbf{A} = \begin{pmatrix} 0 & 1 & 0 & 0 \\ 0 & 0 & 1 & 0 \\ 0 & 0 & 0 & 1 \\ \frac{1}{2} & 2 & \frac{-3}{2} & 0 \end{pmatrix}$.

Example: Solve the initial-value, second order differential equation y" - 2y' + y = e^t for the initial values y(o) = 1 and y'(o) = 1 (note that, for the second-order differential equation, the initial value of y and the initial value of the first derivative of y should be given.)

Solution: The companion system of the problem is $\mathbf{y'}(t) = \begin{pmatrix} 0 & 1 \\ -1 & 2 \end{pmatrix} \mathbf{y}(t) + \begin{pmatrix} 0 \\ e^t \end{pmatrix}$, with the initial values $\mathbf{y}(0) = \begin{pmatrix} 1 \\ 1 \end{pmatrix}$. The solution to the initial-value problem follows the equation **y**(t) = **yp**(t) + **Mn**(t)**y**o, where **yp**(t) is the particular solution, **yp**(t) = **M**(t)**u**(t), **M**(t) is the fundamental matrix, and **Mn**(t) is the normalized fundamental matrix, $\mathbf{Mn}(t) = \mathbf{M}(t)\mathbf{M}^{-1}(0)$. To find the fundamental matrix, solve for the eigensystem of the **A** matrix,

```
In[25]:= Eigensystem[(0 1; -1 2)]
Out[25]= {{1, 1}, {{1, 1}, {0, 0}}}
```

the **A** matrix has one distinct eigenvalue k = 1 and a distinct corresponding eigenvector $\mathbf{u} = \begin{pmatrix} 1 \\ 1 \end{pmatrix}$. Therefore, one solution is $\begin{pmatrix} e^t \\ e^t \end{pmatrix}$. To solve the problem, we need one more distinct eigenvalue and eigenvector. This can be obtained from the equation (**A** - k **I**)**v** = **u** (see the section on the Existence of Definite Solutions). That is, $\left(\begin{pmatrix} 0 & 1 \\ -1 & 2 \end{pmatrix} - \begin{pmatrix} 1 & 0 \\ 0 & 1 \end{pmatrix}\right)\begin{pmatrix} v1 \\ v2 \end{pmatrix} = \begin{pmatrix} 1 \\ 1 \end{pmatrix}$. This results in $\mathbf{v} = \begin{pmatrix} 0 \\ 1 \end{pmatrix}$ and a second solution for the system as $\begin{pmatrix} e^t \\ e^t \end{pmatrix} t + \begin{pmatrix} 0 \\ e^t \end{pmatrix} = \begin{pmatrix} te^t \\ te^t + e^t \end{pmatrix}$. Therefore, the fundamental matrix of the system is $\mathbf{M}(t) = \begin{pmatrix} e^t & te^t \\ e^t & te^t + e^t \end{pmatrix}$. Since **yp(t)** = **M**(t)**U**(t) and **U'**(t) = $\mathbf{M}^{-1}$ **f** (t), find **U'**(t) as,

```
In[26]:= MatrixForm[Inverse[(E^t  tE^t; E^t  tE^t+ E^t)].(0; E^t)]
Out[26]= (-t; 1)
```

therefore, **U'**(t) = $\begin{pmatrix} -t \\ 1 \end{pmatrix}$. To find **U**(t), integrate **U'**(t) in the interval 0 to t to find **U**(t) = $\begin{pmatrix} -\frac{t^2}{2} \\ t \end{pmatrix}$. This will be used to find the particular solution **yp**(t) as **yp**(t) = **M**(t)**U**(t).

```
In[27]:= MatrixForm[(e^t  t e^t; e^t  t e^t + e^t).(-t^2/2; t)]
```

$$\text{Out[27]= } \begin{pmatrix} \frac{e^t t^2}{2} \\ -\frac{1}{2} e^t t^2 + t(e^t + e^t t) \end{pmatrix}$$

To find the initial value solution, first find the normalized matrix **Mn**(t) = $\mathbf{M}(t)\mathbf{M}^{-1}$ (0),

```
In[28]:= MatrixForm[Inverse[(E^t  t E^t; E^t  t E^t+E^t)]] /. {t -> 0}]
Out[28]= (1 0; -1 1)
In[29]:= MatrixForm[(E^t  t E^t; E^t  t E^t+E^t).(1 0; -1 1)]
```

$$\text{Out[29]= } \begin{pmatrix} e^t - e^t t & e^t t \\ -e^t t & e^t + e^t t \end{pmatrix}$$

$\mathbf{M}(t)\mathbf{M}^{-1}(0) = \begin{pmatrix} e^t - te^t & te^t \\ -te^e & e^t + te^t \end{pmatrix}$. To find **y**(t) = **yp**(t) + **Mn**(t)**yo**,

$$\text{In[30]:= MatrixForm}\left[\begin{pmatrix} \frac{e^t t^2}{2} \\ -\frac{1}{2} e^t t^2 + t(e^t + e^t t) \end{pmatrix} + \begin{pmatrix} e^t - t e^t & t e^t \\ -t e^t & e^t + t e^t \end{pmatrix}.\begin{pmatrix} 1 \\ 1 \end{pmatrix}\right]$$

$$\text{Out[30]= } \begin{pmatrix} e^t + \frac{e^t t^2}{2} \\ e^t - \frac{e^t t^2}{2} + t(e^t + e^t t) \end{pmatrix}$$

Therefore, $y(t) = e^t + \frac{1}{2} e^t t^2$.

To solve the same initial-value problem using *Mathematica*,

```
In[31]:= ClearAll[y]
         DSolve[{y''[t] - 2 y'[t] + y[t] == E^t, y[0] == 1, y'[0] == 1}, y[t], t]
```

Out[31]= $\{\{y[t] \to e^t (1 + \frac{t^2}{2})\}\}$

Example: Application to Markets

The market demand and supply of a good produced and sold in a perfectly competitive market are given as Qdt = 100 - .2Pt and Qst = -10 + .3Pet, respectively. P represents the market price and Pe represents the expectation of the market price by the producers. It is assumed that producers form their expectations of market prices by Pet = Pt + .4P't + .2P"t rule, where P't = $\frac{dP}{dt}$ and P"t = $\frac{d^2P}{dt^2}$. In addition, changes in the market price is proportional to excess demand in market with a proportionality factor of .3. That is $\frac{dP}{dt}$ = .3(Qd - Qs). Given that the initial prices are given P(0) = 200, P(1) = 210, solve the dynamic market model and plot the time path of Pt.

Solution: Substitute for Qd and Qs in the price adjustment equation $\frac{dP}{dt}$ = .3(Qd - Qs) to get $\frac{dP}{dt}$ = .3(100 - .2P +10 -.3Pet). Susbstitue for Pet to get .15Pt +1.036P't + .018P"t = 33. This is a second order, nonhomogeneous differential equation. The steady-state value of the price can be derived by setting p' = p" = 0. The steady-state value of the price is P = 220. The differential equation can be solved using *Mathematica* as,

```
In[32]:= Clear[Qd, Qs, p, koala]
         Qd = 100 - .2 p[t]; Qs = -10 + .3 (p[t] + .4 p'[t] + .2 p''[t]);
         koala = DSolve[{p'[t] == .3 (Qd - Qs), p[0] == 200, p[1] == 210}, p[t], t]
```

Out[32]= $\{\{Null^2 (p[t] \to e^{-0.145154 t} (-11.5622 - 8.43783 e^{-57.2652 t} + 220. e^{0.145154 t}))\}\}$

```
In[33]:= p[t_] = p[t] /. koala[[1]];
         Plot[p[t], {t, 0, 50}]
```

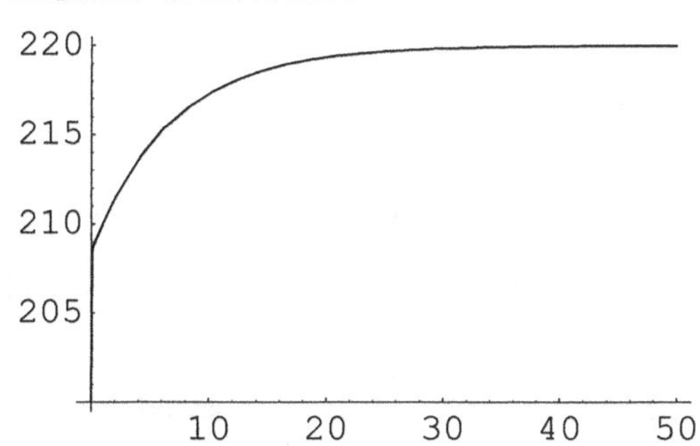

Do More Practice: Change the speed of price adjustment to excess demand from .3 to 3. Keep other assumptions the same. Re-run the program and compare the results.

2- Change the speed of price adjustment to excess demand from .3 to .03. Keep other assumptions the same. Re-run the program and compare the results.

3- Change the Pe equation to Pet = Pt + .6P't + .1P"t. Keep the other assumptions the same. Re-run the program and compare the results.

Example: Application to Markets

In the section for the First Order, Nonhomogeneous Linear Differential Equations, the market demand and supply for a product were given as, Qd = 10 - .2p and Qs = 4 + .1p, respectively. It was assumed that the price adjustment in the market was proportional to the excess demand in market. Now, suppose that the price adjustment in the market depends on the accumulation of the stock of unsold inventories of the product as well as the excess demand. That is, the excess demand in market increases the market prices by the rule $\frac{dp}{dt}$ = .3(Qd - Qs) and the stock of unsold inventories depresses the market prices by the rule $\frac{dp}{dt} = -.08\int_0^t (Qs - Qd)\,dt$. Given the initial market prices of Po = \$6 and P1 = 5.8, find the solution of the differential equation and plot the time path of the price.

Solution: The problem can be formulated as $\frac{dp}{dt} = .3(Qd - Qs) - .08\int_0^t (Qs - Qd)\,dt$. Differentiate the function with respect to t to get $\frac{d^2p}{dt^2} = .3\frac{d(Qd - Qs)}{dt} - .08\,(Qs - Qd)$. Substitute for Qd and Qs from the demand and supply equations and differentiate with respect to t to get P" = -.09P' - .024P + .48, where P" = $\frac{d^2p}{dt^2}$ and P' = $\frac{dp}{dt}$. The steady-state value of P is derived by setting P" = P' = 0. This yields P = 20. To find the adjustment path of P over time solve the second order, nonhomogeneous differential equation above and analyze the time path of Pt. Using *Mathematica,*

```
In[34]:= Clear[p, ox]
         ox = DSolve[{p''[t] == -.09 p'[t] - .024 p[t] + .48, p[0] == 6, p[1] == 5.8}, p[t], t]

Out[34]= {{Null (p[t] → e^(-0.045 t) (20. e^(0.045 t) - 14. Cos[0.14824 t] - 6.81894 Sin
                 [0.14824 t]))}}

In[35]:= p[t_] = p[t] /. ox[[1]];
         Plot[p[t], {t, 0, 100}]
```

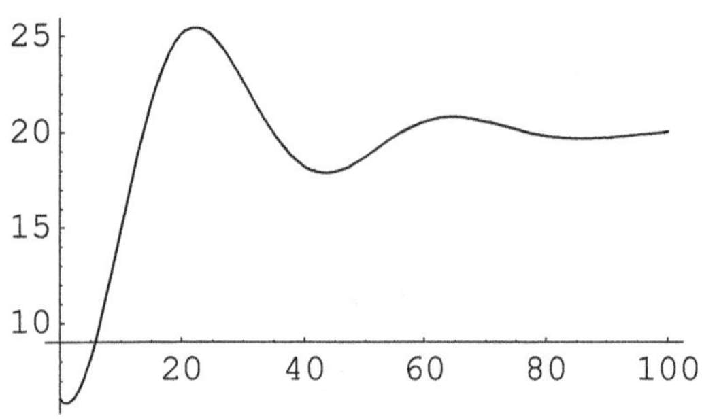

It is obvious from the graph that P converges to its steady state equilibrium value with an oscillatory path.

Do More Practice: Change the speed of price adjustment to excess demand from .3 to 3. Keep other assumptions the same. Re-run the program and compare the results.

2- Change the speed of price adjustment to excess demand from .3 to .03. Change the speed of price adjustment to accumulation of inventories from .08 to .3. Keep other assumptions the same. Re-run the program and compare the results.

3- Change the supply equation to Qs = -4 + 2p and keep the other assumptions of the problem the same. Re-run the program and compare the results.

Problems

1. Write the companion systems of the following differential equations.

a) $y^{(2)} - y^{(1)} - 2y = 0$

b) $3y^{(2)} - 2y^{(1)} - 5y = 0$

c) $y^{(3)} - y = 0$

d) $y^{(3)} - y^{(2)} + 2y^{(1)} - 3y = e^t$

e) $y^{(3)} + 2y^{(2)} - 5y = e^t$

f) $3y^{(4)} - 6y^{(2)} + 3y^{(1)} + y = 0$

2. Solve the following differential equations.

a) $y^{(2)} - 2y^{(1)} + y = e^{2t}$

b) $y^{(2)} - 2y^{(1)} + y = 0$

c) $3y^{(2)} + 2y^{(1)} - y = 0$

d) $y^{(2)} - 4Y = 3e^t$

e) $3y^{(2)} - 2y^{(1)} - 5y = 0$

3. Solve equation a of the problem 2 for the initial values $y(o) = 1$, $y'(o) = 2$.

4. Solve equation b of the problem 2 for the initial values $y(o) = 0$, $y'(o) = 1$.

5. Solve equation c of the problem 2 for the initial values $y(o) = 1$, $y'(o) = 1$.

6. Solve equation d of the problem 2 for the initial values $y(o) = 1$, $y'(o) = 1$.

7. The market demand and supply for a product are given as $Qd = 50 - .2p$ and $Qs = -10 + .3p$, respectively. It is assumed that the price adjustment in the market is proportional to excess demand in market, with a proportionality factor of .4. As well, it is assumed that the price adjustment in market is proportional to the accumulation of the stock of unsold inventories, with a proportionality factor of .1. Given the initial prices of $Po = 100$ and $P1 = 105$, find the steady-state value of the price, solve the dynamic model of the market for Pt, and plot the time path of Pt. Analyze the adjustment process.

VI. Differential Equations and Dynamic Stability of Economic Models

Economic models, in general, consist of a set of behavioral relations and a set of equilibrium conditions. For example, a simple market model consists of two behavioral equations, $Qd = a - bp$, $Qs = c + dp$, and a market-clearing or equilibrium condition, $Qd = Qs$. In these equations, Qd and Qs are quantity demanded and quantity supplied respectively, p is the market price, a, b, c, and d are parameters, where b & d > 0. Solving the system for the equilibrium condition yields $pe = \frac{a-c}{b+d}$ and $Qe = \frac{ad+cb}{b+d}$, where Qe and pe are equilibrium quantity and equilibrium price, respectively.

The equilibrium values derived above represent only the solution to the system of equations. However, the solution is a static one and does not say anything about the time path by which the system moves from one equilibrium to another. The differential equations of this chapter are dynamic models, where time is explicitly included in the models, and the models specify how variables react when they are away from their equilibrium values, or how they move from one equilibrium to another one. The stability analysis of economic models in this section is the study of how the equilibrium values of the variables react if there is any disturbance in the market.

An equilibrium may be stable, unstable, or neutral. An equilibrium is called stable if the values of the variables in question return to the equilibrium following a disturbance. An equilibrium is called unstable if the values of the variables progressively diverge from the equilibrium point as time passes. An equilibrium is neutral if the values of the variables remain unchanged.

In this section we will study the stability of economic models using first-order and higher-order linear differential equations.

Stability Analysis of the First-Order Linear Homogeneous Differential Models: An initial-value first order linear homogenous differential equation y'+ ay = 0, y(0) = yo, has a single (characteristic) root, -a, and the general solution of the equation is, $y(t) = yoe^{-at}$. As t → ∞, it is clear that the value of y will diverge from yo. If a is positive, as t → ∞, y will approach zero. If a is a negative number, as t → ∞, y will grow exponentially away from yo and will approach infinity. Therefore, the stability of an economic model with a first-order linear differential equation will depend on the sign of the (characteristic) root of the differential equation.

Example: Solve the initial-value differential equation y' + 2y = 0, y(0) = 1and analyze the stability of the model.

Solution: Since a = 2 > 0, as t → ∞, the value of y will converge to zero. To show this graphically, solve the problem and plot the time path of y,

```
In[36]:= ClearAll[y, fox]
         fox = DSolve[{y'[t] + 2 y[t] == 0, y[0] == 1}, y[t], t]
```

Out[36]= {{y[t] → e^{-2t}}}

```
In[37]:= y[t_] = y[t] /. fox[[1]]
         Plot[y[t], {t, 0, 5}]
```

Out[37]= e^{-2t}

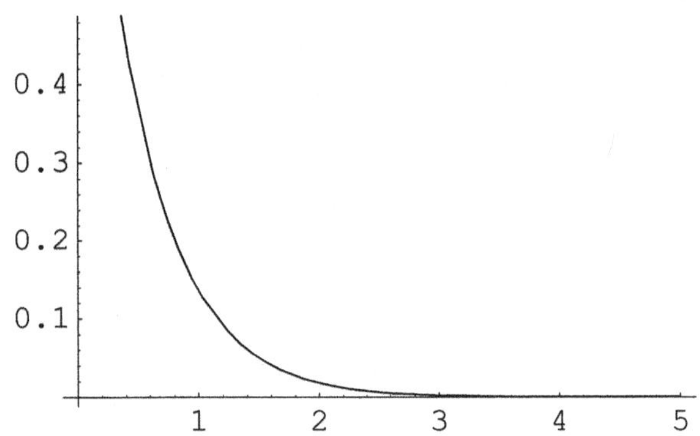

Do More Practice:

Change the value of "a" to -2. Re-run the program and observe changes in solution and the plot.

The conclusion derived in the stability analysis above is true also for the initial-value, first-order nonhomogeneous differential equations of the form y' + ay = b. If the value of "a" in the equation y' + ay = b is positive, as t → ∞, y will approach zero. If a is a negative number, as t → ∞, y will grow exponentially away from yo and will approach infinity. However, if the nonhomogeneous part of the differential equation is a function of t, y' + ay = f(t), then the stability of the model will depend on the form of the f(t) function. For these modes, the best way to analyze the stability of the model is to plot the time path of y and observe its behavior over time.

Example: Solve the initial-value differential equation y' + 2y = 3, y(0) = 1and analyze the stability of the model.

```
In[38]:= ClearAll [y, dog]
         dog = DSolve[{y'[t] + 2 y[t] == 3, y[0] == 1}, y[t], t]
```

Out[38]= {{y[t] → $\frac{1}{2} e^{-2t}(-1 + 3 e^{2t})$}}

```
In[39]:= y[t_] = y[t] /. dog[[1]]
         Plot[y[t], {t, 0, 5}]
```

Out[39]= $\frac{1}{2} e^{-2t}(-1+3e^{2t})$

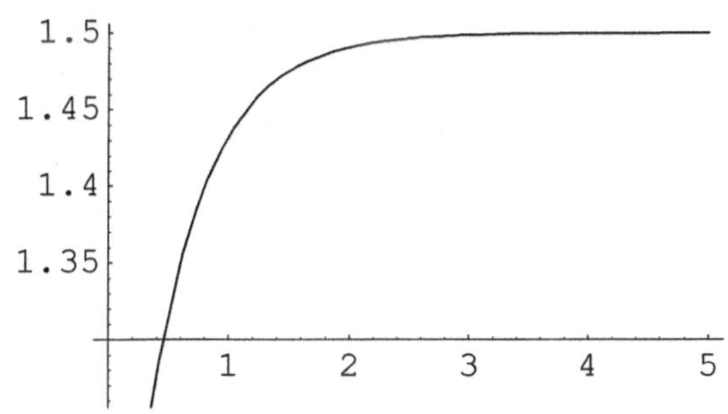

Do More Practice:

1- Change the value of "a" to -2. Re-run the program and observe changes in solution and the plot.

2- Change the value of "b" to -3. Re-run the program and observe changes in solution and the plot.

Example: Solve the initial-value differential equation y' + 2y = 3t, y(0) = 1and analyze the stability of the model.

```
In[40]:= Clear[y, bee]
         bee = DSolve[{y'[t] + 2 y[t] == 3 t, y[0] == 1}, y[t], t]
```

Out[40]= {{Null (y[t] → $\frac{1}{4} e^{-2t}(7-3e^{2t}+6e^{2t}t)$)}}

```
In[41]:= y[t_] = y[t] /. bee[[1]]
         Plot[y[t], {t, 0, 5}]
```

Out[41]= $\frac{1}{4} e^{-2t}(7-3e^{2t}+6e^{2t}t)$

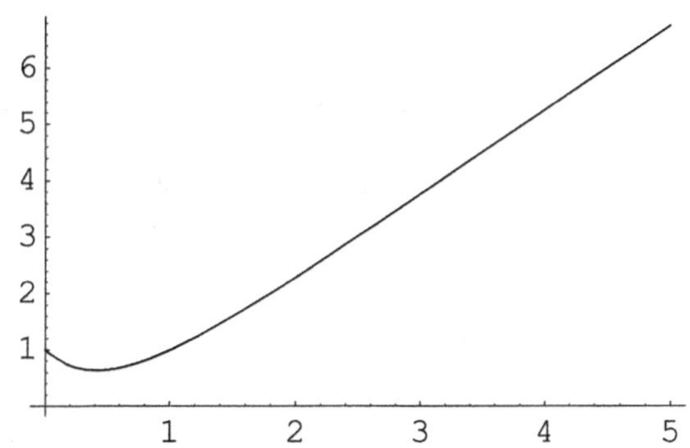

Do More Practice:

1- Change the value of "a" to -2. Re-run the program and observe changes in the solution and the plot.

2- Change the value of f(t) to -3t. Re-run the program and observe changes in the solution and the plot.

3- Change the value of f(t) to $3e^t$. Re-run the program and observe changes in the solution and the plot.

4- Change the value of f(t) to $3e^{-t}$. Re-run the program and observe changes in the solution and the plot.

Example: Application to Markets

Suppose that in the market model above the market price p ≠ pe. Generally, price changes in the market are proportional to excess demand in the market, i.e. $\frac{dp}{dt} = \lambda(Qd - Qs)$, $\lambda > 0$, and $\frac{dp}{dt} = 0$ only if Qd = Qs. Here, λ is the coefficient of adjustment or speed of adjustment . Substituting for Qd and Qs from the demand and supply equations, $\frac{dp}{dt} = \lambda(a - c) - \lambda(b + d)p$, or $\frac{dp}{dt} + \lambda(b + d)p = \lambda(a - c)$. This is a first-order linear differential equation, the solution of which is, $p(t) = (p(o) - \frac{a-c}{b+d})e^{-\lambda(b+d)t} + \frac{a-c}{b+d}$. Since $pe = \frac{a-c}{b+d}$, we can write $p(t) = (p(o) - pe)e^{-\lambda(b+d)t} + \frac{a-c}{b+d}$. Therefore, the dynamic stability of the market requires that, λ(b+d) > 0. Since λ, b, and d are positive by assumption, the stability condition is satisfied as long as demand and supply have the conventional downward sloping and upward sloping shapes, respectively. However, if demand or supply have unconventional slopes, the stability condition would require that b > -d. That is, for the model to be stable, the slope of the supply curve should exceed the slope of the demand curve. If this condition is satisfied, $e^{-\lambda(b+d)t}$ will approach zero and p(t) will approach pe as $t \to \infty$. Otherwise, p(t) will progressively deviate from pe.

Example: Application to Markets

Demand and supply for a product are given as Qd = 10 - .2p and Qs = -2 + p, respectively. A shift in demand has resulted at a market price of p = 8. Suppose that the price changes in the market adjust to excess demand with an adjustment rate of $\frac{dp}{dt} = .04(Qd - Qs)$. Find the equilibrium price and equilibrium quantity. Find the excess demand at p = 8 and the time path of the price adjustment to equilibrium. Analyze the stability of the market.

Solution: To find the equilibrium price and quantity set Qd = Qs and solve for p and Q,

```
In[42]:= ClearAll[p, owl]
         Qd = 10 - .2 p[t]; Qs = -2 + p[t];
         owl = Solve[Qd == Qs, p[t]]

Out[42]= {{Null (p[t] → 10.)}}

In[43]:= Qd /. owl

Out[43]= {8.}
```

Therefore, the equilibrium price and quantity are $10 and 8, respectively. To find the excess demand at p = 8,

```
In[44]:= (Qd - Qs) /. p[t] -> 8

Out[44]= 2.4
```

the excess demand is 2.4 units. To find the time path of the price adjustment to equilibrium, solve the differential equation $\frac{dp}{dt} = .04(Qd - Qs)$, p(o) = 8, and plot the time path of the price,

```
In[45]:= ClearAll[p, owl2]
         owl2 = DSolve[{p'[t] == .04 (Qd - Qs), p[0] == 8}, p[t], t]

Out[45]= {{p[t] → e^(-0.048 t) (-2. + 10. e^(0.048 t))}}

In[46]:= p[t_] = p[t] /. owl2[[1]]
         Plot[p[t], {t, 0, 100}]

Out[46]= e^(-0.048 t) (-2. + 10. e^(0.048 t))
```

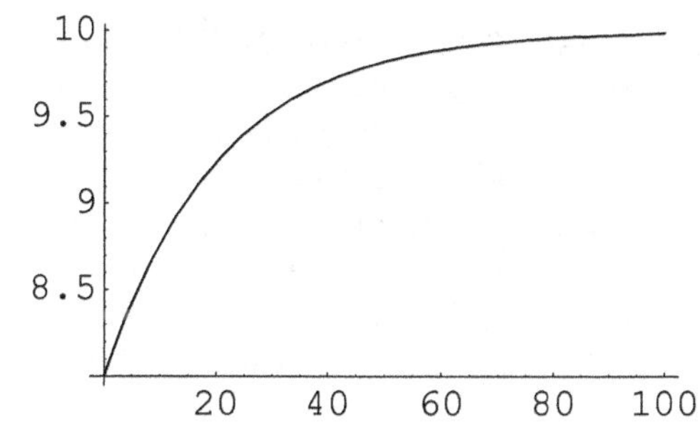

the time path of the price adjustment shows that as $t \to \infty$, the market price approaches the equilibrium price of p = 10.

Do More Practice: Change the speed of adjustment λ in the $\frac{dp}{dt} = \lambda(Qd - Qs)$ from .04 to a larger or smaller number, (.4 or .004). Re-run the program and observe the changes in the solution and the time path of the price adjustment.

2- Change the demand and supply equations of the preceding example to Qd = 10 - .8p and Qs = 4 - .2p. Keep the equation of the adjustment and other assumptions constant. Solve the resulting differential equation and graph the time path of the price adjustment by re-running the program.

3- Change the demand and supply equations of the preceding example to Qd = 10 - .2p and Qs = 16 - .8p. Keep the equation of the adjustment and other assumptions as it is. Re-run the program and compare the solution and the graph to the one in the text.

Stability Analysis of Linear Homogeneous Differential Models of Higher Order: A second-order linear differential equation y'' + ay' + by = 0 has a companion system $\mathbf{y}' = \mathbf{A}\,\mathbf{y} = \begin{pmatrix} 0 & 1 \\ -b & -a \end{pmatrix}\mathbf{y}$. The characteristic roots, or the eigenvalues of the **A** matrix of the system are a solution to the quadratic equation $k^2 + ak + b = 0$ with the roots $k1 = \frac{1}{2}\left(-a + \sqrt{a^2 - 4b}\right)$ and $k2 = \frac{1}{2}\left(-a - \sqrt{a^2 - 4b}\right)$. If k1 and k2 are distinct, the system has two distinct characteristic vectors, **u1** and **u2**, corresponding to the characteristic roots k1 and k2. Also, the system will have two distinct solutions, $\mathbf{y1} = \mathbf{u1}e^{k1t}$ and $\mathbf{y2} = \mathbf{u2}e^{k2t}$. As well, any linear combination of **y1** and **y2** such as, $\mathbf{y} = c1\mathbf{y1} + c2\mathbf{y2} = c1\mathbf{u1}e^{k1t} + c2\mathbf{u2}e^{k2t}$ will also be a solution to the system. The values of c1 and c2, however, are determined by the initial conditions.

If $a^2 - 4b$ of the characteristic roots above is positive, the roots are real and it is clear, from the solution $\mathbf{y} = c1\mathbf{y1} + c2\mathbf{y2} = c1\mathbf{u1}e^{k1t} + c2\mathbf{u2}e^{k2t}$ that the time path of **y** will approach equilibrium, if k1 and k2 are both negative. For the characteristic roots k1 and k2 to be negative, the solutions $k1 = \frac{1}{2}\left(-a + \sqrt{a^2 - 4b}\right)$ and $k2 = \frac{1}{2}\left(-a - \sqrt{a^2 - 4b}\right)$ imply that a and b both have to be positive. Therefore, for the model y'' + ay' + by = 0 to be stable, both a and b have to be positive.

If only one of the characteristic roots is positive while the other one is negative, say for example if k1< 0 and k2 > 0, then the term e^{k1t} in the solution equation will decay to zero as $t \to \infty$, while the term e^{k2t} will grow exponentially away from the equilibrium. In this case, the model has a saddle point.

If $a^2 - 4b$ of the characteristic roots is negative, the roots are complex conjugates. If the real parts of the complex conjugate roots are negative, the time path of y will spiral into equilibrium. The solution and the stability of equations with the complex conjugate roots will be discussed in chapter 14.

In general, for every nth-order linear homogeneous differential equation $y^{(n)} + an\text{-}1y^{(n-1)} + \ldots a1y^{(1)} + aoy = 0$ there is a characteristic equation of the form $k^n + an\text{-}1k^{n-1} + \ldots\ldots + a1k + ao = 0$ the solution of which gives the characteristic roots of the differential equation. The stability of the nth order differential equations requires that all the ai coefficients of the characteristic equation to be positive. If one or more of the ai coefficients are negative, then the system possesses positive roots, and it is dynamically unstable.

Eigenvalues and the Stability Analysis of the Linear Homogeneous Differential Equations: The stability test of the linear differential equations can also be done using the eigenvalue test on the **A** matrix of the system of linear differential equations or on the companion matrix of the companion system of higher order linear differential equations. A system of linear homogeneous differential equations or a higher order homogeneous linear differential will be stable iff the real parts of the eigenvalues of the **A** matrix of the system or the real parts of the eigenvalues of the companion matrix are all negative. If the eigenvalues are real with some positive and some negative, we have a saddle point. If the eigenvalues are complex conjugates with negative real parts, all paths spiral into the equilibrium (stability of the systems with complex conjugate eigenvalues will be studied in chapter 14).

The above condition is equivalent to saying that the necessary and sufficient conditions for the stability of the nth order differential equations are that the trace of the corresponding companion matrix to be negative and the determinant of the corresponding companion matrix to be positive. In many occasions, the trace and determinant conditions are a more convenient way of analyzing the stability of some higher order linear differential economic models.

Example: Analyze the stability of the second-order differential equation $y'' + 5y' + 6y = 0$, $y'(0) = y(0) = 1$.

Solution: The corresponding characteristic equation of the system is $k^2 + 5k + 6 = 0$. Since all the coefficients of the characteristic equation are positive, the characteristic roots are negative and the system is dynamically stable. The characteristic roots of the equation are -2 and -3.

A second way to do the same analysis is to form the companion matrix of the corresponding companion system, $\mathbf{A} = \begin{pmatrix} 0 & 1 \\ -6 & -5 \end{pmatrix}$. The trace of the matrix is tra(**A**) = -5 and the determinant is det(**A**) = 6. Since the trace is negative and the determinant is positive, the necessary and sufficient conditions for the stability of the system are satisfied and the system is dynamically stable.

To solve for y and to analyze the stability of the model using *Mathematica*,

```
In[47]:= ClearAll[y, elk]
         elk = DSolve[{y''[t] + 5 y'[t] + 6 y[t] == 0, y'[0] == y[0] == 1}, y[t], t]

Out[47]= {{y[t] → e^(-3t) (-3 + 4 e^t)}}

In[48]:= y[t_] = y[t] /. elk[[1]];
         Plot[y[t], {t, 0, 10}]
```

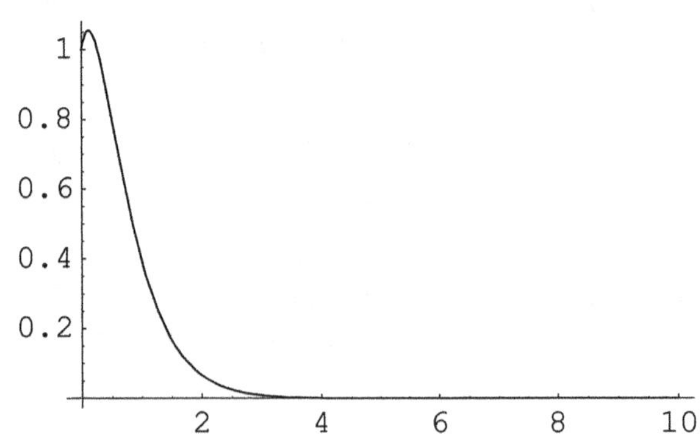

Do More Practice:

1- Change the coefficients form 5 and 6 to 5 and -6 and observe changes in the solution and the plot.

2- Change the coefficients to 3 and 6 and observe changes in the solution and the plot.

Example: Solve the following system of linear homogeneous differential equations and analyze the stability of the model, $\mathbf{Y}' = \begin{pmatrix} 3 & 1 \\ 1 & 2 \end{pmatrix} y$.

Solution: The eigenvalues of the **A** matrix are,

In[49]:= $\text{Eigenvalues}\left[\begin{pmatrix} 3 & 1 \\ 2 & 1 \end{pmatrix}\right]$

Out[49]= $\{(2-\sqrt{3})\,\text{Null},\ (2+\sqrt{3})\,\text{Null}\}$

the eigenvalues are both real and positive. Therefore, the system will be an unstable one.

To solve the problem using *Mathematica*,

```
In[50]:= ClearAll[y1, y2, hen]
         hen = DSolve[{y1'[t] == 3 y1[t] + y2[t], y2'[t] == y1[t] + 2 y2[t],
             y1[0] == y2[0] == 1}, {y1[t], y2[t]}, t]
```

Out[50]= $\{\{\text{y1}[t] \to \frac{1}{10}\left(5\,e^{\frac{1}{2}(5-\sqrt{5})t} - 3\sqrt{5}\,e^{\frac{1}{2}(5-\sqrt{5})t} + 5\,e^{\frac{1}{2}(5+\sqrt{5})t} + 3\sqrt{5}\,e^{\frac{1}{2}(5+\sqrt{5})t}\right),$
$\text{y2}[t] \to \frac{1}{10}\left(5\,e^{\frac{1}{2}(5-\sqrt{5})t} - \sqrt{5}\,e^{\frac{1}{2}(5-\sqrt{5})t} + 5\,e^{\frac{1}{2}(5+\sqrt{5})t} + \sqrt{5}\,e^{\frac{1}{2}(5+\sqrt{5})t}\right)\}\}$

```
In[51]:= y1[t_] = y1[t] /. hen[[1]]
         y2[t_] = y2[t] /. hen[[1]]
```

Out[51]= $\frac{1}{10}\left(5\,e^{\frac{1}{2}(5-\sqrt{5})t} - 3\sqrt{5}\,e^{\frac{1}{2}(5-\sqrt{5})t} + 5\,e^{\frac{1}{2}(5+\sqrt{5})t} + 3\sqrt{5}\,e^{\frac{1}{2}(5+\sqrt{5})t}\right)$

$\frac{1}{10}\left(5\,e^{\frac{1}{2}(5-\sqrt{5})t} - \sqrt{5}\,e^{\frac{1}{2}(5-\sqrt{5})t} + 5\,e^{\frac{1}{2}(5+\sqrt{5})t} + \sqrt{5}\,e^{\frac{1}{2}(5+\sqrt{5})t}\right)$

the plot of y1 and y2 over time is shown below,

```
In[52]:= Plot[y1[t], {t, 0, 5}]
         Plot[y2[t], {t, 0, 5}]
```

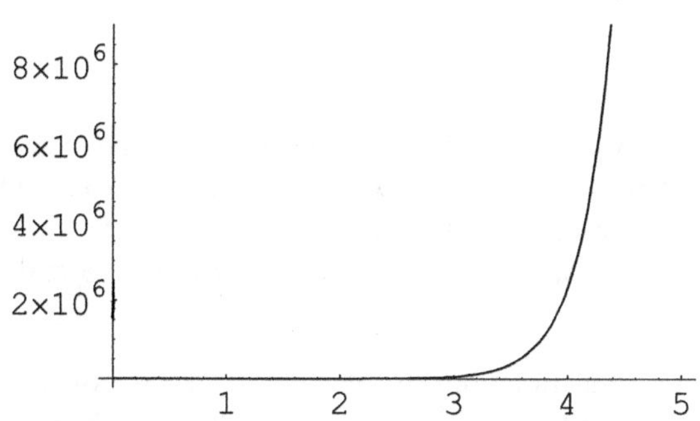

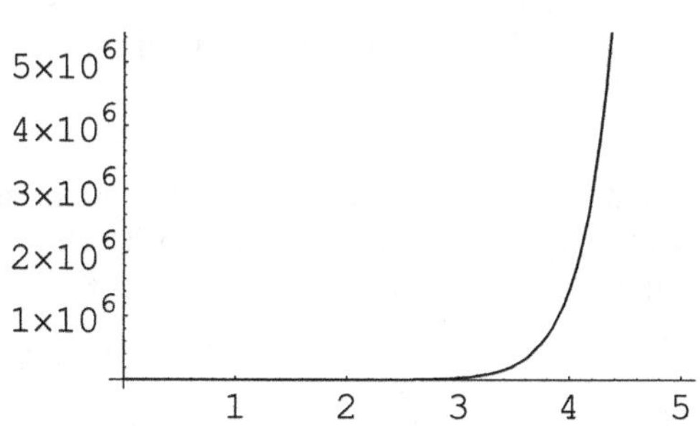

the parametric plot of y1, y2 , and t is shown below,

```
In[53]:= ParametricPlot3D[{t, y1[t], y2[t]}, {t, 0, 10}, Boxed -> False, BoxRatios -> {5, 2, 2},
    AxesLabel -> {t, y1, y2}]
```

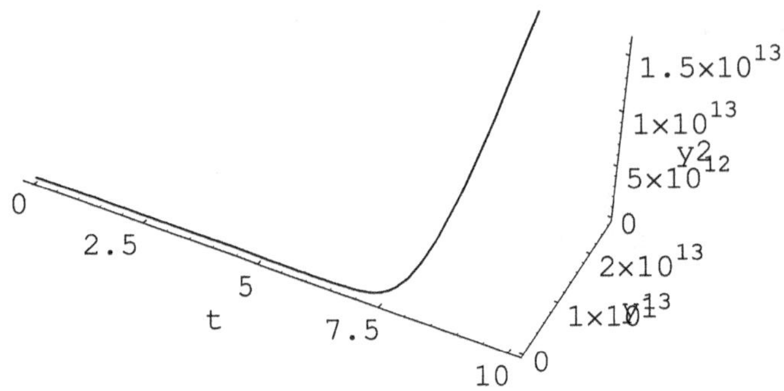

- Graphics3D -

all three graphs show that the system is unstable.

Do More Practice:

1- Change the coefficients of the **A** matrix to $\begin{pmatrix} 1 & 0 \\ 1 & 4 \end{pmatrix}$ and re-run the program. Compare the changes.

2- Change the coefficients of the **A** matrix to $\begin{pmatrix} 0 & 1 \\ 3 & 4 \end{pmatrix}$ and re-run the program. Compare the changes.

3- Change the coefficients of the **A** matrix to $\begin{pmatrix} 1 & -1 \\ 4 & 4 \end{pmatrix}$ and re-run the program. Compare the changes.

Example: Application to IS-LM Model

Suppose that in an IS-LM model of the economy, income(Y) adjusts with an α proportion to the excess demand in the product market and interest rate (r) adjusts with a β proportion to the excess demand for money in the money market. That is, $y' = \alpha(D(y, r) - y)$ and $r' = \beta(L(y, r) - M)$, where D(y, r) is aggregate demand, L(y, r) is demand for money, and M is money supply. Analyze the stability model with respect to the slopes of the IS and LM and the speeds of adjustments in the two markets.

Solution: Linearizing the model (totally differentiating the functions) in the neighborhood of the equilibrium point , y = ye and r = re, yields $dy' = \alpha(Dydy + Drdr - dy)$ and $dr' = \beta(Lydy + Lrdr - dM)$. In the matrix form the system can be written as $\begin{pmatrix} dy' \\ dr' \end{pmatrix} = \mathbf{A}\begin{pmatrix} dy \\ dr \end{pmatrix} = \begin{pmatrix} -\alpha(1-Dy) & \alpha Dr \\ \beta Ly & \beta Lr \end{pmatrix}\begin{pmatrix} dy \\ dr \end{pmatrix}$, where it is assumed that money supply is decided exogenously by the Fed and dM = 0.

The stability of the system requires that the trace of the **A** matrix of the system to be negative and its determinant to be positive. That is $tr(\mathbf{A}) = -\alpha(1 - Dy) + \beta Lr < 0$ and $\det(\mathbf{A}) = -\alpha\beta Lr(1 - Dy) - \alpha\beta LyDr > 0$. The trace condition will be satisfied if (1 - Dy) > 0 and Lr < 0. With the conventional assumptions on the slopes of the IS and LM curves, both of these conditions are satisfied. The determinant condition for the stability requires that $\frac{1-Dy}{Dr} < -\frac{Ly}{Lr}$. Here, $\frac{1-Dy}{Dr}$ is the slope of the IS curve and $-\frac{Ly}{Lr}$ is the slope of the LM curve. The stability of the system requires that the slope of the LM curve exceeds the slope of the IS curve (Note: To find the slope of the IS and LM curves, totally differentiate IS and LM equations at the equilibrium point, where D(Y, r) = y and L(y, r) = M, then find $\frac{dr}{dy}$ for each equation. This will yield $\frac{dr}{dy} = \frac{1-Dy}{Dr}$ as the slope of the IS curve, and $\frac{dr}{dy} = -\frac{Ly}{Lr}$ as the slope of the LM curve.

Example: Application to IS-LM Models

A simple macro model of an economy is represented by the following equations: c = 500 + .9y, i = 100 - 3r, y = c + i, Md = .25y - .8r, and Md = Ms = 400. The first equation expresses consumption (c) as a function of income (y); the second equation expresses investment (i) as a function of interest rate (r); the third equation is the equilibrium condition in the goods market; the fourth equation expresses demand for money (Md) as a function of income and interest rate; and the last equation is the equilibrium condition in the money market. Suppose that the speed of the adjustments in the goods market and money market are proportional to the excess demand in the corresponding markets, with a proportionality factor of .4 for the goods market and a proportionality factor of .6 for the money markets. Given that the initial values of y and r are y(0) = 3000 and r(0) = 10, find the time path of y and r and analyze the stability of the model.

Solution: The equilibrium condition in the goods market requires y = c + i = 600 + .9y - 3r and the equilibrium condition in the money market requires .25y - .8 r = 400. The slope of the IS curve is, $\frac{dr}{dy} = \frac{.10}{-3} = -.033$ and the slope of the LM curve is $\frac{dr}{dy} = \frac{.25}{.8} = .312$. IS and LM have the conventional slopes and the trace condition for the stability of the model is satisfied. The determinant condition for the stability of the model requires that the slope of the LM curve exceeds the slope of the IS curve. This condition is also satisfied. Therefore, the model is a stable model.

To solve the differential equation using *Mathematica*,

```
In[54]:= ClearAll[y, c, i, r, Md, Ms, hog]
         c = 500 + .9 y[t]; i = 100 - 3 r[t]; Md = .25 y[t] - .8 r[t]; Ms = 400;
         hog = DSolve[{y'[t] == .4 (c + i - y[t]), r'[t] == .6 (Md - Ms), r[0] == 10,
             y[0] == 2000}, {y[t], r[t]}, t]
```

Out[54]= $\{\{\text{Null}^2\ (y[t] \to (0. + 0.00366206\,i)\,((53344.9 + 3290.\,i)\,e^{(-0.26-0.362767\,i)\,t} - (53344.9 - 3290.\,i)\,e^{(-0.26+0.362767\,i)\,t} - (0. + 552720.\,i)\,e^{(0.+0.\,i)\,t})),\ \text{Null}^2\ (r[t] \to (0. + 0.000183103\,i)\,((175706. + 334593.\,i)\,e^{(-0.26-0.362767\,i)\,t} - (175706. - 334593.\,i)\,e^{(-0.26+0.362767\,i)\,t} - (0. + 723800.\,i)\,e^{(0.+0.\,i)\,t}))\}\}$

To show the dynamics of the time path to equilibrium,

```
In[55]:= y[t_] = y[t] /. hog[[1]];
         r[t_] = r[t] /. hog[[1]];
```

Out[55]= Null^2

```
In[56]:= ParametricPlot3D[{t, r[t], y[t]}, {t, 0, 20}, Boxed -> False, BoxRatios -> {5, 2, 2},
          AxesLabel -> {t, r, y}]
```

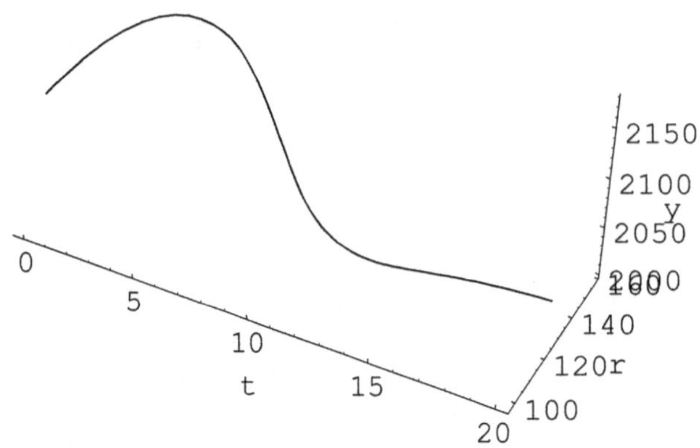

- Graphics3D -

Do More Practice: Change the speed of the adjustments in the goods and money markets from .4 and .6 to .2 and .9, respectively. Re-run the program and observe changes in the solution and the adjustment path.

2- Change the marginal propensity to consume in the consumption function from .9 to .7. Change the coefficient of income in the demand for money equation from .25 to .30. Re-run the program and observe changes in the solution and the adjustment path.

Example: Application to IS-LM Model

Assume that a macro model of an economy is represented by the following equations: c = 350 + .8y, i = 75 - 3r, y = c + i, Md = .25y - .8r, and Md = Ms = 490, where the variables and the relationships are defined as in the previous example. Suppose that the speed of the adjustments in the goods market and money market are proportional to the excess demand in the corresponding markets, with a proportionality factor of .4 for the goods market and a proportionality factor of .6 for the money markets. Given that the initial values of y and r are y(0) = 1500 and r(0) = 9, find the time path of y and r and analyze the stability of the model.

Solution: The equilibrium condition in the goods market requires y = c + i = 425 + .8y - 3r and the equilibrium condition in the money market requires .25y - .8 r = 490. The slope of the IS curve is, $\frac{dr}{dy} = \frac{.2}{-3} = -.066$ and the slope of the LM curve is $\frac{dr}{dy} = \frac{.25}{.8} = .312$. IS and LM have the conventional slopes and the trace condition for the stability of the model is satisfied. The determinant condition for the stability of the model requires that the slope of the LM curve exceeds the slope of the IS curve. This condition is also satisfied. Therefore, the model is a stable model. However, the solution of the system of differential equations using *Mathematica* shows that the solution involves imaginary numbers. Therefore, the time path of variables spirals to equilibrium. To solve the problem using *Mathematica* and to observe the time path to equilibrium, use the following commands,

```
In[57]:= ClearAll[y, c, i, r, Md, Ms, eel]
         c = 350 + .8 y[t]; i = 75 - 3 r[t]; Md = .25 y[t] - .8 r[t]; Ms = 490;
         eel = DSolve[{y'[t] == .4 (c + i - y[t]), r'[t] == .6 (Md - Ms), r[0] == 9,
             y[0] == 1500}, {y[t], r[t]}, t]
```

```
Out[57]= {{y[t] → (0. - 0.021978 i) ((5941.75 - 11125. i) e^((-0.28-0.374166 i) t) -
              (5941.75 + 11125. i) e^((-0.28+0.374166 i) t) +
              (0. + 90500. i) e^((0.+0. i) t)),
          r[t] → (0. - 0.0164835 i) ((5945.49 - 2. i) e^((-0.28-0.374166 i) t) - (5945.49 + 2. i) e^((-0.28+0.374166 i) t) +
              (0. + 550. i) e^((0.+0. i) t))}}
```

the following commands will yield the time path of y(t)

```
In[58]:= Short[y[t_] = y[t] /. eel[[1]]]
         Plot[y[t], {t, 0, 5}]
```

Out[58]= $(0. - 0.021978\, i)((5941.75 - 11125.\, i)\, e^{(-0.28-0.374166\, i) \ll 1 \gg t} - \ll 1 \gg + (0. + 90500.\, i)\, e^{\ll 1 \gg})$

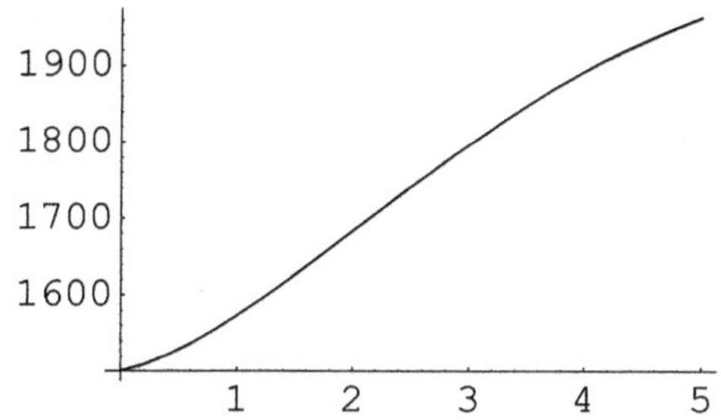

the following commands will yield the time path of r(t)

```
In[59]:= Short[r[t_] = r[t] /. eel[[1]]]
         Plot[r[t], {t, 0, 10}]
```

Out[59]= $(0. - 0.0164835\, i)((5945.49 - 2.\, i)\, e^{(-0.28-0.374166\, i) t} - (\ll 1 \gg)\, e^{\ll 1 \gg} + (0. + 550.\, i)\, e^{(\ll 1 \gg) t})$

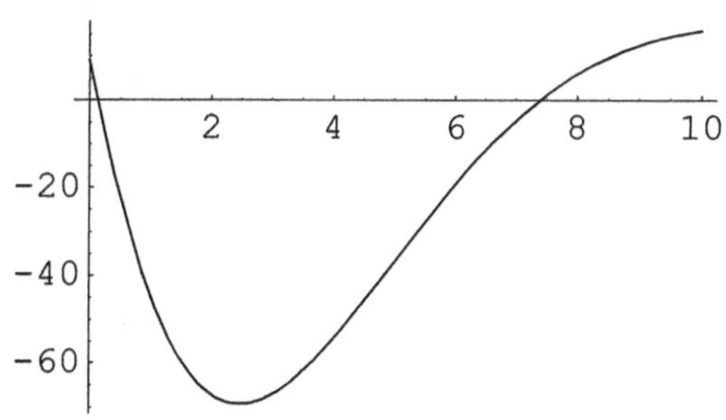


```
In[60]:=
    ParametricPlot3D[{t, r[t], y[t]}, {t, 0, 10}, Boxed -> False, BoxRatios -> {5, 2, 2},
      AxesLabel -> {t, r, y}]
```

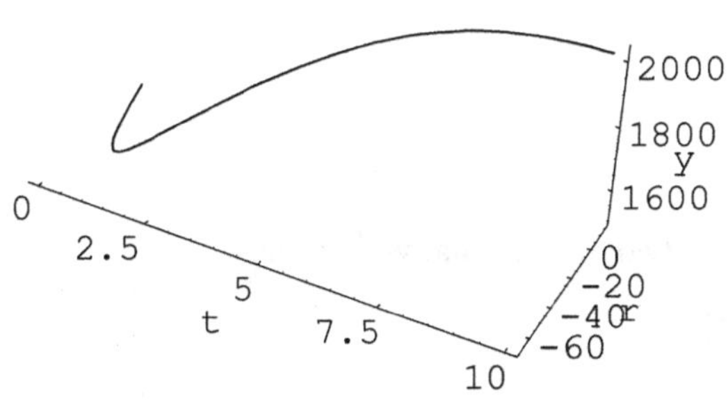

- Graphics3D -

To graph the IS-LM on the standard two dimensional space, use the following commands:

To solve for the IS equation from the equilibrium condition in the goods market,

```
In[61]:= ise = Solve[y == 425 + .8 y - 3 r, r]

Out[61]= {{r → -0.333333 (-425. + 0.2 y)}}
```

To plot the IS,

```
In[62]:= is[y_] = r /. ise[[1]];
         g1 = Plot[is[y], {y, 1800, 2200}]
```

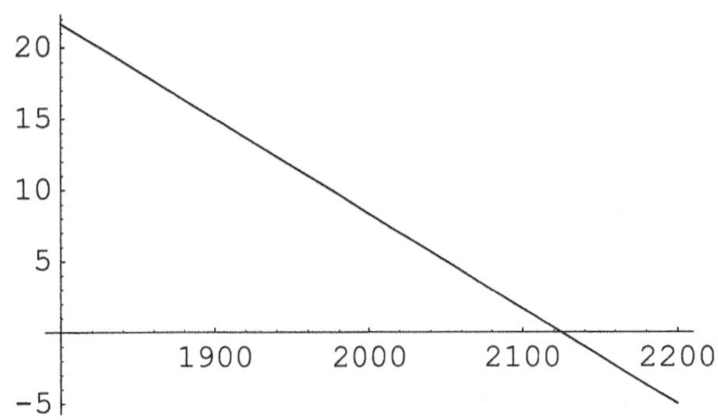

To solve for the LM equation from the equilibrium condition in the money market,

```
In[63]:= lme = Solve[490 == .25 y - .8 r, r]

Out[63]= {{r → -1.25 (490. - 0.25 y)}}
```

To graph the LM,

```
In[64]:= lm[y_] = r /. lme[[1]]
         g2 = Plot[lm[y], {y, 1800, 2200}]

Out[64]= -1.25 (490. - 0.25 y)
```

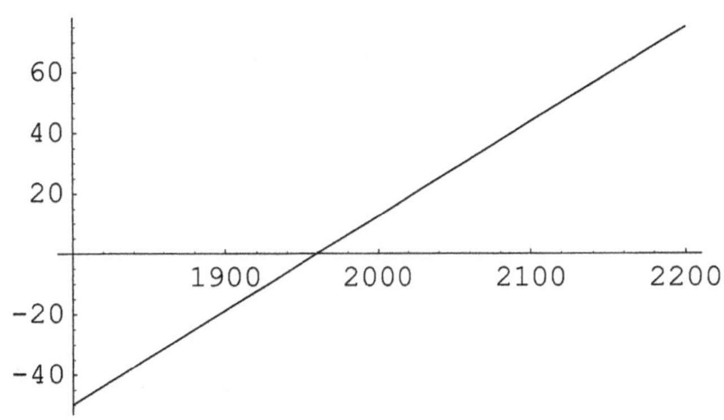

To graph the parametric plot of IS and Lm over t on the two dimensional y - r space,

```
In[65]:= g3 = ParametricPlot[{y[t], r[t]}, {t, 0, 20}]
```

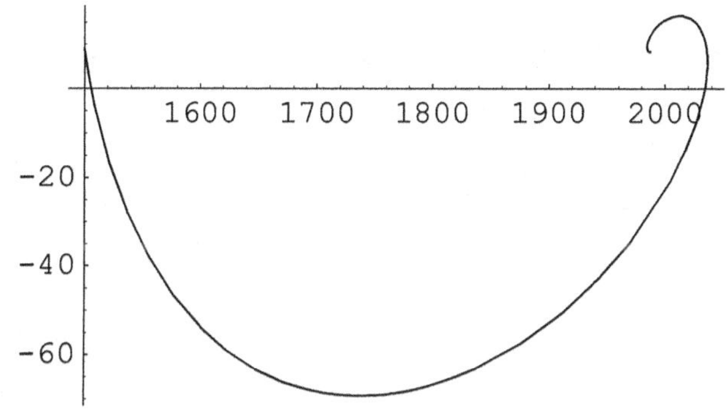

To show the three graphs together on the y - r space,

```
In[66]:= Show[g1, g2, g3]
```

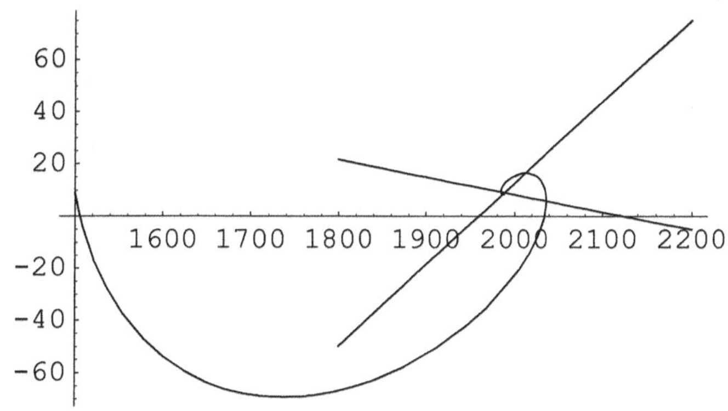

Do More Practice:

1- Change the speed of the adjustments in the goods and money markets from .4 and .6 to .2 and .9, respectively. Re-run the program and observe changes in the solution and the adjustment path.

2- Change the marginal propensity to consume in the consumption function from .8 to .7. Change the coefficient of income in the demand for money equation from .25 to .30. Re-run the program and observe changes in the solution and the adjustment path.

Example:

Solve the following system of linear homogeneous differential equations and analyze the stability of the system: y'1 = y1 + .8y3; y'2 = y1 -.2y2; y'3 = -.3y3.

Solution: The system can be written as $\mathbf{Y'} = \mathbf{AY} = \begin{pmatrix} -1 & 0 & .8 \\ 1 & -.2 & 0 \\ 0 & 0 & -.3 \end{pmatrix}\mathbf{Y}$. The eigenvalues of the **A** matrix of the system are,

In[67]:= Eigenvalues[$\begin{pmatrix} -1 & 0 & 0 \\ 1 & -.2 & 0 \\ 0 & 0 & -.3 \end{pmatrix}$]

Out[67]= {-1., -0.3, -0.2}

the eigenvalues are all real and negative. Therefore, the system will be a stable one. To solve the system and to graph the time path of the variables, type the following commands in *Mathematica*

```
In[68]:= Clear[y1, y2, y3, ray]
         ray = DSolve[{y1'[t] == -y1[t] + .8 y3[t], y2'[t] == y1[t] - .2 y2[t], y3'[t] == -.3 y3[t],
            y1[0] == y2[0] == y3[0] == 1}, {y1[t], y2[t], y3[t]}, t]
```

Out[68]= {{Null (y1[t] → $0.142857\, e^{-1.t}(-1. + 8.\, e^{0.7t})$), Null (y2[t] →
$0.0357143\, e^{-1.t}(5. - 320.\, e^{0.7t} + 343.\, e^{0.8t})$),
Null (y3[t] → $e^{-0.3t}$)}}

```
In[69]:= y1[t_] = y1[t] /. ray[[1]]
            Plot[y1[t], {t, 0, 10}]
```

Out[69]= $0.142857\, e^{-1.t}(-1. + 8.\, e^{0.7t})$

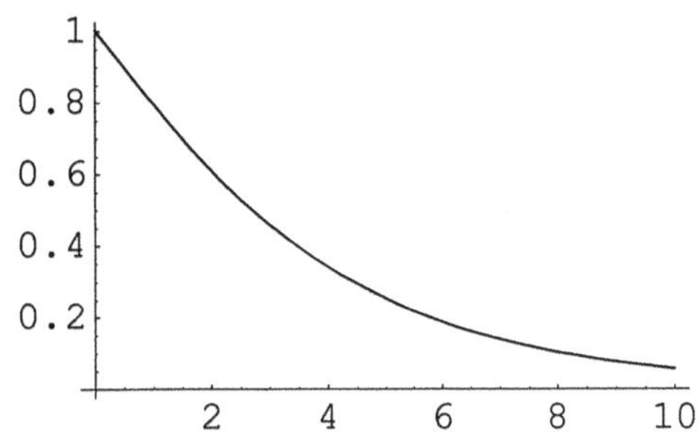

```
In[70] := y2[t_] = y2[t] /. ray[[1]]
      Plot[y2[t], {t, 0, 10}]
```

Out[70]= $0.0357143\, e^{-1.t}(5. - 320.\, e^{0.7t} + 343.\, e^{0.8t})$

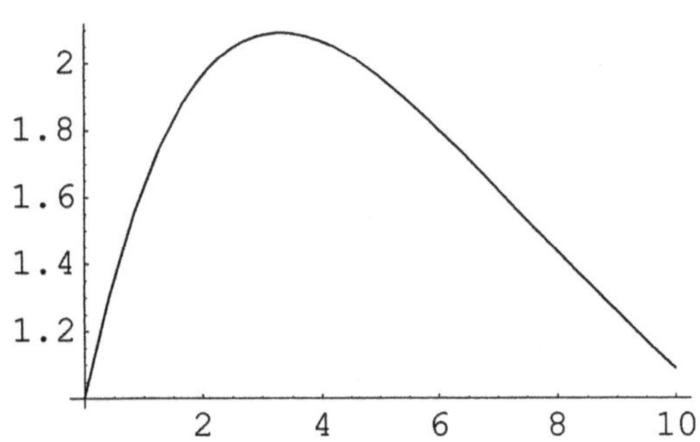

```
In[71]:= y3[t_] = y3[t] /. ray[[1]]
         Plot[y3[t], {t, 0, 10}]
```

Out[71]= $e^{-0.3t}$

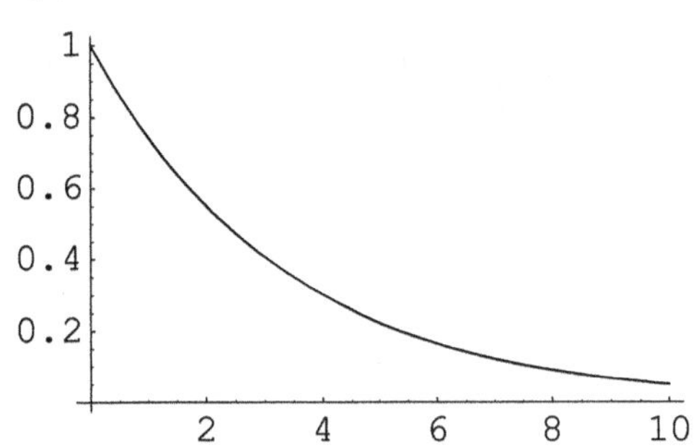

```
In[72] := ParametricPlot3D[{y1[t], y2[t], y3[t]}, {t, 0, 20}, Boxed -> False, BoxRatios -> {5, 2, 2}]
```

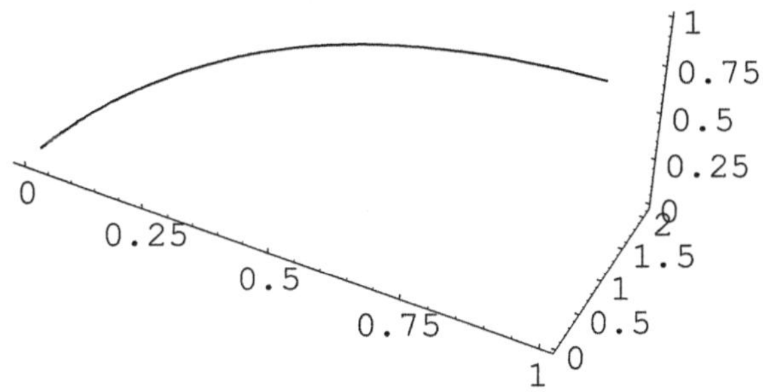

- Graphics3D -

Do More Practice:

1- Change the **A** matrix of the system to $\begin{pmatrix} 1 & 0 & .8 \\ 1 & .2 & 0 \\ 0 & 0 & .3 \end{pmatrix}$ and re-run the program. Compare the results to those in the text.

2- Change the **A** matrix of the system to $\begin{pmatrix} 1 & 0 & .8 \\ 1 & -.2 & 0 \\ 0 & 0 & .3 \end{pmatrix}$ and re-run the program. Compare the results to those in the text.

3- Change the **A** matrix of the system to $\begin{pmatrix} 1 & 0 & .8 \\ 1 & .2 & 0 \\ 1 & 0 & .3 \end{pmatrix}$ and re-run the program. Compare the results to those in the text.

4- Compare the effects of the changes in the eigenvalues of a system on the stability of the system and the time path of the variables using the practices above.

Problems

1. Analyze the stability of the following differential equations

    a) $y' + 2y = 0$

    b) $y' = 2y + 4$

    c) $y'' + 2y' + y = e$

    d) $y'' - 2y' + y = e^t$

    e) $y''' + 6y'' + 11y' + 6y = e^t$

    f) $y''' + 2y'' - y' - 2y = 0$

    g) $y''' - y' + y = 0$

2. Analyze the stability of the markets for the following models

    a) Qd = 100 -2p

       Qs = -20 + 3p

b) Qd = 100 - 2p

Qs = -20 + p

c) Qd = 100 - 4p

Qs = 40 - p

d) Qd = 100 - p

Qs = 160 - 4p

3. A simple macro model of an economy is represented by the following equations: C = 150 + .9y, I = 100 - 4r, y = C + I, Md = .2y - 1.2r, and Md = Ms = 400. The first equation expresses consumption (C) as a function of income (y) ; the second equation expresses investment (I) as a function of interest rate (r); the third equation is the equilibrium condition in the goods market; the fourth equation expresses demand for money (Md) as a function of income and interest rate; and the last equation is the equilibrium condition in the money market. Suppose that the speed of the adjustments in the goods market and money market are proportional to the excess demand in the corresponding markets, with a proportionality factor of .3 for the goods market and a proportionality factor of .8 for the money markets. Given that the initial values of y and r are y(0) = 1500 and r(0) = 10, find the time path of y and r and analyze the stability of the model.

4- Solve and analyze the stability of the following system of linear homogeneous differential equations: y1' = y2 - .8y3; y2' = y1 + y3; y3' = y2 + y3.

5- Solve and analyze the stability of the following system of linear homogeneous differential equations: y1' = 2y1 - y2 + .8y3; y2' = -.6y1 + y3; y3' = 1.1y2 - .3y3.

Appendix 1: Existence of Definite Solution

One very important theory of the system of differential equations is concerned with the variety of solutions and the general solution to the system of differential equations. According to this theory, if **y1**(t) and **y2**(t) are solutions of y'(t) = **A**y(t), then for any arbitrary constants c1 and c2, **y**(t) = c1**y1**(t) + c2y2(t) is also a solution. This equation is called the general solution to the system of differential equations.

Example: Solve the system of equations in the preceding example for the initial values, **y**(o) = $\begin{pmatrix} 1 \\ 9 \end{pmatrix}$.

Solution: The general solution to the problem is, $\begin{pmatrix} y1(t) \\ y2(t) \end{pmatrix} = c1\, e^{-t}\begin{pmatrix} -1 \\ 3 \end{pmatrix} + c2\, e^{3t}\begin{pmatrix} 1 \\ 1 \end{pmatrix}$. To find the solution of the initial-value problem, set t = 0 in the general solution, set $\begin{pmatrix} y1(0) \\ y2(0) \end{pmatrix} = \begin{pmatrix} 1 \\ 9 \end{pmatrix}$, and solve for c1 and c2. That is, $\begin{pmatrix} 1 \\ 9 \end{pmatrix} = \begin{pmatrix} -1 & 1 \\ 3 & 1 \end{pmatrix} . \begin{pmatrix} c1 \\ c2 \end{pmatrix}$. The solution to this system of equations is,

In[73]:= Solve[$\begin{pmatrix} -1 & 1 \\ 3 & 1 \end{pmatrix} . \begin{pmatrix} c1 \\ c2 \end{pmatrix} == \begin{pmatrix} 1 \\ 9 \end{pmatrix}$, {c1, c2}]

Out[73]= {{c1 → 2, c2 → 3}}

This results in the initial-value solutions of y1(t) = $-2e^{-t} + 2e^{3t}$ and y2(t) = $6e^{-t} + 3e^{3t}$.

To solve the same problem using *Mathematica*, type

```
In[74]:= ClearAll[y1, y2, cat]
         cat = DSolve[{y1'[t] == 2 y1[t] + y2[t], y2'[t] == 3 y1[t], y1[0] == 1, y2[0] == 9}, {y1[t], y2[t]}, t]
```

Out[74]= $\{\{y1[t] \to e^{-t}(-2+3e^{4t}),\ y2[t] \to 3e^{-t}(2+e^{4t})\}\}$

Once the problem is solved, use the replacement rule "/." to define the function y1[t] and y2[t] and graph the time path of the variables.

```
In[75]:=y1[t_] = y1[t] /. cat[[1]]
        y2[t_] = y2[t] /. cat[[1]]
        ParametricPlot3D[{y1[t], y2[t], t}, {t, 0, 10}, Boxed -> False,
          BoxRatios -> {4, 2, 2}, AxesLabel -> {y1, y2, t}]
```

Out[75]= $e^{-t}(-2+3e^{4t})$

$3e^{-t}(2+e^{4t})$

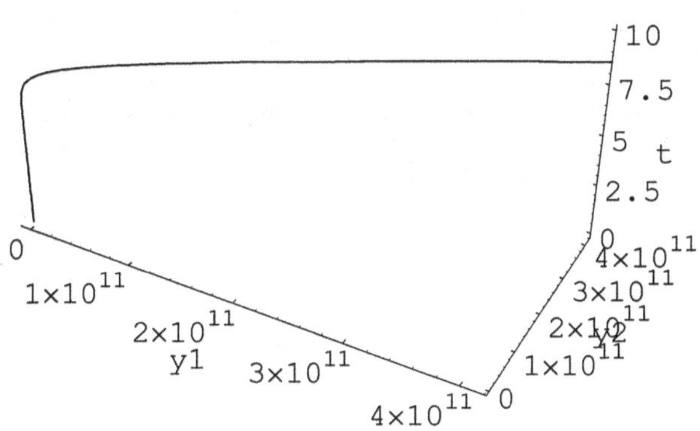

- Graphics3D -

Do More Practice: Change the coefficients of the **A** matrix in the preceding example to $\begin{pmatrix} 1 & 0 \\ 1 & 4 \end{pmatrix}$, find the eigenvalues of the **A** matrix, and re-run the program. Observe the changes in the eigenvalues and their relations to the time path of the solution.

2- Change the coefficients of the **A** matrix in the preceding example to $\begin{pmatrix} 0 & 1 \\ 3 & 4 \end{pmatrix}$, find the eigenvalues of the **A** matrix, and re-run the program. Observe the changes in the eigenvalues and their relations to the time path of the solution.

3- Change the coefficients of the **A** matrix in the preceding example to $\begin{pmatrix} 1 & -1 \\ 4 & 4 \end{pmatrix}$, find the eigenvalues of the **A** matrix, and re-run the program. Observe the changes in the eigenvalues and their relations to the time path of the solution.and redo the example.

A system of n differential equations has a distinct solution for the given initial values if the matrix **A** of the system has n distinct eigenvalues. If the system does not have enough distinct eigenvalues and eigenvectors, the variety of solutions is reduced and the initial-value problem may not be solvable. However, we can search for other solutions that will provide us with a definite solution to the initial value problem. This point will be illustrated through the following example.

Example: Solve the system of differential equations $\mathbf{y}'(t) = \begin{pmatrix} 1 & 3 \\ 0 & 1 \end{pmatrix} \mathbf{y}(t)$ for the initial values $y(0) = \begin{pmatrix} 1 \\ 1 \end{pmatrix}$.

Solution: The eigenvalues and eigenvectors of the **A** matrix are,

```
In[76]:= Eigensystem[(1 3
                      0 1)]

Out[76]= {{1, 1}, {{1, 0}, {0, 0}}}
```

Since the **A** matrix is of order 2x2, to solve the initial-value problem we need two distinct eigenvectors. However, the matrix **A** does not have two distinct eigenvalues, which implies it does not have two distinct eigenvectors. This does not necessarily imply that we can not find two distinct eigenvectors and solve the initial-value problem. If the matrix **A** does not have enough distinct eigenvalues, and there is a deficiency in the number of distinct or linearly independent eigenvectors, we have to search for other solutions that will provide us with enough independent solutions to $\mathbf{y}'(t) = \mathbf{Ay}(t)$ so that it could be solved for the initial values. One possibility is to look for functions of the form, $\mathbf{y}(t) = (\mathbf{u}\,t + \mathbf{v})e^{kt}$ or $\mathbf{y}(t) = (\mathbf{u}\,t^2 + \mathbf{v}\,t + \mathbf{w})e^{kt}$, etc.

Let us explore a solution of the form $\mathbf{y}(t) = (\mathbf{u}\,t + \mathbf{v})e^{kt}$. For $\mathbf{y}(t) = (\mathbf{u}\,t + \mathbf{v})e^{kt}$ to be a solution of $\mathbf{y}'(t) = \mathbf{Ay}(t)$, the differential of $\mathbf{y}(t)$ in $\mathbf{y}(t) = (\mathbf{u}\,t + \mathbf{v})e^{kt}$ should be equal to $\mathbf{Ay}(t)$. That is, $\mathbf{y}'(t) = \mathbf{u}e^{kt} + k(\mathbf{u}t + \mathbf{v})e^{kt} = \mathbf{Ay}(t) = \mathbf{A}(\mathbf{u}te^{kt} + \mathbf{v}e^{kt})$, or $\mathbf{u}e^{kt} + k\mathbf{u}te^{kt} + k\mathbf{v}e^{kt} = \mathbf{Au}te^{kt} + \mathbf{Av}e^{kt}$. By rearranging the terms we get, $(\mathbf{Au} - k\mathbf{u})te^{kt} + (\mathbf{Av} - k\mathbf{v} - \mathbf{u})e^{kt} = \mathbf{0}$. For this equation to be satisfied, the two equations $(\mathbf{A} - k\,\mathbf{I})\mathbf{u} = \mathbf{0}$ and $(\mathbf{A} - k\,\mathbf{I})\mathbf{v} = \mathbf{u}$ should be true. The first equation, $(\mathbf{A} - k\,\mathbf{I})\mathbf{u} = \mathbf{0}$, is the solution that we had before. The second equation, $(\mathbf{A} - k\,\mathbf{I})\mathbf{v} = \mathbf{u}$, provides us with a new solution to the system.

In the example above, the only distinct eigenvalue of the system is $k = 1$, with the corresponding eigenvector $\begin{pmatrix} 1 \\ 0 \end{pmatrix}$. To find another solution we use $(\mathbf{A} - k\,\mathbf{I})\mathbf{v} = \mathbf{u}$. Since we already have k and **u**, we can find **v** from the equation $(\mathbf{A} - k\,\mathbf{I})\mathbf{v} = \mathbf{u}$ by solving $\left(\begin{pmatrix} 1 & 3 \\ 0 & 1 \end{pmatrix} - \begin{pmatrix} 1 & 0 \\ 0 & 1 \end{pmatrix}\right)\begin{pmatrix} v1 \\ v2 \end{pmatrix} = \begin{pmatrix} 1 \\ 0 \end{pmatrix}$,

```
In[77]:= Solve[(0 3   (v1    == (1   , {v1, v2}]
                0 0)   v2)       0)

Solve::svars : Equations may not give solutions for all "solve" variables.

Out[77]= {{v2 -> 1/3}}
```

here, $v2 = \frac{1}{3}$ and v1 can be any arbitrary number, say 1. Therefore, the two solutions are $\mathbf{y1} = e^t\begin{pmatrix} 1 \\ 0 \end{pmatrix}$ and $\mathbf{y2} = te^t\begin{pmatrix} 1 \\ 0 \end{pmatrix} + e^t\begin{pmatrix} 1 \\ \frac{1}{3} \end{pmatrix} = e^t\begin{pmatrix} t+1 \\ \frac{1}{3} \end{pmatrix}$.

The general solution of the problem is, $\mathbf{y}(t) = c1\mathbf{y}1(t) + c2\mathbf{y}2(t) = c1e^t\begin{pmatrix} 1 \\ 0 \end{pmatrix} + c2\,e^t\begin{pmatrix} t+1 \\ \frac{1}{3} \end{pmatrix}$. Solving this equation for the initial values yields, $\begin{pmatrix} 1 \\ 1 \end{pmatrix} = \begin{pmatrix} c1 \\ 0 \end{pmatrix} + \begin{pmatrix} c2 \\ \frac{1}{3}\,c2 \end{pmatrix}$ or, $c2 = 3$ and $c1 = -2$. Therefore. The initial-value problem has a solution $\mathbf{y}(t) = -2\mathbf{y}1(t) + 3\mathbf{y}2(t)$ or $\begin{pmatrix} y1(t) \\ y2(t) \end{pmatrix} = \begin{pmatrix} e^t + 3\,te^t \\ e^t \end{pmatrix}$. Using *Mathematica*, the solution is,

```
In[78]:= Clear[y1, y2, ape]
         ape = DSolve[{y1'[t] == y1[t] + 3 y2[t], y2'[t] == y2[t], y1[0] == 1, y2[0] == 1}, {y1, y2}, t]

Out[78]= {{y1 -> (e^#1 (1 + 3 #1) &), y2 -> (e^#1 &)}}
```

To plot the time path of y1[t] and y2[t], use the replacement rule "/." to define the functions y1[t] and y2[t].

```
In[79]:= y1[t_] = y1[t] /. ape[[1]]
```

Out[79]= $e^t (1 + 3 t)$

```
In[80]:= y2[t_] = y2[t] /. ape[[1]]
```

Out[80]= e^t

```
In[81]:=ParametricPlot3D[{y1[t], y2[t], t}, {t, 0, 10}, Boxed -> False, BoxRatios -> {4, 2, 2},
        AxesLabel -> {y1, y2, t}]
```

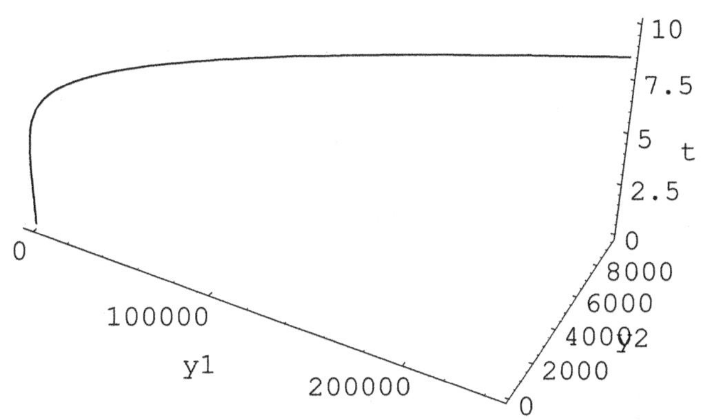

- Graphics3D -

Example: Application to Macroeconomics

The change in national debt of an economy (the budget deficit) is a constant proportion .03 of its GNP and the GNP growth of the economy is 8%. Given the initial GNP is 6000 and the debt is 3000, find the time path of GNP and the debt.

Solution: Lets call the GNP, y and the national debt x. The problem is to solve the system of equations x' = .03y and y' = .08y, for the initial values y(0) = 6000 and x(0) = 4000. The system can be written in matrix form as, $\begin{pmatrix} x' \\ y' \end{pmatrix} = \begin{pmatrix} 0 & .03 \\ 0 & .0.8 \end{pmatrix} \begin{pmatrix} x \\ y \end{pmatrix}$.

```
In[82]:= ClearAll[ x, y, mot]
        mot = DSolve[{x'[t] == .08 y[t], y'[t] == .03 y[t], x[0] == 4000, y[0] == 6000}, {x[t], y[t]}, t]
```

Out[82]= {{Null (x[t] → 4000. (−3. + 4. $e^{0.03 t}$)), Null (y[t] → 6000. $e^{0.03 t}$)}}

The plot of national debt over tine is,

```
In[83]:= x[t_] = x[t] /. mot[[1]]
```

Out[83]= 4000. (−3. + 4. $e^{0.03 t}$)

```
In[84]:= Plot[x[t], {t, 0, 10}]
```

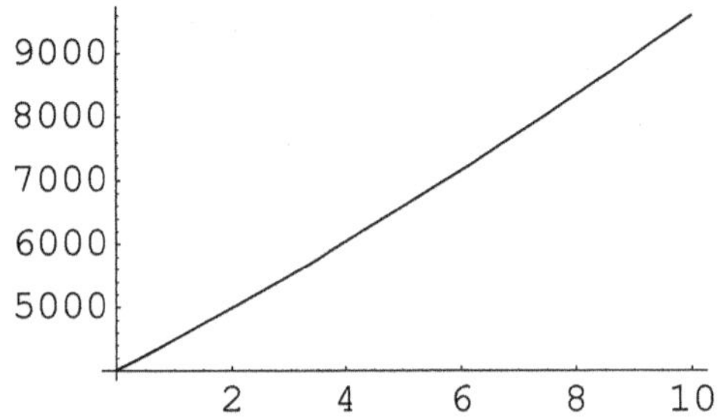

the plot of GNP over time is,

```
In[85]:= y[t_] = y[t] /. mot[[1]]
         Plot[y[t], {t, 0, 10}]
```

Out[85]= $6000.\,e^{0.03\,t}$

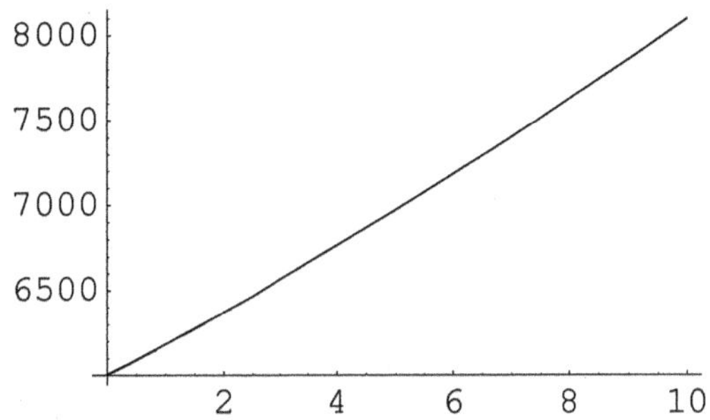

the parametric plot of debt and GDP over time is,

```
In[86]:= ParametricPlot3D[{x[t], y[t], t}, {t, 0, 200}, Boxed -> False, BoxRatios -> {4, 2, 2},
           AxesLabel -> {y1, y2, t}]
```

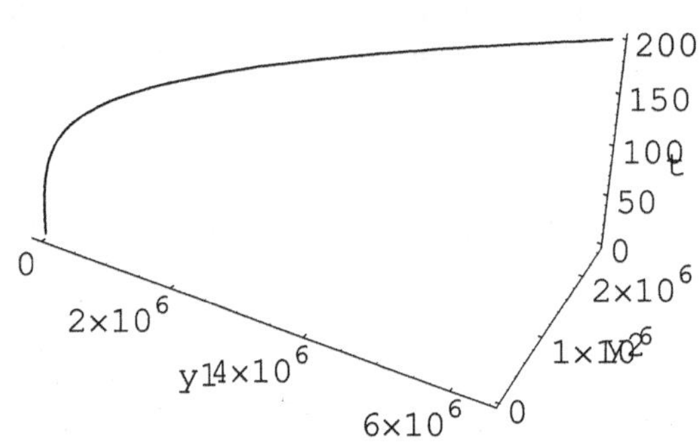

- Graphics3D -

The extent of the affordability of a country's national debt is measured by the ratio of the debt growth to GNP growth. Since the governments pay interest on debt, a ratio less than one implies that the annual interest payment on debt is less than GNP. If the ratio is greater than one, the interest payment on debt may turn greater than GNP and the debt may not be affordable. In this example, the ratio is $\frac{.03}{.08}$ or .375. That is for every \$1 increase in GNP, the debt is increasing by 37.5 cents.

Do More Practice: Change the GNP growth to 8% and the debt growth to 3%. Re-run the program and interpret the solution.

2- Change the GNP growth to 10% and the debt growth to 1%. Re-run the program and interpret the solution.

Example: Solve the following system of differential equations, $\mathbf{y'(t)} = \begin{pmatrix} 0 & -1 & 0 \\ 0 & 0 & 2 \\ 0 & 0 & 0 \end{pmatrix}\mathbf{y(t)}$, for the initial values $\mathbf{y(0)} = \begin{pmatrix} 1 \\ 2 \\ 3 \end{pmatrix}$.

Solution: Find the eigenvalues and eigenvectors of the **A** matrix of the system of equations,

In[87]:= Eigensystem[$\begin{pmatrix} 0 & -1 & 0 \\ 0 & 0 & 2 \\ 0 & 0 & 0 \end{pmatrix}$]

Out[87]= {{0, 0, 0}, {{1, 0, 0}, {0, 0, 0}, {0, 0, 0}}}

The **A** matrix of the system has only one distinct eigenvalue k = 0, with the corresponding eigenvector $\mathbf{u} = \begin{pmatrix} 1 \\ 1 \\ 0 \end{pmatrix}$.

Therefore, one solution is, $y1(t) = \begin{pmatrix} 1 \\ 0 \\ 0 \end{pmatrix}$. To find other solutions, we try a solution of the form, $\mathbf{y2(t) = (ut + v)e^{kt}}$.

As proved in the previous example, for $\mathbf{y2(t) = (ut + v)e^{kt}}$ to be a solution the equation $\mathbf{(A - k\ I)v = u}$ should be satisfied. Since k = 0, $\begin{pmatrix} 0 & -1 & 0 \\ 0 & 0 & 2 \\ 0 & 0 & 0 \end{pmatrix}\begin{pmatrix} v1 \\ v2 \\ v3 \end{pmatrix} = \begin{pmatrix} 1 \\ 0 \\ 0 \end{pmatrix}$. This equation results in the solution set, v2 = -1, v3 = 0, and v1 any arbitrary number, say v1 = 1. That is, $\mathbf{v} = \begin{pmatrix} 1 \\ -1 \\ 0 \end{pmatrix}$. Therefore, the second solution of the system of equations is ,

$$\mathbf{y2(t) = (ut + v)\,e^{kt}} = \begin{pmatrix} 1 \\ 1 \\ 0 \end{pmatrix} t + \begin{pmatrix} 1 \\ -1 \\ 0 \end{pmatrix} = \begin{pmatrix} t+1 \\ t-1 \\ 0 \end{pmatrix}.$$

The system still cannot be solved for the initial values and requires one more solution. Therefore, we search for a solution of the form, $\mathbf{y'3(t) = (ut^2 + vt + w)e^{kt}}$. Differentiating this equation and setting the differential equal to $\mathbf{Ay3(t)}$ yields, $\mathbf{(2ut + v)e^{kt} + k(ut^2 + vt + w)e^{kt} = A(ut^2 + vt + w)e^{kt}}$. By rearranging the terms we get, $\mathbf{(Au - ku)t^2e^{kt} + (Av - kv - 2u)te^{kt} + (Aw - kw - v)e^{kt} = 0}$. For this equation to be zero the following three equations should satisfy:

$\mathbf{Au - ku = 0}$, or $\mathbf{(A - kI)u = 0}$

$\mathbf{Av - kv = 2u}$, or $\mathbf{(A - kI)v = 2u}$

$\mathbf{Aw - kw = v}$, or $\mathbf{(A - kI)w = v}$

The first equation results in the eigenvalue k = 0, with the corresponding eigenvector $\mathbf{u} = \begin{pmatrix} 1 \\ 0 \\ 0 \end{pmatrix}$. The second equation yields, $\begin{pmatrix} 0 & -1 & 0 \\ 0 & 0 & 2 \\ 0 & 0 & 0 \end{pmatrix}\begin{pmatrix} v1 \\ v2 \\ v3 \end{pmatrix} = \begin{pmatrix} 2 \\ 0 \\ 0 \end{pmatrix}$, or $\mathbf{v} = \begin{pmatrix} v1 \\ v2 \\ v3 \end{pmatrix} = \begin{pmatrix} 1 \\ -2 \\ 0 \end{pmatrix}$. The third equation yields $\begin{pmatrix} 0 & -1 & 0 \\ 0 & 0 & 2 \\ 0 & 0 & 0 \end{pmatrix}\begin{pmatrix} w1 \\ w2 \\ w3 \end{pmatrix} =$

$\begin{pmatrix}1\\-2\\0\end{pmatrix}$, or $w = \begin{pmatrix}w1\\w2\\w3\end{pmatrix} = \begin{pmatrix}1\\-1\\-1\end{pmatrix}$. Therefore, the third solution is $\mathbf{y3}(t) = \begin{pmatrix}1\\0\\0\end{pmatrix}t^2 + \begin{pmatrix}1\\-2\\0\end{pmatrix}t + \begin{pmatrix}1\\-1\\-1\end{pmatrix}$.

Now that we have found three linearly independent solutions to the system, we can write the general solution as, $\mathbf{y}(t) = c1\mathbf{y1}(t) + c2\mathbf{y2}(t) + c3\mathbf{y3}(t) = c1\begin{pmatrix}1\\0\\0\end{pmatrix} + c2\left(\begin{pmatrix}1\\0\\0\end{pmatrix}t + \begin{pmatrix}1\\-1\\0\end{pmatrix}\right) + c3\left(\begin{pmatrix}1\\0\\0\end{pmatrix}t^2 + \begin{pmatrix}1\\-2\\0\end{pmatrix}t + \begin{pmatrix}1\\-1\\-1\end{pmatrix}\right)$. Solving this equation for the initial values $\mathbf{y}(0) = \begin{pmatrix}y1(0)\\y2(0)\\y3(0)\end{pmatrix} = \begin{pmatrix}1\\2\\3\end{pmatrix}$ yields c1 = 3, c2 = 1, and c3 = -3. Therefore, the initial-value solution is, $\mathbf{y}(t) = \begin{pmatrix}-3t^2 - 2t + 1\\6t + 2\\3\end{pmatrix}$. Using *Mathematica*, the solution is,

```
In[88]:= ClearAll[y1, y2, y3]
        DSolve[{y1'[t] == 0, y2'[t] == -y2[t], y3'[t] == 2 y3[t], y1[0] == 1, y2[0] == 2,
          y3[0] == 3}, {y1, y2, y3}, t]

Out[88]= {{y1 → (1 &), y2 → (2 e^-#1 &), y3 → (3 e^2#1 &)}}
```

Appendix 2: Solution of the Systems of First Order, Nonhomogeneous Linear Differential Equations

The resemblance of the system of first order, nonhomogeneous linear differential equations $\mathbf{y'}(t) = \mathbf{A}\mathbf{y}(t) + \mathbf{f}(t)$ to the first order, nonhomogeneous linear differential equation $y'(t) = ay(t) + f(t)$ suggests that a similar solution may be applicable here. However, before finding solution, two concepts which are very useful in solving systems of nonhomogeneous differential equations will be discussed below.

Definition 1. Fundamental Matrix A nxn matrix, $\mathbf{M}(t)$, whose columns are linearly independent solutions of the system $\mathbf{y'}(t) = \mathbf{A}\mathbf{y}(t)$ at to is called a fundamental matrix of $\mathbf{y'}(t) = \mathbf{A}\mathbf{y}(t)$ at to.

Example: Find a fundamental matrix for the following system of differential equations $\mathbf{y'}(t) = \begin{pmatrix}1 & -1\\0 & 2\end{pmatrix}\mathbf{y}(t)$.

Solution: The system has two distinct eigenvalues k1 = 1 and K2 = 2, with the corresponding eigenvectors $\mathbf{u1}$ = [$\begin{pmatrix}1\\0\end{pmatrix}$and $\mathbf{u2} = \begin{pmatrix}1\\-1\end{pmatrix}$. Therefore, the solutions of the system of equations are, $\mathbf{y1}(t) = \begin{pmatrix}1\\0\end{pmatrix}e^t$ and $\mathbf{y2}(t) = \begin{pmatrix}1\\-1\end{pmatrix}e^{2t}$. The solution vectors are linearly independent and the fundamental matrix of the system is $\mathbf{y}(t) = \begin{pmatrix}e^t & e^{2t}\\0 & -e^{2t}\end{pmatrix}$, (note that, if the eigenvalues are distinct, the corresponding eigenvectors will be linearly independent.). Clearly, the fundamental matrices are not unique. Any combination of solution vectors which are linearly independent will give a fundamental matrix for the system. It is also easy to prove that $\mathbf{cM}(t)$, where $\mathbf{c}$ is a constant vector, is a solution to the system $\mathbf{y'}(t) = \mathbf{A}\mathbf{y}(t)$.

Definition 2. Normalized Fundamental Matrix: A matrix formed as $\mathbf{Mn}(t) = \mathbf{M}(t)\mathbf{M}^{-1}(to)$ is called a normalized fundamental matrix. From the definition of normalized fundamental matrices, it is easy to see that they are identity matrices at t = 0.

Example: Find the normalized fundamental matrix for the system of differential equations of the preceding example.

Solution: The fundamental matrix of the preceding example is, $\mathbf{M}(t) = \begin{pmatrix}e^t & e^{2t}\\0 & -e^{2t}\end{pmatrix}$. At t = 0, $\mathbf{M}(0) = \begin{pmatrix}1 & 1\\0 & -1\end{pmatrix}$

and, $\mathbf{M}^{-1}(0) = \begin{pmatrix} 1 & 1 \\ 0 & -1 \end{pmatrix}$. The normalized matrix is $\mathbf{Mn}(t) = \mathbf{M}(t)\mathbf{M}^{-1}(0) = \begin{pmatrix} e^t & e^t - e^{2t} \\ 0 & e^{2t} \end{pmatrix}$. This matrix will be an identity matrix at t = 0.

Definitions 1 and 2 help us to find a solution to the nonhomogeneous differential equations. Associated with every nonhomogeneous system is a complementary homogenous system, obtained by setting f(t) = 0. If **M**(t) is the fundamental matrix of the complementary system, we search for a vector **u**(t) such that, y(t) = **M**(t)**u**(t) is a solution to the nonhomogeneous system. If y(t) = **M**(t)**u**(t) is a solution to the system, its differential will correspond to equation **y**'(t) = **A**y(t) + **f**(t). Differential of y(t) = **M**(t)**u**(t) yields, **y**'(t) = **M**'(t)**u**(t) + **u**'(t)**M**(t). Since M(t) is a fundamental matrix of the homogenous system, **M**'(t) = **AM**(t) and **y**'(t) = **M**'(t)**u**(t) + **u**'(t)**M**(t) can be written as, **y**'(t) = **A**(**M**(t)**u**(t)) + **u**'(t)**M**(t) = **A**y(t) + **f**(t). This equation implies that, **M**(t)**u**'(t) = **f**(t), or **u**'(t) = $\mathbf{M}^{-1}$(t)**f**(t). The preceding equation together with the equation **y**(t) = **M**(t)**u**(t) provide a solution to the system. However, this is only one of the many particular solutions of the system. A general solution will be, **y**(t) = **y**p(t) + **y**c(t)C = **M**(t)**u**(t) + **M**(t)C, where C is any vector of constants. The equation **y**(t) = **M**(t)**u**(t) + **M**(t)C is the solution of **y**'(t) = **A**y(t) + **f**(t) for any choice of C.

Example: Find the solution of the following system of nonhomogeneous differential equations **y**'(t) = $\begin{pmatrix} 1 & -1 \\ 0 & 2 \end{pmatrix}$y(t) + $\begin{pmatrix} e^{2t} \\ 1 \end{pmatrix}$.

Solution: To find the fundamental matrix of the complementary homogenous system, as defined in the definition 1 above, find the Eigensystem of the **A** matrix,

```
In[89]:= Eigensystem[(1  -1
                      0   2)]

Out[89]= {{1, 2}, {{1, 0}, {-1, 1}}}
```

the fundamental matrix of the system is $\mathbf{M}(t) = \begin{pmatrix} e^t & -e^{2t} \\ 0 & e^{2t} \end{pmatrix}$. Use the expression $\mathbf{u}'(t) = \mathbf{M}^{-1}(t)\mathbf{f}(t)$ to find **u**(t) of the solution **y**(t) = **y**p(t) + **y**c(t)C = **M**(t)**u**(t) + **M**(t)C. That is, $\mathbf{u}'(t) = \begin{pmatrix} e^{-t} & e^{-t} \\ 0 & e^{-2t} \end{pmatrix}\begin{pmatrix} e^{2t} \\ 1 \end{pmatrix} = \begin{pmatrix} e^t + e^{-t} \\ e^{-2t} \end{pmatrix}$. To find **u**(t), integrate **u**'(t) in the interval zero to t, $\mathbf{u}(t) = \int_0^t \begin{pmatrix} e^t + e^{-t} \\ e^{-2t} \end{pmatrix} dt$,

```
In[90]:= Integrate[{E^t + E^(-t), E^(-2 t)}, {t, 0, t}]

Out[90]= {e^-t (-1 + e^2t), 1/2 - e^-2t/2}
```

The particular solution of the system is, $\mathbf{y}p(t) = \mathbf{M}(t)\mathbf{u}(t) = \frac{1}{2}\begin{pmatrix} e^t & -e^{2t} \\ 0 & e^{2t} \end{pmatrix}\begin{pmatrix} 2e^t - 2e^{-t} \\ 1 - e^{-2t} \end{pmatrix} = \frac{1}{2}(e^{2t} - 1)\begin{pmatrix} 1 \\ 1 \end{pmatrix}$. This solution to the problem is only one of the many particular solutions. A general solution of the problem is, **y**(t) = **y**p(t) + **y**c(t)C = **M**(t)**u**(t) + **M**(t)C = $\frac{1}{2}(e^{2t} - 1)\begin{pmatrix} 1 \\ 1 \end{pmatrix} + \begin{pmatrix} e^t & -e^{2t} \\ 0 & e^{2t} \end{pmatrix}\begin{pmatrix} c1 \\ c2 \end{pmatrix}$ where c1 and c2 are any arbitrary constants.

To solve the same problem using *Mathematica*, use the command "DSolve",

```
In[91]:= ClearAll[y1, y2]
         DSolve[{y1'[t] == y1[t] - y2[t] + E^(2 t), y2'[t] == 2 y2[t] + 1}, {y1[t], y2[t]}, t]

Out[91]= {{y1[t] → 1/2 (-1 + 2 e^2t + 2 e^t C[1] + 2 e^t C[2] - 2 e^2t C[2]), y2[t] → 1/2
            (-1 + 2 e^2t C[2])}}
```

Example: Find the solution of the following system of differential equations, **y**(t) = $\begin{pmatrix} 1 & 0 & 1 \\ 0 & 0 & 1 \\ 0 & 0 & 0 \end{pmatrix} + \begin{pmatrix} 2e^t \\ t \\ 1 \end{pmatrix}$.

Solution: The eigensystem of the **A** matrix is,

```
In[92]:= Eigensystem[{{1, 0, 1}, {0, 0, 1}, {0, 0, 0}}]
Out[92]= {{0, 0, 1}, {{0, 1, 0}, {0, 0, 0}, {1, 0, 0}}}
```

the matrix has two distinct eigenvalues k1 = 0 and K2 = 1, with the corresponding eigenvectors **u**1 = $\begin{pmatrix} 0 \\ 1 \\ 0 \end{pmatrix}$ and **u**2 = $\begin{pmatrix} 1 \\ 0 \\ 0 \end{pmatrix}$. Therefore, the two solutions corresponding to k1 and k2 are, y1(t) = $\begin{pmatrix} 0 \\ 1 \\ 0 \end{pmatrix}$ and **y**2(t) = $\begin{pmatrix} 1 \\ 0 \\ 0 \end{pmatrix} e^t$. To find the general solution of the system, we need one more distinct solution. Let us try a solution of the from **y**3(t) = (**u**t + **v**)e^t (see the section on the existence of definite solution). (**A** - kI)**v** = **u** yields the **v** vector, **v** = $\begin{pmatrix} -1 \\ 1 \\ 1 \end{pmatrix}$ and the third solution is, y3(t) = $\begin{pmatrix} 0 \\ 1 \\ 0 \end{pmatrix} t + \begin{pmatrix} -1 \\ 1 \\ 1 \end{pmatrix} = \begin{pmatrix} -1 \\ t+1 \\ 1 \end{pmatrix}$. The fundamental matrix of the system is **M**(t) = $\begin{pmatrix} 0 & e^t & -1 \\ 1 & 0 & t+1 \\ 0 & 0 & 1 \end{pmatrix}$ and

$$\mathbf{M}^{-1}(t) = \begin{pmatrix} 0 & 1 & -t-1 \\ e^{-t} & 0 & e^{-t} \\ 0 & 0 & 1 \end{pmatrix}$$

and **u**'(t) = $\mathbf{M}^{-1}(t)\mathbf{f}(t) = \begin{pmatrix} 0 & 1 & -t-1 \\ e^{-t} & 0 & e^{-t} \\ 0 & 0 & 1 \end{pmatrix} \begin{pmatrix} 2e^t \\ t \\ 1 \end{pmatrix} = \begin{pmatrix} -1 \\ 2+e^{-t} \\ 1 \end{pmatrix}$. Integrate **u**' (t) in the interval zero to t to get **u**(t),

u(t) = $\begin{pmatrix} -t \\ 2t - e^{-t} + 1 \\ t \end{pmatrix}$. Using this equation we can find **y**(t) as,

y(t) = **M**(t)**u**(t) = $\begin{pmatrix} 0 & e^t & -1 \\ 1 & 0 & t+1 \\ 0 & 0 & 1 \end{pmatrix} \begin{pmatrix} -t \\ 2t - e^{-t} + 1 \\ t \end{pmatrix} = \begin{pmatrix} 2te^t + e^t - t - 1 \\ t^2 \\ t \end{pmatrix}$. This, is only one of the many particular solutions of the system. To find the general solution, **y**(t) = yp(t) + yc(t)C = **M**(t)**u**(t) + **M**(t)C, where C is any arbitrary constant. If the initial value of **y**(to) = **yo** is also given, the initial value problem **y**'(t) = **A**y(t), **y**(to) = **yo** has the unique solution **y**(t) = yp(t) + **Mn**(t)yo, where **Mn**(t) is the normalized fundamental matrix of the complementary system.

Example: Solve the system of differential equations **y**'(t) = $\begin{pmatrix} 1 & -1 \\ 0 & 2 \end{pmatrix}$ y (t) + $\begin{pmatrix} e^{2t} \\ 1 \end{pmatrix}$ for the initial value, y(0) = $\begin{pmatrix} 1 \\ 1 \end{pmatrix}$.

Solution: The solution of the complementary system of the problem is **y**(t) = $\frac{1}{2}(e^{2t} - 1)\begin{pmatrix} 1 \\ 1 \end{pmatrix}$. Normalize the fundamental matrix of the system to get **Mn**(t), **Mn**(t) = $\mathbf{M}(t)\mathbf{M}^{-1}(0) = \begin{pmatrix} e^t & e^{2t} \\ 0 & -e^{2t} \end{pmatrix} \begin{pmatrix} 1 & 1 \\ 0 & -1 \end{pmatrix} = \begin{pmatrix} e^t & e^t - e^{2t} \\ 0 & e^{2t} \end{pmatrix}$.

Therefore, the solution to the initial-value problem is **y**(t) = yp(t) + **Mn**(t)**yo** = $\frac{1}{2}(e^{2t} - 1)\begin{pmatrix} 1 \\ 1 \end{pmatrix}$ + $\begin{pmatrix} e^t & e^t - e^{2t} \\ 0 & e^{2t} \end{pmatrix} \begin{pmatrix} 1 \\ 1 \end{pmatrix} = \frac{1}{2}\begin{pmatrix} 4e^t - e^{2t} - 1 \\ 3e^{2t} - 1 \end{pmatrix}$.

To solve the same problem using *Mathematica*,

```
In[93]:=ClearAll[y1, y2]
        DSolve[{y1'[t] == y1[t] - y2[t] + E^(2 t),
           y2'[t] == 2 y2[t] + 1, y1[0] == 1, y2[0] == 1}, {y1[t], y2[t]}, t]
```

Out[93]= $\{\{y1[t] \to \frac{1}{2}(-1+4e^{t}-e^{2t}),\ y2[t] \to \frac{1}{2}(-1+3e^{2t})\}\}$

Example: Application to Markets

One of the major characteristics of a perfectly competitive industry is that the firms have easy entry or exit to the industry. Suppose that the firms' entry or exit in a perfectly competitive industry (the change in the number of firms, $\frac{dn}{dt}$) is proportional to the gap between the market price (p) and the minimum average total cost (atc), $\frac{dn}{dt} = n' = \lambda(p - atc)$, $\lambda > 0$. That is, if p - atc > 0, the positive economic profit attracts more entry to the industry and results in lower prices in the market. The opposite occurs when p - atc < 0. In addition, assume that changes in the market price is proportional to excess demand in market, $\frac{dP}{dt} = p' = \mu(Qd - Qs)$, $\mu > 0$, where Qd and Qs represent market demand and market supply, respectively. λ and μ represent the speed of adjustment for the number of firms and price, respectively. Given that Qd = 100 - 2p, Qs = 5n, atc = 20, $\lambda = .1$ and $\mu = .2$ formulate the price behavior in the market, solve the resulting system of differential equations, and plot the time path of the price. The initial values of p and n are given as po = 25 and no = 15. Note that, in this problem, it is assumed that the firms in the industry are of the uniform size and that Qs depends on the number of firms in the industry.

Solution: The system of the differential equations can be expressed as p' = .2(100 -2p -5n) and n' = .1(p - 20). In matrix form the system can be written as $\begin{pmatrix} p' \\ n' \end{pmatrix} = \begin{pmatrix} -.4 & -1 \\ .1 & 0 \end{pmatrix}\begin{pmatrix} p \\ n \end{pmatrix} + \begin{pmatrix} 20 \\ -2 \end{pmatrix}$. The steady-state value of p and n are obtained by setting p' = n' = 0. Using *Mathematica*,

In[94]:= ClearAll[Qd, Qs, p, n, chimp]

Solve[$\begin{pmatrix} -.4 & -1 \\ .1 & 0 \end{pmatrix}.\begin{pmatrix} p \\ n \end{pmatrix} + \begin{pmatrix} 20 \\ -2 \end{pmatrix} == \begin{pmatrix} 0 \\ 0 \end{pmatrix}$, {p, n}]

Out[94]= {{Null (p → 20.), Null (n → 12.)}}

the steady-state values of p and n are 20 and 12, respectively. To solve the system ususing *Mathematica*, type

```
In[95] := Qd = 100 - 2 p[t]; Qs = 5 n[t];
chimp = DSolve[{p'[t] == .2 (Qd - Qs), n'[t] == 10 (p[t] - 20), p[0] == 25, n[0] == 15}, {p[t], n[t]}, t]
```

```
In[96] := p[t_] = p[t] /. chimp[[1]];
Plot[p[t], {t, 0, 10}]
```

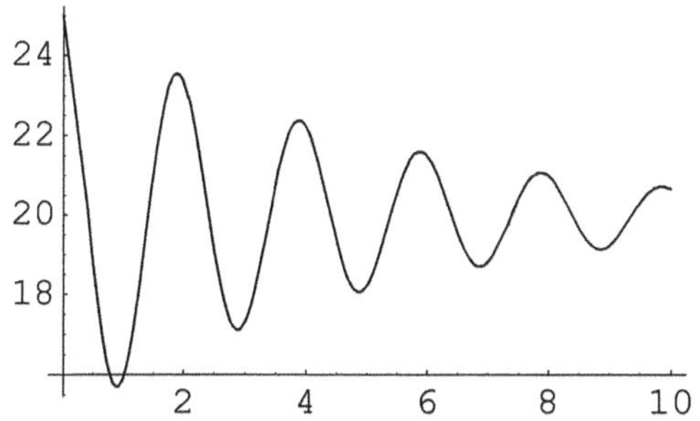

Do More Practice:

1- In the market example of page 333, change the speed of adjustment for price (μ) from .2 to 2, 20 and 200 each time. Hold the speed of adjustment for the number of firms (λ) and other information the same. Solve the problem for each value of the μ and plot the time path of the price. How does the price adjustment process change as the speed of adjustment changes?

2- Keep the speed of adjustment for price (μ) at .2 and change the speed of adjustment for the number of firms (λ) from .1 to 1, 10 and 100. Solve the problem for each value of λ and plot the time path of the prices. How does the price adjustment process change as the speed of adjustment changes?

3- In the market example of page 333, market is initially at disequilibrium, with quantity supplied exceeding quantity demanded by 25 units. Now, let's assume that the market initially has a shortage of 15 units, with Po = \$20 and no = 10. Keep other information the same as are in the example. Solve the problem and plot the time path of the price. How the does price adjustment change as the extent of shortage or surplus in market changes?

CHAPTER 13

DIFFERENCE EQUATIONS

There are many cases in which some economic variable is related to its lagged values by discrete time intervals. Equations that express such a relationship between the present value of a variable and its values in the past are called difference equations. For example, yt = byt-1 is a first-order, homogenous linear difference equation. The problem here is to find the value of Y at time t, given that the value of y at some time in the past is known.

The order of a difference equation is the highest time difference in the equation. For example, yt+1 + 3yt +2 = 0 and yt + 2yt+1 + yt+2 = 0 are first-order and second-order difference equations, respectively. The first two sections of this chapter are concerned with finding solutions of the first-order linear difference equations, followed by the stability analysis of the first-order linear difference equations. The following sections concentrate on the solutions of the second-order linear difference equations and the stability analysis of the second-order difference equations. Finally, we study the solution and stability of the system of linear difference equations.

TOOL BOX

Mathematica **Commands**	**Function**
<<DiscreteMath`RSolve`	Loads Discrete Math Package
RSolve[{y[t] - ay[t-1] ==b, y[0]==yo}, y[t], t]	Solves the difference equation yt-ayt-1=b, Y0= Yo, for Yt.
H[{y1_, y2_}] = {2y1+3y2, y1- 2y2}	Defines any values of y1 and y2 as a set of functions defined in H
f[y1_, y2_, t_] = **NestList**[H, {y1,y2}, n]	Nestlists y1 and y2 of H n times
f[a, b, n]	Finds the first n values visited around y1= a, y2 = b.

I. First-Order Homogenous Linear Difference Equations

Suppose that the value of a variable at time t is a constant proportion b of its value at time t-1. This assumption gives rise to a first-order, homogenous linear difference equation, the general form of which is yt = byt-1. This equation yields the following sequence of values for y: y1 = byo, y2 = by1 = b^2yo, y3 = by2 = b^3yo, ..., yt = yt-1 = b^tyo. It is obvious that, yt = b^tyo satisfies equation yt = byt-1, and is a solution of that equation. If the initial value of y at time t = 0 is also given, then the initial-value problem yt = byt-1, yo = a, has a solution yt = ab^t. This equation satisfies both the initial condition and the difference equation.

To solve the difference equations using *Mathematica*, first load the package "DiscreteMath`RSolve' , then use the command "RSolve". The form of the equations given in the "RSolve" command is similar to the form used in the "DSolve" command of the previous chapter. You can use a single equation, equations with initial conditions or several equations. The solution is returned in the form of a list of replacement rules.

Important Notice: The package "DiscreteMath" should be loaded before any program is run in this chapter. To load DiscreteMath, type <<DiscreteMath`RSolve` and press Shift Enter. If you run a program before running <<DiscreteMath`RSolve`, the program will not run and attempts to load DiscreteMath will also fail. In such cases, close the notebook, exit *Mathematica*, and run <<DiscreteMath`RSolve` before running any program.

Example: Solve the homogeneous difference equation yt = byt-1, y(0) = yo.

Solution: Following the solution rule of the linear homogeneous difference equations, y = yo b^t. Using *Mathematica,* the problem can be solved as,

```
In[1]:= << DiscreteMath`RSolve`
        Clear[y]
        RSolve[{y[t] == b y[t − 1], y[0] == yo}, y[t], t]
        Out[1] = {{y[t] → b^t yo}}
```

Example: Solve the homogeneous difference equation yt = .2yt-1, yo = 7.

Solution: Following the solution rule of the homogeneous difference equations, y = 7 2^t. Using *Mathematica,* the problem can be solved as,

```
In[2]:= Clear[y]
        RSolve[{y[t] == .2 y[t − 1], y[0] == 7}, y[t], t]
        Out[2] = {{Null (y[t] → 7.` 0.2`^t)}}
```

Example: Suppose that an initial amount of A is saved at an interest rate of i. What is the balance in the account at every point at time?

Solution: The problem here is the economic problem of compounding interest, where St = (1 + i)St-1, So = A. Using the solution rule of the initial-value, first-order difference equation, the solution is St = $A(1 + i)^t$.

Example: What will be the value of $10000 invested at an interest rate of 6% compounded annually, after 5, 10, and 15 years?

Solution: Solve the difference equation yt = (1 + .06t)yt-1, y(0) = 10000 and evaluate yt for t = 5, t = 10, and t = 15. Plot the time path of the investment growth.

```
In[3]:= Clear[ y, rose, path1]
        rose = RSolve[{y[t] == 1.06 y[t − 1], y[0] == 10000}, y[t], t];
        y[t_] = y[t] /. rose[[1]]
        Out[3] = 10000.` 1.06`^t
```

```
In[4]:= y[5]
        13382.255776000005`
        y[10]
        17908.476965428545`
        y[15]

        23965.581930996923`
```

to plot the time path of y over t, list the values of the y(t) for a certain range of t and list plot the table of values. (Note: If the coefficient of Yt-1 is positive, y(t) can be graphed directly using "Plot" command. However, if it is negative, the discrete values of y(t) should be evaluated first using "Table" command, then the values can be graphed.)

```
In[5]:=path1=Table[y[t], {t, 0, 10}];

       ListPlot[path1, PlotJoined->True]
```

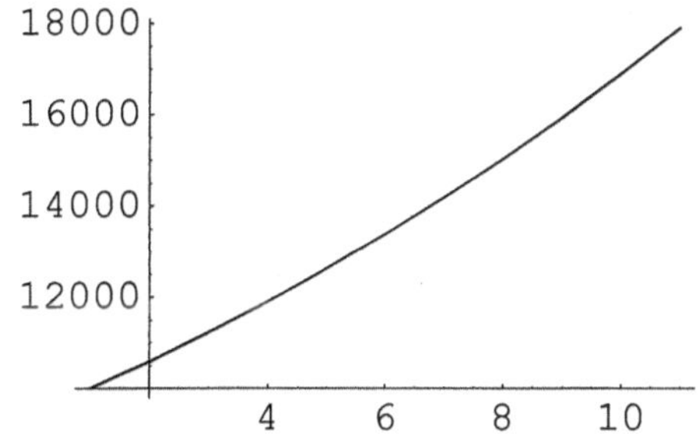

Do More Practice: Change the coefficient of yt-1 to 2. Re-run the program and observe the time path of y(t).

2- Change the coefficient of yt-1 to -2. Re-run the program and observe the time path of y(t).

3- Change the coefficient of yt-1 to -.2. Re-run the program and observe the time path of y(t).

Example: Contrary to Domar's model, which is a continuous time growth model of the economy (see chapter 12), Harrod's model treats time as being discrete (R. F. Harrod, "An Essay In Dynamic theory," Economic Journal, March, 1939; and Toward a Dynamic Economics, New York: McMillan, 1952). The other assumptions of the model, however, are similar to Domar's model. That is, saving (S) is proportional to income (y), St = syt, investment (I) is proportional to change in income, It = c(yt - yt-1), and the equilibrium condition, St = It. The problem is to find the time path of income.

Solving the three equations together yields yt = $\frac{cy_{t-1}}{c-s}$, which is a difference equation. Therefore, given the initial income level yo, the time path of y is determined as yt = yob^t, where b = $\frac{c}{c-s}$.

Example: The actual saving in an economy, St, is a constant proportion .5 of the economy's income, (yt), and the economy's desired investment (It) is a constant proportion 8 of the change in income, DYt = Yt - Yt-1. Assuming that the desired investment is realized find the time path of income, if Y(0) = 6000.

Solution: The problem can be formulated as, St = .5Yt, It = 8(Yt - Yt -1), St = It. Substituting for St and It from the corresponding equations, yields .5Yt = 8(Yt - Yt -1), or Yt = $\frac{8}{7.5}$Yt-1, which is a difference equation with an initial value of Y(0) = 6000. To solve the equation and to graph the time path of Yt, use *Mathematica* as,

```
In[1]:= Clear[ y, sage, path2]
        sage = RSolve[{y[t] == (8/7.5) y[t - 1], y[0] == 6000}, y[t], t];
        y[t_] = y[t] /. sage[[1]]
```

```
In[2] := path2 = Table[y[t], {t, 0, 10}]
ListPlot[path2, PlotJoined -> True]
Out[2] = {6000., 6400., 6826.67, 7281.78,
  7767.23, 8285.04, 8837.38, 9426.55, 10055., 10725.3, 11440.3}
```

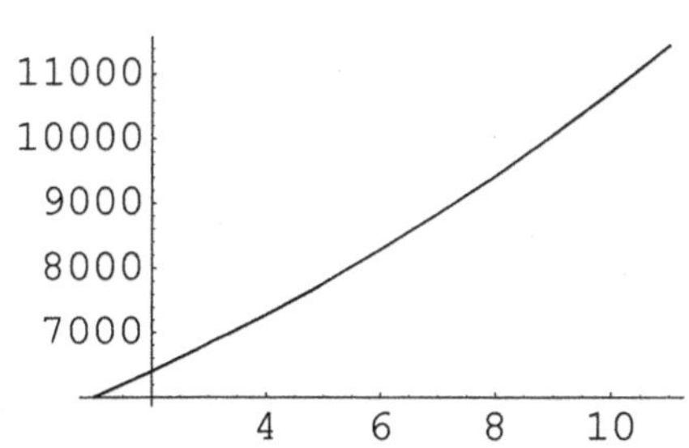

Problems

1. Find the order of the following difference equations.

   a) yt = 3yt-1 + yt-2 + 1

   b) yt = 2yt-2 + 5

   c) yt = 3yt+1 + e^t

   d) yt+1 = yt+2 + 2

2. Solve the following initial-value difference equations.

   a) 4yt + 3yt-1 = 0, yo = 2

   b) yt = yt-1, yo = 10

   c) yt+1 + 2yt = 0, yo = 2

   d) 2yt+1 = yt, yo = 10

   e) yt = -.5yt-1, yo = 10

3. What are the compounded values of $20000 deposited at i = 8%, one year, two years, five years, and ten years from now.

4. Suppose that the actual saving in an economy, St, is a constant proportion .2 of the economy's income, Yt, and the economy's desired investment (It) is a constant proportion 7 of the difference between this year's income, Yt, and the last year's income, Yt-1. Assuming that the desired investment is realized, find the time path of income.

Stability Analysis of the Solutions of First Order, Homogenous Linear Difference Equations:

The solution of the difference equation Yt = bYt-1 is Yt = Yob^t, where Yo and b are known constants. The time path of Y, therefore, depends on the values of Yo and b. The initial value of Y at t = 0, or Yo decides the position of Y at the initial period. The time path of Y, however, depends on the absolute value and the sign of b. Cases that may arise are as follows:

Case 1: b > 1. This gives rise to an divergent nonoscillatory time path.

Example: Graph the time path of Yt = 2Yt-1, Yo = 7.

Solution:
```
In[3]:= Clear[y, okra, path3]
        okra = RSolve[{y[t] == 2 y[t - 1], y[0] == 7}, y[t], t];
        y[t_] = y[t] /. okra[[1]]
```

```
In[4]:= path3 = Table[y[t], {t, 0, 10}];
        ListPlot[path3, PlotJoined -> True]
```

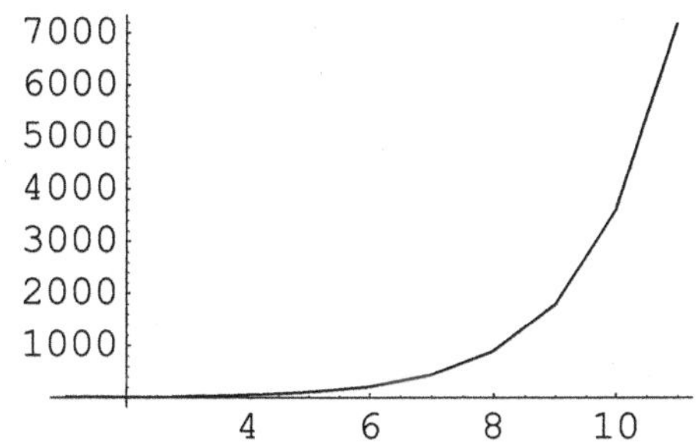

Do More Practice:

1- Change Yo to Yo = -7. Re-run the program and compare the results.

2- Change Yo to Yo = 1. Re-run the program and compare the results.

3- Change Yo yo Yo = -1. Re-run the program and compare the results.

Case 2: b = 1. This gives rise to a constant time path for Y .

Example: Graph the time path of Yt = Yt-1, Yo = 7.

Solution:

```
In[5]:= Clear[y, iris, path4]
        iris = RSolve[{y[t] == y[t - 1], y[0] == 7}, y[t], t]

    Out[5] = {{Null (y[t] → 7)}}
In[6]:= y[t_]=y[t]/.iris[[1]];
path4=Table[y[t], {t, 0, 10}];
ListPlot[path4, PlotJoined->True]
```

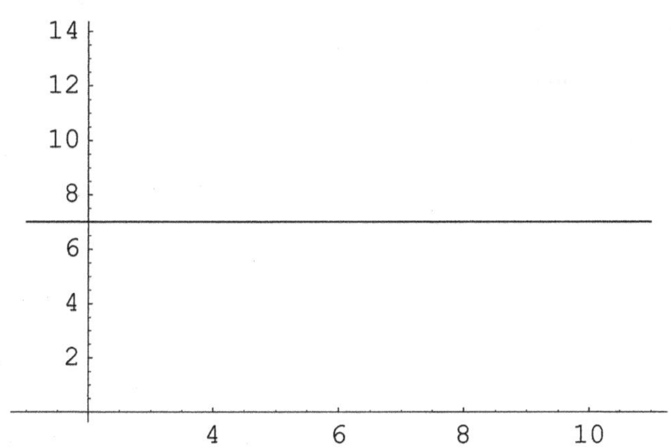

Do More Practice:

1- Change Yo to Yo = -7. Re-run the program and compare the results.

2- Change Yo to Yo = 1. Re-run the program and compare the results.

3- Change Yo yo Yo = -1. Re-run the program and compare the results.

Case 3: 0 < b < 1. This gives rise to a convergent nonoscillatory time path, where the value of y approaches zero as t -> ∞.

Example: Graph the time path of Yt = .5Yt-1, Yo = 7.

Solution:

```
In[6]:= ClearAll[y, fern, path5]
        fern = RSolve[{y[t] == .5 y[t − 1], y[0] == 7}, y[t], t]

        Out[6] = {{Null (y[t] → 7.` 0.5`^t)}}
In[7]:= y[t_]=y[t]/.fern[[1]];
path5=Table[y[t], {t, 0, 10}];
ListPlot[path5, PlotJoined->True]
```

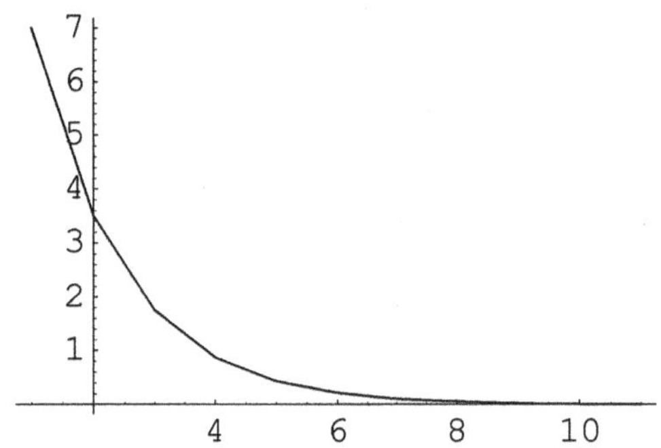

Do More Practice:

1- Change Yo to Yo = -7. Re-run the program and compare the results.

2- Change Yo to Yo = 1. Re-run the program and compare the results.

3- Change Yo yo Yo = -1. Re-run the program and compare the results.

Case 4: -1 < b < 0. This gives rise to a convergent oscillatory time path.

Example: Graph the time path of Yt = -.5Yt-1, Yo = 7.

Solution:

```
In[8]:= ClearAll[y, tulip, path6]
        tulip = RSolve[{y[t] == −.5 y[t − 1], y[0] == 7}, y[t], t]

        Out[8] = {{Null (y[t] → 7.` (−0.5`)^t)}}
In[9]:= y[t_]=y[t]/.tulip[[1]];
path6=Table[y[t], {t, 0, 10}];
ListPlot[path6, PlotJoined->True]
```

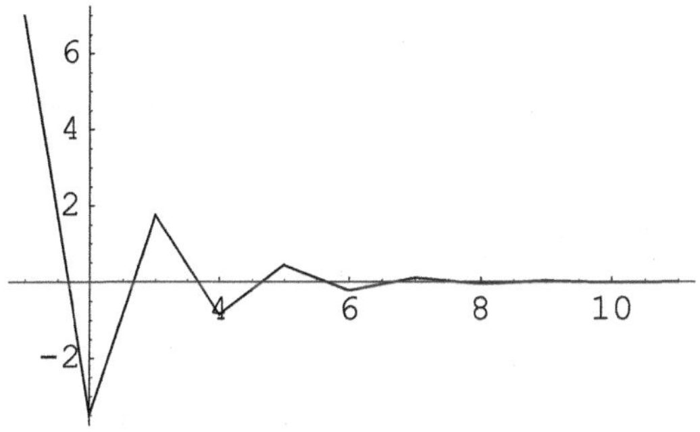

Do More Practice:

1- Change Yo to Yo = -7. Re-run the program and compare the results.

2- Change Yo to Yo = 1. Re-run the program and compare the results.

3- Change Yo yo Yo = -1. Re-run the program and compare the results.

Case 5: b < -1. This gives rise to a divergent oscillatory time path .

Example: Graph the time path of Yt = -2Yt-1, Yo = 7.

Solution:

```
In[10]:= ClearAll[y, cactus, path7]
         cactus = RSolve[{y[t] == -2 y[t - 1], y[0] == 7}, y[t], t]
        Out[10] = {{Null (y[t] → 7 (-2)^t)}}
In[11]:= y[t_]=y[t]/.cactus[[1]];
path7=Table[y[t], {t, 0, 10}];
ListPlot[path7, PlotJoined->True]
```

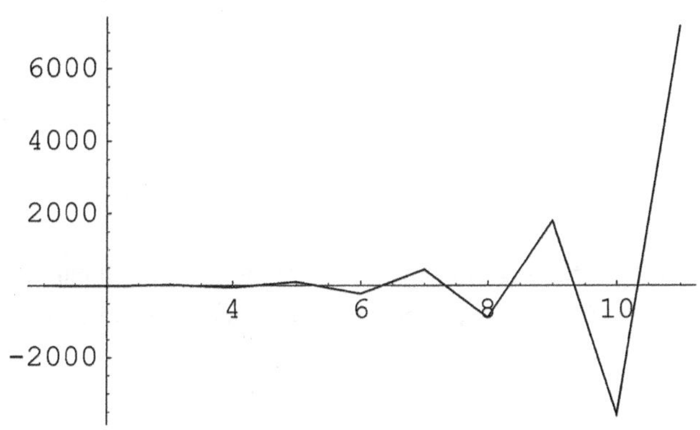

Do More Practice:

1- Change Yo to Yo = -7. Re-run the program and compare the results.

2- Change Yo to Yo = 1. Re-run the program and compare the results.

3- Change Yo yo Yo = -1. Re-run the program and compare the results.

Problems

1. Plot the values of Yt for the following difference equations Discuss the convergence or divergence of the solutions as t ->∞.

a) 4yt + 3yt-1 = 0, yo = 2

b) yt = yt-1, yo = 10

c) yt+1 + 2yt = 0, yo = 2

d) 2yt+1 = yt, yo = 10

e) yt = -.5yt-1, yo = 10

II. First-Order Nonhomogeneous Linear Difference Equations

One major property of every difference equation is that the variable Y of the equation is a function of time. For every first-order nonhomogeneous linear difference equation such as Yt = bYt-1 + c, there is a steady-state or stationary solution for Y such that the dynamic system becomes stationary and Y has no tendency for change. The steady-state value of Y is derived by setting the change is Y equal to zero, ΔYt = 0, or by setting Yt+1 = Yt . That is, for the difference equation Yt = bYt-1 + c, the steady-state value of Y is $Y = \frac{c}{1-b}$, $b \neq 1$.

Starting from any value of Y other than the steady-state value, Y will have a time path that may be convergent to the steady state value or divergent from it. One interesting feature of the difference equations is finding the solution of the system and analyzing the time path of Y.

Consider the following nonhomogeneous difference equation Yt = bYt-1 + c, where c is a constant. The solution here follows the solution of difference equations of the previous section. Equation Yt = bYt-1 + c gives rise to the following sequence of values for y,

$$Y1 = bYo + c$$

$$Y2 = bY1 + c = b(bYo + c) + c = b^2yo + bc + c$$

$$Y3 = bY2 + c = b(b^2Yo + bc + c) + c = b^3Yo + b^2 c + bc + c$$

. . . .

$$Yt = bYt\text{-}1 + c = b^tYo + b^{t-1}c + b^{t-2}c + \ldots + c$$

$$Yt = b^tYo + c(b^{t-1} + b^{t-2} + \ldots + 1)$$

Since the series $(b^{t-1} + b^{t-2} + \ldots + 1)$ adds up to $\frac{(1-b^t)}{1-b}$, the solution of the nonhomogeneous linear difference equation Yt = bYt-1 + c is $Yt = b^tYo + \frac{c(1-b^t)}{1-b}$, if $b \neq 1$. If b = 1, using the same argument, it can be proved that, Yt = Yo + ct.

Example: Application to Markets

The Cobweb model of markets, which is generally true for the agricultural markets, assumes that quantity demanded at time t depends on the market price at t, Qdt = a + bPt, but the quantity supplied at time t depends on the price that prevailed in the previous period. That is, Qst = c + dPt-1. Finally, price adjusts in such away as to clear the market, Qdt = Qst. Find the steady-state equilibrium price and quantity and analyze the stability of the market under these conditions.

Solution: Solving the three equations for the equilibrium condition yields, $Pt = \frac{c-a}{b} + \frac{d}{b}\,Pt\text{-}1$. However, this equation does not reveal what the equilibrium price is. The steady-state equilibrium condition requires that the period-by-period price remains the same, i.e., $Pt = Pt\text{-}1 = Pe = \frac{c-a}{b-d}$. To solve $Pt = \frac{c-a}{b} + \frac{d}{b}\,Pt\text{-}1$, which is a first-order non-homogeneous linear difference equation, apply the solution rule studied above. The solution is $Pt = (\frac{d}{b})^t Po + \frac{c-a}{b} . \frac{1-(\frac{d}{b})^t}{1-\frac{d}{b}} = (\frac{d}{b})^t Po + Pe(1 - (\frac{d}{b})^t)$, or $Pt = Pe + (Po - Pe)(\frac{d}{b})^t$.

In a conventional demand ad supply, $d > 0$, $b < 0$, and $\frac{d}{b}$ is negative. Therefore, the market will exhibit an oscillatory behavior. That is If Po > Pe, price will be above the equilibrium point when t is an even number and below the equilibrium when t is an odd number. The opposite will occur if Po < Pe. it is also clear that the condition for market stability and convergence is $-1 < \frac{d}{b} < 1$. That is, stability of the market requires that the absolute value of the slope of the demand curve exceeds the absolute value of the slope of the supply curve.

Example: Application to Markets

The market demand and supply of a product are given as Qdt = 100 -2Pt and Qst = -8 + 4Pt-1. Given the initial market price of P(0) = 20, find the time path of price adjustment to its steady-state equilibrium.

Solution: The steady-state equilibrium is derived by setting Qdt = Qst, when Pt = Pt-1= P. To find P using *Mathematica*,

```
In[12]:= Clear[p, sumac, path8]
         Solve[100 - 2 p == -8 + 4 p, p]

Out[12]= {{Null (p → 18)}}
```

the steady state equilibrium price is P = 18. To find the time path of price adjustment from P = 20 to P = 18, solve for Pt in terms of Pt-1 by setting Qdt = Qst. This yields Pt = 54 - 2Pt-1, which is a first-order nonhomogeneous difference equation, with initial-value y(0) = 20. To solve this equation using *Mathematica*,

```
In[13]:= sumac = RSolve[{p[t] == 54 - 2 p[t - 1], p[0] == 20}, p[t], t]

Out[13]= {{p[t] → -2 (-9 - 2^t e^(iπt))}}

In[14]:= p[t_] = p[t] /. sumac[[1]];
         path8 = Table[p[t], {t, 0, 10}];
         ListPlot[path8, PlotJoined -> True]
```

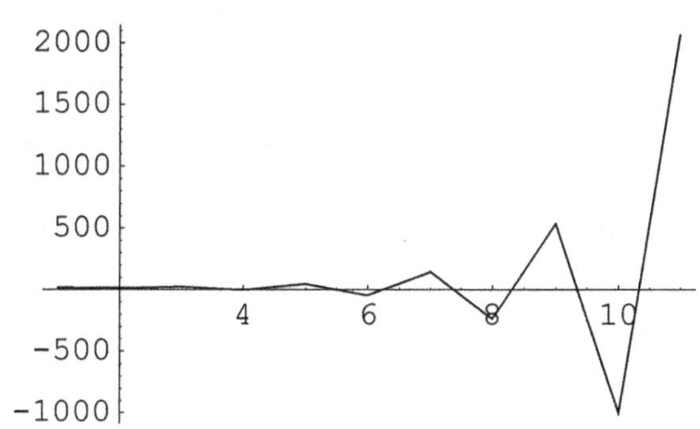

Example: Application to IS-LM Model

Suppose that a dynamic macro model of an economy is expressed as Ct = bYt-1 + Co, It = Io, and Yt = Ct + It, where C is consumption, Y is income, and I is investment. Analyze the time path of income.

Solution: Solving for Yt from the equilibrium equation Yt = Ct + It yields, Yt = bYt-1 + Co + Io. The steady-state equilibrium condition requires that the period-by-period income remain the same. i.e., Yt = Yt-1 = Ye = $\frac{Co + Io}{1 - b}$. However, the time path of Y is determined by the solution of the difference equation Yt = bYt-1 + Co + Io, which is Yt = Yob^t + $\frac{(Co + Io)(1 - b^t)}{1 - b}$= Yo$b^t$+ ye(1 - b^t) = (Yo - Ye)b^t + Ye. Assuming a conventional Keynesian consumption function with b < 1, the solution implies that if there is a displacement in income, the system will approach the equilibrium point gradually as t -> ∞.

Example: Application to Macroeconomic Model

A simple macro model of an economy is expressed as Ct = .75Yt-1 + 500, It = 200, and Yt = Ct + It. Given that the initial income is Y(0) = 3000, graph the time path of income.

Solution: The equilibrium condition yields the initial-value difference equation Yt = .75Yt-1 + 700, Y(0) = 3000. Using *Mathematica*, the solution is,

```
In[15]:= ClearAll[y, catkin, path9]
         catkin = RSolve[{y[t] == .75 y[t - 1] + 700, y[0] == 3000}, y[t], t]
```

```
Out[15]= {{Null (y[t] → 1.33333 (3000. (-2.25 0.75^t + 3. 1.^t) - 2300. If[t ≥ 1, -3. 0.75^t + 3. 1.^t, 0]))}}
```

```
In[16]:= y[t_] = y[t] /. catkin[[1]];
         path9 = Table[y[t], {t, 0, 15}];
         ListPlot[path9, PlotJoined -> True]
```

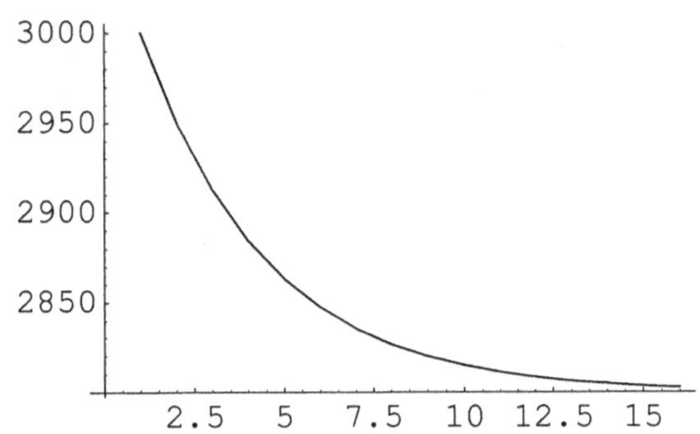

The time path of Y(t) shows that it approaches to its steady-state equilibrium of Y = 2800 in about 15 periods.

DO More Practice:

1- Change the MPC (the coefficient of Yt -1 in the consumption function) from .75 to .9. Keep the other assumptions the same. Re-run the program and compare the results.

2- Change the MPC from .75 to 1.2. Keep the other assumptions the same. Re-run the program and compare the results.

3- Change the initial value from Y(0) = 3000 to Y(0) = 2800. Keep the other assumptions the same. Re-run the program and compare the results.

4- Delete the autonomous spending of 700 in the consumption function. Keep other assumptions the same. Re-run the program and compare the results.

Problems

1. Solve the following difference equations and analyze the stability of each equation.

a) $Yt = Yt\text{-}1 + 5,$ $yo = 2$

b) $5Yt\text{+}1 - Yt = 2,$ $yo = \frac{1}{2}$

c) $3Yt + Yt\text{-}1 = 8,$ $yo = 10$

d) $Yt = 2Yt\text{-}1 + 5,$ $yo = 5$

e) $Yt = -3Yt\text{-}1 - 10,$ $yo = 2$

2. Determine the price behavior of the following market systems and graphically show the price movements.

a) $Qdt = 45 - 5Pt$

$Qst = 10 + 2Pt\text{-}1$

$Po = 7$

b) $Qdt = 200 - 3Pt$

$Qst = -10 + 4Pt\text{-}1$

$Po = 28$

c) $Qdt = 150 - 2Pt$

$Qst = 10 + 5Pt\text{-}1$

$Po = 15$

d) $Qdt = 80 - 4Pt$

$Qst = -40 + 2Pt\text{-}1$

$Po = 15$

3. Suppose that the price adjustment in shock market follows the rule $Pt\text{+}1 = Pt - .3(Qst - Qdt)$. Given the following market models find the time path of Pt and determine the stability of the models.

a) $Qdt = 21 - 2Pt$

$Qst = -3 + 6Pt$

b) $Qdt = 100 - 5Pt$

$Qst = -20 + 3Pt$

4. A macro model of an economy is expressed as: $Ct = .75Yt\text{-}1 + 1000$, $It = 100$, and $Yt = Ct + It$, where C is consumption, I is investment, and Y is income. Find the time path of income and graphically show the adjustment of income from $Yo = 5000$ to equilibrium.

III. Second-Order Homogenous Linear Difference Equations

The general form of a second order, homogenous linear difference equation is Yt + aYt-1 + bYt-2 = 0 where a and b are arbitrary constants. The solution of the first order, homogenous difference equation of the previous section suggests that a solution of the form, Yt = k^t may be applicable here also. Substitute Yt = k^t and its lagged equivalent into the difference equation Yt + aYt-1 + bYt-2 = 0 to obtain, $k^t + ak^{t-1} + bk^{t-2} = 0$. This equation is called the root equation or auxiliary equation. Dividing both sides of the root equation by k^{t-2} reduces the root equation to a quadratic equation $k^2 + ak + b = 0$. This equation has the roots k1 = $\frac{-a + \sqrt{a^2 - 4b}}{2}$ and

k2 = $\frac{-a + \sqrt{a^2 - 4b}}{2}$. Through direct substitution, it is easy to show that y1 = $k1^t$, y2 = $k2^t$, or any linear combination of y1 and y2, such as Yt = c1Y1 + c2Y2 satisfies the difference equation Yt + aYt-1 + bYt-2 = 0 and is a solution of that equation. However, depending upon the value and the sign of a^2 - 4b in the solution of the root equation, three distinct cases can occur:

Case1 a^2 - 4b - 4b > 0: Under this assumption, k1 and k2 are unequal real roots and y1 = $k1^t$ or y2 = $k2^t$, or any linear combination of y1 and y2 such as Yt = c1Y1 + c2Y2 = c1$k1^t$+ c2$k2^t$ are solutions to the equation.

Example: Solve the second order, homogenous difference equation Yt + Yt-1 - 6Yt-2 = 0.

Solution: Form the auxiliary equation, $k^2 + k - 6 = 0$. This equation has two unequal real roots, k1 = 2 and k2 = -3. Therefore, y1 = 2^t or y2 = -3^t, or any linear combination of y1 and y2 will be a solution of the equation.

Using *Mathematica*, the solution can be found as,

```
In[17]:= Clear[y]
         RSolve[y[t] + y[t - 1] - 6 y[t - 2] == 0, y[t], t]
```

Out[17] = $\{\{\text{Null}\ (y[t] \rightarrow \frac{1}{5}(-3)^{2+t}(-C[1] + 2C[2]) + \frac{1}{5}2^{2+t}(C[1] + 3C[2]))\}\}$

If the initial values of Yt at times t = 0 and t = 1 are also given, then the initial-value problem Yt + Yt-1 - 6Yt-2 = 0, Yo = 6, Y1 = 2 has a solution, Yt = c1$k1^t$ + c2$k2^t$ = c1$(2)^t$ + c2$(-3)^t$. Substituting for the initial values yields the two equations, c1 + c2 = 6 and 2c1 -3c2 =2, the solution of which yields, c1 = 4 and c2 =2. Therefore, Yt = 4$(2)^t$ + 2$(-3)^t$ is a solution to the initial-value problem. It is easy to show that this solution satisfies both the initial value and the difference equation.

To solve the same initial-value problem using *Mathematica,* use the RSolve command as,

```
In[18]:= Clear [y]
         RSolve[{y[t] + y[t - 1] - 6 y[t - 2] == 0, y[0] == 6, y[1] == 2}, y[t], t]
```

Out[18]= $\{\{\text{Null}\ (y[t] \rightarrow 2((-3)^t + 2^{1+t}))\}\}$

Case 2 a^2 - 4b = 0: Under this assumption, k1 = k2 = k = -$\frac{a}{2}$ and Y1 = k^t is a solution to the difference equation. However, to solve for the initial-value problem, we need to search for another solution. It is easy to show that Y2 = tk^t is also a solution of the difference equation (this can easily be proved by substituting tk^t into the difference equation and showing that it satisfies the equation). Therefore, any linear combination of y1 and y2, such as Yt = c1Y1 +c2Y2 = c1k^t + c2tk^t is also a solution of the problem.

Example: Solve the second order, homogenous difference equation Yt + 4Yt-1 + 4Yt-2 = 0 for the initial values Yo = 2, Y1 = 0.

Solution: Form the auxiliary equation, $k^2 + 4k + 4 = 0$. This equation has only one root, k = -2. Therefore, the solution to the difference equation is y1 = $(-2)^t$. However, to solve for the initial values, we need one more solution. We try the solution Y2 = $t(-2)^t$. Since this solution also satisfies the difference equation, a linear combination of Y1 and Y2, Yt = c1Y1 + c2Y2 = c1 $(-2)^t$+ c2t$(-2)^t$ will also satisfy the difference equation.

Solving for the initial values yields, c1 = 2 and c2 = -2. Therefore, the solution of the initial-value problem is Yt $= 2(-2)^t - 2t(-2)^t$. By direct substitution, it can be shown that this solution satisfies both the initial values and the difference equation.

To solve the same problem using *Mathematica*,

```
In[19]:= Clear[y]
         RSolve[{y[t] + 4 y[t - 1] + 4 y[t - 2] == 0, y[0] == 2, y[1] == 0}, y[t], t]

Out[19] = {{Null (y[t] → -2^(1+t) e^(iπt) (-1 + t))}}
```

Case 3 $a^2 - 4b < 0$: Under this assumption, k1 and k2 are two distinct complex conjugate roots. Solutions of difference equations with complex conjugate roots are discussed in chapter 14.

Example: Application to Macroeconomics

The acceleration principle states that the net investment in the economy during the time period t (It) depends on the rate of change of aggregate income, It = v(Yt-1 - Yt-2), where v is a constant coefficient. This principle, together with the consumption model Ct = bYt-1can be merged into a simple multiplier-accelerator model of the economy. Solving the two equations above for the equilibrium condition Yt = Ct + It, yields Yt = (b+v)Yt-1 -vYt-2. This equation is a second-order, homogeneous difference equation that can be solved to find the time path of Yt. The solution and the stability of the model follows the three cases discussed above.

Example: Application to Macroeconomics

Suppose that in the example above investment adjusts to change in income at a constant proportion 8, (v = 8), and that marginal propensity to consume, b = .9. Given the initial values Yo = 6000 and Y1 = 6200, find the time path of Y.

Solution: The assumptions above yields the second-order difference equation Yt = 8.9Yt-1 - 8Yt-2 with the initial values Yo = 6000 and Y1 = 6200. Using *Mathematica*, the solution and the time path of Yt can be found as.

```
In[20]:= ClearAll[y, lily, graph1]
         lily = RSolve[{y[t] == 8.9 y[t - 1] - 8 y[t - 2], y[0] == 6000, y[1] == 6200}, y[t], t]

Out[20]= {{Null (y[t] → 5983.57 1.01452^t + 16.4257 7.88548^t)}}

In[21]:= y[t_] = y[t] /. lily[[1]];
         graph1 = Table[y[t], {t, 0, 5}];
         ListPlot[graph1, PlotJoined -> True]
```

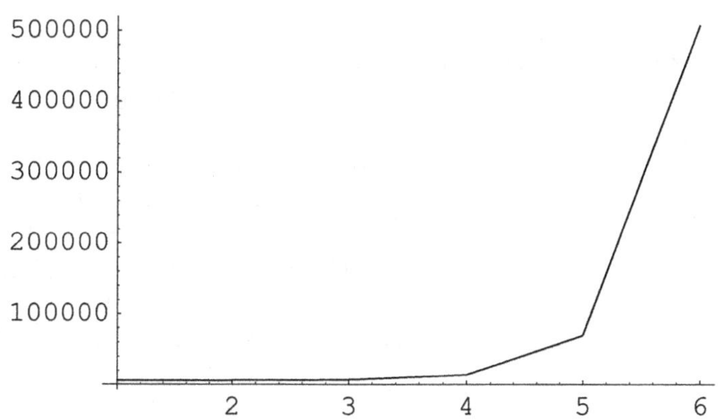

Do More Practice:

1- Change b to .75. Keep other assumptions the same. Re-run the program and compare the results.

2- Change v to 2. Keep other assumptions the same. Re-run the program and compare the results.

3- Change v to .7 and b to .9. Keep other assumptions the same. Re-run the program and compare the results.

4- Change v to .9 and b to .7. Keep other assumptions the same. Re-run the program and compare the results.

Problems

1. Solve the following difference equations and graph the time path of Yt..

a)	Yt + 3Yt-1 + 2Yt-2 = 0,	Yo =2,	Y1 =3
b)	Yt + Yt-1 - 2Yt-2 = 0,	Yo = 1,	Y1 = -1
c)	Yt - 4Yt-2 = 0,	Yo = 1,	Y1 = 1
d)	6Yt + 5Yt-1 + Yt-2 = 0,	Yo = 0,	Y1 = 1
e)	9Yt - 12Yt-1 + 4yT-2 = 0,	Yo =1,	Y1 = 0
f)	Yt + 8Yt-1 + 16Yt-2 = 0,	Yo = 2,	Y1 = -4

2. A simple Keynesian macro-model of the economy is described as: It = 4.07(Yt-1 - Yt-2), Ct = .73Yt-1, and Yt = Ct + It where I is investment, C is consumption and Y is income. Given that Yo = 4000 and y1 = 4150, find the time path of income and graph it.

IV. Second Order Nonhomogeneous Linear Difference Equations

The solution of a second-order, nonhomogeneous linear difference equation of the form Yt + aYt-1 + bYt-2 + c = 0 consists of two parts: The solution of the complementary homogenous part, Yc, and the particular solution, Yp, of the nonhomogeneous difference equation. The general solution of the difference equation is the sum of the solution to the complementary homogenous equation and the particular solution. That is, Yt = Yc + Yp. To obtain the particular solution of the nonhomogeneous part, try a function f(t) = z such that z satisfies the difference equation for all t values.

Example: Solve the second-order, nonhomogeneous linear difference equation Yt - 2Yt-1 - 3Yt-2 + 8 = 0 for the initial values Yo = 3 and Y1 = 2.

Solution: The complementary homogenous equation Yt - 2Yt-1 - 3Yt-2 = 0 has the auxiliary equation $k^2 - 2k - 3 = 0$, with the roots k1 = 3 and k2 = -1. Therefore, the solution of the complementary homogenous part is Yc = $c1\,(3)^t + c2\,(-1)^t$. To obtain a particular solution, we try f(t) = z, such that z satisfies the difference equation, for all t. That is, z - 2z - 3z + 8 = 0, or z = 2. Therefore, the particular solution is Yp = 2. The general solution of the difference equation is Yt = Yc + Yp = $c1\,(3)^t + c2\,(-1)^t + 2$. Given the initial values, c1 and c2 can be found from the general solution of the difference equation. Substitute for Yo and y1 to get 3 = c1 + c2 + 2 and 2 = 3c1 - c2 + 2, respectively. Solving these two equations together yields, c1 = $\frac{1}{4}$ and c2 = $\frac{3}{4}$. The solution of the initial-value, nonhomogeneous difference equation is Yt = $\frac{1}{4}\,3^t + \frac{3}{4}\,(-1)^t + 2$. It can be shown, through direct substitution, that this solution satisfies both the initial conditions and the difference equation. To solve the same problem using *Mathematica*, input as follow

```
In[22]:= Clear[y]
        RSolve[{y[t] – 2 y[t – 1] – 3 y[t – 2] + 8 == 0, y[0] == 3, y[1] == 2}, y[t], t]
        Out[22] = {{Null (y[t] → 1/4 (8 + 3 (–1)^t + 3^t))}}
```

Example: Application to Macroeconomics

The multiplier-accelerator model of the previous section can be extended to a second order nonhomogeneous difference equation by including autonomous spending and/or by including government spending. Suppose that the investment at time t (It) is defined as a constant 4 proportion of the change in income, It = 4(Yt-1 - Yt-2); Consumption at time t is defined as Ct = .80Yt-1 + 400, and the government spending is Gt = 200. Given the equilibrium condition Yt = Ct + It + Gt and the initial values Yo = 5000 and Y1 = 5100 find Yt and graph the time path of Yt.

Solution: Solving the equations of the model for the equilibrium condition yields Yt - 4.8Yt-1+ 4Yt-2 - 600 = 0 with the initial values Yo = 5000 and Y1 = 5100. This is an initial-value, second-order, nonhomogeneous difference equation that can be solved for Yt. Using *Mathematica* the solution is,

```
In[23]:= ClearAll[y, fuchsia, graph2]
        fuchsia = RSolve[{y[t] – 4.8 y[t – 1] + 4 y[t – 2] – 600 == 0, y[0] == 5000, y[1] == 5100}, y[t], t];
```

```
In[24]:= y[t_]=y[t]/.fuchsia[[1]];
graph2=Table[y[t], {t, 0, 5}];
ListPlot[graph2, PlotJoined->True]
```

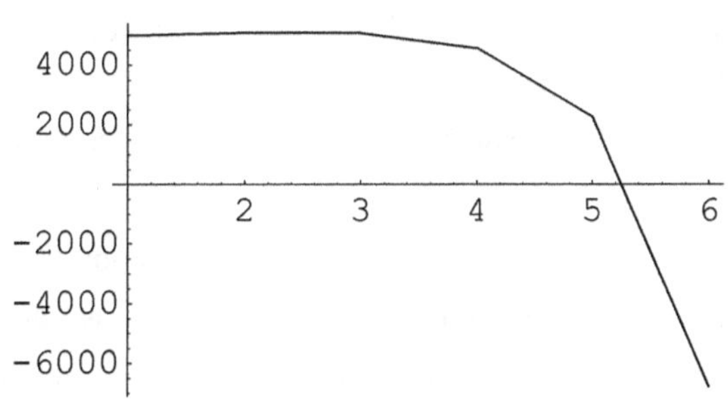

Do More Practice:

1- Change b to .75. Keep other assumptions the same. Re-run the program and compare the results.

2- Change v to 2. Keep other assumptions the same. Re-run the program and compare the results.

3- Change v to .7 and b to .9. Keep other assumptions the same. Re-run the program and compare the results.

4- Change v to .9 and b to .7. Keep other assumptions the same. Re-run the program and compare the results.

Problems

1. Solve the following difference equations and graph the time path of Yt.

   a) Yt - Yt-1 - 2Yt-2 + 4 = 0, Yo =0, Y1 = 1

   b) Yt - 4Yt-2 + 6 = 0, Yo = 10, Y1 = 2

   c) Yt+2 - 7Yt+1 + 10Yt + 12 = 0, Yo = 10, Y1 = 5

   d) Yt+2 - 9Yt + 8 = 0 Yo = 2, Y1 = 3

   e) Yt + 4Yt-1 + Yt-2 =18, Yo = 1, Y1 = 2

   f) Yt - 4Yt-1 + 4Yt-2 -3 = 0, Yo = 0, Y1 = 2

2. A simple Keynesian macro-model of an economy is described as It = 5.08(Yt-1 - Yt-2), Ct = .8Yt-1 + 200, Yt = Ct + It + Gt. Given that Yo = 3000 and Y1 = 3200, find the time path of income when Go = 150.

Stability Analysis of the solutions of Second-Order, Linear Difference Equations: The general solution of a second-order, nonhomogeneous difference equation is Yt = $c1\,k1^t + c2k2^t + c$, where c1, c2 and c are constants and k1 and k2 are the roots of auxiliary equation corresponding to the homogenous part of the difference equation. This solution generates a time path for Yt as t varies from 0 to ∞. In this solution, the only values that change as t changes are $k1^t$ and $k2^t$. Therefore, changes in Yt depend on the sign and absolute values of k1 and k2.

A common way to study the time path of Yt is to transform Yt = $c1\,k1^t + c2k2^t + c$ to Yt = $k1^t(c1 + c2\left(\frac{k1}{k2}\right)^t) + c$ by factoring out the root that has the greater absolute value. Since it is assumed that |k1| > |k2|, $\left(\frac{k1}{k2}\right)^t$ approaches zero as t approaches ∞. Therefore, the time path of xYt can be determined by concentrating on the behavior of $k1^t$, or the greater root only.

If k1 and k2 are real roots, (k1≠ k2), Yt will have a number of different time paths depending upon the sign and magnitude of k1 and k2. In general, if the absolute value of the largest root is smaller than one, as t approaches ∞, Yt will approach its long run equilibrium value. Otherwise, Yt will move further away from equilibrium. Some of those cases are studied in the following examples.

Example: The case where k1 and k2 > 1.

Solve the difference equation Yt - 5Yt-1 + 6Yt-2 - 10 = 0, Yo = 8 and Y1 = 13.

Solution: The solution of the difference equation is, Yt = $2^t + 2(3)^t + 5$. Factor out 3^t to transform the solution to $3^t(\left(\frac{2}{3}\right)^t + 2) + 5$. As t approaches ∞, $\left(\frac{2}{3}\right)^t$ approaches zero, while 3^t approaches ∞. Therefore, Yt will explode gradually in a positive direction. The solution and the graph using *Mathematica* illustrate the time path of Yt.

```
In[25]:= ClearAll[y, corymb, graph3]
         corymb = RSolve[{y[t] - 5 y[t - 1] + 6 y[t - 2] - 10 == 0, y[0] == 8, y[1] == 13}, y[t], t]

         Out[25] = {{Null (y[t] → 5 + 2^t + 2 3^t)}}
```

```
In[26]:= y[t_]=y[t]/.corymb[[1]];
graph3=Table[y[t], {t, 0, 10}];
ListPlot[graph3, PlotJoined->True]
```

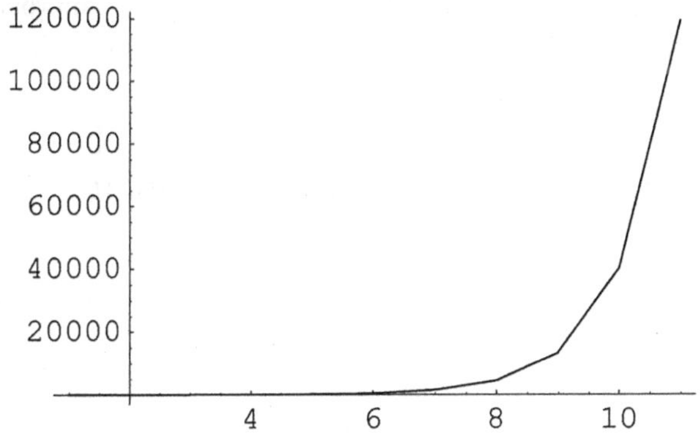

Example: The case where 0 < k1 & K2 < 1.

Solve the difference equation 6Yt - 5Yt-1 + Yt-2 - 4 = 0, Yo = 10, Y1 = 5 and analyze the time path of Yt.

Solution: The solution of the difference equation is $Yt = 2\left(\frac{1}{2}\right)^t + 6\left(\frac{1}{3}\right)^t + 2$. Factor out $\left(\frac{1}{2}\right)^t$ to transform the solution to $Yt = \left(\frac{1}{2}\right)^t\left(2 + 6\left(\frac{2}{3}\right)^t\right) + 2$. As t approaches ∞, both $\left(\frac{1}{2}\right)^t$ and $\left(\frac{2}{3}\right)^t$ approach zero. Therefore, Yt gradually approaches its long run equilibrium level, 2. The solution and the graph using *Mathematica* illustrate the time path of Yt.

```
In[27]:= ClearAll[y, dafodil, graph4]
         dafodil = RSolve[{6 y[t] - 5 y[t - 1] + y[t - 2] - 4 == 0, y[0] == 10, y[1] == 5}, y[t], t]
         Out[27] = {{Null (y[t] → 2^(1-t) 3^(-t) (3 2^t + 3^t + 6^t))}}
```

```
In[28]:= y[t_]=y[t]/.dafodil[[1]];
graph4=Table[y[t], {t, 0, 5}];
ListPlot[graph4, PlotJoined->True]
```

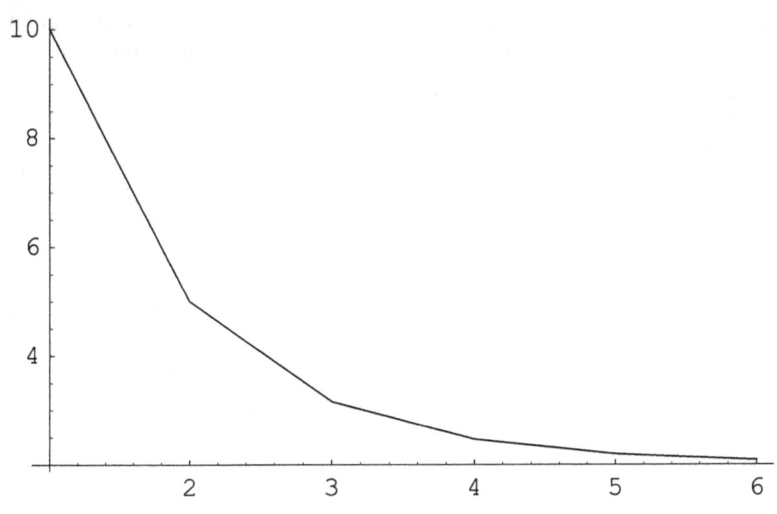

Example: The case where k1 & k2 < -1.

Solve the difference equation Yt + 5Yt-1 + 6Yt-2 - 24 = 0, Yo = 5, Y1 = -5.

Solution: The solution of the difference equation is $Yt = 2(-2)^t + (-3)^t + 2$. The solution can be written as $Yt = (-3)^t(1 + (\frac{2}{3})^t) + 2$. As t approaches ∞, $(\frac{2}{3})^t$ approaches zero and $(-3)^t$ approaches ∞. Therefore, Yt will explode with an oscillatory path. The solution and the graph using *Mathematica* illustrate the time path of Yt.

```
In[29]:= ClearAll[y, mint, graph5]
         mint = RSolve[{y[t] + 5 y[t - 1] + 6 y[t - 2] - 24 == 0, y[0] == 5, y[1] == -5}, y[t], t]
```

Out[1] = $\{\{\text{Null}\,(y[t] \to 2 + (-3)^t + (-1)^t 2^{1+t})\}\}$

```
In[30]:= y[t_]=y[t]/.mint[[1]];
graph5=Table[y[t], {t, 0, 10}];
ListPlot[graph5, PlotJoined->True]
```

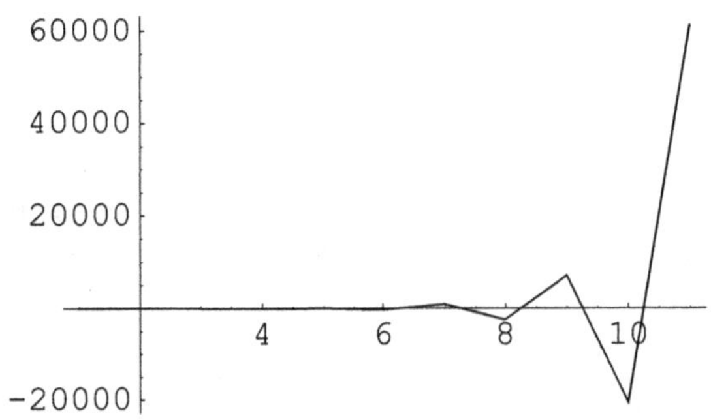

Example: The case where -1 < k1 & K2 < 0.

Solve the difference equation 6Yt + 7Yt-1 + 2Yt-2 + 30 = 0, Yo = 3, Y1 = -5, and analyze the time path of Yt.

Solution: The solution of the difference equation is $Yt = 3(\frac{-2}{3})^t + 2(\frac{-1}{2})^t - 2$. Factor out $(\frac{-2}{3})^t$ to get $Yt = (\frac{-2}{3})^t(3 + 2(\frac{3}{4})^t) - 2$. As t approaches ∞, both $(\frac{-2}{3})^t$ and $(\frac{3}{4})^t$ approach zero and Yt approaches its long run equilibrium, -2. However, since the roots are negative numbers, Yt approaches -2 with an oscillatory path around the equilibrium value. The solution and the graph using *Mathematica* illustrate the time path of Yt.

```
In[31]:= ClearAll[y, poppy, graph6]
         poppy = RSolve[{6 y[t] + 7 y[t - 1] + 2 y[t - 2] + 30 == 0, y[0] == 3, y[1] == -5}, y[t], t]
```

Output[31] = $\{\{\text{Null}\,(y[t] \to 3\,(-\frac{2}{3} + ((\frac{2}{3})^t + \frac{2^{1-t}}{3})\,e^{i\pi t}))\}\}$

```
In[32]:= y[t_]=y[t]/.poppy[[1]];
graph6=Table[y[t], {t, 0, 10}];
ListPlot[graph6, PlotJoined->True]
```

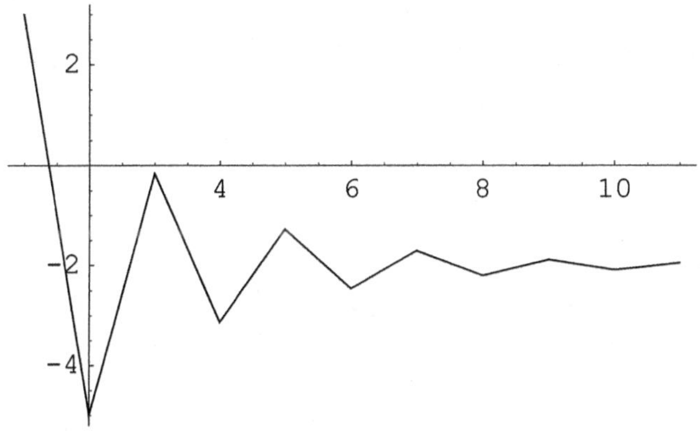

Example: The case where k1 = k2 > 1.

solve the difference equation Yt - 4Yt-1 + 4Yt-2 -2 = 0, Yo = 3, Y1 = 10 and plot the time path of Yt.

Solution: The solution of the difference equation is Yt = $2^t + 3t2^t + 2$. The solution can be written as Yt = $2^t(1 + 3t) + 2$. As t approaches ∞, Yt will gradually explode in the positive direction. The solution and the graph using *Mathematica* illustrate the time path of Yt.

```
In[33]:= ClearAll[y, basil, graph7]
        basil = RSolve[{y[t] – 4 y[t – 1] + 4 y[t – 2] – 2 == 0, y[0] == 3, y[1] == 10}, y[t], t]

        Out[33] = {{Null (y[t] → 2 + 2^t + 3 2^t t)}}
In[34]:= y[t_]=y[t]/.basil[[1]];
graph7=Table[y[t], {t, 0, 10}];
ListPlot[graph7, PlotJoined->True]
```

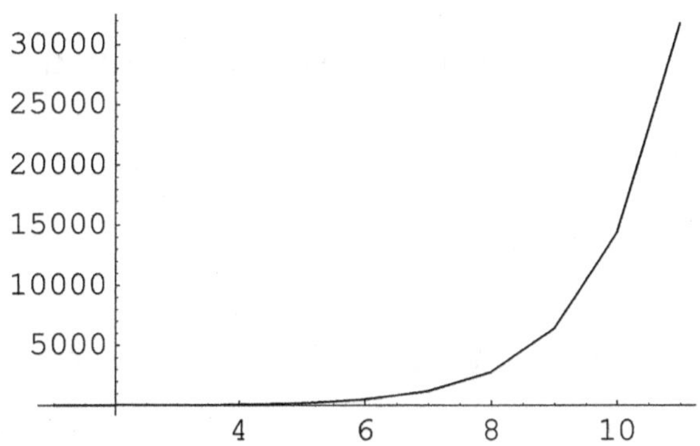

Example: The case where 0 < k1= k2 < 1.

Solve the difference equation 4Yt - 4Yt-1 + Yt-2 + 2 = 0, Yo = 4, Y1 = 5, and graph the time path of Yt.

Solution: The solution of the equation is Yt = $2((\frac{1}{2})^t + 4t(\frac{1}{2})^t) + 2$. The solution can be written as Yt = $(\frac{1}{2})^t(2 + 4t) + 2$. As t approaches ∞, Yt will approach its long run equilibrium, -2. The solution and the graph using *Mathematica* illustrate the time path of Yt.

```
In[35]:= ClearAll[y, bay, graph8]
         bay = RSolve[{4 y[t] - 4 y[t - 1] + y[t - 2] + 2 == 0, y[0] == 4, y[1] == 5}, y[t], t]

         Out[35] = {{Null (y[t] → 2^(1-t) (3 - 2^t + 4 t))}}
```

```
In[36] := y[t_] = y[t] /. bay[[1]]
graph8 = Table[y[t], {t, 0, 10}]
ListPlot[graph8, PlotJoined -> True]
Out[36] = 2^(1-t) (3 - 2^t + 4 t) {4, 5, 7/2, 7/4, 3/8, -9/16, -37/32, -97/64, -221/128, -473/256, -981/512}
```

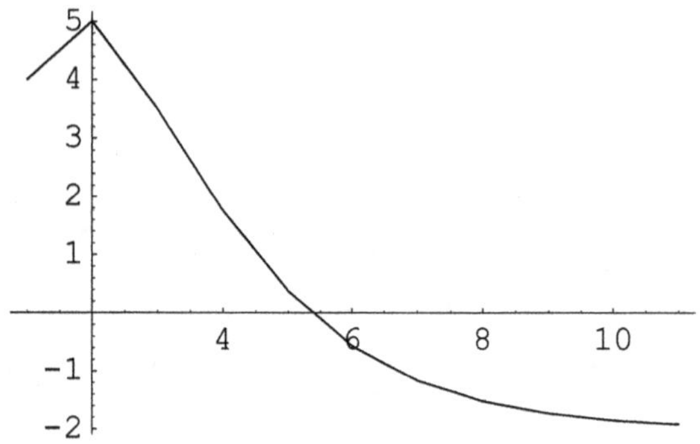

Example: The case where -1 < k1=k2 < 0.

Solve the difference equation 4Yt + 4Yt-1 + Yt-2 - 9 = 0, Yo = 3, Y1 = -2, and graph the time path of Yt.

Solution: The solution of the difference equation is $Yt = 2\left(\frac{-1}{2}\right)^t + 4t\left(\frac{-1}{2}\right)^t + 1$. The solution can be written as $Yt = \left(\frac{-1}{2}\right)^t(2 + 4t) + 1$. As t approaches ∞, $\left(\frac{-1}{2}\right)^t$ approaches zero and Yt approaches its long run equilibrium, 1. The solution and the graph using *Mathematica* illustrate the time path of Yt.

```
In[37]:= ClearAll[y, thyme, graph9]
         thyme = RSolve[{4 y[t] + 4 y[t - 1] + y[t - 2] - 9 == 0, y[0] == 3, y[1] == -2}, y[t], t]

         Out[37] = {{y[t] → 2^(-t) (2 (-1)^t + 2^t + 4 (-1)^t t)}}
```

```
In[38] := y[t_] = y[t] /. thyme[[1]]
graph9 = Table[y[t], {t, 0, 10}]
ListPlot[graph9, PlotJoined -> True]
Out[38] = 2^(-t) (2 (-1)^t + 2^t + 4 (-1)^t t) {3, -2, 7/2, -3/4, 17/8, 5/16, 45/32, 49/64, 145/128, 237/256, 533/512}
```

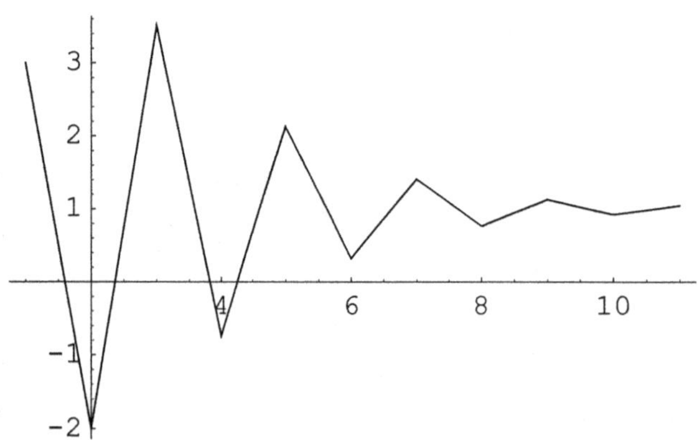

Example: Application to Markets

The demand and supply for a product are expressed as Qdt = 20 - Pt and Qst = -4 + 1.1Pet, respectively. Pt is the market price of the product and Pet is the expected price. Given that expectations of future prices by the producers follow the simple adaptive expectations rule, Pet = $\frac{9}{11}$Pt-1 + $\frac{2}{11}$Pt-2, solve the market model and analyze the time path of Pt, if the initial prices are Po = 10 and P1 = 8.

Solution: The market equilibrium requires that Pt = 24 - .6Pet. Substituting for Pet yields the second-order difference equation Pt = 24 - .9Pt-1 - .2Pt-2. Given the initial values Po = 10 and P1 = 8, this equation can be solved to trace its time path. Using *Mathematica*,

```
In[39]:=ClearAll[p, fennel, tplot1]
        fennel = RSolve[{p[t] == 24 - .9 p[t - 1] - .2 p[t - 2], p[0] == 10, p[1] == 8}, p[t], t];

In[40]:=p[t_]=p[t]/.fennel[[1]];
tplot1=Table[p[t], {t, 0, 10}];
ListPlot[tplot1, PlotJoined->True]
```

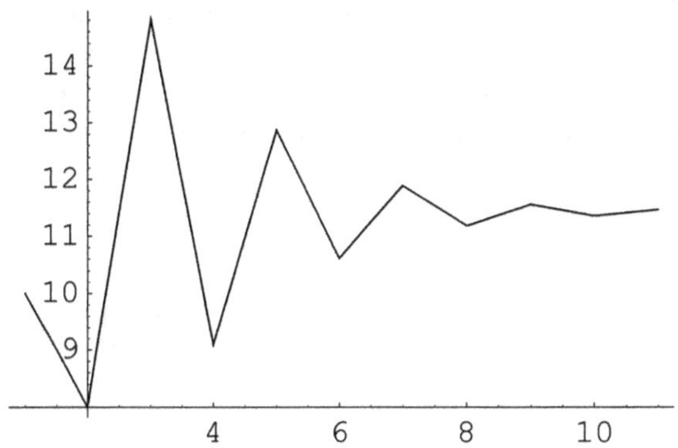

Do More Practice

1- Suppose that the expectations of the future prices follows the rule Pet = .5Pt + .3Pt-1 + .2Pt-2. R-run the program and compare the results.

2- Suppose that demand and supply are given as Qdt = 20 - 1.2Pt and Qst = -4 + Pet, respectively. The other assumptions stay the same as in example. Re-run the program and compare the results.

Problems

1. Solve the following second-order linear difference equations and graph the values of Yt for t = 0 to t = 10. Analyze the time path of Yt as t approaches ∞.

a)	4Yt - Yt-2 -9 = 0,	Yo = 11,	Y1 = 1
b)	Yt + Yt-1 - 6Yt-2 + 4 = 0,	Yo =6,	Y1 = 1
c)	Yt + 4Yt-1 + 4Yt-2 -18 = 0,	Yo = 3,	Y1 = -4
d)	10Yt - 23Yt-1 + 12Yt-2 + 3 = 0,	Yo = 6,	Y1 = 6.1

e)	$2Y_t - Y_{t-1} - 6Y_{t-2} + 4 = 0$,	$Y_0 = 6$,	$Y_1 = 4$
f)	$2Y_t + 7Y_{t-1} + 6Y_{t-2} + 30 = 0$,	$Y_0 = .5$,	$Y_1 = -6$
g)	$6Y_t + Y_{t-1} - Y_{t-2} + 12 = 0$,	$Y_0 = 3$,	$Y_1 = -\frac{17}{6}$
h)	$4Y_t - 12Y_{t-1} + 9Y_{t-2} + 2 = 0$,	$Y_0 = 0$,	$Y_1 = 4$

2. The demand and supply equations of a product are given as, $Qd_t = 10 - P_t$ and $Qs_t = -8 + .8P_{et}$ where, P_t is the market price and P_{et} is the expected price. Given that expectations of future prices by the producers follow the simple adaptive expectations model, $P_{et} = \frac{7}{8}P_{t-1} + \frac{1}{8}P_{t-2}$, solve the market model and analyze the time path of P_t, if the initial prices are $P_0 = 15$ and $P_1 = 8.1$.

3. The demand and supply equations for a product are given as $Qd_t = 30 - 2P_t$ and $Qs_t = -20 + 8P_{et}$ where, P_{et} is the expected market price. If the expectations of future prices by the producers follow the simple adaptive expectations rule, $P_{et} = \frac{5}{8}P_{t-1} + \frac{3}{8}P_{t-2}$, solve the market model. Given the initial prices $P_0 = 6.5$ and $P_1 = 3$. Analyze the time path of prices. Will the market price approach its long run equilibrium?

4. Consider a simple macro model of economy where $Y_t = C_t + I_t$, $C_t = 100 + .8Y_{t-1}$, and $I_t = 10(Y_{t-1} - Y_{t-2})$. The first equation is the equilibrium condition and expresses income at time t (Y_t) equal to consumption at time t (C_t) and investment at time t {I_t). The second equation expresses consumption as a function of lagged income and the third equation expresses investment as a function of the change in lagged income. Given that the initial income levels are $Y_0 = 450$ and $Y_1 = 480$, find the time path of Y_t and analyze the stability of the system.

V. System of First-Order Linear Difference Equations

Economic variables may be related to each other through a system of several discrete time dynamic inter-related variables and equations. For example, a discrete time dynamic IS-LM model of the economy may be expressed by the system of equations $\Delta Yt = \alpha[D(Y, r) - Y]$ and $\Delta rt = \beta[L(Y,r) - M]$, where Y is the income or output, r is the interest rate, M is the money supply, and α and β are the coefficients of adjustment. The first equation represents equilibrium in the output market, IS. It expresses changes in output, $\Delta Yt = Yt+1 - Yt$, as α proportion of the gap between aggregate demand, D(Y, r), and aggregate output, Y. The second equation, $\frac{dr}{dt}$, represents the equilibrium in the money market, LM. It expresses changes in the interest rate as a β proportion of the gap between the demand for money, L(Y,r), and the money supply, M.

The laws that govern the behavior of variables in this model and the system can be estimated by a system of linear difference equations. That is, a model which assumes that the change of each variable is a linear combination of other variables.

System of First-Order Linear Homogeneous Difference Equations: The general form of a system of first-order linear homogeneous difference equation is $\mathbf{\Delta Y}t = \mathbf{AY}t$ where, $\mathbf{\Delta Y}(t)$ is a vector with elements $\begin{pmatrix} \Delta Y1t \\ \Delta Y2t \end{pmatrix}$, **A** is a matrix of constant coefficients, and **Y**(t) is the vector with elements $\begin{pmatrix} Y1t \\ Y2t \end{pmatrix}$.

Equation $\mathbf{\Delta Y}t = \mathbf{AY}t$ may be written as $\begin{pmatrix} Y1t+1 - Y1t \\ Y2t+1 - Y2t \end{pmatrix} = \mathbf{A}\begin{pmatrix} Y1t \\ Y2t \end{pmatrix}$ or $\begin{pmatrix} Y1t+1 \\ Y2t+1 \end{pmatrix} = \mathbf{A}\begin{pmatrix} Y1t \\ Y2t \end{pmatrix} + \mathbf{I2}\begin{pmatrix} Y1t \\ Y2T \end{pmatrix} = (\mathbf{A} + \mathbf{I2})\begin{pmatrix} Y1t \\ Y2t \end{pmatrix}$, where **I2** is a 2x2 identity matrix. Representing $\begin{pmatrix} Y1t+1 \\ Y2t+1 \end{pmatrix}$ by **Y**t+1 and $\begin{pmatrix} Y1t \\ Y2t \end{pmatrix}$ by **Y**t, the model can be written in matrix form as **Y**t+1 = (**A** + **I2**)**Y**t. The resemblance of this equation to Yt+1 = kYt suggest that a trial solution of the form Yt+1 = Yok^t may be applicable to the system of difference equations. Let us study the possibility by searching for a constant vector **u** and a number k such that, **Y**t+1 = $\mathbf{u}k^t$ satisfies **Y**t+1 = (**A** + **I2**)**Y**t. Since **Y**t+1 = $\mathbf{u}k^t$, for **Y** to be a solution it is necessary and sufficient that $\mathbf{u}k^t = (\mathbf{A} + \mathbf{I2})\mathbf{u}k^{t-1}$. This equation implies that, $(\mathbf{A} + \mathbf{I2})\mathbf{u} - k\mathbf{u} = ((\mathbf{A} + \mathbf{I2}) - k\mathbf{I})\mathbf{u} = \mathbf{0}$. Therefore, k is the eigenvalue and **u** is the eigenvector of the matrix (**A** + **I2**) (see, Chapter 5, Special Topics in Matrix Algebra).

Example: Solve the following system of linear homogeneous difference equations, $\Delta Y1t = 2Y1t + Y2t$ and $\Delta Y2t = 3Y1t$, $Y1o = 1$ $Y2o = 1$.

Solution: The system in the matrix form is, $\begin{pmatrix} \Delta y1t \\ \Delta y2t \end{pmatrix} = \begin{pmatrix} 2 & 1 \\ 3 & 0 \end{pmatrix} \begin{pmatrix} y1t \\ y2t \end{pmatrix}$. Find the eigenvalues and eigenvectors of **A** + **I2**= $\begin{pmatrix} 3 & 1 \\ 3 & 1 \end{pmatrix}$,

```
In[41]:=Eigensystem[(3 1
                     3 1)]
```

```
Out[41] = {{0, 4}, {{-1, 3}, {1, 1}}}
```

the system has eigenvalues 0 and 4, with the corresponding eigenvectors $\begin{pmatrix} -1 \\ 3 \end{pmatrix}$ and $\begin{pmatrix} 1 \\ 1 \end{pmatrix}$, respectively. Therefore, the general solution of the system is $\begin{pmatrix} Y1t+1 \\ Y2t+1 \end{pmatrix} = c1\begin{pmatrix} -1 \\ 3 \end{pmatrix}0^t + c2\begin{pmatrix} 1 \\ 1 \end{pmatrix}4^t$, where c1 and 2 are two constants to be decided by the initial conditions. Substitution for the initial values Y1o = Y2o = 1 results in $Y1t+1 = 4^t$ and $Y2t+1 = 4^t$. To solve the same problem with *Mathematica* type,

```
In[42]:=Clear[y1, y2]
        RSolve[{y1[t + 1] == 3 y1[t] + y2[t], y2[t + 1] == y2[t] + 3 y1[t], y1[0] == 1,
          y2[0] == 1}, {y1[t], y2[t]}, t]
```

Out[42]= {{y1[t] → 4^t, y2[t] → 4^t}}

Example : Solve the system of differencen equationsΔ**Yt** = $\begin{pmatrix} 1 & 3 \\ 0 & 1 \end{pmatrix}$**Yt** for the of Y1, Y2, and the relationship initial values **Y0** = $\begin{pmatrix} 1 \\ 1 \end{pmatrix}$.Plot the time path between y1 and y2 over time.

Solution: **A** + **I**2 matrix of the system is $\begin{pmatrix} 2 & 3 \\ 0 & 2 \end{pmatrix}$. The system can be written as $\begin{pmatrix} Y1t+1 \\ Y2t+1 \end{pmatrix}$ = $\begin{pmatrix} 2 & 3 \\ 0 & 2 \end{pmatrix}\begin{pmatrix} Y1t \\ Y2t \end{pmatrix}$. Using the M*athematica* command RSolve the problem can be solved as,

```
In[43]:=Clear[y1, y2, parsley, oregano, tplot2, tplot3]
        A = (1 3
             0 1);
        H[{y1_ , y2_}] = (A + IdentityMatrix[2]).{y1, y2}
```

Out[43]= {2 y1 + 3 y2, 2 y2}

```
In[44]:=oregano =
          RSolve[{y1[t + 1] == 2 y1[t] + 3 y2[t], y2[t + 1] == 2 y2[t], y1[0] == y2[0] == 1}, {y1[t], y2[t]}, t]
```

Out[44]= {{y1[t] → 2^{-1+t} (2 + 3 t), y2[t] → 2^t}}

to find the time path of Y1 and Y2,

```
In[45]:=y1[t_] = y1[t] /. oregano[[1]];
        tplot2 = Table[y1[t], {t, 0, 6}];
        ListPlot[tplot2, PlotJoined -> True]
```

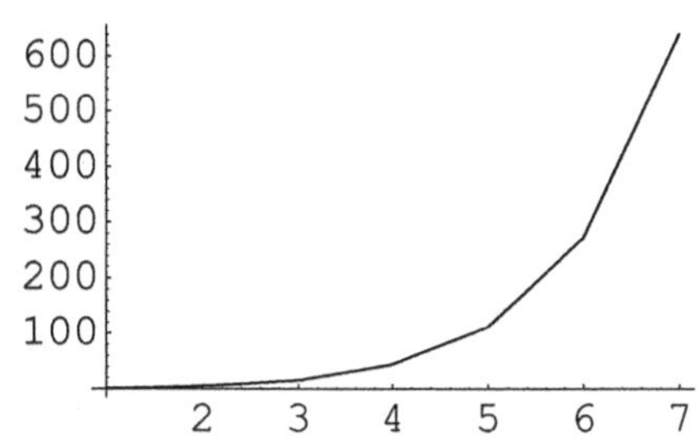

to find the time path of y2,

```
In[46]:=y2[t_] = y2[t] /. oregano[[1]];
        tplot3 = Table[y2[t], {t, 0, 6}];
        ListPlot[tplot3, PlotJoined -> True]
```

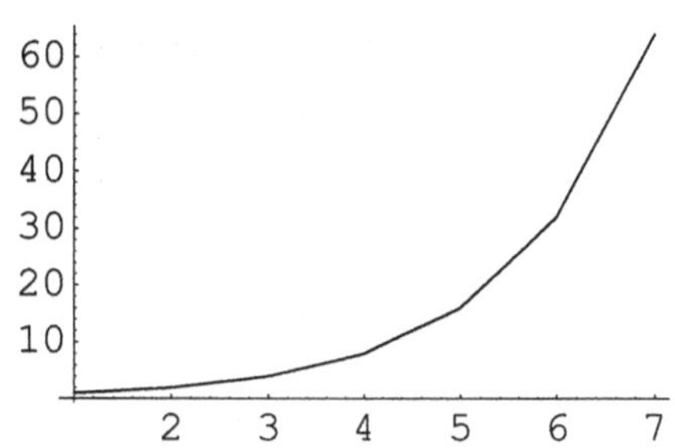

The two graphs above represent the time path of y1 and y2 individually. To graph the relationship between y1 and y2 in the y1-y2 space, use the "NestList" command of *Mathematica*, as shown below. This graph shows the dynamic relationship between y1 and y2 over time.

```
In[47]:=f[y1_, y2_, t_] := NestList[H, {y1, y2}, 10];
        parsley = f[1, 1, 10];
        ListPlot[parsley, PlotStyle -> PointSize[.02]]
```

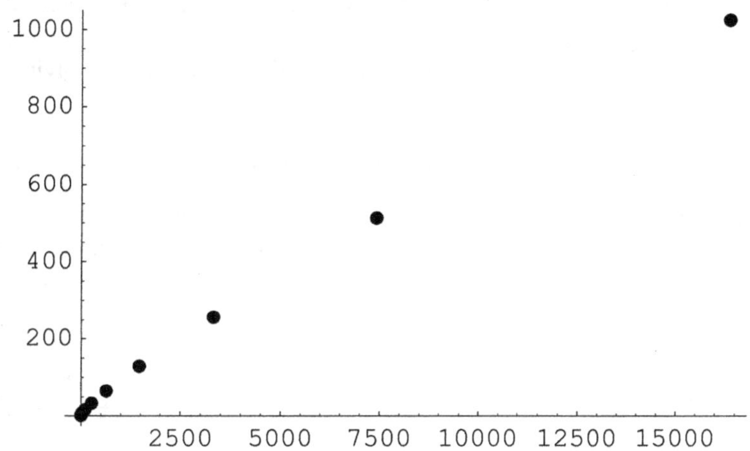

Do More Practice:

1- Change the coefficients of the **A** matrix in the preceding example to $\begin{pmatrix} 1 & 0 \\ 1 & 4 \end{pmatrix}$, find the eigenvalues of the **A** + **I2** matrix, and re-run the program. Observe the changes in the eigenvalues and their relations to the solution.

2- Change the coefficients of the **A** matrix in the preceding example to $\begin{pmatrix} 0 & 1 \\ 3 & 4 \end{pmatrix}$, find the eigenvalues of the **A** + **I2** matrix, and re-run the program. Observe the changes in the eigenvalues and their relations to the solution.

3- Change the coefficients of the **A** matrix in the preceding example to $\begin{pmatrix} 1 & -1 \\ 4 & 4 \end{pmatrix}$, find the eigenvalues of the **A** + **I2** matrix, and re-run the program. Observe the changes in the eigenvalues and their relations to the solution.

Problems

1. For the following matrices, find a solution to $\mathbf{\Delta Y}t = \mathbf{AY}t$.

a) $\mathbf{A} = \begin{pmatrix} 1 & -3 \\ 0 & 1 \end{pmatrix}$ b) $\mathbf{A} = \begin{pmatrix} 2 & 0 \\ 1 & 0 \end{pmatrix}$

c) $\mathbf{A} = \begin{pmatrix} 1 & 2 \\ 2 & 1 \end{pmatrix}$ d) $\mathbf{A} = \begin{pmatrix} 2 & 3 \\ 1 & 0 \end{pmatrix}$

e) $\mathbf{A} = \begin{pmatrix} 0 & 1 \\ \frac{5}{3} & \frac{2}{3} \end{pmatrix}$ f) $\mathbf{A} = \begin{pmatrix} 1 & 4 & 1 \\ 0 & 2 & 3 \\ 0 & 0 & 3 \end{pmatrix}$

g) $\mathbf{A} = \begin{pmatrix} 1 & 0 & 1 \\ 0 & 2 & 1 \\ 4 & 0 & 1 \end{pmatrix}$ h) $\mathbf{A} = \begin{pmatrix} 1 & 1 & 0 \\ 0 & 0 & 1 \\ 0 & 1 & 1 \end{pmatrix}$

2. Solve problems a, b, c, and d of the problem 1 for the initial value, $\mathbf{y}(0) = \begin{pmatrix} 1 \\ 2 \end{pmatrix}$. Plot the time path of Y1t and Y2t.

3. Solve problems a and b of the problem 1 for the initial value, $\mathbf{y}(0) = \begin{pmatrix} 0 \\ 3 \end{pmatrix}$. Plot the time path of Y1t and Y2t.

System of First-Order Nonhomogeneous Linear Difference Equations: The general form of a system of first-order nonhomogeneous linear difference equations is $\mathbf{\Delta Y}$t = $\mathbf{AY}$t + $\mathbf{B}$ where, $\mathbf{\Delta Y}$t = $\mathbf{AY}$t is the homogenous part of the system and $\mathbf{B}$ is a vector of constants. If the $\mathbf{A}$ matrix is non-singular matrix, then the steady-state equilibrium for the system exists and can be derived by setting $\mathbf{\Delta Y}$t = $\mathbf{0}$ and solving for Y1t and Y2t. That is, $\begin{pmatrix} Y1t \\ Y2t \end{pmatrix} = \mathbf{A}^{-1}.\ \mathbf{B}$. The solution of the system, however, follows the solution of the complementary homogeneous part, discussed in the previous section, plus the particular solution due to nonhomogeneous part.

Example: Solve the following system of nonhomogeneous linear difference equations, ΔY1t = -.4Y1t + .2Y2t + 2.8 and ΔY2t = .8Y1t - .7Y2t - 3.8, Y1o = Y2o = 8. Plot the time path of Y1t, Y2t, and the relationship between Y1t and Y2t over time.

Solution: The system in the matrix form is, $\begin{pmatrix} \Delta y1t \\ \Delta y2t \end{pmatrix} = \begin{pmatrix} -.4 & .2 \\ .8 & -.7 \end{pmatrix} \begin{pmatrix} y1t \\ y2t \end{pmatrix} + \begin{pmatrix} 2.8 \\ -3.8 \end{pmatrix}$. The steady-state equilibrium condition can be derived by setting $\begin{pmatrix} \Delta y1t \\ \Delta y2t \end{pmatrix} = \begin{pmatrix} 0 \\ 0 \end{pmatrix}$, and solving for $\begin{pmatrix} y1t \\ y2t \end{pmatrix}$. Using *Mathematica*,

```
In[48]:=Clear[y1, y2, paprika, rice, tplot4, tplot5, tplot6]
        A = ( -.4   .2 );  B = (  2.8 );  y = ( y1t );
            (  .8  -.7 )       ( -3.8 )       ( y2t )
        Solve[A.y + B == 0, {y1t, y2t}]

Out[48]= {{y1t → 10., y2t → 6.}}
```

the steady-state equilibrium of the system is at the Y1 and Y2 values of 10 and 6, respectively. To solve the difference equation, find $\mathbf{A} + \mathbf{I}2$ matrix, and solve for $\begin{pmatrix} Y1t+1 \\ Y2t+1 \end{pmatrix}$ in terms of $\begin{pmatrix} Y1t \\ Y2t \end{pmatrix}$. That is, $\begin{pmatrix} Y1t+1 \\ Y2t+1 \end{pmatrix} = (\mathbf{A} + \mathbf{I}2)\begin{pmatrix} Y1t \\ Y2t \end{pmatrix} + \mathbf{B}$. Using Mathematica,

```
In[49]:=H[{y1_ , y2_}] = (A + IdentityMatrix[2]).{y1, y2} + B

Out[49]= {{2.8 + 0.6 y1 + 0.2 y2}, {-3.8 + 0.8 y1 + 0.3 y2}}
```

use the resulting equations to write the "RSolve" command as,

```
In[50]:=paprika = RSolve[{y1[t + 1] == 2.8 + .6 y1[t] + .2 y2[t],
              y2[t + 1] == -3.8 + .8 y1[t] + .3 y2[t], y1[0] == y2[0] == 8}, {y1[t], y2[t]}, t]
```

to graph the time path of Y1,

```
In[51]:=y1[t_] = y1[t] /. paprika[[1]]
        tplot4 = Table[y1[t], {t, 0, 20}];
        ListPlot[tplot4, PlotJoined -> True]
```

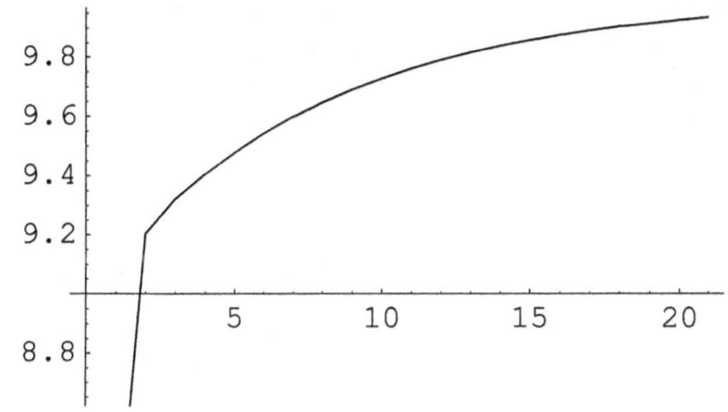

it is obvious that as t → ∞, Y1 converges to its steady-state equilibrium 10. To graph the time path of Y2t,

```
In[52]:=y2[t_] = y2[t] /. paprika[[1]];
        tplot5 = Table[y2[t], {t, 0, 20}];
        ListPlot[tplot5, PlotJoined -> True]
```

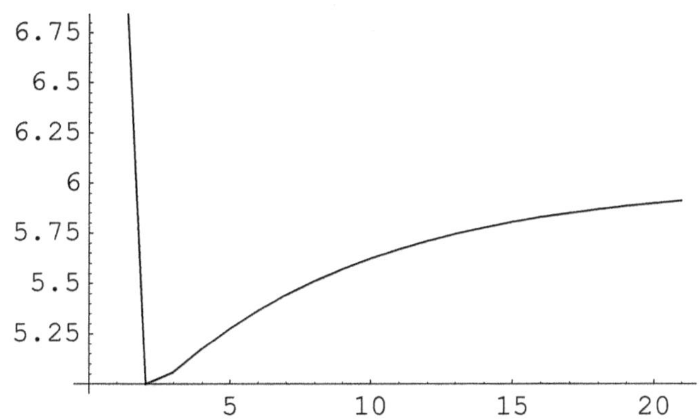

it is obvious that as t → ∞, Y2 approaches its steady-state equilibrium 6. To graph the first 100 points visited over time in adjustment towards the equilibrium point, use the following commands,

```
In[53]:=H[{y1_, y2_}] = {2.8 + .6 y1 + .2 y2, -3.6 + .8 y1 + .3 y2};
        f[y1_, y2_, t_] := NestList[H, {y1, y2}, 100];
        rice = f[1, 1, 100];
        tplot6 = ListPlot[rice, PlotStyle -> PointSize[.02]]
```

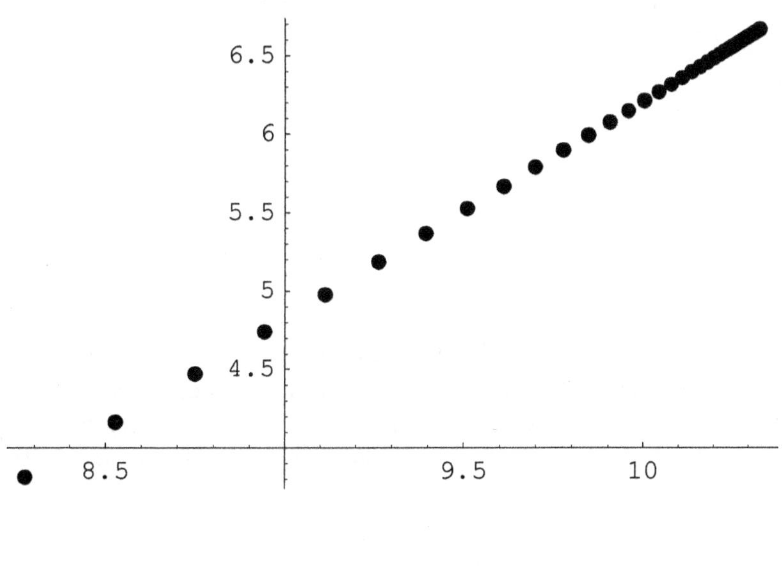

It is also possible to graph the two equations of the system in the y1-y2 space and study the dynamics of the adjustment over time from one point to another or the dynamics of the adjustment to steady-state equilibrium. To do this, find y2 in terms of y1 for the two equations of the system by setting $\Delta Y1t = \Delta Y2t = 0$ and plot them together with the previous graph, as shown below,

```
In[54]:=Clear[y1, y2, eq1, eq2, tplot7]
        eq1 = Solve[-.4 y1 + .2 y2 + 2.8 == 0, y2]
Out[54]= {{y2 → -5. (2.8 - 0.4 y1)}}
In[55]:=eq2 = Solve[.8 y1 - .7 y2 - 3.8 == 0, y2]
Out[55]= {{y2 → 1.42857 (-3.8 + 0.8 y1)}}
In[56]:=tplot7 = Plot[{-5 (2.8 - .4 y1), 1.42857 (-3.8 + .8 y1)}, {y1, 0, 20}]
        Show[tplot6, tplot7]
```

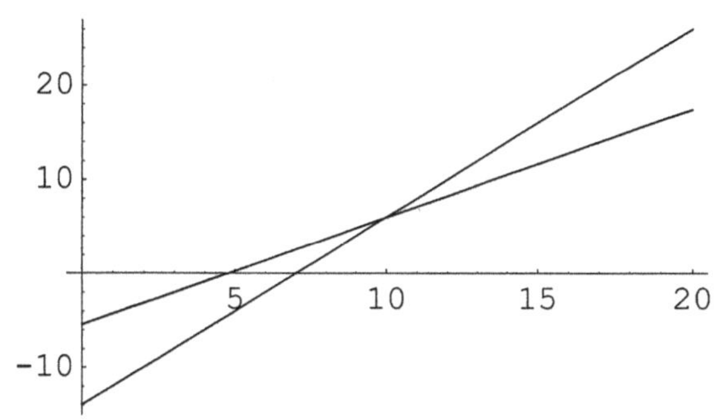

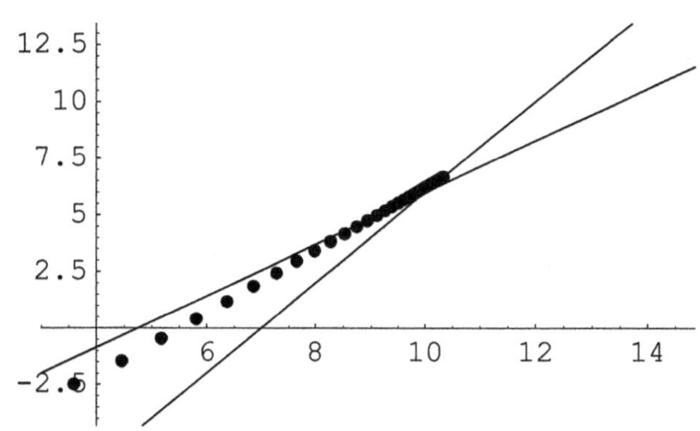

Example: Application to IS-LM Model

A simple macro model of an economy is represented by the following equations: C = 400 + .8Y, I = 100 - 8r, Y = C + I, Md = .25Y - .9r, and Md = Ms = 500. The first equation expresses consumption (C) as a function of income (y); the second equation expresses investment (I) as a function of interest rate (r); the third equation is the equilibrium condition in the goods market; the fourth equation expresses demand for money (Md) as a function of income and interest rate; and the last equation is the equilibrium condition in the money market. Suppose that the change in income in the goods market and the change in interest rate in the money market are proportional to excess demand in the corresponding markets, with a proportionality factor of .4 for the goods market and a proportionality factor of .5 for the money markets. Given that the initial values of y and r are Y(0) = 2200 and r(0) = 10, find the time path of Y and r and analyze the dynamic relationship between Y and r over time.

Solution: The change in income (ΔYt) can be expressed in terms of the excess demand in the goods market as ΔYt = .4[500 + .8Yt - 8rt - Yt] or ΔYt = -.08Yt - 3.2rt + 200 and the change in interest rate (Δrt) can be expressed in terms of the excess demand in the money market as Δrt = .5(.25Yt - .9rt - 500) or Δrt = .125Yt - .45rt - 250. The system can be written in matrix form as $\begin{pmatrix} \Delta Yt \\ \Delta rt \end{pmatrix} = \begin{pmatrix} -.08 & -3.2 \\ .125 & -.45 \end{pmatrix} \begin{pmatrix} Yt \\ rt \end{pmatrix} + \begin{pmatrix} 200 \\ -250 \end{pmatrix}$. By setting $\begin{pmatrix} \Delta Yt \\ \Delta rt \end{pmatrix}$ = **0**, the long run equilibrium can be derived as,

```
In[57]:=Clear[y, r, yucca, tplot8, tplot9]
        Solve[{-.08 y - 3.2 r + 200 == 0, .125 y - .45 r - 250 == 0}, {y, r}]
Out[57]= {{y → 2041.28, r → 11.4679}}
```

the steady-state equilibrium income and interest rate are 2,041.28 and 11.47, respectively. To study the dynamic adjustment of income and interest rate from the initial levels to the equilibrium levels, solve the system of difference equations $\begin{pmatrix} \Delta Yt \\ \Delta rt \end{pmatrix} = \begin{pmatrix} -.08 & -3.2 \\ .125 & -.45 \end{pmatrix} \begin{pmatrix} Yt \\ rt \end{pmatrix} + \begin{pmatrix} 200 \\ -250 \end{pmatrix}$. To solve the system, find **A** + **I**2 matrix, and solve for $\begin{pmatrix} Yt+1 \\ rt+1 \end{pmatrix}$ in terms of $\begin{pmatrix} Yt \\ rt \end{pmatrix}$. That is, $\begin{pmatrix} Yt+1 \\ rt+1 \end{pmatrix} = (\mathbf{A} + \mathbf{I}2) \begin{pmatrix} Yt \\ rt \end{pmatrix} + \mathbf{B}$. Using Mathematica,

```
In[58] := A = ( -.08  -3.2 ); B = ( 200 );
              ( .125  -.45 )       ( -250 )
H[{y_ , r_}] = (A + IdentityMatrix[2]).{y, r} + B
Out[58] = {{200 - 3.2 r + 0.92 y}, {-250 + 0.55 r + 0.125 y}}
```

use the resulting equtions to write the "RSolve" command as,

```
In[59]:= yucca=RSolve[{y[t+1]==200 -3.2r[t] + .92y[t],r[t+1]
    ==-250 +.55r[t]+.125y[t], y[0]==2100, r[0]==10},
   {y[t], r[t]}, t];
```

to graph the time path of Y, input as follows

```
In[60]:=y[t_] = y[t] /. yucca[[1]];
        tplot8 = Table[y[t], {t, 0, 100}];
        ListPlot[tplot8, PlotJoined -> True]
```

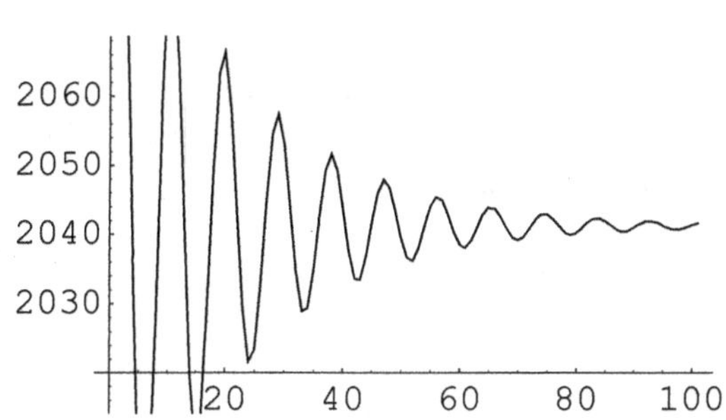

it is obvious that as t→∞, Y approaches its steady-state equilibrium value 2041. To graph the time path of r,

```
In[61]:=r[t_]=r[t]/.yucca[[1]];
tplot9=Table[r[t], {t, 0,100}];
ListPlot[tplot9, PlotJoined->True]
```

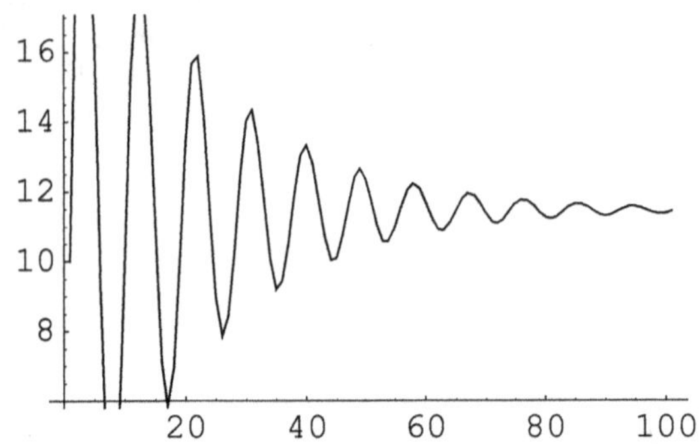

it is obvious that as t→∞, r approaches its steady-state equilibrium value 11.47. To graph the first 100 points visited in the dynamic adjustment from one point to the equilibrium point use the following commands,

```
In[62]:=H[{y_, r_}] := {200 - 3.2 r + .92 y, -250 + .55 r + .125 y}
     f[y_, r_, t_] := NestList[H, {y, r}, 100]
     vanilla = f[2500, 8, 100];

   pl1 = ListPlot[vanilla, PlotStyle -> PointSize[.02], PlotRange -> All]
```

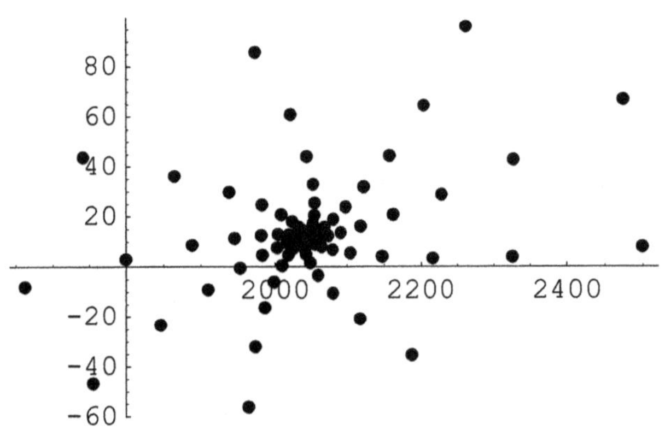

It is also possible to graph the IS and LM curves in y-r space and study the dynamics of the adjustment over time from one point to another or to the equilibrium point. To do this, find r in terms of y for IS and LM curves and plot them. Then, show the lisplot above and the IS-LM graphs together as shown below,

```
In[63]:=
     IS = Solve[500 + .8 y - 8 r == y, r]
     LM = Solve[.25 y - .9 r == 500, r]
     Out[63] =  {{r → 0.125` (500.` - 0.19999999999999996` y)}}
     {{r → 1.1111111111111112` (-500.` + 0.25` y)}}
```

In[64]:=
```
pl2 = Plot[{.125 (500 - .2 y), 1.111 (-500 + .25 y)},
    {y, 2000, 2200}, PlotRange -> All]
```

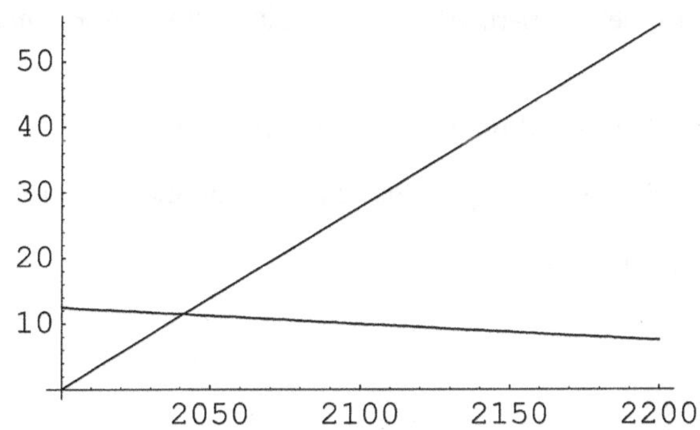

In[65]:=
```
Show[pl1, pl2]
```

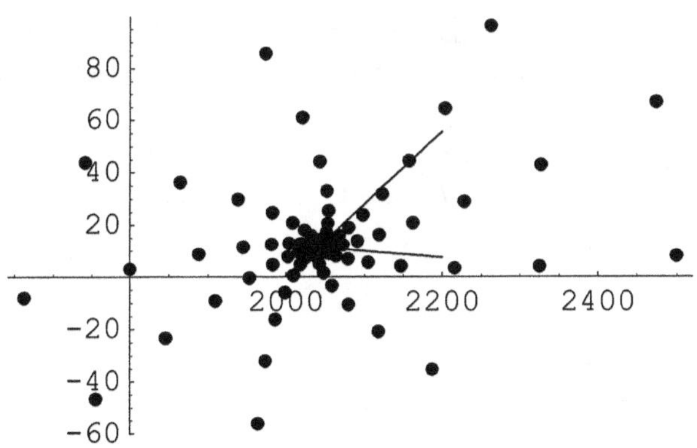

it is obvious from this graph that the equilibrium point is asymptotically stable and as time passes the points converge to the steady-state equilibrium point. The stability of the system of difference equations is discussed in the following section.

Do More Practice:

1- Change the speed of the adjustments in the goods and money markets from .4 and .5 to .2 and .9, respectively. Re-run the program and observe changes in the solution and the adjustment path.

2- Change the speed of the adjustments in the goods and money markets from .4 and .5 to .9 and .2, respectively. Re-run the program and observe changes in the solution and the adjustment path.

3- Suppose in the IS-LM model of the example investment is independent of the interest rate. Re-run the program under this assumption and compare the results.

4- Change the model as C = .7Y + 400, I = 100 - r, Y = C + I, Md = .3Y - 1.2r, and Md = Ms = 480. Keep the speed of adjustments in the goods and money markets the same as in the example. Re-run the program and observe changes in the solution and the adjustment path.

Stability Analysis of the System of Linear Difference Equations: The eigenvalues of the **A** + **I** matrix of the system of linear difference equations reveal much about the time path of the variables of the system and overall stability of the equilibrium point. The equilibrium point of a system of linear difference equations will asymptotically be stable if and only if the absolute value of the real parts of the eigenvalues of the **A** + **I** matrix are less than one.

Example: Analyze the stability of the equilibrium point of the IS-LM model of the preceding example.

Solution: The **A** matrix of the system of the difference equations of the preceding example is $\mathbf{A} = \begin{pmatrix} -.08 & -3.2 \\ .125 & -.45 \end{pmatrix}$. Find the eigenvalues of the **A** + **I**2 matrix,

```
In[66]:=Clear[A]
    A = ( -.08  -3.2
          .125  -.45 );
    Eigenvalues[A + IdentityMatrix[2]]
    Out[66] = {0.7350000000000001` + 0.604793353138078` i, 0.7350000000000001` -
        0.604793353138078` i}
```

the real parts of the eigenvalues of the A + I2 matrix is .735 and the system is asymptotically stable.

Example: Analyze the stability of the system of difference equations ΔY1t = .3Y1t - .1Y2t - 10, ΔY2t = -Y1t +1.2Y2t + 12, Y1(0) = 50, Y2(0) = 25. Solve the system and show the time path of each variables and the relationship between y1 and y2 over time.

Solution: The system can be written in the matrix form as $\begin{pmatrix} \Delta Y1t \\ \Delta Y2t \end{pmatrix} = \begin{pmatrix} .3 & -.1 \\ -1 & 1.2 \end{pmatrix}\begin{pmatrix} Y1t \\ Y2t \end{pmatrix} + \begin{pmatrix} -10 \\ 12 \end{pmatrix}$. To analyze the stability of the system, find the eigenvalues of the **A** + **I**2 matrix,

```
In[67]:=Clear[A, B]
    A = ( .3  -.1
          -1  1.2 );  B = ( -10
                            12 );
    Eigenvalues[A + IdentityMatrix[2]]
    Out[67] = {2.3000000000000003`, 1.2000000000000002`}

    In[68] := Clear[y1, y2, tendrils, sessile, pl3, pl4, pl5, pl6]
    H[{y1_ , y2_}] = (A + IdentityMatrix[2]).{y1, y2} + B
    Out[68] = {{-10 + 1.3` y1 - 0.1` y2}, {12 - y1 + 2.2` y2}}
```

use the resulting equations to write the "RSolve" command,

```
In[69]:=tendrils = RSolve[{y1[t + 1] == -10 + 1.3 y1[t] - .1 y2[t],
        y2[t + 1] == 12 - y1[t] + 2.2 y2[t], y1[0] == 50, y2[0] == 25}, {y1[t], y2[t]}, t]
```

to graph the time path of Y1, use the *Mathematica* commands as,

```
In[70]:=y1[t_]=y1[t]/.tendrils[[1]];

pl3=Table[y1[t], {t, 0, 10}];

ListPlot[pl3, PlotJoined->True]
```

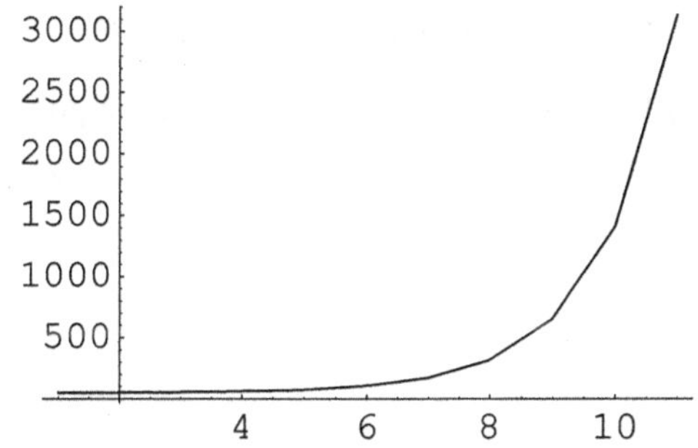

to graph the time path of Y2,

```
In[71]:=y2[t_] = y2[t] /. tendrils[[1]];
        pl4 = Table[y2[t], {t, 0, 10}];
        ListPlot[pl4, PlotJoined -> True]
```

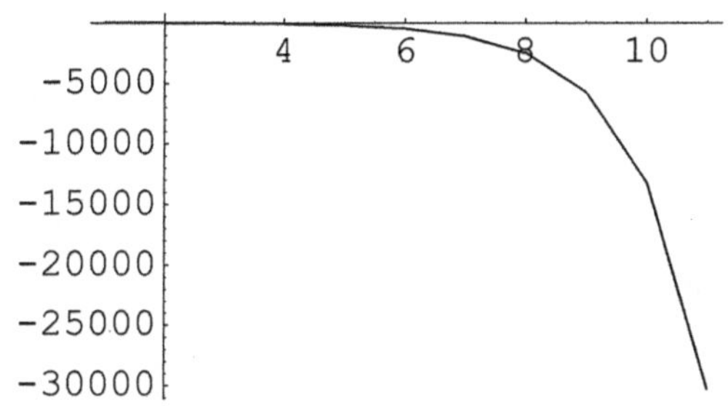

To graph Y1t against Y2t over time use the following commands,

```
In[72]:=H[{y1_, y2_}] := {-10 + 1.3 y1 - .1 y2, 12 - y1 + 2.2 y2}
        f[y1_, y2_, t_] := NestList[H, {y1, y2}, 5]

        sessile = f[40, 20, 5];
        pl5 = ListPlot[sessile, PlotStyle -> PointSize[.02], PlotRange -> All]
```

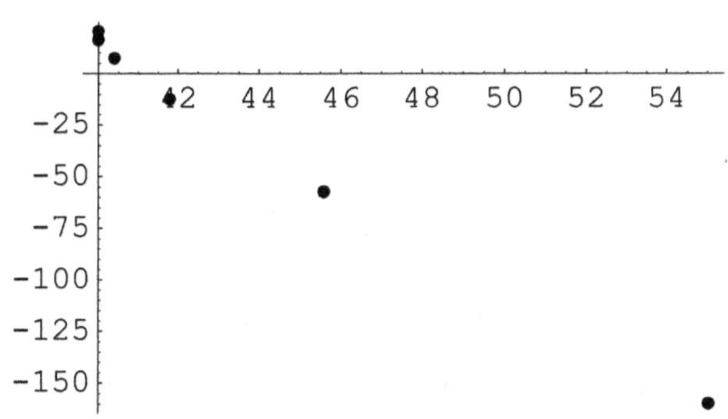

It is also possible to graph the equations of the system for their steady-state equilibrium by setting $\begin{pmatrix} \Delta Y1t \\ \Delta Y2t \end{pmatrix} = \begin{pmatrix} 0 \\ 0 \end{pmatrix}$ and study the adjustment over time from any point on the Y1-Y2 space towards the equilibrium point.

```
In[73]:=
    Clear[IS, LM]
    IS = Solve[-10 + .3 y1 - .1 y2 == 0, y2]
    LM = Solve[12 - y1 + 1.2 y2 == 0, y2]
    Out[73] = {{y2 → 10.` (-10.` + 0.3` y1)}}
    {{y2 → -0.8333333333333334` (12.` - 1.` y1)}}
```

```
In[74]:=pl6 = Plot[{10 (-10 + .3 y1), -.83 (12 - y1)}, {y1, 10, 60}, PlotRange -> All]
    Show[pl5, pl6]
```

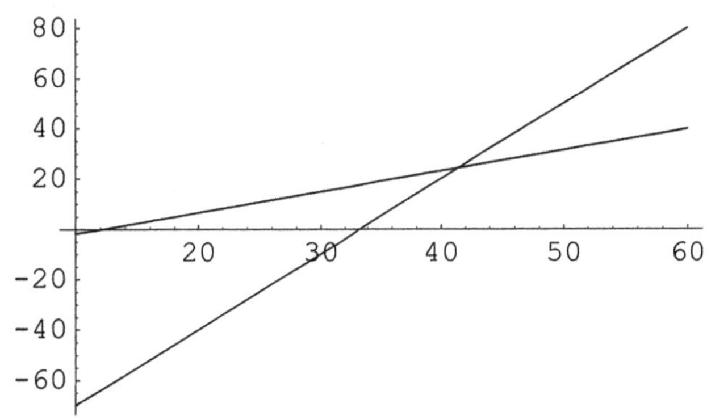

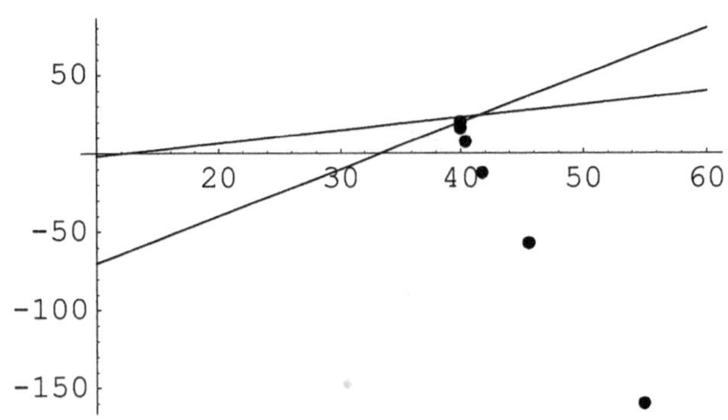

the graphs of Y1, Y2, and the dynamic relations between Y1 and Y2 indicate that the variables and the model are all divergent.

Problems

1. For the following system of difference equations

$$\begin{pmatrix} \Delta Y1t \\ \Delta Y2t \end{pmatrix} = \begin{pmatrix} -.3 & 0 \\ 1.2 & .8 \end{pmatrix} \begin{pmatrix} Y1t \\ Y2t \end{pmatrix} + \begin{pmatrix} 9 \\ -40 \end{pmatrix}, \quad \begin{pmatrix} Y1\ (0) \\ Y2\ (0) \end{pmatrix} = \begin{pmatrix} 5 \\ 30 \end{pmatrix}$$

a. find the steady-state equilibrium point.

b. solve the system of difference equations and graph the time path of Y1 and Y2.

2. For the following system of difference equations

$$\begin{pmatrix} \Delta Y1t \\ \Delta Y2t \end{pmatrix} = \begin{pmatrix} -.3 & -1 \\ 1.5 & .2 \end{pmatrix} \begin{pmatrix} Y1t \\ Y2t \end{pmatrix} + \begin{pmatrix} 40 \\ -30 \end{pmatrix}, \quad \begin{pmatrix} Y1\ (0) \\ Y2\ (0) \end{pmatrix} = \begin{pmatrix} 12 \\ 35 \end{pmatrix}$$

a. find the steady-state equilibrium point.

b. solve the system of difference equations and graph the time path of Y1 and Y2.

3. For the following system of difference equations

$$\begin{pmatrix} \Delta Y1t \\ \Delta Y2t \end{pmatrix} = \begin{pmatrix} -1.2 & .8 \\ .5 & .2 \end{pmatrix} \begin{pmatrix} Y1t \\ Y2t \end{pmatrix} + \begin{pmatrix} -5 \\ -10 \end{pmatrix}, \quad \begin{pmatrix} Y1\ (0) \\ Y2\ (0) \end{pmatrix} = \begin{pmatrix} 10 \\ 20 \end{pmatrix}$$

a. find the steady-state equilibrium point.

b. solve the system of difference equations and graph the time path of Y1 and Y2.

4. For the following system of difference equations

$$\begin{pmatrix} \Delta Y1t \\ \Delta Y2t \end{pmatrix} = \begin{pmatrix} 1 & -1 \\ .5 & 2 \end{pmatrix} \begin{pmatrix} Y1t \\ Y2t \end{pmatrix} + \begin{pmatrix} -24 \\ -20 \end{pmatrix}, \quad \begin{pmatrix} Y1\ (0) \\ Y2\ (0) \end{pmatrix} = \begin{pmatrix} 16 \\ 5 \end{pmatrix}$$

a. find the steady-state equilibrium point.

b. solve the system of difference equations and graph the time path of Y1 and Y2.

c. graph Y1 and Y2 on the Y1-Y2 space and 50 points around the steady-state equilibrium.

d. analyze the stability of the system.

4. For the following system of difference equations

$$\begin{pmatrix} \Delta Y1t \\ \Delta Y2t \end{pmatrix} = \begin{pmatrix} -1 & 1 \\ .5 & 2 \end{pmatrix} \begin{pmatrix} Y1t \\ Y2t \end{pmatrix} + \begin{pmatrix} -24 \\ -20 \end{pmatrix}, \quad \begin{pmatrix} Y1\ (0) \\ Y2\ (0) \end{pmatrix} = \begin{pmatrix} 6 \\ 2 \end{pmatrix}$$

a. find the steady-state equilibrium point.

b. solve the system of difference equations and graph the time path of Y1 and Y2.

c. graph Y1 and Y2 on the Y1-Y2 space and 50 points around the steady-state equilibrium.

d. analyze the stability of the system.

2. A simple macro model of an economy is represented by the following equations: $C = 800 + .9Y - 2r$, $I = 200 - 4r$, $Y = C + I$, $Md = .4Y - r$, and $Md = Ms = 3800$. The first equation expresses consumption (C) as a function of income (y) and interest rate (r); the second equation expresses investment (I) as a function of interest rate; the third equation is the equilibrium condition in the goods market; the fourth equation expresses demand for money (Md) as a function of income and interest rate; and the last equation is the equilibrium condition in the money market. Suppose that the change in income in the goods market and the change in interest rate in the money market are proportional to excess demand in the corresponding markets, with a proportionality factor of .5 for the goods market and a proportionality factor of .8 for the money markets. Given that the initial values of y and r are $Y(0) = 10000$ and $r(0) = 8$, find the time path of Y and r and analyze the stability of the model and the dynamic relationship between Y and r over time.

CHAPTER 14

COMPLEX NUMBERS AND SOLUTION OF DIFFERENTIAL AND DIFFERENCE EQUATIONS

A complex number is represented by a +bi, where a and b are real numbers and i is the imaginary unit, having the property $i^2 = -1$. In the preceding two chapters, we intentionally avoided equations with complex roots. However, in solving differential and difference equations, it is quite often the case that the roots are complex numbers. Therefore, this chapter will concentrate on complex numbers and the solutions of differential and difference equations having complex roots.

TOOL BOX

Mathematica Commands	Function
DSolve[y' [t] == ay[t] + b, y[t], t]	Solves the differential equation for y(t).
DSolve[{y' [t] == ay[t] + b, y[0] ==yo}, y[t], t]	Solves the initial-value differential equation for y(t).
RSolve[{y[t] - ay[t-1] ==b, y[0]==yo}, y[t], t]	Solves the difference equation yt-ayt-1=b, Y0= Yo, for Yt.

I. Complex Number

A complex number can be represented in two forms, rectangular form and polar form. The rectangular form of the complex number is a + bi. Now, suppose that we use the rectangular coordinate system to represent an ordered pair of numbers, a and b, along the horizontal axis and vertical axis as in the following figure (for definition of rectangular and polar coordinate systems and transformation from one to other, see chapter one, section Mathematical Relations)

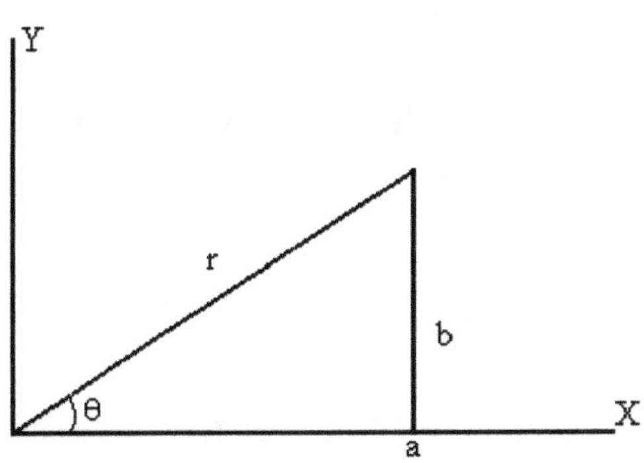

Using trigonometry, $\sin\theta = \frac{b}{r}$ or $b = r\sin\theta$, and $\cos\theta = \frac{a}{r}$ or $a = r\cos\theta$. Therefore, $a + bi = r\cos\theta + ir\sin\theta = r(\cos\theta + i\sin\theta)$ where, by the Pythagorean theorem, the length of OA or r is $r = \sqrt{a^2 + b^2}$. The form $r(\cos\theta + i\sin\theta)$ is called the polar form. The polar form contains r and the sine and cosine functions of the angle θ, all obtained from a and b.

The algebra of the complex numbers is exactly the same as the real numbers. However, since the complex numbers can be written in two different forms, it should be noted that, generally, addition and subtraction involving complex numbers are more conveniently carried out in the rectangular form, while multiplication and division in the polar form.

Example: Find $(a + bi)^2$, $(a + bi)^3$, and $(a + bi)^n$.

Solution: The polar form of a + bi is $r(\cos\theta + i\sin\theta)$. Therefore, $(a + bi)^2 = r^2(\cos\theta^2 - \sin\theta^2 + 2i\sin\theta\cos\theta)$. Since, $\sin 2\theta = 2\sin\theta\cos\theta$ and $\cos 2\theta = \cos\theta^2 - \sin\theta^2$, $(a + bi)^2 = r^2(\cos 2\theta + i\sin 2\theta)$. Using the same method, it can be shown that $(a + bi)^3 = r^3(\cos 3\theta + i\sin 3\theta)$, and by generalization, $(a + bi)^n = r^n(\cos n\theta + i\sin n\theta)$.

II. Imaginary Roots and Solution of Differential Equations

The general solution of the differential equations was discussed in chapter 12. Here, the solution can be extended to the case of equations with imaginary roots.

Suppose that for the system of differential equations y'(t) = Ay(t), the characteristic root of the A matrix, k, is a complex root with the corresponding eigenvector u. One solution of the equation is $y(t) = ue^{kt}$. The solution, however, will consist of a real part and an imaginary part. It can be shown that the real part and the coefficient of the imaginary unit are each a solution to the system.

Example: Solve the system of differential equations $y'(t) = \begin{pmatrix} 1 & -1 \\ 2 & -1 \end{pmatrix} y(t)$.

Solution: The A matrix of the system has two eigenvalues, k1 = i and k2 = -i. The eigenvector corresponding to k1 = i is $u1 = \begin{pmatrix} 1+i \\ 2 \end{pmatrix}$. Therefore, one solution is $y1(t) = \begin{pmatrix} i \\ 1 \end{pmatrix} e^{it}$. However, since $e^{it} = \cos t + i\sin t$, (by the Euler theorem $e^{(a+ib)t} = e^{at(\cos bt + i\sin bt)}$), the solution $y(t) = \begin{pmatrix} 1+i \\ 2 \end{pmatrix} e^{it} = \begin{pmatrix} 1+i \\ 2 \end{pmatrix}(\cos t + i\sin t) = \begin{pmatrix} \cos t + i\sin t + i\cos t - \sin t \\ 2\cos t + 2i\sin t \end{pmatrix} = \begin{pmatrix} \cos t - \sin t \\ 2\cos t \end{pmatrix} + i\begin{pmatrix} \cos t + \sin t \\ 2\sin t \end{pmatrix}$. The real part of the solution is $\begin{pmatrix} \cos t - \sin t \\ 2\cos t \end{pmatrix}$ and the coefficient of the imaginary unit is $\begin{pmatrix} \cos t + \sin t \\ 2\sin t \end{pmatrix}$. Therefore, $\begin{pmatrix} \cos t - \sin t \\ 2\cos t \end{pmatrix}$ and $\begin{pmatrix} \cos t + \sin t \\ 2\sin t \end{pmatrix}$ or any linear combination of the two solutions is a solution to the system. This can easily be proved through direct substitution.

For k2 = -i, the corresponding eigenvector is $u2 = \begin{pmatrix} 1-i \\ 2 \end{pmatrix}$ and the solution is $y2(t) = \begin{pmatrix} 1-i \\ 2 \end{pmatrix} e^{-it} = \begin{pmatrix} 1-i \\ 2 \end{pmatrix}(\cos(-t) + i\sin(-t)) = \begin{pmatrix} 1-i \\ 2 \end{pmatrix}(\cos t - i\sin t) = \begin{pmatrix} \cos t - i\sin t - i\cos t + \sin t \\ 2\cos t - 2i\sin t \end{pmatrix} = \begin{pmatrix} \cos t + \sin t \\ 2\cos t \end{pmatrix} + i\begin{pmatrix} -\cos t - \sin t \\ -2\sin t \end{pmatrix}$. Therefore, $\begin{pmatrix} \cos t + \sin t \\ 2\cos t \end{pmatrix}$ and $\begin{pmatrix} -\cos t - \sin t \\ -2\sin t \end{pmatrix}$ or any linear combination of the two solutions is a solution to the system. To solve the same problem using Mathematica, use "DSolve" command,

```
In[1]:= Clear[y1, y2]
        DSolve[{y1'[t] == y1[t] - y2[t], y2'[t] == 2 y1[t] - y2[t]}, {y1[t], y2[t]}, t]

Out[1]= {{y1[t] → -C[2] Sin[t] + C[1] (Cos[t] + Sin[t]), y2[t] → C[2] Cos[t] + 2 C[1] Sin[t] - C[2] Sin[t]}}
```

Example: Solve y" + 9y = sin 3t for the initial values y(0) = 1, y'(0) = 0 and plot the time path of y(t).

Solution: Let y' = x. This implies that, y" = x' = -9y + sin3t. The two equations yields a system of differential equations that, in matrix form, can be written as $\begin{pmatrix} y' \\ x' \end{pmatrix} = \begin{pmatrix} 0 & 1 \\ -9 & 0 \end{pmatrix}\begin{pmatrix} y \\ x \end{pmatrix} + \begin{pmatrix} 0 \\ \sin 3t \end{pmatrix}$. The matrix A of the complementary homogeneous part of the system has two eigenvalues, k1 = 3i and k2 = -3i, with the corresponding eigenvectors, $u1 = \begin{pmatrix} -i \\ 3 \end{pmatrix}$ and $u2 = \begin{pmatrix} i \\ 3 \end{pmatrix}$. The solution of the complementary homogeneous part of the system corresponding to k1 is $yc1(t) = \begin{pmatrix} -i \\ 3 \end{pmatrix} e^{3it} = \begin{pmatrix} -i \\ 3 \end{pmatrix}(\cos 3t + i\sin 3t) = \begin{pmatrix} -i\cos 3t + \sin 3t \\ 3\cos 3t + 3i\sin 3t \end{pmatrix} = \begin{pmatrix} \sin 3t \\ 3\cos 3t \end{pmatrix} + i\begin{pmatrix} -\cos 3t \\ 3\sin 3t \end{pmatrix}$. Note that we have used the trigonometric facts that $\sin 2\theta = 2\sin\theta\cos\theta$ and $\cos 2\theta = 2\cos^2\theta - 1 = 1 - 2\sin^2\theta$. Therefore, $\begin{pmatrix}\sin 3t\\ 3\cos 3t\end{pmatrix}$ or $\begin{pmatrix}-\cos 3t\\ 3\sin 3t\end{pmatrix}$ or any linear combination of the two solutions is a solution to the complementary homogenous part of the system.

The solution of the complementary homogeneous part of the system corresponding to k2 is yc2(t) = $\begin{pmatrix}i\\ 3\end{pmatrix}e^{-3it}$ = $\begin{pmatrix}i\\ 3\end{pmatrix}(\cos -3t + i\sin -3t)$ = $\begin{pmatrix}i\\ 3\end{pmatrix}(\cos 3t - i\sin 3t)$ = $\begin{pmatrix}i\cos 3t + \sin 3t\\ 3\cos 3t - 3i\sin 3t\end{pmatrix} = \begin{pmatrix}\sin 3t\\ 3\cos 3t\end{pmatrix} + i\begin{pmatrix}\cos 3t\\ -3\sin 3t\end{pmatrix}$. Therefore, $\begin{pmatrix}\sin 3t\\ 3\cos 3t\end{pmatrix}$ or $\begin{pmatrix}\cos 3t\\ -3\sin 3t\end{pmatrix}$ or any linear combinations of the two solutions are a solution to the complementary homogenous part of the system.

The general solution of the system follows the solution steps described in chapter 11, according to which y(t) = yp(t) + ycC = M(t)u(t) + Mn(t)x(o), where yp is the particular solution, yc is the solution of the complementary homogenous part, M(t) is the fundamental matrix of the complementary homogeneous system, and C is any arbitrary constant (see, chapter 11). Taking two of the above solutions, the fundamental matrix is formed as M(t) = $\begin{pmatrix}\cos 3t & \sin 3t\\ -3\sin 3t & 3\cos 3t\end{pmatrix}$ with $\mathbf{M}(t)^{-1} = \frac{1}{3}\begin{pmatrix}3\cos 3t & -\sin 3t\\ 3\sin 3t & \cos 3t\end{pmatrix}$ and $\mathbf{M}(o)^{-1} = \frac{1}{3}\begin{pmatrix}3 & 0\\ 0 & 1\end{pmatrix}$. Therefore, the normalized fundamental matrix is $Mn(t) = M(t)\mathbf{M}(o)^{-1} = \begin{pmatrix}\cos 3t & \frac{1}{3}\sin 3t\\ -3\sin 3t & \cos 3t\end{pmatrix}$. To find the particular solution, yp(t) = M(t)u(t), where $u'(t) = M(t)^{-1}f(t) = \frac{1}{3}\begin{pmatrix}3\cos 3t & -\sin 3t\\ 3\sin 3t & \cos 3t\end{pmatrix}\begin{pmatrix}0\\ \sin 3t\end{pmatrix} = \frac{1}{3}\begin{pmatrix}-\sin^2 3t\\ \sin 3t\cos 3t\end{pmatrix}$. Integrate u'(t) in the interval 0 to t to get $u(t) = \int_0^t \mathbf{u}'(t)\,dt = \frac{1}{6}\begin{pmatrix}\frac{1}{6}\sin 6t - t\\ -\frac{1}{6}\cos 6t + \frac{1}{6}\end{pmatrix}$. Therefore, yp(t) = $\frac{1}{6}\begin{pmatrix}\cos 3t & \sin 3t\\ -3\sin 3t & 3\cos 3t\end{pmatrix}\begin{pmatrix}\frac{1}{6}\sin 6t - t\\ -\frac{1}{6}\cos 6t + \frac{1}{6}\end{pmatrix} = \frac{1}{18}\begin{pmatrix}-3t\cos 3t + \sin 3t\\ 9t\sin 3t\end{pmatrix}$. The general solution y(t) = yp(t) + $Mn(t)xo = \frac{1}{18}\begin{pmatrix}-3t\cos 3t + \sin 3t\\ 9\sin 3t\end{pmatrix} + \begin{pmatrix}\cos 3t & \frac{1}{3}\sin 3t\\ -3\sin 3t & \cos 3t\end{pmatrix}\begin{pmatrix}1\\ 0\end{pmatrix} = \begin{pmatrix}-\frac{1}{6}t\cos 3t + \frac{1}{18}\sin 3t + \cos 3t\\ \frac{1}{2}t\sin 3t - 3\sin 3t\end{pmatrix}$.

Therefore, $y(t) = -\frac{1}{6}t\cos 3t + \frac{1}{18}\sin 3t + \cos 3t$ is the solution of the problem. The solution satisfies both the differential equation and the initial conditions.

To solve the same problem using Mathematica,

```
In[2]:= Clear[y1, y2, sol]
        sol = DSolve[{y''[t] + 9 y[t] == Sin[3 t], y[0] == 1, y'[0] == 0}, y[t], t]

Out[2]= {{Null (y[t] → 1/6 (6 Cos[3 t] − t Cos[3 t] + 1/3 Sin[3 t]))}}

In[3]:= y[t_] = y[t] /. sol[[1]];
        Plot[y[t], {t, 0, 20}]
```

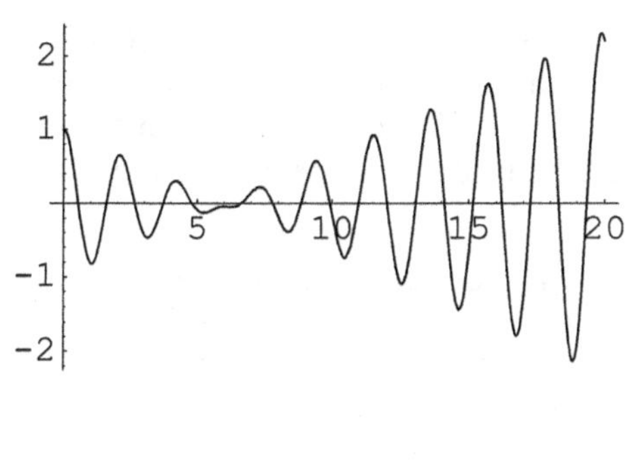

Problems

1. Solve the following system of equations

a) $x'(t) = \begin{pmatrix} 0 & 2 \\ -2 & 0 \end{pmatrix} \mathbf{x}(t)$

b) $x'(t) = \begin{pmatrix} 1 & 3 \\ -3 & 1 \end{pmatrix} x(t)$

c) $x'(t) = \begin{pmatrix} 1 & 5 \\ -2 & 1 \end{pmatrix} \mathbf{x}(t)$

d) $x'(t) = \begin{pmatrix} 1 & 0 & 0 \\ 0 & 0 & 1 \\ 0 & -1 & 0 \end{pmatrix} x(t)$

e) $x'(t) = \begin{pmatrix} 1 & 1 & 0 \\ 0 & 0 & -2 \\ 0 & 2 & 0 \end{pmatrix} x(t) + \begin{pmatrix} 0 \\ 0 \\ \sin 2t \end{pmatrix}$

2. Solve the following initial-value problems

a) $y'' + y = 0, \quad y(0) = 0, \quad y'(0) = 1$

b) $y'' + 2y = \sin t, \quad y(0) = 0, \ y'(0) = 1$

c) $y'' + 2y' + 5y = \cos 2t - 4\sin 2t, \quad y(0) = 1, \quad y'(0) = 0$

d) $y''' + y'' + 4y = 0, \quad y(0)=1, \ y'(0) = 0, \ y''(0) = -4$

e) $y''' - y'' + 3y' + 5y = 2\sin 2t + 9\cos 2t, \quad y(0) = 1, \ y'(0) = 0$, and $y''(0) = -4$

III. Imaginary Roots and Solution of Difference Equations

The general solution of difference equations was discussed in chapter 13. Here, the solution will be extended to the case of equations with imaginary roots.

Given the second order, homogenous linear difference equation Yt + aYt-1 + bYt-2 = 0, the reduced auxiliary equation of the difference equation is $k^2 + ak + b = 0$. Suppose that the roots k1 andk2 that satisfy the auxiliary equation are complex conjugate numbers (The two numbers (a + bi) and (a - bi) are called complex conjugate numbers. The main property of the complex conjugate numbers is that they add and multiply to real numbers (the sum of the two numbers here is 2a and the product is $a^2 + b^2$). If k1 and k2 are the roots of the auxiliary equation, the general solution of the difference equation is Yt = c1Y1 + c2Y2, where Y1 is the solution corresponding to k1 ($Y1 = k1^t$), y2 is the solution corresponding to k2, ($Y2 = k2^t$), and c1 and c2 are constants. Since k1 and k2 are complex conjugates, let k1 = a + bi and k2 = a - bi. The polar forms of the roots are k1 = a + bi = r(cos θ + isin θ) and k2 = a - bi = r(cos θ - isin θ) where, $r = \sqrt{a^2 + b^2}$ and $\theta = \tan^{-1}(\frac{a}{b})$. Therefore, the general solution of the difference equation is $Yt = c1(r\cos\theta + ri\sin\theta)^t + c2(r\cos\theta - ri\sin\theta)^t$.

By expanding and collecting the terms, the solution can be written as $Yt = r^t[(c1 + c2)\cos \theta t + i(c1 - c2)\sin \theta t]$. It can be shown, by direct substitution, that the real part of this equation, $r^t(c1 + c2)\cos \theta t$, or the coefficient of the imaginary unit of the general solution, $r^t(c1 - c2) \sin \theta t$, or any linear combination of the two is a solution to the difference equation. Assuming C1 = c1 + c2 and C2 = c1 - c2, the general form of the solution can be written as $Yt = C1r^t\sin \theta t + C2r^t\cos \theta t$.

Example: Solve the second-order, homogenous linear difference equation Yt - 2Yt-1 + 2Yt-2 = 0, Yo = 1, Y1 = 2, and analyze the time path of Yt.

Solution: The reduced auxiliary equation of the difference equation is, $k^2 - 2k + 2 = 0$. This equation has the roots $k1 = 1 + i$ and $k2 = 1 - i$. Therefore, the general solution is $Yt = C1r^t \sin \theta t + C2r^t \cos \theta t$, where $r = \sqrt{a^2 + b^2}$, and $\theta = \tan^{-1}(\frac{a}{b})$. Here, a = 1, b = 1, $r = \sqrt{2}$, and $\theta = \tan^{-1}(1) = 45^o = \frac{\pi}{4}$. Thus, the general solution is $Yt = C1(\sqrt{2})^t \sin 45t + C2(\sqrt{2})^t \cos 45t$. To find C1 and C2, substitute for the initial values of Y for t = 0 and t = 1 to obtain C1 = 1 and C2 = 1. The initial-value solution is $Yt = (\sqrt{2})^t(\sin 45t + \cos 45t)$. This solution satisfies both the difference equation and the initial values.

The solution also represents a time path for Yt which produces a set of repetitive oscillations, the amplitude of which increases with the passage of time. The solution $Yt = (\sqrt{2})^t(\sin 45t + \cos 45t)$ consists of two terms. The term(sin 45t + cos45t) produces a set of repetitive oscillations of uniform amplitude at the intervals $t = \frac{360}{\theta} = \frac{360}{45} = 8$, as shown in the following graph,

```
In[4]:= Plot[Sin[Pi t/4] + Cos[Pi t/4], {t, 0, 20}]
```

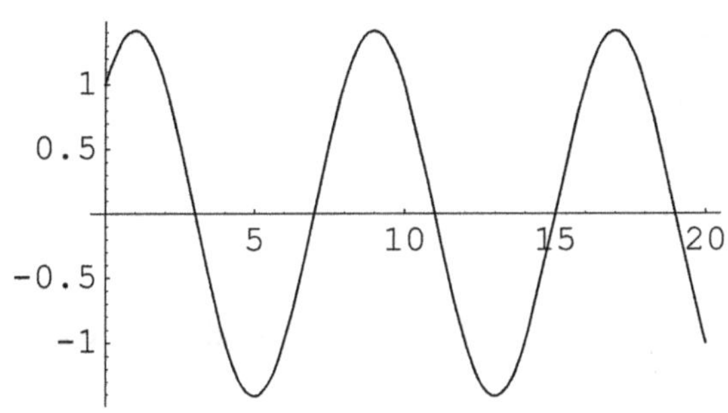

while the term $(\sqrt{2})^t$ explodes as t increases. The graph of $(\sqrt{2})^t$ is shown below.

```
In[5]:= Plot[(Sqrt[2])^t, {t, 0, 20}]
```

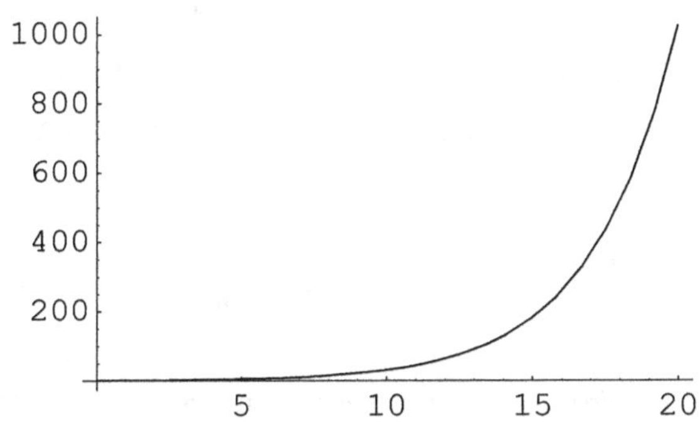

Therefore, $Yt = (\sqrt{2})^t(\sin 45t + \cos 45t)$ will have an oscillatory time path with an increasing amplitude. The graph of the solution to the difference equation is shown below

```
In[6]:= Plot[(Sqrt[2])^t (Sin[Pi t/4] + Cos[Pi t/4]), {t, 0, 20}]
```

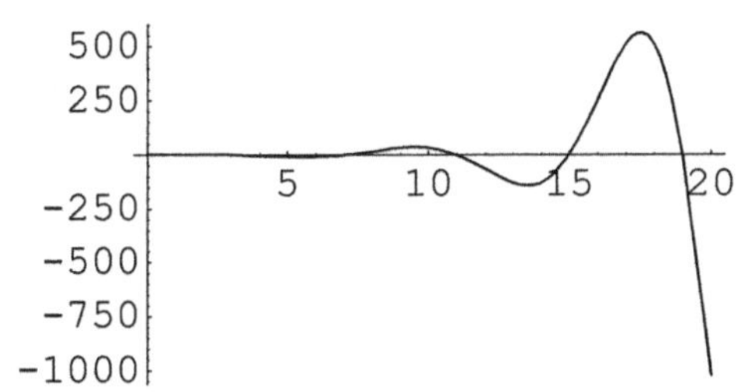

To solve the same problem using Mathematica and to plot the solution, use the following command

```
In[7]:= << DiscreteMath`RSolve`

        Clear[y, sol1]
        sol1 = RSolve[{y[t] - 2 y[t - 1] + 2 y[t - 2] == 0, y[0] == 1, y[1] == 2}, y[t], t]
```

Out[7]= $\{\{y[t] \to 2^{\frac{1}{2}+\frac{t}{2}} \, \text{Sin}[\frac{1}{4}\pi(1+t)]\}\}$

```
In[8]:= y[t_] = y[t] /. sol1[[1]];
        Plot[y[t], {t, 0, 20}]
```

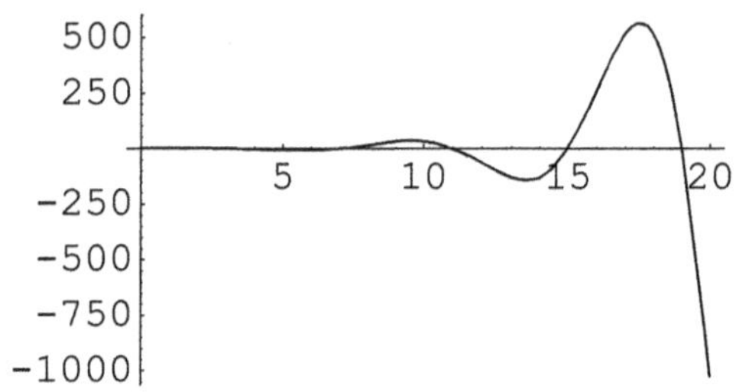

Since $\sin(\frac{\pi}{4}+\frac{t}{4}) = \sin\frac{\pi}{4}\cos\frac{t}{4} + \cos\frac{\pi}{4}\sin\frac{t}{4}$, it can easily be shown that the Mathematica's solution is the same as the solution derived from the auxiliary equation.

Example: Solve the second-order, homogenous linear difference equation 2Yt - 2Yt-1 + yt-2 = 0, Yo = 2, Y1 = 3, and analyze the time path of Yt.

Solution: The reduced auxiliary equation of the difference equation is $2k^2 - 2k + 1 = 0$. This equation has the roots $k1 = \frac{1+i}{2}$ and $k2 = \frac{1-i}{2}$. Therefore, $a = \frac{1}{2}$, $b = \frac{1}{2}$, $r = \frac{\sqrt{2}}{2}$, and $\theta = \tan^{-1}(1) = 45^o = \frac{\pi}{4}$. Thus, the general solution of the problem is, $Yt = C1(\frac{\sqrt{2}}{2})^t \sin 45t + C2(\frac{\sqrt{2}}{2})^t \cos 45t$. To find C1 and C2, substitute for the initial values of Y for t = 0 and t = 1 to obtain C1 = 4 and C2 = 2. Therefore, the initial-value solution is $Yt = 4(\frac{\sqrt{2}}{2})^t \sin 45t + 2(\frac{\sqrt{2}}{2})^t \cos 45t$. This solution satisfies both the difference equation and the initial values.

To find the time path of xYt as t changes, factor out $2(\frac{\sqrt{2}}{2})^t$ to get $Yt = 2(\frac{\sqrt{2}}{2})^t$ (2sin 45t + cos 45t). The term (2sin 45t + cos 45t) has a repetitive oscillatory time path of uniform amplitude at intervals $t = \frac{360}{\theta} = \frac{360}{45} = 8$. However, the term $2(\frac{\sqrt{2}}{2})^t$ diminishes as t increases. Therefore, $Yt = 4(\frac{\sqrt{2}}{2})^t \sin 45t + 2(\frac{\sqrt{2}}{2})^t \cos 45t$ will have an oscillatory time path with a decreasing amplitude. Yt approaches zero as t approaches ∞. The following figures show the plots of 2sin 45t + cos 45t, $2(\frac{\sqrt{2}}{2})^t$, and $Yt = 2(\frac{\sqrt{2}}{2})^t$ (2sing 45t + cos 45t), respectively.

```
In[9]:= Plot[2 Sin[Pi t/4] + Cos[Pi t/4], {t, 0, 20}]
```

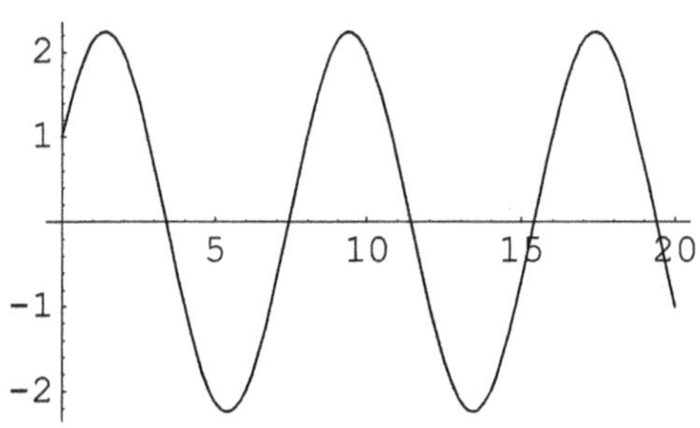

```
In[10]:= Plot[2 (.5 Sqrt[2])^t, {t, 0, 20}]
```

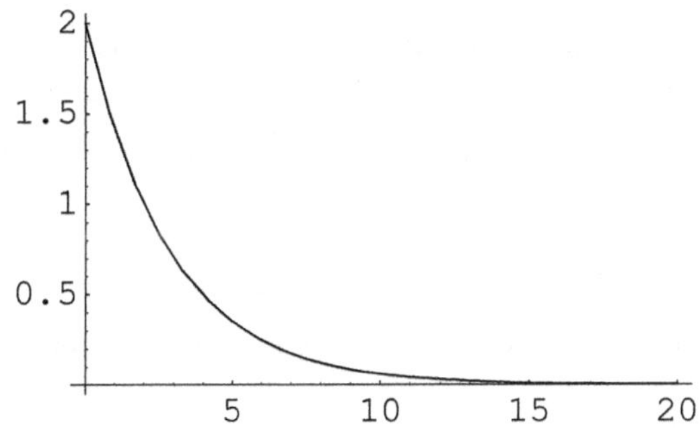

```
In[11]:= Plot[2 (.5 Sqrt[2])^t (2 Sin[Pi t/4] + Cos[Pi t/4]), {t, 0, 20}]
```

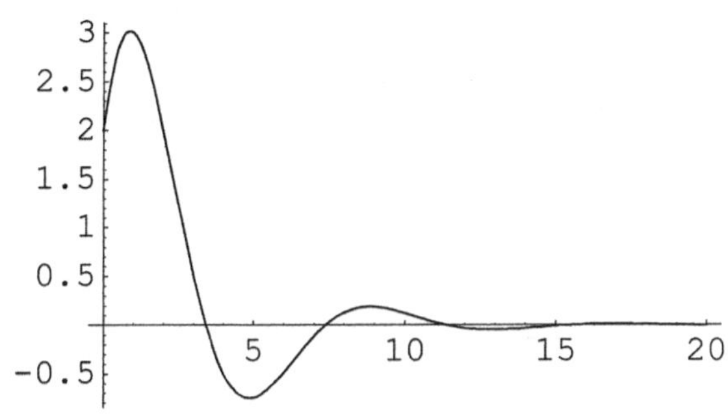

To solve the same problem using Mathematica and to plot the time path of Yt use the following commands

```
In[12]:= ClearAll[y, sol2]
         sol2 = RSolve[{2 y[t] - 2 y[t - 1] + y[t - 2] == 0, y[0] == 2, y[1] == 3}, y[t], t]
```

Out[12]= $\{\{\text{Null}\,(y[t] \to 2\,\text{If}[t \geq 0,\ 2^{\frac{1}{2}(-3-t)}\, e^{-\frac{1}{4} i(-1+t)\pi}\, ((3+i) - (1+3i)\, e^{\frac{i t \pi}{2}}),\ 0])\}\}$

```
In[13]:= y[t_] = y[t] /. sol2[[1]];
         Plot[y[t], {t, 0, 20}]
```

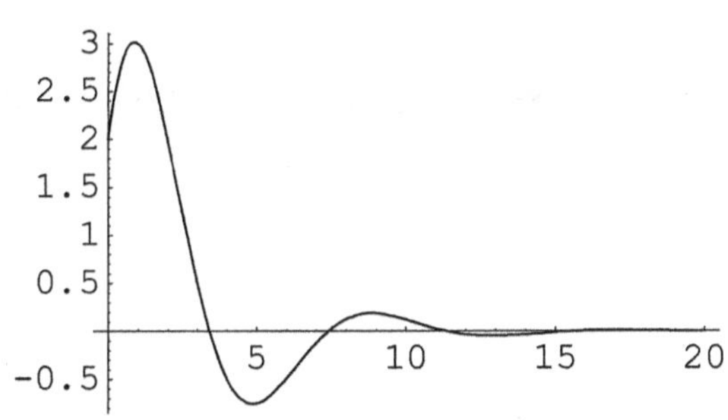

Although the Mathematica's solution looks different than the solution using the auxiliary equation, it can easily be shown that the solutions are in fact the same.

Example: Solve the second-order, nonhomogeneous linear difference equation 2Yt -2Yt-1 + Yt-2 = 5, Yo = 2, Y1 = 3, and analyze the time path of Yt.

Solution: The complementary homogenous equation 2Yt -2Yt-1 + Yt-2 = 0 has the reduced auxiliary equation

$2k^2 - 2k + 1 = 0$. This equation has the roots $k1 = \frac{1+i}{2}$ and $k2 = \frac{1-i}{2}$. Therefore, $a = \frac{1}{2}$, $b = \frac{1}{2}$, $r = \frac{\sqrt{2}}{2}$, and $\theta = \tan^{-1}(1) = 45^{\circ} = \frac{\pi}{4}$. Thus, the general solution of the complementary homogenous equation is $Yct = C1\left(\frac{\sqrt{2}}{2}\right)^t \sin 45t + C2\left(\frac{\sqrt{2}}{2}\right)^t \cos 45t$. To obtain a particular solution to the difference equation, we try f(t) = z, such that z satisfies the non-homogeneous difference equation, for all t. That is 2z - 2z + z = 5, or z = 5. Therefore, the particular solution is Yp = 5, and the general solution to the difference equation is Yt = Yc + Yp = $Yt = C1\left(\frac{\sqrt{2}}{2}\right)^t \sin 45t + C2\left(\frac{\sqrt{2}}{2}\right)^t \cos 45t + 5$.

Substitute the initial values of Y for t = 0 and t = 1 to obtain C1 = -1, C2 = -3. Thus, the solution of the initial-value problem is $Yt = -\left(\frac{\sqrt{2}}{2}\right)^t \sin 45\,t - 3\left(\frac{\sqrt{2}}{2}\right)^t \cos 45\,t + 5$. This solution satisfies both the nonhomogeneous difference equation and the initial values.

The time path of Yt in this example is similar to the time path of the complementary homogeneous part of Yt in the previous example. However, here Yt approaches 5 as t approaches ∞. The following graph shows the time path of Yt.

```
In[14]:= Plot[-(.5 Sqrt[2])^t Sin[Pi t/4] - 3 (.5 Sqrt[2])^t Cos[Pi t/4] + 5, {t, 0, 20}]
```

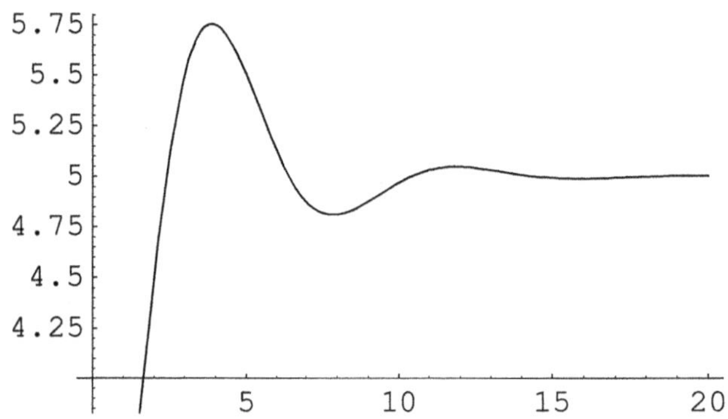

To solve the same problem using Mathematica use the following commands

```
In[15]:= ClearAll[y, sol3]
         sol3 = RSolve[{2 y[t] - 2 y[t - 1] + y[t - 2] == 5, y[0] == 2, y[1] == 3}, y[t], t]
```

Out[15]= $\{\{\text{Null}(y[t] \to \text{If}[t \geq 0, 2^{\frac{1}{2}(-1-t)} e^{-\frac{1}{4} i(-5+t)\pi} ((2-i) - (5-5i) 2^{t/2} e^{\frac{it\pi}{4}} + (1-2i) e^{\frac{it\pi}{2}}), 0])\}\}$

```
In[16]:= y[t_] = y[t] /. sol3[[1]];
         Plot[y[t], {t, 0, 20}]
```

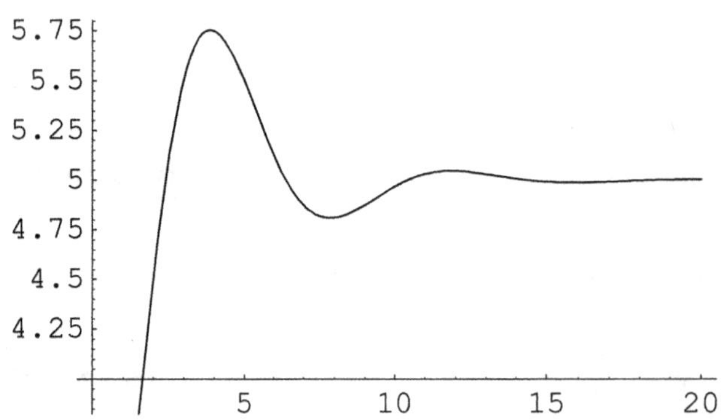

Problems

1. Solve the following difference equations and analyze the time path of Yt.

a) $Y_t - 2Y_{t-1} + 2Y_{t-2} = 0$, $Y_0 = 1$, $Y_1 = 2$

b) $Y_t - 2Y_{t-1} + 4Y_{t-2} = 0$, $Y_0 = 3$, $Y_1 = 1$

c) $3Y_t - 6Y_{t-1} + 4Y_{t-2} = 0$, $Y_0 = 1$, $Y_1 = 3$

d) $3Y_t - 7Y_{t-1} + 4Y_{t-2} = 0$, $Y_0 = 5$, $Y_1 = 17$

e) $2Y_t + Y_{t-2} = 0$, $Y_0 = 3$, $Y_1 = 4$

f) $Y_t - 4Y_{t-1} + 4Y_{t-2} = 0$, $Y_0 = 1$, $Y_1 = 2$

g) $Y_t - 7Y_{t-1} + 12Y_{t-2} = 0$, $Y_0 = 2$, $Y_1 = 4$

h) $Y_t + 9Y_{t-2} = 0$, $Y_0 = 10$, $Y_1 = 12$

2. Solve the following nonhomogeneous difference equations and analyze the time path of Yt.

a) $Y_t - 2Y_{t-1} + 2Y_{t-2} = 3$, $Y_0 = 1$, $Y_1 = 2$

b) $2Y_t + Y_{t-2} = 6$, $Y_0 = 3$, $Y_1 = 4$

c) $Y_t - 7Y_{t-1} + 12Y_{t-2} = 18$, $Y_0 = 2$, $Y_1 = 4$

d) $Y_t + 2Y_{t-1} + 2Y_{t-2} = 10$, $Y_0 = 1$, $Y_1 = 2$

www.ingramcontent.com/pod-product-compliance
Lightning Source LLC
LaVergne TN
LVHW080632011026
849160LV00014B/1194